3RD WORLD CONFERENCE ON ENGINEERING EDUCATION

ORGANISING COMMITTEE

Conference Chairman:
Eur Ing Professor T.V. Duggan
University of Portsmouth, UK

Academic Conveners:
Dr B.E. Mulhall
University of Surrey, UK
G.P. White
University of Portsmouth, UK

Secretary and Treasurer:
Dr M.R.I. Purvis
University of Portsmouth, UK

Co-Chairman (SEFI):
Professor B.L. Button
Nottingham Polytechnic, UK

Co-Chairman (University of Surrey):
Professor A. Walker
University of Surrey, UK

Conference Executive and Secretariat:
Mrs Christine Asher (Conference Secretariat)
University of Portsmouth, UK
Mrs Drusilla Moody
Conference Executive, Portsmouth, UK
Miss Denise Olway (Conference Secretariat)
University of Portsmouth, UK

Members of the Local Organising Committee:
Eur Ing Dr J.M. Bement
University of Portsmouth, UK
Professor J.T. Boardman
University of Portsmouth, UK
C. Douthwaite
University of Surrey, UK
D. Dring (Exhibition Director)
University of Leeds, UK
J.G.W. Huffell
Consultant, Portsmouth, UK
Dr D.E.P. Jenkins
Past President, SEFI, UK
Professor P.B. Morice
University of Southampton, UK
Professor B.S. Plumb
University of Plymouth, UK
Dr Z.J. Pudlowski (Secretary ILG-EE)
University of Sydney, Australia
Professor R.A. Smith
University of Sheffield, UK

L. Sucharov
Computational Mechanics Publications, Southampton, UK

International Liaison Group for Engineering Education and International Advisory Committee:
Professor M.S. Agarwal
Indian Institute of Technology, Bombay, India
Professor Ludmila Amani
New York Institute of Technology, USA
Professor G. Augusti
Universita di Roma, Italy
Mr H. Bedalian
Balfour Beatty Construction Ltd, UK
Professor J.T. Boardman
University of Portsmouth, UK
Professor B.L. Button (Co-Chairman SEFI)
Nottingham Polytechnic, UK
Professor Benzhu Chen
Shenyang Institute of Aeronautical Engineering, PR China
Dr S.R. Cheshier
Southern College of Technology, Georgia, USA
Professor T. Cole
University of Sydney, Australia
Dr B.J. Cory
Imperial College of Science and Technology, UK
Professor P. Darvall
Monash University, Melbourne, Australia
K. Davies
IBM Europe, Belgium
Professor Dr Ing K. Detert
Universitat Gesamthochschule Siegen, Germany
Professor A. Dewedar
Suez Canal University, Ismailia, Egypt
C. Douthwaite
University of Surrey, UK
D. Dring (Exhibition Director)
University of Leeds, UK
Eur Ing Professor T.V. Duggan (Chairman)
University of Portsmouth, UK
Dr H. Falk
Royal Institute of Technology, Stockholm, Sweden
Professor G. Frade
Ecole des Mines de Paris, France
Professor Marcus Giorgetti
Escola de Engenharia de Sao Carlos, Brazil
Dr S. Goodland
Imperial College of Science and Technology, UK

Professor G.W. Heinke
University of Toronto, Canada
J.G.W. Huffell
Consultant, Portsmouth, UK
Dr D.E.P. Jenkins
Past President, SEFI, UK
Dr G. Jordan
Marconi Research Centre, Chelmsford, UK
Professor G. Kardos
Carleton University, Ottawa, Canada
Professor V. de Kosinsky
EEC Relations Executive, University of Liege, Belgium
Professor C. Kuo
University of Strathclyde, UK
Professor B.E Lee
University of Portsmouth, UK
Professor T.P. Leung
Hong Kong Polytechnic, Hong Kong
Professor J.C. Levy
Consultant, UK
Professor T. Lipski
Politechnika Gdanska, Gdansk, Poland
Professor T. Marek
Jagiellonian University, Kracow, Poland
Mr Markku Markkula
Helsinki University of Technology, Finland
Professor V. Martino
Ecole Nationale d'Ingenieurs de St Etienne, France
Professor A. Melezinek
Universitat fur Bildungswissenschaften, Austria
Professor H K Messerle (Chairman ILG-EE)
University of Sydney, Australia
Professor P.B. Morice
University of Southampton, UK
Dr B.E. Mulhall (Academic Convener)
University of Surrey, UK
Professor E. Onate
Technical University of Catalunya, Barcelona, Spain
Professor A.I. Petrenko
Kiev Polytechnical Institute, USSR
Professor E.R. Petty (President, SEFI)
University of Limerick, Ireland
Professor B.S. Plumb
University of Plymouth, UK
Mr E. Prosser
The European Commission, Brussels, Belgium
Dr Z.J. Pudlowski (Secretary ILG-EE)
University of Sydney, Australia

Dr. M.R.I. Purvis
University of Portsmouth, UK
Dr. W.N. Roebuck
University of Sydney, Australia
Professor E. Rovida
Polytechnic of Milan, Italy
Dr Leighton E. Sissom
President, American Society for Engineering Education, USA
Professor R.A. Smith
The University of Sheffield, UK
Dr B. Street
Air Products plc, UK
L. Sucharov
Computational Mechanics Publications, Southampton, UK
Dr A.L. Tilmans
Kansas College of Technology, USA
Professor S. Waks
Israel Institute of Technology, Israel
Professor M.S. Wald
Fachhochschule Hamburg, Germany
Professor A. Walker (Co-Chairman, University of Surrey)
University of Surrey, UK
Professor Wei
Southeast University, Nanjan, PR China
G.P. White (Academic Convener)
University of Portsmouth, UK
Professor Dr G. Wolf
Technikum Winterthur Ingenieurschule, Switzerland
Professor W.A. Woods
Queen Mary College, University of London, UK
Professor Y. Yamazaki
Toyohashi University of Technology, Japan
G.G. Zahler
Consultant, UK

Members of Organising Committee

Back row L-R: Cliff Douthwaite, Derek Dring, Peter Morice, Mike Purvis, Lance Sucharov, Mike Bement
Front row L-R: Denise Olway, Brian Mulhall, Graham White, Terry Duggan, Bryan Button, Drusilla Moody, Christine Asher

3rd World Conference on Engineering Education

Vol 1: International, Quality and Environmental Issues

Proceedings of the 3rd World Conference on Engineering Education, held in Portsmouth, UK, during 20-25 September 1992

Editor: T.V. Duggan

Computational Mechanics Publications
Southampton Boston

T.V. Duggan
Faculty of Engineering
University of Portsmouth
Anglesea Building
Anglesea Road
Portsmouth PO1 3DJ
UK

British Library Cataloguing in Publication Data

A Catalogue record for this book is available
from the British Library

ISBN 1-85312-190-8 Computational Mechanics Publications, Southampton
ISBN 1-56252-118-7 Computational Mechanics Publications, Boston, USA

Set
ISBN 1-85312-189-4 Computational Mechanics Publications, Southampton
ISBN 1-56252-117-9 Computational Mechanics Publications, Boston, USA

Library of Congress Catalog Card Number 92-81591

This work is subject to copyright. All rights are reserved, whether the whole or part of the material is concerned, specifically the rights of translation, reprinting, re-use of illustrations, recitation, broadcasting, reproduction on microfilms or in other ways, and storage in data banks.

© Computational Mechanics Publications 1992

Printed and bound by Bell & Bain Ltd, Scotland

The use of registered names, trademarks, etc., in this publication does not imply, even in the absence of a specific statement, that such names are exempt from the relevant protective laws and regulations and therefore free for general use.

PREFACE

This major World Conference on Engineering Education, held in Portsmouth, UK (20-25 September, 1992), is the third in the series. It follows the World Conference on Education in Applied Engineering and Engineering Technology held in Cologne, in the Federal Republic of Germany, in April 1984, and the World Conference on Engineering Education for Advancing Technology, held in Sydney, Australia, in February 1989. It was organised by the University of Portsmouth in partnership with the University of Surrey, on behalf of the International Liaison Group for Engineering Education, combined with the Annual Conference of the Société Européenne pour la Formation des Ingenieurs (SEFI).

These volumes represent the edited versions of most of the papers presented at the conference. They have been collected into theme areas covering a very wide spectrum of interests in engineering education. These include international issues, quality aspects, courses and teaching issues, widening access and provision, academic-industrial links, and subject specific topics such as environmental engineering, computers in teaching, projects, design and manufacture.

The main theme of the conference was that of Engineering Education for the 21st Century, and the enthusiasm with which this theme has been grasped is very encouraging. The response to the call for papers was overwhelming, and it is evident from the papers published in these volumes that the subject of Engineering Education is now well established as a subject area in its own right.

There are many who have assisted with the preparation of the conference and the publication of these papers. Whilst it might be invidious to mention individuals, I would like to thank especially all those members of the Organising Committee who have contributed in any way, but in particular the Conference Secretariat and Executive who have handled most of the administration so effectively; the Academic Conveners who largely put the programme together; and the Publisher, Computational Mechanics, for their care and attention in ensuring the highest quality of these books. I should also like to express my appreciation to the International Liaison Group for recommending that this Conference take place in Portsmouth in 1992. Finally, I wish to express my personal appreciation, as well as that of the Organising Committee, to the University of Portsmouth for its support and encouragement throughout the planning and delivery of this event, and also to all those who have given support or assisted in any way in sponsoring the conference. The quality of papers presented, the innovation and experiences shared will, I hope, make the Conference and these proceedings a milestone in the history of Engineering Education.

Terry Duggan
Conference Chairman

CONTENTS

SECTION 1: KEYNOTE ADDRESS

The Changing Nature of Engineering Education and The Shape of Things to Come *T.V. Duggan*	3
SEFI: Its Policies and Role in Engineering Education *E.R. Petty*	15
Engineering Education: Serving the Customer? *B.W. Manley*	21
Engineering Education in the (Soviet) Commonwealth of Independent States (CIS) *L.E. Sissom*	29
Engineering Education in Poland: Facing the United Europe *A. Filipkowski*	41
Strategy for Engineering Education Development for Hong Kong in the 21st Century *T.P. Leung*	53
Comparison of Engineering Education between U.S.A. and Japan - Focusing on Undergraduate Education *H. Ohashi*	61
An International Approach to Career Development *J.A. Lorriman*	69
Engineering as a Liberal Art *K. Corfield*	75

SECTION 2: INTERNATIONAL ISSUES

Engineering Education and Training in a Rapidly Industrialising Malaysia *A.A. Abang Abdullah*	87
The Mechanical Engineering Curriculum in Finland and the U.S.A. *T.E. Leinonen*	93
New Trends, Old Schemes and Accreditation for Engineering Schools in Latin America *M.S. Navarro, R.P. Jetton*	99

Technology Transfer Through Postgraduate Training: 105
Effectiveness of Current Practice and Feasibility of
New Form of Cooperation
M.S.J. Hashmi

The Role of Engineering Technology Education in Technology 111
Transfer between the United States and Developing Countries
W.R. Hager, J. Baguant

The Impact of the German Model Gesamthochschule on the 117
Academic Education of Engineers
K. Detert

Attributes for the Baccalaureate Engineer: What are the 123
Desires of Industry?
D.L. Evans, D.L. Shunk

Technology Assessment: An International Issue 129
M.A. Brown

Engineering Education in Shantou University 135
T. Jiang

CADCAM in Developing Countries 141
M.A. Eason

Developments in Engineering Education at Cambridge 147
University
S.C. Palmer

The Future of Engineering Education in Canada 153
A. Meisen, K.F. Williams

The Development of Postgraduate Studies at Xavier 159
University, Cagayan de Oro, Philippines (A Link
Programme Supported by the British Council)
M. Cui, M. Mendoza, T.J. Oliver

An Exploration of the Tendency for Development and Reform 165
of Higher Engineering Education in China
X.M. Zhang

"CIM in Orbit" - Masters Course in CIM (Computer 171
Integrated Manufacturing). The Surrey Experience
P.G. Ranky

Planning, Development and Implementation of a Joint European Masters Degree: A Test Case for Cooperation under Erasmus Scheme *M.S.J. Hashmi*	177
Towards Internationally Recognised Engineering First Degrees *M.S.J. Hashmi*	183
International Cooperation in Engineering Technology Education: An Interim Evaluation *G.J. van Woudenberg, J.P. Rey*	189
Masters Degrees for Professional Practice and Management *S.K. Al Naib*	195
Internationalising the Technical Curriculum *S.M. Kazem, E.L. Widener*	201
Experiences in Providing Educational Support for a Developing Country *W.A. Barraclough, M.D. Bramhall, Z.A. Famokun, R.G.Harris*	207
Some Experiences of Curriculum Development and Technology Transfer between the U.K. and Developing Countries *T.I. Pritchard, J.M. Bullingham, D.G. Rivers*	213
Working Together to Update the Engineering Profession - The East-West Distance Education Project *M. Markkula, A. Hagström*	219
Trans-European Cooperation Through International Student Seminars *A.J. Miller, H.W. Holz*	225
A New Journal to Assist in Teaching and Research in Engineering for International Development *P.H. Oosthuizen, J. Jeswiet*	231
Establishment of an MSc Programme in Indonesia *J.S. Younger, A.D. May, T. Soegondo*	237
International Technical Education Program *R. Massengale*	243

Tempus Projects "Computer Aided Learning and Simulation Technologies" and "New Curricula and Courses in Theoretical Engineering Education" at the Czech Technical University K. Květoň, M. Starý	249
Collaborative Railway Roller Rig Project S.D. Iwnicki, Z.Y. Shen	255
Simultaneous Degree Programs in Engineering and a Second Language for American Students J.S. DiGregorio, T. Krauthammer, K.M. Grossman, W.R. Hager, M.E. Keune, H.J. Sommer III	261
Foreign Language Learning for Engineers D.J. Croome, G.K. Cook, S.E. Poole	267
Languages for Engineers at Nottingham Polytechnic E. Stewart, A. Jones	273
Engineering Degrees and Foreign Languages B.E. Mulhall	279
Language Teaching as a Key Element in Engineering Education or Why Should Languages be Mandatory in All Engineering Courses? R. Meillier	285
Foreign Language Learning for Students of Engineering: Some Theoretical and Practical Considerations P. Hand	291
Case Studies for Use in Teaching Engineering for International Development P.H. Oosthuizen	297
Engineering Education in the Global Context. Working Together M.L. Watkins	303
An Expansive Technique for Promoting International Industrial Exchanges H.E. Newman	309
Efficiency Through Mobility: The European Networks V. de Kosinsky	315

Engineering Education in Developing Countries A.H. Pe, A. Sevillano, E. Tadulan	321
Equivalence Questions in Engineering Courses - The Effect of the Student Exchange Process C.A. Walker	327
Summer Engineering Program for U.S. Students in London J.W. Lucey, E.W. Jerger	333
Transition Programs in Canadian Engineering Faculties P.M. Wright, J.D. McCowan, J.D. Ford	339
Development and Reorganization of Engineering Studies in Italy G. Augusti	345
Engineering Education in the Philippines S.P. Claridge, E.L. Tadulan	351
Educational Programs to Combat the Serious Lack of Professional Engineers in China's Village and Town Industry H. Wang, Z. Li	357
COMNET V. de Kosinsky	363
The Teaching of English as a Second Language for Engineers M. Rigal, R. Suarez	367

SECTION 3: QUALITY ASPECTS

How Effective is the Teaching of Engineering? M. Acar	375
"As Wise as We are Smart" Engineering - An Education in the Abstract G.A. Hartley	381
A Non-Vocational Approach to Development of Engineering First Degrees H. Cawte	387
Quality in Engineering Education: Changing the Culture - Time for a New Paradigm S.R. Cheshier	393

Quality Measures in Engineering Education in Australia and the U.K. *R.K. Duggins*	399
The Function of Examinations *R.H. Dadd*	405
Encouraging Ingenuity in Civil Engineering Students *S.G.D. Johnston, C. Williams*	411
What about the Teacher? *R.G.S. Matthew, D.G. Hughes, R.D. Gregory, L. Thorley*	417
The Future Pattern of First Degree Courses in Engineering in the U.K. *J.J. Sparkes*	423
Improving Student Learning: Some Contextual Dimensions *B.L. Button, R.M. Metcalfe, I.P. Solomonides*	429
Enhancing Engineering Education using Experiences in Medical College *Z. Xiao, B. Chen*	435
The Chemical Engineer in Society *S. Ruhemann*	439
Learning by Doing: WPI Engineer-in-Society Projects in London *L. Schachterle, R.D. Langman, W.R. Grogan, M.L. Watkins, G.L. Watson*	445
Engineers, Societies and Sustainable Activity *J.R. Duffell*	451
Teaching Engineers to Break the Rules is Rational and Desirable *M. Hancock*	457
The Role of the Study of History in the Formation of Engineers *W. Addis*	463
Psychology Teaching in the Undergraduate Electrical Engineering Curriculum: Prospects and Practice *J. MacDonald, D. Van Laar*	469

Formation of Engineers for the 21st Century *W.J. Plumbridge*	475
Formation of the New Engineer *D.G. Elms*	481
Future Engineers, Are We Trying to Attract the Right People? *K. Travers*	487
Optimisation of Fundamental, Applied and Humanitarian Principles and their Supporting Structures as the Basis of Future Engineers' Training *A.A. Minayev, E.S. Traube*	493
Changes in the Management of Engineering Education within a New and Complex U.K. Environment *D.A. Sanders, G.E. Tewkesbury, D.C. Robinson, D.G. Sherman*	499
Commercial Applications of Co-ordinate Measurement Equipment for the Manufacture of Three Dimensional Sculptured Surfaces *W.W. McKnight, D. Crossen*	505
Testing and Advisory Centre for Industrial Trucks *J. Mather*	511
CAE/CAD/CAM Education for Industry *B.T. Cheok*	517
The Marketing of Higher Education Engineering Courses *R.A. Otter*	525
Commerce in the HE Sector *R. Fletcher*	531

SECTION 4: ENVIRONMENTAL ENGINEERING

Environmental Education in Engineering Courses *R. Van Der Vorst, F. Schmid*	539
Environmental Engineering - Bridging the Educational Divide *D.J. Blackwood, S. Sarkar*	545

Impact of Environmental Issues on Engineering Education *T.V. Duggan, K. McIvor, M.R.I. Purvis*	551
Educating Engineers for Positive Environmental Action *D.J. Hardy*	557
Engineering Education in the Philippines with Emphasis upon Energy and the Environment *J. Mabaylan, A.H. Pe, M.R.I. Purvis*	563
The Establishment and Operation of a MSc in Energy and Environmental Systems *C.U. Chisholm, S. Burek*	569

SECTION 1: KEYNOTE ADDRESSES

The Changing Nature of Engineering Education and The Shape of Things to Come

T.V. Duggan

Faculty of Engineering, University of Portsmouth

Abstract
The changing nature and demands of new technological developments, the interdisciplinary nature of engineering and the ability to apply concurrent engineering more effectively, present engineering educators with new challenges. Whilst some of these may be met by extending the portfolio of existing provision, it is suggested that a more radical approach is required, articulating technician engineer, undergraduate and postgraduate provision through the development of integrated modularised programmes, and including a structured approach to continuing education and training in partnership with industry. To meet such challenges requires greater flexibility, not only to satisfy the needs of engineering education, but the corporate needs of academic institutions and the more general needs of industry, the service sector and society as a whole.
Keywords: Engineering education, Modularisation, Total professional development, Faculty structure.

1   Introduction

Engineering education at all levels is undergoing momentous change throughout the world, and nowhere is this more apparent than in the United Kingdom. There are many reasons for this, stemming largely from the rate of change of technology, but of particular importance are the greater complexity of systems of all kinds, and the information technology revolution created by the ubiquity of the computer. There is also greater awareness and concern about environmental issues, not only from the point of view of correcting the damage to the environment which has already taken place, but, of even greater importance, the necessity to produce environmentally sound designs which consider requirement to retirement needs. Now, more than at any previous time in history, the engineering community has both a responsibility and an opportunity to contribute substantially to improving the quality of life on a global scale. Engineers should be concerned not only with the application and management of science and technology in creating new benefits for mankind, but with the socio-economic factors involved and the interaction between engineering, the environment and human systems. It is suggested that, if this is accepted, then a much more radical approach to engineering education and training is required.

It is perhaps surprising that engineering educators at the forefront of knowledge in their specialised areas and often responsible for bringing about technological change, are in many instances those who have the greatest difficulty in accepting the fundamental changes which are required in engineering education to satisfy the long term needs of industry and society. Could this be because they perceive such changes as diluting the intellectual standing of their own particular specialism in the formation of engineers? Without recognising the necessity for, and benefits associated with, an holistic approach which includes continuous education and training (CET) or, as it is more frequently referred to in the UK, continuous professional development (CPD), it is likely that such attitudes will continue to be maintained. It is amazing that the total quality concept, which has received almost universal acceptance elsewhere, appears not to have been accepted generally amongst engineering educators, and there are many who still see the initial formation of engineers as an end in itself.

2 Factors Influencing Engineering Education

Engineering is concerned with a methodical approach to the solution of broad based problems, and in all major projects a systems approach is required. Consequently, it no longer seems logical at the stage of initial formation to retain the narrow specialisations of the past and simply offer conventional "stand alone" single honours courses in civil engineering, mechanical engineering, electronic and electrical engineering, and so on. Nor should the separatism of technician engineering courses and professional engineering courses, which has been the practice, be accepted as desirable without significant justification. Yet very few engineering faculties have developed comprehensive broad based integrated modular programmes capable of satisfying current and future requirements and, or so it would appear, none with complete articulation from technician engineer education through to postgraduate level. The almost universal situation is that courses for technician or incorporated engineers, undergraduate engineering courses aimed at satisfying the educational requirements for initial formation of professional engineers and postgraduate courses are usually considered to be almost completely separate activities, rather than as a continuum. The present situation, certainly in the United Kingdom, results in an unbalanced output, with too few technician engineers being produced and registered relative to the number of graduate professional engineers. Furthermore, in the past the professional engineering institutions in the UK have placed far too much emphasis on the entry requirements to degree courses, on the classification of degree obtained by graduates, and on the condition that it is necessary to include substantial elements of subject specific curricula if the courses are to be accredited by the appropriate professional institution. Whilst substantial progress has been made, it seems that the interdisciplinary nature of engineering has not yet been sufficiently accepted by many engineering educators and some of the professional engineering institutions to break down this traditional thinking, and until this inertia is converted into momentum, the desired changes are going to be difficult to achieve. These attitudes do little or nothing to widening of access and provision, and simply continue to contribute to the fragmentation of the engineering profession in the UK.

What is needed is a systems approach to engineering education. The identification of the goals; the skills required and the skills available to achieve those goals; the mismatch between them; the educational needs; how these are to be satisfied; and an evaluation of the results, with appropriate feedback (Fig 1).

The most important aspect of this simple systems approach is that of performing the needs analysis. In producing the specification for engineering education the purposeful inputs required to achieve the desired objectives must be identified, the interaction between them must be considered, and the associated uncontrollable inputs and undesired outputs must be assessed. There is nothing new about this, it is just one elementary early stage in the morphology of the design process, Duggan (1970), and yet engineering educators rarely seem to apply a sensitivity analysis to the design of their courses.

All major projects which involve the production of an artefact commence with the recognition of an identified need, and chronologically progress sequentially through a series of steps which proceed from a feasibility study, through to preliminary and detail design, and then to manufacture, distribution, consumption and retirement (Asimow, 1962). These chronologically sequential steps can, to a large extent now be carried out in parallel, a process being frequently referred to as concurrent engineering. There is nothing new about concurrent engineering either; the design process has always included a consideration of the subsequent stages of manufacture, distribution, consumption and retirement. What is new, however, is the ability to simulate these activities at the design stage more effectively using sophisticated software. All engineering projects need to include a consideration of at least three important questions. Is the proposed system physically realisable; is it financially viable; and is it economically worthwhile? Other questions also need to be addressed, and of particular relevance are those relating to producing environmentally sound designs. If the identified needs are wrong, it matters little how successful subsequent stages are, or whether concurrent engineering is applied, since the project as a whole will be a failure. This also applies to engineering courses.

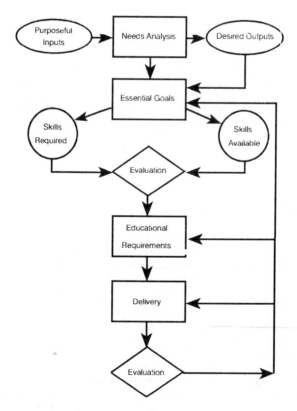

Fig. 1   Systems Approach to Education

Is this what has happened to engineering education, or has the design morphology not been applied at all or so long ago that it has become obsolete? Is it recognised that concurrent engineering now enables the morphology of design, manufacture and development to proceed in parallel, thus minimising the triumph of development over design? The information technology revolution has enabled much wider availability of information to be achieved, increased the reliability of design, reduced the need for development and enabled more effective decision making to be accomplished.

In a recent address to an assembly of senior industrialists and academics, Sir John Fairclough (1992), Chairman of the Engineering Council and former Chief Scientific Adviser to the Cabinet Office in the UK, questioned whether the evolutionary process associated with the formation of experienced engineers in the United Kingdom has kept pace with changing demands. He suggests that if it is accepted that development and manufacture (D & M) are now more tightly coupled than research and development (R & D), then this should be reflected in education courses, and this appears not to be the case.

Perhaps the greatest test of all which should be applied to engineering education, and indeed to education in general, is whether or not it provides the right basis for promoting lifelong learning. Since the half life of an engineer's knowledge is now about five years, and considerably less in the very high technology areas, the concept of providing a basis for promoting lifelong learning is essential in the formation of engineers. It is also fundamental in the design of undergraduate courses to recognise that this initial formation is but one stage in the continuous process of total professional development.

Many factors affecting engineering education, have been discussed eg. Finniston et al (1989) and Duggan (1989a,1989b,1990,1991), and some of these are specific to particular countries. But whatever the system of education, it is certain that engineering education for the twenty first century will need to be very different from that of the present and the past. Courses will need to address environmental issues and include consideration of how to handle people, develop market strategies, how to sell and cost products, and to recognise the existence of concurrent engineering. Systems engineers, capable of working across disciplines and exercising interdisciplinary skills will be the order of the day, and all engineers will participate in the process of lifelong learning. Whilst named degrees with specific pathways will almost certainly survive and continue to satisfy specific needs, the emphasis will be on modular courses, carefully designed to provide flexible pathways, and integrated with other courses to incorporate those aspects associated with the humanities, social sciences, cultural studies, business and finance and environmental studies. Indeed, engineering is, and should be seen, as much as a liberal art as it is a science, providing a sound basis of education for life itself. All other non-engineering courses are likely to benefit from the wider availability of information and the developments in technology, and it is difficult to perceive how any educational provision in the future can exist without the inclusion of some aspects of engineering and technology.

3 Future Engineering Education

What should the future shape of engineering education look like, and what are the identified impediments to achieving it? The answers to these questions depend very much upon how the needs are identified in the first place, what desired outputs are defined as essential in order to satisfy these needs and what purposeful inputs are required to achieve the desired outputs. A total systems approach is necessary which recognises the existence of various sub-systems within the whole. For example, engineering itself represents only one aspect of the socio-economic environment and the socio-production-consumption cycle. The initial formation of engineers is but part of the process of lifelong learning, and so undergraduate courses should not be seen as an end in themselves, but rather as an element in the process of continuing education and training.

The question must be asked as to whether it is reasonable or desirable to have separate pathways for professional (chartered) engineers and technician or incorporated engineers. If so, should there be appropriate ladders and bridges to enable mobility between the two? At a meeting of the twelve member states of the European Community, organised by the Commission des Titres D'Ingenieur, Paris (1990), the delegates re-confirmed their view that two separate pathways are desirable, and a summary of the reasons for this view has been presented elsewhere, Duggan (1991). However, Sir John Fairclough (1992) has suggested that this evolved process should be re-assessed, and consideration given to articulating technician or incorporated engineer education with that of the education and training of professional engineers. He also argues that a more general foundation for engineering qualification is required, and puts forward the idea that the first two years for all engineering degree courses should be largely common, with the award of a general engineering degree at the conclusion of this two year foundation. There would seem to be no reason why this general degree should not replace existing technician engineering courses, designed appropriately to satisfy the challenges of widening access and provision. It would be expected that whilst some people would stay on to complete an honours or masters degree programme, others would start work, with the opportunity to continue their professional development in the workplace and advance their academic qualifications by taking modules appropriate and pertinent to their chosen career. The links between such a new two year unclassified engineering degree for technician engineers and the longer honours degree necessary to satisfy the academic requirements for professional engineers would need to be carefully considered. The articulation will also need to recognise the wide variety and ability of students entering such a programme.

There is certainly no doubt that future engineering courses in the UK will need to be very different from the present if they are to continue to satisfy identified needs and be attractive to young people. Greater choice and more flexible programmes are desirable and will no doubt emerge. The idea of a two year general engineering degree as a launch pad providing a suitable entry point for a career in engineering has considerable merit, particularly if this could also provide a means of fully integrating technician engineers into the profession. Leaving aside the important fact that engineers generally have many valuable transferable skills, it is at least questionable whether the United Kingdom is really producing too many chartered professional engineers and not enough registered technician engineers. There is some evidence that this is, in fact, the case and that many graduate engineers find themselves working as technician engineers. The consequence of this is that it can be demotivating by not matching expectations to aspirations, and it additionally leads to a shortage of talent in key technician activities. The proposed articulation between technician engineers and professional engineers would do much to overcome this situation. Furthermore, it would also overcome a major drawback in the present system, which does not encourage people to start their careers as a technician engineer because of the difficulties of achieving professional engineering status as their career develops.

There is a real need to rationalise vocational qualifications and to establish a framework for the assessment of competence based learning. In the UK this task is being undertaken by the National Council for Vocational Qualifications (NCVQ) set up in 1986. The work of NCVQ has not, however, figured highly in the thinking and planning of higher education institutions and has had little impact on the universities. It is now becoming clear that this national framework is likely to lead to fundamental and far reaching reform, integrating academic courses with workplace experience. The standards of competence upon which National Vocational Qualifications (NVQs) are based are derived by lead bodies and reflect:

> (a) the underpinning knowledge and understanding required for effective performance in employment;
>
> (b) the ability to transfer competence from place to place and context to context; and
>
> (c) the ability to respond positively to foreseeable changes in technology, working methods, markets and employment patterns and practices.

It is suggested by Marks (1991), that "the NVQ statement of competence is light years away from the style and format of traditional syllabuses, not only in its emphasis on outcomes but, more particularly, in its specification of observable performance, and in the precision, consistency and refinement with which that performance is specified"

Similar frameworks for competence based learning have been developed in many countries, and these are bound to have an effect on engineering courses, their relationship with workplace experience, and on the equivalencies which are assessed and agreed internationally.

4 Possible Future Faculty Structures

If the future philosophy of engineering education is to be based upon the principles and strategy outlined in the previous section, it has implications on engineering courses and the way in which engineering departments, schools and faculties are organised. It will also very significantly affect the role and influence of the professional engineering institutions and the way in which courses are accredited. No longer will there be strong arguments for retaining separate departments of civil engineering, electrical and electronic engineering, mechanical engineering, chemical engineering, and so on, at least not on the grounds of undergraduate provision. It follows from this that the accreditation processes currently undertaken by the professional engineering institutions would need to be revised, with the process shifting from individual courses to the schools themselves. There would appear to be substantial resistance to this shift, and it is likely that, in the

UK, the Engineering Council will need to play a key role if such a shift is to be brought about.

It has been argued elsewhere, Duggan (1989a) that the traditional academic departmental and faculty structures in Higher Education Institutions have inhibited many of the interdisciplinary developments perceived as desirable, and some changes have already been proposed and implemented. However, as the need to expand educational provision develops still further, and modularisation and credit accumulation and transfer (CAT) become established, it is both timely and desirable to take the arguments further.

One model which enables the integration of the various engineering and other specialisms to be achieved on an institutional wide basis is that of a matrix structure, and some institutions have already either adopted such a structure or attempted to move in this direction. It is important to stress that a total systems approach is required if the optimum benefits are to be achieved. The management structure should be as flat as possible, with the minimum number of management levels required to achieve the management function.

In order to provide the maximum flexibility of modularisation and credit transfer across a whole institution, all barriers must be removed and easy pathways made available. This suggests a three-dimensional model which can best be thought of as a globe, with the lines of longitude representing programme areas, the lines of latitude the variety of subject specialisms, and the different time zones (and cultures) representing faculties.

The two-dimensional model shown in Figure 2 represents a typical single faculty structure. The vertical lines (lines of longitude) represent programme areas and could, for example, be specifically identified as courses such as Civil Engineering, Electronic and Electrical Engineering, Engineering and Engineering Systems, Manufacturing Systems Engineering, Engineering with Business Studies, and so on. It is likely that broader pathways would be identified and preferred, justifying the description of programme areas rather than courses. Each programme area would be the responsibility of a Programme Area Leader (PAL).

The horizontal lines (lines of latitude) represent the various subject specialisms, each subject area being the responsibility of a Subject Area Leader (SAL). Thus, the subject areas can travel around the globe and thereby be integrated where appropriate with other pathways and programme areas in each of the faculties or "time zones".

This matrix system, which has been rotated through ninety degrees compared with most traditional matrix structures, would provide an effective integration of subject specialisms and programme areas throughout the whole institution. In addition to the overwhelming advantages this would have specifically in engineering, it would also enable full advantage to be gained from the implementation of modularisation and credit accumulation and transfer. In addition to the traditional pathways, it would also provide a wonderful opportunity for a wide variety of new and exciting pathways, with very substantial gains in understanding different cultures and developing stronger and wider ranging interdisciplinary research.

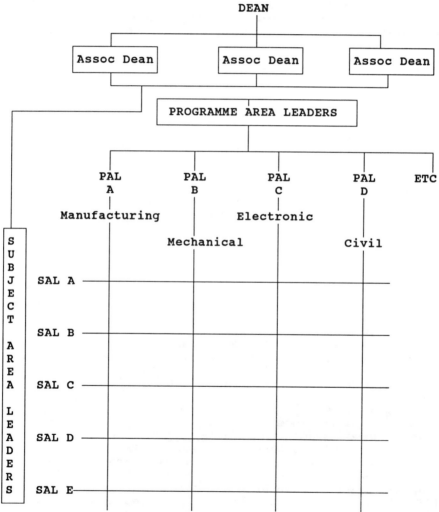

Fig 2 Proposed Faculty Structure for Modularised CATS

The next step on would be that of networking with other institutions and industries. But there is the potential for international networking, and the possibilities are limited only by one's imagination. A truly systems approach on a grand scale! But why not?

For those who are concerned about maintaining the status quo, the proposed matrix structure is unlikely to have any appeal. However, although the two-dimensional representation shown is not intended to illustrate the management structure, it is nevertheless obvious that this represents a substantially simpler structure than most which currently exist, providing greater flexibility and opportunity, and more effective integration. Figure 2 indicates Associate Deans with specific portfolio but these, like the programme areas shown, are simply illustrative.

There are several advantages associated with this three dimensional model. Firstly it enables existing programme areas to be retained and strengthened without placing constraints on new developments; secondly it enables successful faculty structures to be maintained and further developed, providing for more effective management whilst encouraging greater faculty and global integration; and thirdly it offers considerable opportunities for staff development. Additionally, in the longer term it would probably produce substantial savings. It is certainly not proposed that all higher education institutions should adopt this model, but several already have done so and it does justify serious consideration.

5 Concluding Remarks

This paper offers a number of suggestions regarding the future development and shape of engineering education. In particular some of the major factors influencing engineering education have been identified, specifically those associated with the complexity of systems of all kinds, the influence of information technology brought about as a consequence of the ubiquity of the computer and the necessity for a more interdisciplinary approach. The educational process associated with the formation of engineers should be seen as one element in a continuous process of total professional development. The way in which future courses might develop have been discussed, with an emphasis on the articulation and integration of technician engineering education and training and postgraduate studies. It is also suggested that competence based learning is likely to influence future developments in vocational courses. An approach is advocated which encourages life long learning and this leads to the introduction of modularised credit rated programmes, enabling a flexible approach to be used. Finally a proposed structural model is suggested which would enable a total systems concept to be applied in developing engineering education as part of a global approach.

But what about the content of engineering courses? What should they contain? And should prior experiential learning (PEL) be accepted as an integral part of courses? Should all courses be more closely integrated with industry, and should industry provide in-house training and education as a contributory part of academic qualifications? These are just some of the questions which this conference will address, and no doubt the sharing of experiences will enable all delegates to benefit and formulate ideas which will take us forward. Forward to the twenty first century!

6 References

Duggan, T.V. (1970) **Applied Engineering Design and Analysis**, Iliffe Books, London.

Asimow, M. (1962) **An Introduction to Design**, Prentice Hall, Englewood Cliffs, N.J.

Fairclough, J. (1992) Competitiveness for International Survival, **Conference on Continuing Professional Development,** CRAC, Castle Park, Cambridge.

Finniston, H. M., Duggan, T.V. and Bement, J.M. (1989) Integrated Engineering and its Influence on the Future of Engineering Education in the UK, **Int J Appl Engrg Ed,** 5, 135-145.

Duggan, T.V. (1989) How Technology will Change Engineering Education, **Int J Appl Eng Ed**, 5, 753-758.

Duggan, T.V. (1989) Future International Cooperation in Engineering Education, **Int J Appl Eng Ed**, 5, 687-690.

Duggan, T.V (1990) Trends and Attitudes to Change in Engineering Education, **Proc ASEE Annual Conf,** Toronto.

Duggan, T.V. (1991) An Overview of Engineering Education in Europe, **Australasian J of Engng Ed,** 2, 155-166.

Commission des Titres D'ingenieur, (1990), **European Conference on Assessment and Accreditation of Engineering Training and Qualifications,** Paris.

Marks, R. (1991) Implications of National Vocational Qualifications for the Polytechnic Sector, **Project Report for the Committee of Directors of Polytechnics,** CDP,Whitfield Street, London.

SEFI: Its Policies and Role in Engineering Education

E.R. Petty

Department of Materials Science and Technology, University of Limerick

Abstract

The background to the development of The European Society for Engineering Education (SEFI) is briefly outlined. This is followed by the Society's Aims and the methods used to achieve them.

A review of the success of the Working Groups together with special SEFI initiatives and links with other appropriate organizations illustrate SEFI's view of its role in the development of engineering education at all levels in a rapidly changing Europe.

Keywords: SEFI, Policies, Role, Organization, Initiatives.

1. History

SEFI was born in 1973. "Confinement" began on 19 January at the Katholieke Universiteit Leuven, Belgium and the birth was registered later that year at the First Annual Conference at Chatenay-Malabry (France). It was fathered by some thirty representatives of engineering education institutions from ten European countries.

With adolescence it has now nearly 300 institutional members from 22 countries. Annual Conferences have been held in 11 of these countries ranging from Napoli to Copenhagen and from Dublin to Helsinki. Presidents have come from each of these countries.

The Velvet Revolution has broadened the membership and scope of our activities into Central and Eastern Europe. Thus the Society can now truly call itself European.

Membership is mainly on an institutional basis which allows any academic of the institution the rights of full membership of SEFI. There are some 250 Universities enroled at present. Other members include International Organizations, National Societies, industries and individual members.

2. Aims

The Aims of SEFI as outlined in the Statutes are:-

a) To contribute to the development and to the improvement of engineering education as well as to the improvement of the position of the engineering professionals.

b) To provide appropriate services and to promote information about engineering education.

c) To improve communication and exchange between teachers, researchers and students in the different European countries.

d) To develop cooperation between educational engineering institutions and establishments of higher technical education.

e) To promote cooperation between industry and those engaged in engineering education.

d) To act as an "interlocutor" between its members and other societies or organisations.

These are general aims; we have decided recently to specify a Mission Statement which will be reviewed and updated annually. This Statement includes the following Objectives we have set currently:

3. Objectives

a) To assist the institutions of higher engineering education by participating in European and World developments of engineering education.

b) To support and promote new initiatives in engineering education.

c) To provide the necessary information to academic institutions, individual engineers and industry which provides for industry's requirements in terms of recruitment and further quality courses.

d) To support continuing engineering education to meet industry's identified needs and support specific local courses utilizing experts from industry and academic institutions.

e) To provide a network of contacts which allows for informal improvement within academic institutions to the benefit of their students and, subsequently, to industry.

f) To provide leadership in consultation with the Commission of the European Community in developing European solutions to relevant, identified demographic needs.

g) To ensure that industry is aware of and can capitalize on academic research.

h) To provide leadership in the development of management skills in technologists, and technology appreciation and utilization for managers in industry.

i) To stimulate research into the needs of the profession as a whole and relevant to industry.

j) To monitor and measure required industrial standards for engineers.

k) To support small and medium sized enterprises in planning the future requirements and enabling them to fully participate in engineering education and in the benefits of new technologies.

We are moving on most of these fronts but, of course, are limited by finance and manpower. In order to make the most of the available resources we have structured SEFI so that certain people or groups with enthusiasm for specific initiatives are given responsibility for these initiatives.

4. Structure of SEFI:

The Society is run by an Administrative Council of 21 members elected for a three-year mandate. One third of the Council is renewed annually at each General Assembly.

The Bureau, a smaller planning group, forms an executive body with the President; members of the Bureau communicate or meet more frequently than the Council. A Secretary-General and Assistant Secretary-General operate from the European Cultural Foundation office in Brussels.

The main work on aspects of engineering education is done by Working Groups which consist of 10 to 20 self-selecting members interested in a specialist area/topic. At present there are 8 Working Groups; these are:-

- Curriculum Development (CDG)
- Continuing Engineering Education (CEE)
- Computer Assisted Engineering Education (CAEE)
- International Exchange (IE)
- Mathematical Education in Engineering (MEE)
- Women in Engineering (WEG)
- Engineering Education and Developing Countries (EEDC)
- Environmental Engineering (EEG)

Working Groups organize meetings, seminars, workshops, etc and publish reports and monographs. A selection of activities is given below:

RECENT ACTIVITIES OF WORKING GROUPS:

1991

Berlin 14/16 March Curriculum Development in European Engineering Education (CDG)

Warsaw 4/5 April European Co-operation in E.E. in East and West (E/W.T.F.)

Budapest 10/13 April Sixth European Seminar on Mathematics in E.E. (M.W.G.)

Copenhagen 6/11 August Women in E.E. (W.E.G.)

Marseille 9/13 Sept Seminar and Exhibition of W.G. on Women into Engineering (W.E.G.)

1992

Miskolc 8/10 April Women into Engineering (W.E.G.)

Nottingham 26/28 March Computers in E.E. (C.D.G. & C.A.E.E.)

Lisbon 28/30 April 2nd European Forum on C.E.E.

Helsinki 2/5 June 5th World Conference on C.E.E.

Miskolc 25/28 June Generalist and Specialist E.E. (C.D.G.)

There is also a major and very active Task Force on East/West interaction which has obtained support from TEMPUS, as well as governmental and industrial support.

Two committees look after SEFI publications, see below.

5. Special Initiatives and Links

As well as the specific activities of the Working Groups, a number of special initiatives have been set up, often in conjunction with other organizations. A selection of these include:

- ETAP - European Thesis Abroad Programme helps to arrange for students to prepare their diploma theses in another institution in another country. It is run by the I.E. Working Group.
- EURASCOPE - European Advanced Short Courses Programme for Engineers. A SEFI initiative.
- EUROPACE - European Programme in Advanced Continuing Education. Industrial Consortium producing satellite-delivered videos and support material. SEFI has been involved from its inception.
- ETMI - European Technology Management Initiative. A joint effort between SEFI and three other organizations.
- ICEE - Information Centre on Engineers in Europe. A pooling of information by SEFI and FEANI on engineering education.
- IACEE - International Association for Continuing Engineering Education. SEFI was one of the Co-founders.

SEFI has also co-operated to greater or lesser extents with many other organizations such as ASEE, IGIP, UNESCO, CESAER, etc. As we develop, I should expect all these links to strengthen.

6. Publications

A society with the avowed Aims expressed above to 'promote information' and 'improve communication' must, of necessity, produce and disseminate material in various published formats. SEFI has developed a range of such publications which include:-

- European Journal for Engineering Education. This scientific journal appears quarterly.
- SEFI News, a members information magazine, appears bimonthly.
- SEFI Guide. A compendium of European Engineering courses is enlarged and up-dated every 3 to 4 years. A Guide for Eastern Europe is also presently being prepared.
- Working Group Reports and Monographs often produced from material presented at their seminars.
- Proceedings of Annual Conferences.

- Documents and Cahiers are occasional publications reviewing or reflecting work being done in E.E.

7. The Future

There are a number of important strategic issues facing European Engineering Education at present. These include:

- Europe's declining and ageing population.
- The move by this reducing number of students away from the 'hard' disciplines - science and engineering.
- The increasing rate of obsolescence of knowledge and skills.
- Increasing student and staff exchange under ERASMUS and TEMPUS.
- European recognition of professional qualifications allowing freer movement of engineers.
- The growing recognition that society and the economy need generalists as well as specialists.
- The impact of a more open Central and Eastern Europe.

"These factors indicate a blurring of the boundaries between general and professional education" (1) and make flexible modular curricula inevitable.

SEFI has begun to address many of these issues and is preparing to shift up a gear in order to improve its level and rate of response.

8. Conclusion

As SEFI approaches its twentieth Birthday, it can be very proud of its achievements during its 'teens'. The present status of these achievements has been reviewed and tasks for the next stage of development are outlined.

References

1. (1991) Memorandum on Higher Education in the European Community. Memorandum from the Commission to the Council, 5 Nov 91 (COM (91) 349 final), p.15.

Engineering Education: Serving the Customer?
B.W. Manley
The Institution of Electrical Engineers, London

Abstract
Engineering education must be designed to serve the needs of the customer; but who is the customer? Is it the school-leaver searching for an experience that will match his aspirations, or industry seeking to fill its ranks with appropriate skills?
 Engineering courses in higher education set out to fulfil both ends. Are the customers satisfied? There is ample evidence that they are not. Students in many countries are turning away from engineering, and industry - at least in Britain - does not look primarily to engineers to fill its senior positions.
 The paper identifies and explores the mismatches that may exist between Industry need and Course purpose, as well as student perception and ability.

1. Introduction.

 Britain, like most other European countries, spends about 5% of its gross national product in educating its young people. The re-investment of such a large fraction of the product of wealth generating industry and commerce ought to be done with a clear sense of national priority, an evident relevance to the continued economic success of the country, and a careful eye upon its effectiveness.
 While these are questions which must primarily be resolved within the context of each nation state, the same problems are faced in many countries, and the degree of interdependence between countries today leads inevitably to a convergence in the answers which are reached.
 Within the world of engineering education there are indeed common problems. Many countries are finding the attraction of engineering decreasing among the young at a time when there is a need to increase the number of engineers. All are facing the problem of a growing pressure to enlarge the syllabus. Few countries are either sufficiently wealthy, or have the will to provide the means to spend all that would be wished by the educational practitioners to meet all these pressures.
 In Britain we are certainly facing all of these dilemmas. The number of young people applying to study engineering in polytechnic or university reduced continuously between 1983 and 1990. There is evidence that the trend may have now reversed, but this is significantly influenced by the

economic recession, and the consequential rapid overall growth in the number choosing to stay in education, so deferring their entry into the job market.

There has been a continuing pressure to extend the syllabus. Employers are concerned to see a broader engineering content, and the inclusion of business and economic elements. The need for mastery of more than one language in business today is well appreciated, and increasingly features in engineering courses.

There is little doubt that these are important features, particularly within the context of a Europe providing wider opportunities to the graduate engineer. However, the continuing expansion of an already crowded course is certainly a factor contributing to the falling number attracted to engineering, and to the high proportion dropping-out of engineering courses.

All this is taking place against a background in which higher education in Britain is under increasing financial pressures, and there are great dangers that standards will fall as a result. To avoid that we must know what the customer wants; then we can decide what is important, and what can be sacrificed.

2. Who is the customer?

Who then is the customer? Is it the young person deciding upon what career to embark, whether to go into higher education, and what to study? Is it industry and commerce, anxious to achieve excellence in its engineering functions? Is it the nation, recognizing that engineering industry creates the national wealth, wanting engineering knowledge to pervade society, and engineers to provide the infra-structure of society?

In fact, each of these form a part of the customer network that needs to be satisfied. Higher education has a task within that, but cannot alone correct a system that may be out of balance. Thus, if high quality young people are not attracted in sufficient quantity into courses leading to honours degrees in engineering, it is due, to a great extent, to society's and industry's perception of the value of the chartered engineer. It will not be corrected by increasing or even maintaining the supply of lower quality engineering graduates. That will indeed confirm those perceptions, and compound the problem.

The better response would be to maintain or raise standards, and accept fewer onto such courses. Although the market operates slowly, the most certain way of changing the perceptions of industry is by reducing the quantity, and increasing the quality of the Chartered Engineer. The problem faced by the universities in following such a course is to keep the other customer satisfied - society, as represented by government, anxious only to see increasing numbers in higher education.

There is also a tension between the other two customers, the student and the employer.

The students expectations are:

- To gain the qualification that will establish him in a career providing security, interest, money, and opportunities for high position.

- To pursue an interest in engineering matters that will provide him with the opportunity to extend that interest into a career.

The employers expectations are to attract:

- A few high calibre, flexible entrants who will be expected to progress into the management of engineering.

- Many skilled entrants able to undertake a broad range of technical tasks, and bring them to fruition.

It is evident that not all those students who expect high position will satisfy the employers' criteria, nor will all who wish to be engaged on the most challenging of engineering tasks find many of the jobs with which they are confronted to be satisfying.

Thus, one of the dilemmas faced by higher education is to help satisfy the different criteria held by the various customers.

3. Are the customers satisfied?

There is ample evidence that none of the customers are fully satisfied today:

3.1 The young person considering a career

- Applications to study engineering have been falling (IEE, 1991). The young person is less attracted to engineering as a profession, either through a lack of understanding, or a rejection of what it offers (Woolnough, 1991).

- In Britain, engineering does not attract the highest calibre university entrants (Smithers and Robinson, 1991).

- The apparent disenchantment of many graduates with their first experience of working in engineering industry. This implies a mismatch between the expectation engendered during undergraduate years, and the actuality of the work-place. Some 44% of sponsored final year undergraduates surveyed in Britain said that their experience of industry had put them off a career in engineering (Industry Ventures, 1991).

- Women in particular do not see engineering as providing a satisfying job, and see it as male orientated. Participation rates are low throughout Europe (Van den Berghe, 1986).

- The non-completion rates for engineering courses are 25% in polytechnics and 18% for universities. The reasons include a disenchantment with engineering and a belief that it is boring

(Keys and Wardman, 1991).

3.2 The employer anxious to achieve excellence in his engineering function.

- The often expressed wish for more broadly based engineers, rather than the specialised graduates of today (Engineering Council, 1987).

- The widely held view that the engineer is a poor communicator, and generally not favoured for the most senior management jobs in industry and commerce.

- The relatively low salaries paid to engineers. In 1990 the gross annual salary of a newly-qualified engineer in Britain was among the lowest in Europe (Egor International, 1991).

- The frequent employment of graduate engineers as technicians (Cassels, 1990).

3.3 The nation, recognizing that engineering industry creates the national wealth.

- The evidence is that engineering is seen as an enemy of the environment by much of the media.

- In Britain, the engineer is equated with the mechanic and not perceived to provide the key to national wealth creation.

- Engineering is seen to be a relatively unimportant sector in society. Some 70% of 4th year secondary school students surveyed believed a lawyer to be more important than an engineer (Manpower 2000, 1990).

- The government places no priority upon education for engineering. It provides funding for places in higher education only in response to the demands of the eighteen year-old, not with any judgement of national need.

If higher education is to play its part in improving the level of satisfaction, it must try to define more explicitly with employers what kind of graduates are needed. While representing only one of the customers, industry provides the lever that can make engineering again into a career that attracts the young.

4. Defining the need.

Engineers are required throughout industry and commerce, and their functions include:

Innovation   Research
                   Development and design

Implementation   Production management
                   Manufacturing systems
                   Project management

Services   Technical marketing
                   Maintenance and service
                   Quality management
                   Test and measurement

Engineers are also, of course, involved in general management functions, but access is usually through one or other of these sectors.

The formation of an engineer will differ according to function, and the nature of undergraduate programmes must reflect this wide range. The needs for each function differ; a reasonable catalogue of the attributes which an engineer will possess in varying degrees might include:

Engineering knowledge   -   Knowledge of specific engineering technology
                         -   Wide knowledge of engineering disciplines

Engineering creativity   -   Ability to apply theory to practice
                         -   Creative and imaginative skills
                         -   Ability to solve problems

Personal qualities   -   Ability to cooperate and work in teams
                         -   Ability to communicate effectively orally and in writing
                         -   Ability to lead and manage
                         -   Determination to achieve

Commercial awareness   -   Business and economic understanding
                         -   Financial knowledge
                         -   Commercial acumen

In qualitative terms, we might ascribe the relative importance of the various attributes to the different functions;

	KNOWLEDGE	CREATIVITY	COMMERCIAL	PERSONAL
INNOVATION	High	High	Medium	Medium
IMPLEMENTATION	Medium	High	High	High
SERVICES	Medium	Medium	High	High

Although certain functions may thus call for particularly high levels of specific abilities, an engineer in whatever function thus needs a wide range of skills and attributes, reaching far beyond the specific boundaries of a particular engineering specialism. Of course, these are not all acquired within the limits of an undergraduate degree course at university; most are developed in the course of a career. However, the base upon which through-career learning is built must be laid at that early stage.

We must not make the mistake of believing that there is a hierarchy of quality defined by the three functional sectors. The engineer providing the services that underpin successful engineering industry must be of as high a quality as the engineer engaged in innovation and implementation of new products. There will also be engineers at all levels within a sector, but the technician is not a low quality chartered engineer. This seems to be a misapprehension in much of engineering industry in Britain, and contributes significantly to the poor regard in which the engineer is held.

5. Achieving the ends.

In the world of industry today, the emphasis is upon the achievement of high quality in all aspects of performance. To achieve that requires engineers of the highest quality, and today in Britain we are failing to attract the most able young people into the profession. As a consequence industry does not thrive, jobs are lost, and the need for engineers appears to erode - and as a consequence even fewer high quality young people are attracted. We are in a vicious circle, and we must find a way to halt the regression, and that can be done only by restoring the attraction of engineering to the most able.

Perhaps the major contribution that higher education has made to the problem has been to seek uniformity in the engineering degree. The driving force has been the wish to see courses accredited by the professional engineering institutions, so providing a direct route to chartered engineer status. This apparent lack of differentiation offers no distinctive attraction to the most able.

The chartered professional institutions have compounded the problem by the natural wish to see their numerical strength increase. They have carefully followed the Engineering Council's requirements, and have not lowered standards. Indeed, accreditation has helped the development of best practice throughout. They have, however, been willing accessories in a process which has resulted in a rather narrow range of possible degree courses throughout the universities and polytechnics, most concentrating upon imparting the maximum amount of engineering knowledge in the minimum time.

The professional institutions, together with the Engineering Council, nonetheless have an important role in correcting the problem. It is now necessary to develop new qualifying routes for the Chartered Engineer and the Incorporated Engineer which recognize their differing needs, and provide bridges at all career stages from one to the other.

The response of the universities must complement those changes. Access must be wide and flexible, so as to discourage early specialisation, and encourage the development of broad syllabuses in the secondary schools. The initial courses must establish the balance among the attributes alluded to earlier, recognising that personal skills and commercial awareness are important partners of engineering knowledge.

Specialism should be deferred at least until the third and, preferably, the fourth years of the undergraduate course. The acquisition of specialist knowledge should be structured so as to be readily accessible to those already in work. This implies courses comprising short modules that can be accumulated towards a Masters degree, and ideally, an NVQ. This approach will also help provide the bridges that will give the opportunity to those in work to progress in their careers, and in the profession of engineering.

Industry needs more Incorporated Engineers than Chartered Engineers. The universities must recognize that, and structure their courses to provide both outcomes. However, it should not be presumed that the Incorporated engineer course is the first part of a course aimed at producing a Chartered engineer.

Finally, we would do well to consider what it is we are trying to achieve. Perhaps it has been best expressed by Paul Gray, the former President of the Massachusetts Institute of Technology:

> "The goal is to equip undergraduates with attitudes, habits of mind, and approaches to learning that will assure a lifetime of technical competence, social contribution, and personal fulfilment"

If we succeed in doing that, we will have served all the customers well.

6. References.

IEE (1991) Schooling our Resources. Report of Task Group on Schools Activities.

Woolnough, B.E. (1991) The Making of Engineers and Scientists. Oxford University Department of Educational Studies.

Smithers, A. and Robinson, P. (1991) Beyond Compulsory Schooling. The Council for Industry and Higher Education.

Industry Ventures (1991) Survey into the Attitudes and Intentions of Final Year Undergraduate Engineers.

Van den Berghe (1986) Engineering Manpower, a comparative study on the employment of graduate engineers in the western world, Unesco.

Keys, W. and Wardman, M. (1991) Research into Engineering Education. National Foundation for Educational Research.

Egor International Limited (1991) The Training and Employment of Engineers in Europe.

Cassels, Sir John (1990) Education, Training and Competitive Advantage. The Ninth Hitachi Lecture, University of Sussex.

Manpower 2000 (1990) Images, Education and Recruitment in Engineering. Southern Science Forum, University of Southampton.

Engineering Education in the [Soviet] Commonwealth of Independent States (CIS)

L.E. Sissom

American Society for Engineering Education

Abstract

This paper gives a review of the education system of the former Soviet Union (USSR) and the aspirations of the new Commonwealth of Independent States (CIS) as they relate to engineering education. It examines key aspects of their engineering education--curricula, laboratories and philosophy. Cited are activities, anticipated pedagogical changes and proposed interface with the rest of the world.

Engineering educators of the CIS are most interested in having their degrees (diplomas) recognized for entry into the profession--workplace or graduate school--in the international arena. Under development is an accreditation (evaluation) system which will assure that graduates meet minimum criteria. The proposed accreditation system was first envisioned as an "escape" from the state-controlled program specificity, permitting more input from professionals rather than from bureaucrats. It is now viewed as the ideal way to develop an engineering program--with professionals defining the constituency of the profession.

Keywords: accreditation, evaluation, Soviet Union, Commonwealth of Independent States, engineering education

1 Introduction

In 1991, the Soviet Association of Engineering Universities and the American Society for Engineering Education (ASEE) signed an agreement on reciprocal cooperation. Both parties pledged joint efforts to cooperate in the furtherance of engineering education. Specifically identified were joint research efforts, faculty and student exchange, joint colloquia and a mechanism for addressing change. The last item came to be the most important one because of the tremendous changes that have occurred in such a short time since these efforts were initiated. The Agreement of Cooperation came after a delegation of American engineering educators visited the Soviet Union in October 1990, with a reciprocal visit of soviet educators to the U. S. in November 1990.

Efforts intensified in November 1991 when a delegation of ten American engineering educators evaluated 32 Soviet engineering institutions (listed in Appendix A). All of the participants were seasoned evaluators for the Accreditation Board for Engineering and Technology (ABET), the accreditation agency for engineering and engineering technology in the United States. In most instances, at least two evaluators visited each of the schools. They evaluated the faculty, students, curricula, administration and commitment to the programs. The evaluators prepared reports on the individual institutions, and a joint report was prepared as an overview. Results of the evaluation were discussed in a Soviet-U. S. conference in Washington, DC in the spring of 1992.

There is currently a crisis in CIS engineering education, reflecting the woes of the current economic crisis. Engineering is not regarded as highly as it was just a few years ago. Included among the reasons for this loss of prestige are, Venda (1991):

♦ Salary, standard of living and social prestige are no longer directly related to intellectual abilities, creativity and productivity. Some vocational workers (such as bus drivers) sometime earn more than engineers.

♦ Promoters and bureaucrats often earn more than engineers, making engineers want to pursue these "functioner" roles.

♦ Admission to graduate schools has become more dependent upon union organization, party membership, etc. than upon intellectual ability and creativity. Practically all university presidents, institute rectors and other key administrators, including in many instances the chairs, in the higher education community are former union and party activists. Professional educators have essentially disappeared from the bureaucratic ministries of education, leading to a lowering of the intelligentsia.

♦ A declining percentage of resources has been going into education, with the military complex getting about 50 percent of the total, but that too is now declining.

All of these factors are expected to change now although some may get worse before improving.

2 Education in the Commonwealth of Independent States (CIS)

For many years the soviet education system progressed from the general to the highly specialized, generally divided as follows.

♦ Kindergarten: Up to age 6

♦ Elementary School: Grades 1-4

♦ Middle School: Grades 5-8

♦ Secondary School: Grades 9-10 (Note: The terms "high school" and "higher school" are usually reserved for education beyond the secondary school.)

♦ Professional-Technical Schools (for worker professions): 1 to 3 years after Middle or Secondary School

♦ Technical Schools (for technicians): 4 years after Middle or Secondary School

♦ Universities/Institutes (for "higher" education): 5.5 years after Secondary School

In the last three years, there has been an effort to extend secondary schools through grade 11. All institutions evaluated by the U. S. delegation now claim to require that their engineering education is based upon an 11-grade secondary school preparation.

Before the independence of Estonia, Latvia and Lituania, the Soviet system of higher education included 895 universities and institutes, Venda (1991). **Universities** are broad-based institutions which provide a variety of fundamentals; a few include engineering. Graduates of the universities usually work in research centers as scholars and in the secondary schools as teachers. **Institutes** provide education for specialists in industry, culture, agriculture, medicine and specialized scientific centers. Universities and institutes bear the name of the city in which they are located.

Institutes are of two types: (1) polytechnics (with many different faculties) and (2) narrow specialized institutes (e.g., aircraft instrument making). Many of the leading engineering institutes are polytechnics, e.g. Moscow Technical University (named after Bauman) and Kiev Polytechnic Institute. Although most of the leading engineering schools have specialized names, such as the Leningrad (St. Petersburg) Electrical Technical Institute (LETI), they have a wide range of engineering specialties.

Students and faculty have access to libraries at their own schools in addition to the library of the Academy of Science. Schools with an enrollment of around 5,000 generally have libraries with about a million volumes. A very small percentage of the holdings are in English and other western languages, limiting access to the world literature. This is a crucial point in specialized research where they may be "re-inventing the wheel." (It is little wonder that the Russians claim that the radio was invented by Popov while most of the remainder of the world claims it was done by Marconi!)

About 5 million students pursue higher education in the CIS, in over 450 specialties and more than 1,000 narrow fields of occupational training. The average length of education is 9.8 years, compared to 14 years in the United States and other developed nations, Venda (1991).

3 Academies of Science

Before the metamorphosis of the USSR into the CIS, there were four academies in the USSR, viz.

- ✦ Academy of Science
- ✦ Academy of Pedagogical Science
- ✦ Academy of Medical Science
- ✦ Academy of Agricultural Science

While all of the academies were designated "science," they included engineers, primarily in the first one. Membership, full and corresponding (associate), in the academies is very honorable and well paid. Many are scientific bureaucrats rather than scientists, however.

In addition to the countrywide academies, each Republic, except for Russia, had its own "Republican" Academy of Science as well as other academies. The Russian federation is now subsuming the Soviet Academy. It is highly likely now that the academies of the individual states will become more active and fill the void left by the somewhat undefined character of the former soviet academies.

4 Structure of Engineering Education in the CIS

In the immediate past upon completion of an engineering program of study, a graduate was granted a diploma (not a degree), the only professional recognition accorded graduates. The "diploma engineer" is viewed in about the same way as those from the German *hochschules*. Most diplomas represent highly-specialized preparation in a very narrow area, including at least one semester of clinical practice in a regional enterprise (industry) within the student's field. The level of specialization is also reflected in the diploma work (senior project) which is undertaken after successfully passing a comprehensive examination (at some institutes). Now, there is an effort in many of the engineering institutes to develop multi-level degree programs, similar to those in the United States--baccalaureate, masters and doctorate--although the equivalence will be questionable until the secondary school preparation can be thoroughly evaluated.

At the present time, all of the schools evaluated award a diploma (engineer) upon completion of 5.5 years of satisfactory study beyond 11 years in secondary school. [Some of the institutes have just begun or will begin admitting students for Bachelor of Science (B.S.) degrees to be awarded after four years of study A few of the institutes will award masters degrees at the end of 5.5 years, a practice which is still being questioned in most circles.] In some institutes, the completion of an additional half year, with concentration in a research area, will result in the graduate being given the designation of "research engineer."

There are two general levels of graduate study: *aspirantura* and *doctorantura*. Candidate of Science **aspirants** pursue graduate study for at least three years beyond the diploma at an engineering institute or university/polytechnic, graduating upon publicly defending a research thesis. Graduates are designated *kandidat* of Science. The CIS institutions claim that *kandidat* degrees are equivalent to the American Ph.D.'s although there is widespread disagreement on this, even by some within the CIS.

Although granted to a very small percentage of those who seek it, a second graduate "degree," Doctor of Science (D.Sc.), may be awarded to a *doctorant* upon completion of a second thesis (dissertation)--of major significance with very important theoretical, experimental and practical results--after a *kandidat* attains prominence in his/her field, usually at the age of 40-60. [The Doctor of Science, D.Sc., degree is approximately equivalent to the U. S. Ph.D. plus being a Fellow (professional distinction) in a professional society.] Recipients are granted significantly higher salaries, status and rank. In some institutes, only those with the D.Sc. degree may be promoted to full professor.

The Schematic of Higher Education in the Appendix shows the relationship among the USSR diploma/degrees, the proposed CIS degrees and those of the USA.

5 Engineering Programs of Study

The curricula are quite specialized but current. In general, mathematics and basic sciences are excellent. Engineering sciences are often intermingled with other components of the curricula such that they are difficult to identify. To illustrate, thermodynamics as taught in American institutions does not exist; however, elements of it are integrated into other courses--e.g., heat treatment in material science and combustion in internal combustion engines. Many of the engineering sciences, fluid mechanics and vibrations, e.g., are integrated in special projects, laboratory exercises and design projects.

Despite the curricula being rather rigidly imposed by the Ministry of Higher Education and Science, the categorization of their components are identified quite differently by the individual institutes. Here are some example categorizations taken from the published syllabi of three institutes (based upon clock hours in class), compared with the criteria prescribed by the U. S. Accreditation Board for Engineering and Technology (ABET, 1991).

	ABET	MATI	TPI	UAI
Mathematics-Basic Sciences	25.0%	25.0%	21.0%	16.0%
Engineering Sciences	25.0	25.7	38.0	25.0
Engineering Design	12.5	31.5	14.0	30.0
Humanities-Social Sciences	12.5	11.9	21.0	14.0
Other	25.0	5.9	6.0	15.0
	100.0	100.0	100.0	100.0

MATI   Moscow Aviation Technical Institute
TPI   Tula Polytechnic Institute
UFA   Ufa Aviation Institute

The overall curricula provide integrated educational experiences directed toward the development of the ability to apply pertinent knowledge to the identification and solution of practical problems in the designated areas of engineering specialization. The CIS schools require more time in class and under the tutelage of the faculty than that required in the USA (maybe double), but there is considerably less homework.

The CIS faculty are probably more accessible to students since there is less pressure for them to "earn their keep" via funded research and development programs. There are faculty, however, who lack initiative and do not take pride in their work just as there are in all parts of the world. With everything "belonging to the state," it is easy for the faculty to get caught up in their own very narrow interests to the exclusion of all other components of the desired broad-based education. It is highly probable that this is as much the reason for their specialized curricula as the state specificity. It has been naturally easier in their bureaucratic system for the faculty to conform to the specialist path, especially where that path coincides with the state prescribed emphasis areas, such as the military complex or the space program. The students truly believe that they have the best rockets in the world while their cars break down or while they can't get gasoline for them.

In general, the computers available to CIS students are somewhat obsolete by western standards--and obsolete even by soviet standards in some institutes. But there are bright spots. For example, the Ufa Aviation Institute has one microcomputer for every 10 students, compared to about one for every 50 students throughout the entire CIS. A small number of them are 286-type units with an even smaller number of 386 types. There were some isolated 486 machines in about 20 percent of the institutes visited.

While not common, there were a few Sun-type workstations (UNIX-based) built in Russia with Asian components. At the Leningrad Institute for Aircraft Instrument Making, for example, there were two such units with 24 terminals tied to them via a local area network. Furthermore, there was a impressive array of software for them, especially in the CAD/CAM area. An electronic mail (E-mail) node is available at some institutes but not readily accessible to anyone even where available.

Perhaps more important than the availability of computers, however, was the integration of computer use into the curricula. Such was the case in every institute visited. All of the students appeared to be computer literate by the time they were in their second year of study. In many of the institutes, there were early admission programs where secondary school 10th graders took classes on campus two or three days per week, notably in the area of computer programming and mathematics.

Some of the institutes' laboratories were truly outstanding--e.g., a physics lab at the Ufa Aviation Institute. Several laboratories, which looked otherwise dull and dismal, had some very sophisticated equipment in them--e.g., scanning electron microscopes (SEM), mass spectrometers, Raman spectroscopy equipment, robots, and outstanding manufacturing equipment.

An outstanding example of laboratories was found at the Ufa Aviation Institute. Specialized laboratories which were evaluated include: Robotics; Material Handling; CAD; Surface Finishing; Machining Center; Flexible Manufacturing; Signature Analysis; Metal Processing; Manufacturing; Casting; Material Finishing; Cybernetics; Avionics; Engines and Power Plants; and Strength of Materials.

As another example, the Tula Polytechnic Institute had over 100 machines in its Machine Tool Laboratory, including lathes, grinders, milling machines, gear hobbers, etc. While some of them were over 30 years old, they were still functional. And there were some modern numerical control (NC) and computer numerical control (CNC) units which were being used regularly. Institute personnel were working with both hydraulic and electrical units, and they had set up a small-scale flexible manufacturing system.

There was such an emphasis on manufacturing throughout the institutes that it is worthy of special note.

6 Independent Enterprises (Manufacturing)

Practically all of the institutes had some commercial operations as a part of the institutes. Spin-off enterprises (industries), which become self-supporting, are strongly encouraged. The Leningrad Electrotechnical Institute (LETI) has 25 independent companies within it. In some cases, the enterprises pay a special levy to the institutes in lieu of taxes. Faculty members have excellent contacts with industries and interact directly in the design and production of military and commercial products. Branch centers of some institutes are located within industries, facilitating the formal education of employed workers and students who are engaged in clinical practice.

Science parks have been established in some of the major cities, e.g., in St. Petersburg with the cooperation of several of the institutes led by the Leningrad Electrotechnical Institute (LETI). The St. Petersburg science park will concentrate primarily on environmental issues. It has free trade zone status, enabling exportation for hard currency. Moscow's science park, called International Technology Village and supported strongly by the Moscow Aviation Technical Institute (MATI), is seeking foreign enterprises, ideally American and Finnish companies.

As an example of longevity, the Aviation-Mechanical Department of the Moscow Aviation Technical Institute (MATI) has an "Automatized Systems Technology" enterprise which has been functional for over a decade. It is registered as a participant in external economic activities. It offers services both domestically, e.g., to the Ministry of Science and Higher Education, and abroad, e.g., to private industry in England.

The mechanical engineering faculty of the Tula Polytechnic Institute (TPI) earns about a million rubles per year (November 1991 valuation) with its manufacturing program, selling both domestically and abroad. Products are primarily precision components. It appears that most of the well-educated faculty (*kandidats*, e.g.) are spending a large amount of time in the manufacturing enterprise, actually operating the machines. Most western nations would view this as an unwise use of specialized talent. Other departments of TPI are similarly engaged in commercial operations although none as extensive as that in mechanical engineering.

There are outstanding manufacturing facilities in some of the institutes. While they are devoted to commercial applications, students participate in the operation and gain invaluable experience--experience not acquired by most western students while in school. At the Ufa Aviation Institute (UAI), for example, Institute personnel are currently hot forming rotors, etc. for compressors used in domestic and foreign industries. The operation, generating temperatures over 1,000 degrees Celsius, is very sophisticated and equal to that seen in private industry.

7 Organization of Schools and Faculties

Universities, polytechnics and institutes are presided over by rectors. Most of them rose to their positions from being active in the Communist Party. Some were "assigned" to schools which were out of their desired location. Since their appointments are political, a great deal of their activities are political, some serving on the former Supreme Soviet, the principal governing body of the nation (equivalent to the Senate in the United States).

Vice rectors direct the principal administrative units on campus. The First Vice Rector is the chief academic officer, equivalent to the Provost or Dean of Faculties in the United States. The numbers of vice rectors depend upon the size of the schools and the extent of the rector's external commitments. They, too, are often political appointees. In general, however, the American evaluators found their credentials to be in keeping with their assigned responsibilities.

Schools or colleges are generally called faculties, presided over by a dean or chair, depending upon the local organization set in place by the rector and his chief advisors. Oftentimes there are sub-faculties, sometimes called chairs, somewhat equivalent to U. S. engineering departments. Then there are multiple specialities--options or concentrations.

Faculty rank parallels that in western institutions although having different titles. The beginning faculty member is an **Assistant** (not assistant professor), analogous to the U. S. instructor. A number of assistants hold only a diploma, i.e., no graduate work, although many of them become *aspirants* for the Candidate of Science degree (*kandidat*). With satisfactory progress toward the *kandidat* and good performance, the assistant is promoted after two to five years to **Senior Teacher**, analogous to assistant professor in the United States. Senior teachers are given more responsibilities for laboratories and course content, rarely permitted for the assistant.

Upon attaining the *kandidat* and with excellent performance as a senior teacher, the faculty member may be promoted to **Dosent**, analogous to the U. S. associate professor. Dosents account for over 50 percent of the faculty members--ranging from 53 to 80 percent where numbers were made available. Dosents rarely become faculty chairs although there was one notable example at the Moscow Aviation Technical Institute (MATI). The highest faculty rank, **Professor**, is attained by very few--usually around 10 percent, ranging from 5 to 12 percent in the institutes where data were published. Professors normally have their own faculty, or sub-faculty, within their speciality. Oftentimes, professors are made Head of the Chair or Dean in their discipline. The rank of professor is highly regarded and sought. They receive higher salaries and are accorded higher status. It is rare that a faculty member is promoted to the rank of professor without having a Doctor of Science degree (*doktorant*). The number of *doktorants* is estimated to be around 6,000.

In addition to the faculty members, each school has a large number of supporting personnel--often from 30 to 50 percent of the total, viz, research faculty, engineers, a variety of technicians and production workers.

8 Proposed Education Reform

In March 1987, guidelines for remodeling the system of higher and specialized secondary education were adopted by the USSR Council of Ministers. It was partially a response to this reform that led to grade 11 now being expected as the beginning point for engineering education. Despite the metamorphosis which has taken place since then, a major portion of the proposed reforms has been assimilated and adopted by the independent states.

In early 1991, the Russian State Committee for Science and Higher Schools issued plans for higher education reform, citing social, economic and administrative deficiencies of the existing system, Balzer (1991). From an education standpoint, narrow specialization was criticized in favor of the well-rounded educational experience. It is likely now that this reform plan will be accepted throughout the CIS since the goal is to decentralize the decision-making and regulation processes and to adopt the UNESCO international standards for reporting and evaluation.

The new system will drop the distinction between higher and specialized secondary education in favor of a multilevel system of higher education, similar to that in the United States. The intent is to make higher education accessible to everyone in accordance with their abilities. Included in the plan is a system of evaluating the quality of all types of educational activities. The goal is to prepare practitioners who will be able to cope with the changing character of the economy and labor force. On-the-job continuing education is an important aspect of the plan.

The proposed higher education system, designed to build upon 11 grades in the secondary school, will consist of four levels with broad access to the first level and competitive access at the other levels. The levels are:

✦ **Incomplete Higher Education (Lower Division)**: Two years of broad fundamentals which will prepare graduates for further study at a higher level.

✦ **Basic Higher Education (Upper Division)**: Two years of professional training in one of the areas of science, technology or culture following the "incomplete higher education." Graduates will receive a diploma, recognizing competence in the particular field of study, and a Baccalaureate degree. The diploma will enable the graduate to continue study at the third level in his/her field.

✦ **Complete (or Specialized) Higher Education**: At least two years of study based upon the envisioned future of the selected specialty. Programs will include clinical internships in enterprises (industries) related to the program of study, ideally where the student will work upon completing the program. Graduates will be awarded diplomas, certifying their status as specialists. Those completing the required research will be awarded a *Magister* degree.

✦ **Graduate Study**: At least three years of additional study and research for those competitively selected. Graduates will be awarded a *Kandidat* degree.

These degrees are shown in the Schematic of Higher Education in the Appendix, compared with the USSR diploma/degrees and those of the USA.

Upon comparison of these proposed reforms with the existing engineering education system, Section 4 above, there are a number of questions, primarily centered upon the overlapping and fuzzy degree designations. Will the proposed *magister* degree replace the old *kandidat* degree? It is obvious that the proposed *kandidat* degree will reflect competence somewhere between the old *kandidat* degree and the old Doctor of Science degree. Will the old Doctor of Science degree fade into oblivion? Will those who hold the Candidate of Science (*Kandidat*) degrees--more than 30,000--demand that their degrees be accepted as equivalent to the American Ph.D.s? These are issues which cannot be answered at the present time. They are unanswerable until the new system is in place, if fully instituted, and has produced graduates.

Scheduled to be started September 1, 1992, the proposed system will function alongside the existing system in those regions which choose to adopt it. It is doubtful that such freedom of selection will result in anything short of chaos. Of special concern is the individual student who cannot be expected to understand the options available.

9 Conclusions

The nuclear disaster at Chernobyl blew the first hole in the iron curtain although the Soviet system's flaws had been surfacing throughout the unsuccessful war in Afghanistan. The Armenian earthquake of December 1988 revealed the shoddiness of Soviet architecture, the weakness of the infrastructure and the low quality of emergency medical aid (e.g., glass bottles of blood plasma with rags for stoppers). The Soviet Union was exposed as a third-world country with a space program and nuclear weapons, Lourie (1991). *Glasnost* (openness) and *perestroika* (reconstruction) sounded encouraging when first introduced by Mikhail Gorbachev. Can the new Commonwealth of Independent States bring order to the crippled giant whose natural resources exceed those of any other country?

The infrastructure is in shambles. The communication system is very poor. Even though international communication is quite dependent upon the fax machine, some schools do not even have one that will interface with international telephone systems due to lack of circuits and related hardware. Most schools have to send faxes to major cities, such as Moscow or St. Petersburg, and have them relayed. Fax communication out of the country normally requires a minimum of one day waiting in the queue.

Distribution of goods is even worse than the communication system. Agrarian economies of the west have recognized the need to move farm products to the regional markets and have developed transportation systems to accommodate the need. That has not been done in the CIS. For example, pickup trucks which are so common in the western nations, are almost nonexistent in the CIS. There are few refrigerated large trucks for moving perishable products. Furthermore, there is no scheduling system which depends upon good communications and computers.

The collective farm system has produced a citizenry which is not interested in feeding anyone except themselves. The "lack of ownership" is a major psychological problem, affecting pride in addition to production. The transportation system is not very dependable, with the exception of a few aspects of the rail and air system, although they are now operating in the face of an adverse supply of fuel and replacement parts.

The most glaring deficiencies are in management, entrepreneurial skills and understanding of fundamental economics. These deficiencies are being addressed but in an uncoordinated fashion. The International Management Center at the Leningrad Electrotechnical Institute (LETI) is an example of excellent attention to these deficiencies although available to only a few students and itself having inadequate faculty.

Graduates of the CIS engineering schools are well prepared to practice engineering in their specialized disciplines. Compared with western schools, the fundamentals in mathematics and basic sciences are very good; engineering sciences are marginal; engineering design is above average; and humanities-social sciences are satisfactory in recently-revised curricula. The educational experience is limited in breadth, however, being highly specialized. Clinical experiences and diploma works (senior projects) are excellent.

Students are well motivated, articulate and have lofty goals. Enrollment by females has been dropping sharply because they are not readily accepted in the workplace as equal to the males.

In some of the institutes, laboratory equipment is in a poor state of repair. With a few exceptions, facilities are cramped, dark, dismal and poorly maintained. Some space is poorly utilized--in hallways and less-than-optimum arrangement of equipment. The faculty has done an amazingly good job under adverse conditions, however.

Assuming a rational conversion from military to civilian production, the curricula available in the engineering schools are well suited to the country's needs. Very few engineering school personnel have recognized that the same technology used in producing missiles is not required in making consumer products. Such changes in the internal philosophy will result in a short-term decline in the need for engineers and technologists.

Publication by the faculty is driven more by motivation than by necessity. The publish-or-perish syndrome does not exist in the CIS.

Overall, engineering education is being provided remarkably well under the circumstances currently being faced within the Commonwealth of Independent States. It is commendable that they are preparing their graduates for recognition throughout the world--in both graduate schools and the marketplace.

References

Accreditation Board for Engineering and Technology, **1991 ABET Accreditation Yearbook**.

Balzer, Harley D. (1991) "Education and Technology After Perestroika: Can Schooling Make a Difference?" **Proceedings**, NATO Conference on Technology and Transition in the USSR, 17-20 September 1991.

Lourie, Richard (Ed.) **Predicting Russia's Future: How 1000 Years of History are Shaping the 1990s**, Whittle Communications, 1991.

Venda, Val F. (1991) "Prospects of the US-SU Collaboration in Engineering Education," Unpublished Consulting Report for ASEE, 5 August 1991.

Acknowledgment

While acknowledging the valuable information provided by other members of the American team which evaluated the CIS engineering schools, the content of this paper is the sole responsibility of the author. The dynamic changes in the former Soviet Union may result in conditions which are different from some of those described in this paper. The author would appreciate receiving comments on further changes in those matters addressed herein.

Appendix

Schools Evaluated by American Delegation

Moscow Electronic Machine Building Institute
Moscow State Technical University (Bauman Institute)
Kiev Automobile-Highway Institute
Kiev Polytechnic University
Nikolaev Shipbuilding Institute
Moscow Car Building Institute
Moscow Aircraft Technology Institute
Moscow Automobile-Highway Institute
Tula Polytechnic Institute
Ufa Aviation Institute
Leningrad Institute of Aircraft Devices
Leningrad Electrotechnical Institute
Moscow Machine Tool Institute
Leningrad State Marine Technical University
Leningrad Mining Institute
Leningrad State Technical University
Sverdlovak Urals Polytechnic Institute
Sverdlovak Mining Institute
Moscow Institute of Radioelectronics
Moscow Power Institute
Leningrad Mechanical Institute
Minsk: Byelorussia Technical University
Minsk Institute of Radioengineering
Moscow Physics Technical Institute
Moscow Institute of Electronic Machine Building
Moscow Institute of Civil Aviation Engineering
Krasnoyarsk Polytechnic Institute
Krasnoyarsk Institute of Space Techniques
Tomsk Institute of Automated Control Systems and Radioelectronics
Omsk Polytechnic Institute
Moscow Oil and Gas Institute
Cheluabinsk Polytechnic Institute

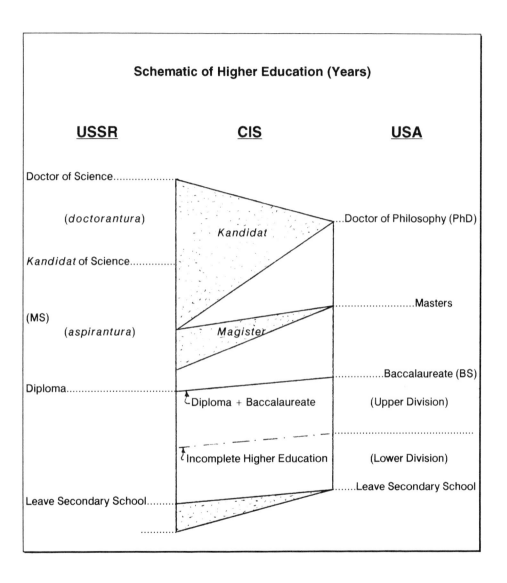

Engineering Education in Poland: Facing the United Europe

A. Filipkowski

Warsaw University of Technology

Abstract
The paper describes the problems which must be solved to approach the Polish engineering education to the Western-European standards. Joining the European Community challenges us to create engineering manpower capable to work in the United Europe. The existing educational system in Poland is described. On that basis it is pointed out what should be changed in curricula, equipment, industry co-operation and continuing education. The role of EC organizations and programs in internationalization of Polish engineering education is emphasized. Conclusions show that the problem is difficult but possible to solve.
Keywords: Engineering Education, European Education, Curricula, United Europe.

1 Introduction

The fall of communism in Poland and starting to shift the economy towards the market rules, have put a great and difficult challenge to education. During 44 years in the period 1945 to 1989 the Polish system of education was under domination of communistic ideology and its development was different than in Western countries. The academic staff made many efforts to oppose against these tendencies and the Polish education was the most autonomous among such called "socialistic" countries. Nevertheless the effects of indoctrination, limitations in international co-operation and curricula imported from Soviet Union had some impact on the level of academic education in Poland.

With this heritage we now face new problems. It is a truism that the success of any economical, political and intellectual reconstruction depends mostly on the elites available in the country. The Polish elites are now limited in number and some on the level not meeting our needs. We must create them as soon as possible and that is the main challenge for the Polish education. To educate the new decision-making personnel, understanding the meaning of the "European dimension", takes more time, than building a factory or changing the banking system. We should therefore make a great effort to reconstruct our university education.

Before more detailed discussion, a short review of the university education in Poland will be presented, with emphasis on the engineering education. The higher education is performed by various schools which are divided on the basis of broad area of disciplines. Engineering education is in specialized schools called "Politechnika" (University of Technology - UT) or "Wyższa Szkoła Inżynierska" (Higher School of Engineering - HSE). Other schools are concerned with sciences (Universities), economy (Higher Schools of Business and Economics), pedagogy, agriculture, theology, art, etc.

UT's and HSE's in the number of 20 are governed by the Ministry of National Education. Students can be enrolled after completing their secondary education and obtaining the certificate called "matura". The primary and secondary education lasts basically 12 years and students entering the University are 19 years old. Vocational education lasts 13 years and pupils receive the "matura" and a technician certificate. They get their profession at that moment and can be employed as technical staff, but many of them continue their education on the university level.

The main type of engineering higher education in Poland is based on one 5-year course leading to the degree called "magister-inżynier", corresponding to the M.Sc. (eng). Students taking these courses are not divided into undergraduates and graduates.

Many universities offer also 3-year or 4-year courses leading to the degree "inżynier", which means engineer, and corresponding to B.Sc. (eng). Students graduated in those schools are prepared to professional engineering activity and not to research and science. For students graduated on that level, additional postgraduate 2-year courses are offered for "magister-inżynier" degree. Those courses are mostly in the evening, part time type.

Full time studies leading to the doctor degree exist, but so far are not very popular in Poland. They are organized mainly for foreign doctoral students. Most doctor degrees are gained in Poland by persons with "magister-inżynier" degree working professionally in research and as university teachers submitting a dissertation to an authorized scientific council. A second doctor degree exists in Poland called "doktor habilitowany", which can be compared with the British D.Sc. It is awarded to persons with doctor degree having substantial achievements in research and who submit a dissertation with original contribution in a broad area of science. Only persons with that degree can promote doctors.

Persons with "doktor habilitowany" degree can apply for the position or title of a professor. The title of a professor is awarded by the President of State for people with outstanding achievements in science and education. The position of a professor has two levels: lower - extraordinary professor - given by the Rector of the University and higher - ordinary professor, for those with the title - given by the Minister of National Education.

In September 1990 a new Higher Education Bill was passed by the Polish Parliament. Big Universities, with a well recognized level of education, received a high degree of autonomy. They can establish their own bylaws and have a great freedom in modernizing the curricula and the organization of studies. The UT's and HSE's are owned by the State and receive each year financial resources from the State budget.

Main part of financial resources is for teaching activities, some are for research. The University can also have contracts with industry and with foreign entities in the area of research.

The diagram of the sequence of engineering education in Poland is shown in fig. 1.

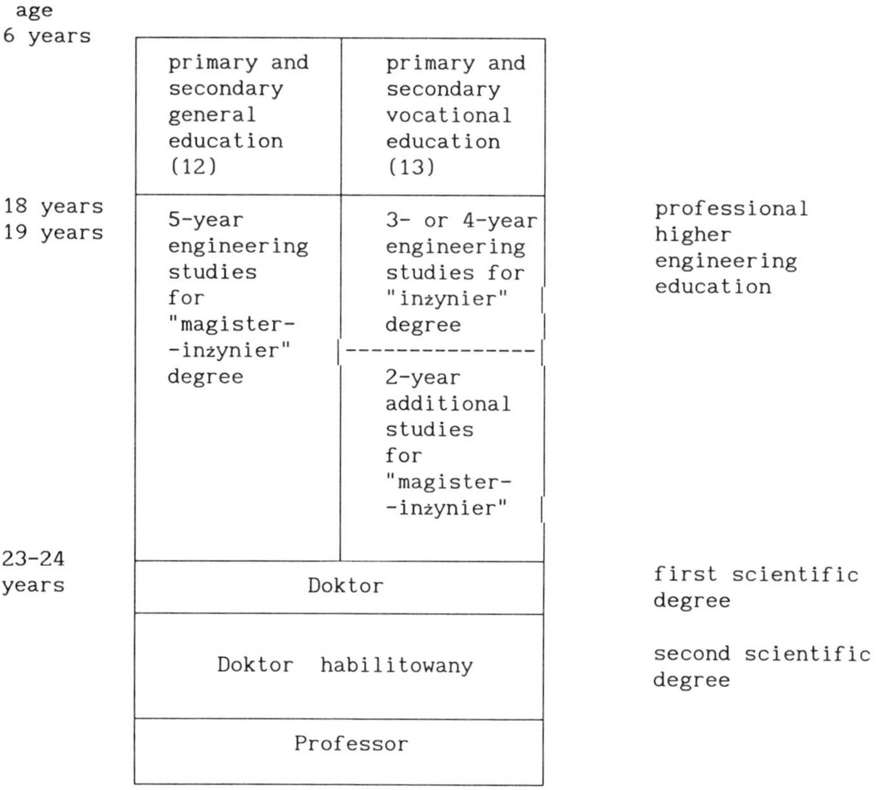

age 6 years			
	primary and secondary general education (12)	primary and secondary vocational education (13)	
18 years 19 years	5-year engineering studies for "magister--inżynier" degree	3- or 4-year engineering studies for "inżynier" degree	professional higher engineering education
		2-year additional studies for "magister--inżynier"	
23-24 years	Doktor		first scientific degree
	Doktor habilitowany		second scientific degree
	Professor		

Fig.1. Simplified scheme of the paths leading to engineering degrees in Poland

2 Curricula: do they fit to European standards?

The engineering education in Poland is on a good theoretical level but practical training of students is limited by the poor and rather obsolete laboratory equipment. Our alumni when starting to work in Western countries, have usually sufficient educational background to meet the employers' requirements. The level of mathematics and physics is at Polish high schools and technical universities rather high and that gives to our students a good theoretical background.

The level of engineering courses is also good, many academic teachers have spent some time abroad, mainly in the US and have some experience in modern education methods. The problem is therefore not in the quality of courses but in the curricula and equipment.

As it was said earlier, the basic type of engineering education is a 5-year course leading do M.Sc. degree. This was found to be too expensive and not meeting the needs for educated engineering personnel. The 5-year studies give a knowledge suitable for R & D activities rather than for engineering services and work in factories. The 3- or 4-year courses are mostly given by small provincial HSE's and do not fill the needs for that kind of education. The Main Council for Higher Education, being the highest advisory body in university studies, prepared a new proposal being based on two obligatory kinds of education, performed by each University of Technology: 3-year course (leading to B.Sc. degree) with practical orientation and 5-year course (with possible 3 + 2 version), leading to M.Sc. degree, with theoretical and research orientation. Some other options of this idea are also discussed, where one or two years are common for both kinds of education, then the student chooses the B.Sc. (shorter) or M.Sc. (longer) stream of studies. This system would have two advantages:
1. More young people would have the possibility of university studies.
2. The system would be more compatible with the Western.

The first advantage is considered to be extremely important. The rate of scholarization (ratio of students enrolled to the number of young people in the academic age) is in Poland only 8%, compared with 30% in developed countries. Our aim is to obtain this rate as high as 20% at least. Giving the possibility of completing shorter, more practical studies will certainly help in obtaining that goal.

The second problem is the specialization. In engineering there are now 26 areas and the student, when applying for enrollment, must decide which one he chooses. They are listed in table 1.

Table 1. Areas of engineering education in Poland.
1 Architecture
2 Automation and Robotics
3 Basic technical problems
4 Biocybernetics and biomedical engineering
5 Biotechnology
6 Buildings
7 Chemical engineering
8 Chemical technology
9 Computer engineering
10 Environmental engineering
11 Electronics
12 Electrotechnics
13 Geodesy and Cartography
14 Material engineering
15 Mechanical engineering
16 Metalurgy
17 Mining and geology
18 Nuclear engineering
19 Oceanotechnology
20 Organization and industrial management

21 Poligraphy
22 Sanitary engineering
23 Technical education
24 Telecommunications
25 Textile engineering
26 Transport

Unfortunately it is not the end of making engineering studies more specific. Each faculty and each department can make more divisions on higher years of studies and finally the student leaves the university being sometimes a specialist in a very narrow field. This is a heritage after the centralistic system, when the number of different specialists were planned in advance in the ministries. We are aware that this now must be changed, but the resistance against changes is very strong. As the first step the Main Council for Higher Education proposed a new list with 20 areas, in which some (No.No. 3, 4, 18, 21, 23, 26 in tab. 1) are cancelled, some are combined or changed. This is still much too many. Facing the united Europe the Polish engineering education must shift towards more general and flexible knowledge for our students, giving them the possibility of broad choice of jobs available for them after completing their studies.

The third problem is training in languages. For years Russian was the dominant language taught, now the students must learn at least one of the western languages. The level is still too low. In my opinion an engineer with M.Sc. degree should know 2 foreign languages, among them English, French or German fluently. This is a *sine qua non* condition for our engineering curricula to fit the European standards.

The fourth problem is the humanistic training of our students. During 45 years they were not taught but indoctrinated in marxism-leninism and communistic political economy. At all universities institutes were established specially for conducting those subjects. Now new curricula are prepared for training our students in market economy, management, philosophy, sociology etc., but there is no experience and limited teaching personnel. Particularly in that field there is a broad interest from Western countries and our universities get a great help both in financial support and in academic staff.

The four problems listed concern the changes which should be done in the education system in Poland to fit better our curricula to what is required and taught in EC countries. These requirements can be called the "European standards". Based on that, detailed changes should be done in syllabi of particular subjects and courses. An action is now done in most of our UT's and HSE's to decrease the number of hours the students must spend in classes. So far it was 33 hours per week, now some schools or departments have 25, aiming towards 20. At the same time much more individual home work of the student is demanded, including problem solving, literature reading etc. Some departments have introduced flexible individual way of studies, allowing the sudents to make their own choice of courses and duration of studies (subject of course to some limitations, like prerequisites). This new style of teaching needs changes in syllabi, courses must give to the students more general and flexible knowledge.

It is obvious that these tendencies face resistance from older, more conservative academic staff, who is well used to the old style of work. The reform is proceeding slowly, but with good hope for the future.

3 Equipment

The engineering studies should have a broad practical background and that needs laboratory equipment. Its quality at Polish universities is very diverse, but in general it is rather obsolete and exploited. Before 1989 the hard currency was very expensive in Poland and it was centrally distributed by the Ministry to the universities. So, very often among many old and obsolete instruments one could find a single modern western apparatus of very high performance. That, of course, could not change the general low level of our equipment.

After 1989, when the Polish currency has been convertible, the economic recession brought severe limitations to the university budgets. And again buying good equipment is very limited. Easy access to rather cheap personal computers of the IBM PC type from Far East shifted our expenditures toward computerizing of laboratories. Now academic staff and students have easy access to PC's, usually connected in LAN's or sometimes in large university of even international networks. Software, being often illegal before 1990, now is being officially purchased. Students and academic staff have been fascinated with computer technology and now most of them are highly skilled in using them.

Unfortunately computers can only simulate the real world and do not create it. The university laboratories still face a severe lack of hardware, like modern measuring apparatus, technological, chemical and mechanical equipment etc. As an example I can give the semiconductor technology. Our students during their project exercises can design electronic integrated circuits but have no facilities to perform them. In that particular case there is a co-operation with the Eurochip organization and the diskettes with our students' projects are sent to Grenoble in France. Later the encapsulated integrated circuits come back to Poland and are measured by the students. Most of the circuits are working, showing that the design skills of our students are good, but unfortunately, we can not train them in all aspects of modern technology.

The issue of equipment is very important in making our engineering education compatible with that in Western Europe. Unfortunately its solution is much more expensive than changing the curricula and probably will not be done in the near future. The economic recession in Poland and the budgetary deficit push the investment expenditures for universities down the list of needs. Optimal decisions are needed in choosing the most important items to be purchased for the limited funds we have. Some help in that matter from Western countries is also expected.

The possibilities of using the equipment owned by the Polish industry is discussed later.

4 Internationalization of engineering studies in Poland

The discussion in sections (2) and (3) has shown that the engineering education in Poland is not yet fitted to the European standards but the problem is well recognized and many activities have started. In that matter the Task Force: Human Resources, Education, Training and Youth of the European Community Commission is very active. The role of the programme TEMPUS (Trans-European Mobility Programme for University Studies), which has started in September 1990 can not be overestimated.

One of the most important issues in internationalization of university studies, not only in engineering, is the possibility of credit transfer. The students should have the possibility to change the country of their studies and to have all credits accepted and recognized at another university. This is a very difficult problem to solve and even in the EC countries the initiative called ECTS (European Community Course Credit Transfer System) is limited to some universities and only to 5 areas of studies (management, history, medicine, chemistry and mathematics). Further efforts must be done as it is the most crucial point to get the true, European way of education. So far in Poland some success in that issue has been obtained in physics.

What is the position of Poland in that approach? The four problems listed in section (2) should be solved before we start to think about including Poland to the European Credit Transfer action. The following conditions should be met:
1. Good understanding and recognition of undergraduate and graduate student conception.
2. Curricula should be compatible, giving a similar level of practice and abstraction.
3. Polish students must know European languages, at Polish universities courses should be available in commonly known languages, at least in English.
4. The infrastructure (student halls, cafeterias, entertainments etc.) should meet the European standards.

A great role in the internationalization is the programme TEMPUS. In Poland in the second year of its existence about 50 JEP's (Joint European Projects) are running, most of them in engineering education. Action 1 (JEP's) and Action 2 (mobility) have the greatest impact on the restructurization of engineering education in Poland. Retraining of our academic staff, equipment supplies and mutual visits of teachers and students bring important effects of creating new, modernized curricula, better understanding of differences and similarities, building the "European dimension" on both sides of the co-operating parties.

The first two years of TEMPUS have shown not only its advantages but also some drawbacks. It is financed from the foundation PHARE (Poland and Hungary Assistance for Reconstruction in Economy), being governed by the X Directorate of the EC Commission and the national governments devote a very limited part of it for TEMPUS. As the effect of this, only about 10% of well prepared JEP proposals are accepted for realization and the range of influence on Polish university education is therefore limited. Observing how beneficial have been the effects of TEMPUS projects running so far, I think that the way of financing and selecting the projects should be changed and the existence of TEMPUS should not be limited in time. Many other educational programs existing in EC countries should be extended to Central European Countries. Among them the most important are COMETT (COMmunity programme on Education and Training in Technology) and ESPRIT. The first gives the possibility for students to be trained in industry, which is particularly important for Polish students of engineering because of shortage in equipment and poor technological level of our plants. The second concerns research activities at universities. It is well known that engineering education can not have a high standard if the teachers are not involved in up-to-date research.

International co-operation, leading also to the "European dimension" of the university studies is emphasized through participation in international organizations and associations. Polish UT's and HSE's are members of SEFI, HEURAS, IACEE, CRE, UATI etc. It gives many useful contacts all over Europe, includes Polish universities in the international lobby of higher education. For example, a study visit organized by HEURAS in 1991 in Belgium opened many new trends and visions for engineering education in Poland. Approaching of engineering education to European standards made necessary creating of international co-operation offices at the Polish universities. These units should help in formal and administrative management of international relations and accelerate our re-entering to Europe.

5 Engineering education versus industry

In developed countries the co-operation between industry and technical universities is bilateral: the university educates man-power for industry and creates innovations in technology, while industry gives to the universities the possibility of practical training and shows the paths for research and development. Many examples of such fruitful co-operation can be pointed out, IBM educational programme is one of the most spectacular.

In Poland the economic recession for several years has stopped this co-operation, which, frankly speaking, never had been effective. Big, state owned factories, managed by communistic rules, have had no possibilities of establishing a high level of modern machines and equipment. The communistic system on the other hand created an antieconomic and ineffective organizational structure, with overextended bureaucracy. When communism collapsed and the international free market has been opened, it occured that Polish industry (particularly that using high technology, like electronics, chemistry, fine mechanics etc.) produces worse and more expensive goods than those obtainable in the West. The difficult and lengthy procedure of privatization and restructuralization of Polish industry is now going on. In that transition period the role of universities is rather limited and the access of our staff and students to modern technology is almost none.

A new area in which the role of universities can be important are the small - and medium - size private enterprises. In my opinion this should be the nucleus introducing high technology, new concepts and industrial innovation in our country. But without excessive help during the developing period only swindle short-term, maximum-profit-oriented enterprises can exist. In that problem I see an important role of universities in creating science parks, business incubators, supplying the small- and medium-size enterprises with new ideas and inventions. First business incubators and science parks with large international co-operation have already been created and the prognostics are rather optimistic. We feel that some help from Western countries in this field is crucial.

Small and medium enterprises in the development period can hardly serve for industrial training of university students in Poland. Our approach to European standards of education on the other hand needs urgently an easy access to modern technology. Without these contacts the education of our young engineers will be incomplete. The European Community created a programme COMETT, with its goal to train students in technology. So far Poland has a very limited access to this programme, the only way is through TEMPUS. It is our aim to open that possibility for our country. Selected students from Poland should spend several months in modern, well equipped factories in Western Europe. They would later have the knowledge of how to manage the enterprise on an international level. Industrial training is, in my opinion, a *sine qua non* condition for preparing our engineers for effective joining the international co-operation in the United Europe.

6 Continuing education

Engineers educated in Poland during the last 45 years present a considerable man-power, but usually their education has not been updated.

The continuing education was a forgotten area. Some courses on computers or weekend training at universities did not concern more than several percents of our engineering personnel. In Western Europe the universities use about 60% of their educational activity for continuing education, in Poland only 2-5%. The effects of this situation are now wretched. Most of our engineers aged over 40, who have spent their whole professional life in Polish industry, are not ready to join the modern, restructured plants and factories. They must be retrained.

The necessity of developing the continuing education is now starting to be well understood in Poland. Realization is not easy. In Western countries the continuing education fees are usually covered by the employers. They are not low, ranging several thousand dollars for one week training. The Polish enterprises have no funds for that, the universities also can not organize such courses free. The consciousness that continuing education is so important is not yet existing both in employers' and employees' minds.

Most universities of technology in Poland now have persons responsible for continuing education, but that does not yet solve the problem. Three items are needed: money, trainers and trainees. The lack of money is a general disease now in Poland. The trainers are more easily available, nevertheless they should be retrained in modern technological trends and in continuing education methodology, preferable in the West. The trainees will be available, when the courses will be at reasonable prices.

We hope to get help in continuous education from the Western Europe. The International Association for Continuing Engineering Education already has members in Poland. But so far the EC programs like EUROPRO have not been available for Central- and East-European countries. We should join these programs if we want to join the European standards in engineering.

Introducing continuing education in Poland should have two steps:
- retraining of trainers in good continuing education centers in Western Europe and participation of our trainees in Western courses. Financial support for these activities would be necessary, or at least reduced fees.
- starting our own continuing education in Poland available for a broad scope of engineers at reasonable prices, updating their knowledge in engineering, managing, economics etc.

With this goal reached the older generation of engineers in Poland will be able to join the technological challenge facing Poland entering in to the united Europe.

7 University Management

Last but not least the engineering education must be based on a well organized infra-structure of the University. Education means not only teachers, students, curricula and equipment, but also buildings, energy supply, administration and maintenance. The communism left the managing system of the Polish universities in a disastrous state. The administrative personnel is numerous, units responsible for maintenance work on crazy economical basis, energy consumption is too large. At the same time the economic recession in our country and the budget shortage bring a severe reduction in financial resources available by the universities. The existing system of management must be completely changed.

At most Polish universities these changes are slowly done. In many opinions - too slowly. But the problem is not easy. The financial, banking and accounting system must be changed and it takes time. The university administration should be fully computerized in network systems, with software compatible for all offices. This goal has been partly achieved, but more must be done to have the possibility of meaningful reduction of administrative personnel. The maintenance units should be private and working on market based economy. We know what should be done, but we often don't know how to do it. And now after two years of life in a free country there is still much to do in that field.

The restructuring of our university management system has been done so far on the basis of our own ideas with a very limited experience and knowledge of how it is done elsewhere. We expect to get some help in that area from the West: we urgently need training visits of our managers at the EC universities and advise from experts coming to Poland.

8 Conclusions

Joining the United Europe puts to the Polish engineering education a very difficult challenge to create man-power prepared for restructurization of Polish industry and research. The universities of technology are ready to accept this challenge, the academic staff and students are in general open to changes approaching us in Europe. But the communistic heritage makes it necessary to overcome enormous difficulties, which needs great effort and ingenuity. Not only our curricula and equipment should be fitted to European standards, but also our connections with industry, management of schools and re-education of older engineering personnel. To solve these problems which are facing us in parallel and simultaneously we need intensive help from Western European countries, particularly in programs aimed for modernizing our education and in training and retraining our academic staff, students and engineers.

The author wishes to thank Mr Yves Beernaert, the Secretary General of HEURAS and SEFI, who first introduced to me the problems of engineering education in EC countries and has stimulated my enthusiasm for approaching the European dimension in engineering education in Poland.

Strategy for Engineering Education Development for Hong Kong in the 21st Century

T.P. Leung
Department of Mechanical and Marine Engineering, Hong Kong Polytechnic

Abstract

Hong Kong is in political transition and with a sudden expansion of student numbers for degree entry, the design of engineering courses to suit Hong Kong's future need is important. This paper intends to analyse the strengths and weaknesses of engineering education in the context of Hong Kong, taking into account cultural, historical and geographical factors and the future industrial development and services to be provided by Hong Kong. Design of courses and the maintenance of standards to meet the future need will be analysed. The role of the Hong Kong Government and the Hong Kong Institution of Engineers as the professional assessment body will also be presented. This paper emphasizes the need to have an integrated approach in dealing with engineering education in view of the very complex societal structure and ever-changing needs of a city like Hong Kong.

Keywords: Political Transition, Special Administrative Region, Language, Course Structure, Engineers in Society, Engineering Training, Maintenance of Standards, Hong Kong Institution of Engineers.

1 Introduction

Education is an activity of major economic significance. It is concerned with developing and upgrading human resources which are indispensable for the economic development. To discuss the engineering education development for Hong Kong in the next century, we must understand the past, present and the possible future economic situations of Hong Kong before any strategy for education can be developed.

Hong Kong is situated at the mouth of the Pearl River in South China, and was colonised by Britain in 1840. On 19 December, 1984, the Chinese and British Governments signed the Joint Declaration on the Question of Hong Kong , affirming that the Government of the People's Republic of China (PRC) will resume the exercise of sovereignty over Hong Kong with effect from 1 July 1997. China agrees that Hong Kong will under the guiding principle of 'one country, two systems' become a Special Administrative Region (SAR). The socialist system and policies of China will not be practised in

Hong Kong, and the capitalist way of life of Hong Kong is guaranteed for the following 50 years. Thus the year 1997 and the 50 years thereafter are seen by the people in Hong Kong as a challenge and opportunity.

Economically, up to and just after World War II, Hong Kong was basically an entrepot for China. The formation of the PRC in 1949 had a major effect on Hong Kong and with the subsequent UN embargo on trade with China, Hong Kong shifted its economy to light manufacturing, establishing first a textile and garment industry and later plastics and consumer electronics industries. In the decade commencing in 1970, Hong Kong further diversified into financial services. Since 1978, with the 'open door' policy of China, the entrepot business of Hong Kong has revived, and today Hong Kong has a complex economy of three interrelated sectors--diversified manufacturing and manufacturing services, financial and business services, and entrepot services; and is the tenth trading economy in the world.

Hong Kong has a population of about six million and since Hong Kong has no natural resources, her people are her most valuable asset. The Hong Kong Government understands the importance of education for supporting Hong Kong's survival and has built up a rather successful pseudo-British education system in Hong Kong.

2 Challenge and Opportunities in the Next Century

In manufacturing, Hong Kong is facing competition from Taiwan and Korea for high value-added products, and from Thailand and Malaysia in labour-intensive products. In financial services, competition is from Australia and Singapore.

Some people in Hong Kong view the transition in 1997 with doubt and uncertainty, and there is a brain drain problem. The year 1997 also presents Hong Kong with an opportunity to develop a special relationship with China. As a result of China's open door policy, Hong Kong manufacturers have set up joint-ventures and out-processing activities in China, shifting assembly and other labour-intensive operations to China, especially in the Pearl River Delta region. Hong Kong entrepreneurs have employed around three million workers there, while keeping the high value-added aspect of manufacturing in Hong Kong. Hong Kong has grown to become a 'manufacturing service centre' which provides supporting services to production plants in China in the form of up-front services such as marketing, design and sourcing, as well as rear-end services such as warehousing, quality control, trade financing and shipping.

Observations since 1984 indicate that China is sincere in allowing Hong Kong to maintain her way of life as Hong Kong is useful in this respect. Hong Kong serves as an important gateway for China to the outside world. Hong Kong will continue to be useful to China if Hong Kong can maintain the present prosperity. The Hong Kong economy holds the key to her future.

Some businessmen have studied ways of building prosperity in Hong Kong in the future. They proposed that Hong Kong will need to remain

an international open city, and there is also a need for improving the
present economic infrastructure. They proposed strategies to develop
human resources and technology. The key elements in their strategies
are a skilled and adaptable workforce, accessible technology,
entrepreneurial management and capital availability, (SRI, 1989).
Areas of technology towards which Hong Kong can advance are also
identified by them as well as by other groups, (Kao and Young, 1991).

3 Bases of Strategy for Engineering Education Development

The bases of planning for the future engineering education can be
derived from the above discussion, whilst taking into account of other
constraints.
(1) Hong Kong will continue to be an international city.
(2) Hong Kong will have a special relationship with China after
1997 as an SAR.
(3) Hong Kong will grow to become a manufacturing service centre in
the region.
(4) Hong Kong needs to develop new high value-added products and
services to face competition from other Asia-Pacific regions and even
from China.

4 Proposed Strategy for Future Engineering Education Development

Traditionally engineering education in Hong Kong has been modelled on
the British system. Presently there are three universities and two
polytechnics offering degree courses in engineering, all with A-level
entry. There are now about 8500 first year first degree places,
sufficient to cover about 10% of the 17-20 age group. In 1994-95 the
places are planned to increase to 15,000, representing no less than
18% of the group. Of the places available , no less than 25% will be
in Engineering. In addition there will be about 4500 sub-degree
places at Higher Diploma and Higher Certificate level for the training
of higher technicians and technician engineers, many of whom have the
capability of being able to be upgraded into professional engineers
through further education. The increase in opportunities for higher
education is meant not only as a counter measure for the brain drain
but also to prepare for the increasing demand for manpower for Hong
Kong and the manufacturing bases in China. The Hong Kong community
places high value and importance on education, which in turn has its
deep roots in Chinese tradition, where traditionally the scholar had
the highest social status . In modern days, learning is still keenly
sought after and highly respected in Hong Kong. The engineering
courses offered must be able to attract the best young people and to
prepare them for the future.

4.1 Language
To remain international, language will be an important factor. At
present all engineering courses in the tertiary institutions are
conducted in English. It is expected that English will continue to be

the medium of instruction as English is the most widely used language in technology, trade and finance. However proficiency in Chinese will also be emphasized since the trend will be that the industry of Hong Kong will be more and more linked to that of China. The present day educators in Hong Kong are worrying at the students' ever-lowering standards in the use of languages, English and Chinese included. Ways of maintaining the language standards will form an important part of the future engineering courses. With the formation of trading blocs in America and Europe, and the continued dominance by the Japanese economy in Asia-Pacific , the study of other languages such as German, French and Japanese will be encouraged. This is a way to enhance Hong Kong's international links.

4.2 New Areas of Technology

The areas of technology identified to be essential for the continued prosperity of Hong Kong (SRI 1989; Kao and Young 1991) are industrial automation, precision engineering, information technology, biotechnology, materials technology, and environmental technology. The first two areas are important for upgrading the technological capabilities of the manufacturing industries. The last four areas are, with the current technological bases and advantages of Hong Kong, niche products of which may be developed to foster the further growth of the Hong Kong economy.

The trend for development of technology is to go interdisiplinary. The existing engineering education system, being modelled after the British, has the advantage that the courses offered are being of international standard, satisfying the standards set by professional engineering institutions and the academic accreditation body (CNAA). However one may feel that the general subject requirements set by the professional bodies are quite rigid, so that the engineering courses tend to be 'structured', and a course with a mix of cross-discipline and interdisciplinary subjects may have difficulties in gaining the recognition of these bodies. With the transition in 1997, Hong Kong will possibly be given a freer hand in tailoring courses to suit her economy strategy. It is expected that the future course will have only a very small number of core subjects and with plenty of modular subjects for selection. In fact the modular trend is already entrenched in the MSc courses in Industrial Automation, and Precision Engineering offered by the Hong Kong Polytechnic. Hong Kong has one advantage that she has the 'Hong Kong Institution of Engineers' (HKIE) which is a single statutory body to represent the engineering profession composed of 13 engineering disciplines and set standards of performance and conduct. The qualification process of becoming a member of HKIE is of international standard and HKIE has a mutual recognition agreement with the UK Institution of Civil Engineers. It is expected that a single statutory engineering body will allow the engineering course content to be more flexible and easily adapted to the ever developing technological scene.

4.3 Engineering Training

Though the core subjects required for study may be reduced, there is one area where training may be reinforced. This is the Basic Workshop

Practice Training, which is equivalent to the Engineering Applications I in many of the UK professional engineer training requirement. In general Hong Kong does not have many large companies which have sufficient facilities to provide systematic training in this respect. The basic training is essential for the formation of engineers, and this would then need to be provided in education and training institutions so that a graduate engineer can fit in with the engineering world directly right after graduation. Such an Industrial Centre has been set up in the Hong Kong Polytechnic which provides a total of up to four months of training in traditional and modern processes to students over their three years of study. It is expected that engineering training will be maintained in this way in the future.

4.4 Engineering Management and Engineers in Society

Engineering management is also an area which needs to be developed, not only to train engineers to manage the future high technology, but also to provide managerial support for the manufacturing bases in China. Hong Kong people are well known for their diligence, adaptability, ambition and entrepreneurship. In the English speaking world, it is a general observation that an engineer can seldom get to the very top positions of large corporations, these positions are usually taken up by lawyers or accountants. The social status of engineers is not that high when compared with some other professions. Engineering courses are in general academically more demanding than other professions or disciplines. A modern engineering graduate is however well-rounded and can easily adapt to all types of job environment. In fact fresh engineering graduates in Hong Kong are often recruited by trading and business firms for their training and especially their computer literacy. However there seems to be one important point missing from the present engineering education, engineering students are not instilled with a vision to become leaders of society. It is astonishing that when we compare a graduate engineer with a graduate social worker to find the latter has the vision and confidence of becoming a leader of society. The present day engineer is perhaps too much indulged in his engineering capabilities and does not have the wide outlook to face the society and the world. The wealth of present world has been created in a large part by past engineering feats. There should be a revamp and upgrading of the content of the courses now generally known as Engineering Management, Engineers in Society or the like to train future engineers to have a wide outlook, entrepreneurial spirit and a vision of becoming leaders of society in creating wealth.

4.5 Continuing Education and Training

What Hong Kong needs is a skillful and adaptable workforce. Continuing education and training (CET) for this workforce will be important to provide them with the knowledge they have not learnt in their formal education and also to upgrade what they learnt. The tertiary institutions in Hong Kong are all expanding their provision for CET work. Undergraduates in the institutions will be trained so that they would become life-long learners through formal CET provided or self-learning.

4.6 Research and Development

Hong Kong needs to develop new high value-added products to face ever demanding competition. With the quickening of the technological pace of the world, Hong Kong can no longer rely on purely acquiring technology from the outside world when it is fully developed, otherwise Hong Kong will always lag behind due to the ever-shortening of the product cycle. Foreseeing this trend, there are now greatly enhanced public funds for research activities to the tertiary education institutions through the University and Polytechnic Grant Committee, and the Industry and Technology Development Council of the Government. The private sector is also beginning to take a keen interest, particularly in research and development work of the applied category. The Hong Kong Polytechnic now has a number of industry sponsored project and research projects through the Teaching Company Schemes.

Links with China enhances R&D work in Hong Kong, since China has a very strong R&D workforce. Hong Kong tertiary institutions now have over 100 joint research projects with China, and these are mainly carried out in Hong Kong. Research degree places are now filled by qualified research students from Hong Kong and China. Co-operation with China in R&D also involves developing and commercializing research results initially completed in China. It is expected that academic exchanges with China and the rest of the world will be more frequent so as to maintain the international outlook of Hong Kong and upgrading the R&D standards of Hong Kong.

4.7 Quality Assurance

To maintain Hong Kong's international competitiveness, the future engineering education in Hong Kong must also be of international standard. This is to be guaranteed by the Hong Kong Council for Academic Accreditation (HKCAA) and the Hong Kong Institution of Engineers (HKIE). The HKCAA is modelled after the CNAA of UK, and consists of local and overseas academics as well as local community leaders. The present qualifying processes of HKIE are modelled after UK professional engineering institutions, but HKIE is seeking to have more international links so that its membership will be recognised internationally. The presence of overseas academics and the international links of HKIE will help Hong Kong to maintain the high quality and standard of engineering education.

5 Conclusion

Hong Kong is facing a major transition in 1997. The development of engineering education in the next century must be matched with the anticipated economic, social and political developments. The future course will be designed so that the graduates will be wide in outlook, well versed in English and other foreign languages, and will have the vision to become leaders of society. The unique entrepreneurial spirit of Hong Kong is to be promoted. The course structure should be flexible so as to facilitate quick assimilation of new technologies, and continuing education and training will become an important part of the engineering education. Basic engineering training will be acquired by the engineers while they are studying as undergraduates. R&D will become important and co-operation with China in this respect will be close. The quality of the education will be maintained to international standard through liaison with and participation in the outside world.

References

Kao, Charles K and Young, Kenneth (1991) **Technology Road Maps for Hong Kong**.

SRI International (1989) **Building Prosperity: A Five-Part Economic Strategy For Hong Kong's Future**, a report prepared for the Hong Kong Economic Survey Limited.

Comparison of Engineering Education between U.S.A. and Japan - Focusing on Undergraduate Education

H. Ohashi

Department of Mechanical Engineering, Kogakuin University

Abstract
Features of Japanese engineering education at undergraduate level are characterized by the comparison with those of U.S.A. in terms of admission, study motivation, courses, graduation thesis, accreditation, retention rate, job finding, expectation of employers and role of professors. The study motivation and intensity are generally lower in Japan but graduation thesis in senior year seems effective to enhance students' ability of cooperative work and knowledge synthesis. Japanese employers evaluate fresh graduates as raw material which is to be wrought and shaped after employment, while US employers evaluate them as preliminary finished product which can be used immediately after employment.
Keywords: Engineering, Education, Undergraduate, Comparison, U.S.A., Japan

1 Introduction

Japanese school systems at secondary and tertiary level were completely remodeled in 1949 copying exactly those of U.S.A. University education was then expanded from three to four years and at the same time general education on humanities, social and natural sciences (minimum 36 credits), language education (8 credits) and physical education (4 credits) were added on major education (76 credits) as an indispensable part of university education. Japanese universities, which had concentrated in the education of specialized major in pre-reform period, were urged to build organizations and faculties to conduct the new assignment in a relatively short preparation time.

Because of the above mentioned historical circumstances, the spirit of general education has been transplanted superficially in Japan and both university authorities and students have assigned less priority and endeavor to general education. Consequently the first two years allotted principally to general education have become rather a relaxed period during which students compensate gloomy years of entrance preparation for universities of better ranking and this academic climate has been spoiling the study moral of students throughout the college life.

Japanese primary and secondary educations up to high schools have established an internationally acknowledged reputation for their high quality, well-balanced education with a high average level of student achievement. To the contrast, the higher education system of Japan is considered as the weakest part of the entire system. Among the most discussed issues are the entrance examination, the quality of undergraduate education, rigidities in the university-based research system, and the limited opportunity for graduate and continuing education (cf. Japanese Education Today, 1987).

Japanese industries have been steadily developing over the past 30 years and have established a firm standing in international competitiveness. This remarkable performance has been supported partly by the massive supply of fresh engineers from universities and colleges. If the engineering education in Japanese universities have been so ineffective as the university education has been evaluated in general, Japanese industries must have badly experienced the lack of qualified young engineers who are the central work force for the development of highly automated production lines and ever increasing R&D tasks.

There must be a kind of subtle understanding between employers and education of Japan so that universities supply engineers who may be not well finished at the point of graduation but are willing and flexible enough to be trained after the employment in accordance with the employers' needs for specific engineering assignment.

As already mentioned the Japanese education system is the duplicate of US system. There is, therefore, little difference in the organizations, degree-requirements and curricula of both systems as far as we compare them from written documents. When we check the contents of typical text books of a specific course used in Japan and U.S.A. for example, they are admirably similar in items and depth.

The present paper is to have a closer look at the undergraduate education of both countries and to pick up essential and structural differences which emerge from the superficial similarity.

2 Comparison of Undergraduate Education

Undergraduate education of Japan can be described or ironically summarized by the famous words of Edward Fiske (1983);

"American students, by and large, take examination to get out of school, Japanese take them to get in."

or of Edwin Reischauer (1986);

"The squandering of four years at the college level on poor teaching and very little study seems an incredible waste of time for a nation so passionately devoted to efficiency."

Table 1 summarizes the difference of undergraduate education of both countries from various aspects, that is, admission, study motivation, courses, accreditation, graduation thesis, retention rate, job finding, expectation of employers and role of professors. These are mostly personal view of the author and there could be another analysis and comparison both from American and Japanese sides.

Table 2 illustrates two examples of courses which typical mechanical engineering students of the 3rd year take in the first semester. Examples are drawn from the Johns Hopkins University and the University of Tokyo and show the case of students who finish minimum credits required for graduation.

As the example indicates, US students concentrate in relatively small number of courses in a semester, while Japanese take quite a number of courses simultaneously. Heavy credit requirement for Japanese students in the 3rd year is caused by the fact that they concentrate principally in graduation thesis (10 credits) in the 4th year and consequently they have to finish most of required credits for graduation practically in three years. Graduation thesis is compulsory in majority of Japanese engineering education and its positive educational effect is widely acknowledged. It is also true, however, that the introduction of thesis work results in overcrowded courses and deteriorating study accomplishment. The contrast seen in Table 2 could represent the character of engineering education of both countries.

3 Engineering Education in Life-Time Career Development

In Japanese system, entrance examination is indeed tough but once a student overcomes the barrier, his way to the graduation is relatively easy. He may take quite a variety of courses covering from big bang to artificial reality just to satisfy his curiosity about something new, but the understanding remains mostly superficial just like the case of watching TV programs.

Students afford to share abundant time for leisure and sports activities so that the four years of study are virtually a period of enjoying the blooming youth and to recharge energy for the severe battle among corporate warriors in which they will be involved after the graduation.

Industries regard the fresh graduates as raw material which can be shaped by the On- and Off-the-Job Training after the employment and do not lay much value on students who have definite objective and intention in a specific field. Japanese industries used to prefer it fat pupae which could metamorphose to any insects companies want. The faculties of Japanese universities have had excuse for their less demanding education that they are offering a quiet restful period for these pupae.

In the U.S.A. most of recruits are intended to fulfill specific job vacancies with corresponding specialities. Therefore, US graduates are often requested to state in the job application in what aspects they are distinct and special. They should be preliminary finished products at the point of graduation. All these aspects including life-time career development in companies are metaphorically illustrated in Figure 1.

When we look closer, there found several basic differences in the concept of education which have their roots in traditions, social structures, status systems, etc. of both nations. Japanese engineering education has been successful in the sense that it has been supplying an adaptive and productive engineering force for industries, in large part because Japanese way of education is effective not in fostering study accomplishment but in enhancing wide basic knowledge, disciplined work habits and group cooperativeness through the combination of courses, practices and graduation thesis work.

Such a feature of Japanese education has been positively accepted by industries and employers, but opposite opinions are gradually emerging among industries that the past system is not adequate to stimulate the creativity of engineers, the very sought qualification for pioneering in the development of new products and concepts. Large part of the strength of Japanese industries has relied on the high motivation due to group consciousness but the trend is shifting definitely toward individual creativity. Japanese engineering education must reform itself to cope with this changing demand.

The problems of higher education system are widely recognized in Japan and they have been long the subject of extensive scrutiny in and out of government. Rigidity, excessive uniformity and lack of choice have been pointed out as the most serious cause of the illness. National Standard of University Qualification was amended last year in such way that total credits of 124 were prescribed as the minimum condition for graduation while separate credit requirements for general, language and major education were removed. Every university has started to look for a uniqueness of education and this movement could result in a firm step toward the improvement of undergraduate education.

4 Concluding Remarks

Undergraduate educations of U.S.A. and Japan are compared each other. Although they are similar as far as organizations and curricula are concerned, there found several basic differences which are rooted in the traditions, social structures and status systems of both nations. The target of Japanese engineering education started to shift gradually from the supply of flexible, disciplined engineers to the development of individual feature and creativity.

References

U.S. Study of Education in Japan (January 1987) edited by Dorman, C.H., U.S. Department of Education.

Fiske, E.B. (1983) Japan's schools: Exam Ordeal Rules Each Student's Destiny, New York Times, July 12, p. A1.

Reischauer, E. (1986) Introduction, The Japanese school: Lessons for Industrial America, by Duke, B., New York, Praeger, p. xviii.

Table 1 Comparison of undergraduate education

Japan	U.S.A.
Admission	
Judged by scores of National Test Center for University Admission and/or entrance examination of individual university. Personal judgment is regarded as unfair and only scores of paper test serve as sacred scale for admission.	Multilateral judgment by SAT/ACT scores, recommendation, academic record at high school, interview, etc. Much account is given to personal views and evaluations.
Study motivation	
Very low. Students are suddenly depressurized from the stress of entrance exam and start to enjoy non-academic activities which they had to forgo during grueling period of entrance preparation. All know that credentials are given for the successful entrance, not for exit.	Students begin study with fresh willingness. They are highly self-motivated since everyone realizes that better academic achievement implies better opportunity and promotion in the career.
Courses	
Students take 10 to 15 courses a week, each course once (90 min) a week. Passive attitude in class with few question and involvement. Frequent cut class. Result of exam at the semester end determines pass or fail. Poor guidance for systematic learning with few prerequisite. One-way classes with silent listening; Poor understanding = poor student's ability	Take 4 to 6 courses a week, each 1 to 3 times a week. Much home assignment to improve understanding, supported by office hour and teaching assistants. Effective learning through systematic arrangement of prerequisite. Two-way classes with much question and assignment; Poor understanding = poor professor's performance
Graduation thesis	
Compulsory for graduation. Students are intensively involved in the thesis work and learn team work together with graduate students. They experience a flavor of academic research for the first time and find learning is exciting, but often too late. Students are in a sense free labor force for research activity.	Not common. Student may participate in project work on paid basis.

Table 1 Comparison of undergraduate education, continued

Japan	U.S.A.
Accreditation	
Before the foundation of a university, Ministry of Education inspects whether all requirements (mostly quality and quantity of faculties and facilities) are met according to the regulation. 　If once permitted, the license keeps valid forever. No guarantee on the quality of education but complete freedom of curriculum design. 　Insensitive to accreditation because universities accept certified students.	Periodical accreditation by ABET guarantees the quality of education. 　Freedom of curriculum design is somehow restricted. 　Accreditation is important because universities certify students.
Retention rate	
High, 90 to 95%. 　Pass entrance exam + 4 years = diploma except for those who are extraordinarily lazy. 　Diploma indicates nothing about scholastic marks or ranking.	Moderate, 70 to 80%. 　Diploma is a certificate of achievement with possible indication of honor.
Job finding	
Department heads are almost employment mediating officers. 　Strong ties with industries through the mediation of recruit activities.	Fundamentally individual activities of students with the assistance of placement officers. 　Professors help students by writing recommendation letters.
Expectation of employers	
Little. 　They will be better trained in company after employment.	High. 　They will be treated as rookie specialists with limited experience.
Role of professors	
Produce manpower with brand of specific university. 　Keep human-relations with graduates for longer time and serve as consultant for business and private matters.	Provide students with knowledge and skill. 　Search for funds to help students by scholarship or laboratory job, etc. 　Highly skilled in writing attractive proposals and applications.

Table 2 Comparison of courses taken by mechanical engineering students of the 3rd year in the first semester

University of Tokyo		Johns Hopkins University	
Course	Credit	Course	Credit
Mathematics 2	3	Fluids I	3
Numerical Analysis	1.5	Heat Transfer	3
Machine Design	1.5	Circuits	1
Production Technology	1.5	Electrical Eng. Lab	3
Machine Dynamics	1.5	Technical elective	3
Automatic Control 1	1.5	Humanities or Social Sci. elective	3
Structural System Eng.	1.5		
Electrical Eng. 2	1.5		
Fluid Mechanics 2	1.5		
Mechatronics 1	1.5		
Heat Transfer	1.5		
Experiment & Instrumentation	1.5		
Software Eng. 2	1.5		
Mechanical Eng. Seminar 1	2		
Design & Drawing Practice 2	3		
Mechanical Eng. Lab. 1	1		
Subtotal	27	Subtotal	16

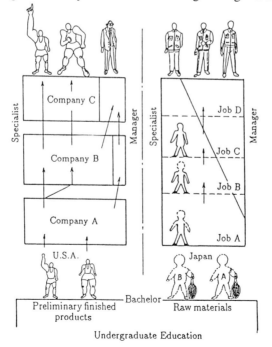

Figure 1 Comparison of life-time engineering career

An International Approach to Career Development

J.A. Lorriman

The Centre for Consultancy, Leicester

Abstract

This paper presents the key findings from an international Anglo-Japanese research project and relates these to a number of important career development initiatives in which the author has been involved. In addition, this paper puts forward a framework for an International Modular Masters degree within the aegis of the International Association for Continuing Engineering Education (IACEE). Above all, it is argued that there must be both much more structure in the career development of technical staff, and much greater international mutual learning, which is a major theme of a book the author is currently writing for Oxford University Press on the lessons to be learnt from the Japanese approach to education, training and management.

1 International Research Project

In 1985 the author had the privilege of being awarded a Winston Churchill Travelling Fellowship to visit Japan for six weeks (Lorriman, 1985 and Lorriman 1986). The purpose was to compare the approach to the education and training of electronics and software engineers in Japan with that in the UK. Much valuable assistance and advice was given by the late Professor Keith Thurley, Professor of Personnel Management at the London School of Economics (LSE), who also participated in many of the visits.

Keith Thurley had been involved in research in various British electronics companies and extended this research into GPT; at the same time the author was appointed Nancy Seear Research Fellow in the Personnel Management Department of the LSE.

It became obvious that the three key problems in British companies in providing effective Continuing Professional Development are:

1. Inadequate ongoing technical and management training.

2. Insufficient career development.

3. Giving engineers enough challenge in their jobs. They are both misutilised and underutilised.

In 1987 a two year research project was initiated by the author together with Keith Thurley. It involved some 50 engineers in electronics companies in both the UK and Japan. The purpose was to study in depth the training experienced by a limited number of engineers over a six-month period. The documentation was based on the Institution of Electrical Engineers' Professional Development Record (PDR), which will be described later in this paper.

The importance of these results is that it illustrates a vitally important advantage possessed by Japanese companies; their approach to the skills development of their staff gives them a breadth and depth of talent in their employees which provides enormous flexibility. This is a key strategic advantage in the world marketplace.

Some of the key results are shown in Tables 1, 2 and 3:

Table 1: Utility of Knowledge/Skills Acquired Through On-the-Job Training (O-J-T)

	Japan (%)	UK (%)
For present tasks	20.0	58.2
For future tasks	45.2	25.4
Total number of O-J-T incidents	83	67

There is an interesting comparison of the approaches to On-the-Job and Off-the-Job Training in the two countries:

Table 2: Methods of On-the-Job Training

	Japan (%)	UK (%)
Formal		
Individual	0.0	0.0
With one another	2.5	1.5
Group	38.8	3.0
Spontaneous		
Individual	21.1	79.1
With one another	15.0	8.9
Group	22.5	7.5

Table 3: Methods of Off-the-Job Training

	Japan (%)	UK (%)
Internal seminar	53.6	52.2
Internal course	10.7	17.4
External seminar	28.6	13.0
External course	7.1	17.4

Absolutely key in the Japanese approach is the personal responsibility accepted by their managers for the development of skills by their staff. These statistics also demonstrate that there is need for much more long-term orientated learning by technical staff, that there is greater room for group learning within the organisation itself and that more effective external seminars and courses can be developed, including such approaches as an International Modular Masters degree framework.

2 A Framework for an International Modular Masters Degree

The IEE has its own modular MSc approach, the Continuing Education Modular Masters, and SEFI also has its international post-graduate modular EuroPro degree.

GEC, the largest electronics company in the UK, has agreed with a number of Universities, Polytechnics and Management Colleges to organise a modular post-graduate system in which there is mutual accreditation of modules towards an MBA or MSc within a scheme known as the Professional Development Consortium. Open University courses are also accepted within this agreement, and the intention is to extend it to include Universities elsewhere in Europe.

In the USA there is the satellite based National Technological University, and modules from this system, together with the European equivalent, EuroPace, are being introduced into the Commonwealth of Independent States.

These are just a few examples of many such schemes and there seems little doubt that this modular and international approach will become more and more popular as companies increase the international career development opportunities for their graduates. The author has therefore been asked by IACEE to chair a Continuing Professional Development Working Group to implement a framework whereby all these different modular Masters degrees can be made truly interchangeable and international. This was announced at the 5th World Conference on Continuing Engineering Education in Helsinki in May 1992 and the objective is that this international framework should effectively be in place for the Sixth World Conference in 1995, with most of the major modular Master degree schemes agreeing to participate by then.

The proposed framework for this International Modular Masters is:

- It would be launched, publicised, co-ordinated and licensed by IACEE.

- Each Founding Organisation will co-ordinate the approval of participating Masters degree courses by appointing a Steering Committee to approve, license, monitor and publicise the IMM in their territories.

- Participating providers of modules within the IMM will have to be prepared to accredit all modules awarded by other IMM providers licensed by IACEE.

- Students enrolling on the scheme will do so by registering with a local licensed IMM provider.

- The scope and duration of the studies would be based on the EuroPro degree, ie equivalent to a total of 40 weeks (one year) of full time study. The duration of study would be between 2 and 4 years, with 6 years as the maximum.
- The required initial qualification, like EuroPro, might be a minimum 3 year recognised engineering degree. In addition, the participant would have at least 3 years professional experience.

The benefits to participating students would be the opportunity to extend the modules available from their existing modular Master degree courses with modules from other schemes and other countries. The benefits to the providing universities and polytechnics would clearly be more students, and employers will no doubt benefit from broader developed engineers in an international context.

3 The Key Elements of Successful Continuing Professional Development

Most CPD schemes around the world are far less effective than they might be, simply because there is no focus or coherence. This will only be the case if there are three essential and coherent components:

3.1 The Commitment of Individuals

Only through systematic analysis of their learning and career development needs and experiences will individuals be able to achieve real fulfilment of their abilities. The only effective way in which this can be done is by the use of Personal Development Plans, such as the Institution of Electrical Engineers' Professional Development Record. The evidence is that, for example, large Japanese companies, using this approach, are over twice as productive as smaller companies, which are much less systematic in developing the skills and careers of their employees.

3.2 The Manager as Coach

It is absolutely essential that managers are given the skills and confidence to coach effectively. Unfortunately, most managers outside Japan have never learnt the skills of coaching, mainly because they have never been coached themselves. They therefore neither understand the enormous power it can give in releasing the real potential in their staff, nor do they feel confident enough to do it.

The approach used with outstanding success by the author's consultancy is Achievement Coaching. Unlike most 'Management Coaching', the technique used starts from the employee's agenda and directly complements the use of Personal Development Plans. Managers are taught to coach by being coached themselves and very often the change in their management style is quite remarkable, as are the results they are able to elicit from their staff, whose trust and respect for them rises enormously. One of the Case Studies in The Engineering Council's 'CPD - the Practical Guide to Good Practice', which was written by the author of this paper and has been made available free to the 300,000 registered technicians and professional engineers in the UK (The Engineering Council, 1991), is the American company Digital. Ken Olsen, the company's President and Founder, has made it very clear that the role of managers is not to achieve objectives themselves, but rather to enable their staff to achieve. Staff are encouraged to make decisions; the outcome, say Digital, is either a correct decision or else a powerful learning experience.

3.3 The Organisation as a Learning Environment

Most potential learning from formal training courses is much less valuable than it could be. The reason is simply that the environment at the workplace is not supportive of learning and the course itself is either not properly identified by the manager for its relevance to the employee or it is not effectively followed up. In addition to the individual and management commitment to learning, there must be as many ways as possible sought to learn within the organisation. Every opportunity must be taken to foster mutual learning, whether verbally or written, one to one or in groups. Time has to be allocated as a priority to this activity, for this is the only way most organisations will successfully utilise the real talent and knowhow within that potentially gives them their commercial advantage over their competitors.

4 The Professional Development Record

Last year the author completed his three year term as Chairman of the Professional Development Committee of the Institution of Electrical Engineers (IEE), which with 120,000 members is the largest Learned Society in Europe. In 1986, in response to a Discussion Document from the Engineering Council, the IEE developed a document at the author's suggestion which it called the Professional Development Record (PDR). It is a document designed to help employees to structure their experiences, both on and off the job, as well as their career development objectives.

The IEE piloted the PDR with 200 engineers in GPT in 1987 and launched it nationally in April 1988. By the end of 1990 the 5,000th had been sold and that figure is now over 8,000.

5 The Cost of Staff Turnover

Working in conjunction with the Institute of Manpower Studies, the author developed a cost model of the direct costs when a member of technical staff leaves the company. This is important in terms of asking ourselves how high the cost will be if we do not take appropriate action to provide engineers with much more satisfying and challenging careers.

This model takes the following factors into account:

- Loss of efficiency while the employee is looking for another opportunity.
- Loss of efficiency during the notice period.
- Loss of resource while a replacement is found and other employees cover for

the leaver.
- Recruitment expenses.
- Training of replacement employee.
- Learning time for new employee to become fully effective.
- Diversion of other staff.

The total costs involved are thereby estimated at a minimum of £20,000 each time an employee leaves.

During the the first two years of GPT, the author's previous company, as a result of introducing personal development plans and better structured career development, the turnover rate of the 4,000 graduates was reduced from 10.9% to 5.4%, which represented a very considerable financial saving to the company of over £4 million pa.

As was shown earlier in this paper, the evidence from research indicates strongly that the major reason for dissatisfaction, and therefore staff mobility, amongst graduates is lack of training and career development. It is therefore extremely important to quantify the financial benefits of investment in this way in terms of the savings in lower turnover costs.

These and many other lessons are described in a book which the author and Professor Takashi Kenjo of Japan's University of Industrial Technology will have published early next year by Oxford University Press (Kenjo and Lorriman, 1993).

6 References

Lorriman, J. (1985) The education and training of technicians and engineers in Japan, Report on Winston Churchill Travelling Fellowship 1985.

Lorriman, J. (1986) Ichiban - the Japanese approach to engineering education, Electronics and Power, August 1986, The Institution of Electrical Engineers.

The Engineering Council (1991) Continuing Professional Development - the Practical Guide to Good Practice, 10 Maltravers Street, London WC2R 3ER.

Kenjo, T. and Lorriman, J. (1993) Japan's Winning Margins - Education, Training and Management, Oxford University Press.

Engineering as a Liberal Art
K. Corfield
Former Chairman and Chief Executive, STC plc

Abstract
The great achievements of engineering and the quality of
intellect required for their accomplishment are called in
evidence of the claim for engineering to be regarded as
one of the Liberal Arts. By being so perceived it is
claimed that engineering would attract a greater proportion of the best intellects in the population. This, and
the fact that the discipline of engineering would be more
widespread among those practising other callings enhance
the respect in which engineering is held. Comparisons are
made with the countries in which this situation more
nearly exists. To bring about this desirable end proposals
are made for change in the training and education of
engineers and a simplification in their qualification
process aimed at removing the confusion which presently
exists in the lay population as to what constitutes an
engineer. The universal qualification of a first degree
would be followed by additional training and a higher
degree for those aspiring to the highest professional
standing. There would be a single accreditation body
empowered by the State to award and protect the status of
professional engineers based primarily on academic success
supplemented by submissions on training by employers and
Institutions. Education both broader and deeper than at
present practised is seen to be the critical factor in
bringing about the objectives propounded.
Keywords: Engineering a liberal art, Professional
Qualification, National accreditation, German Engineers

1 Introduction

It was Baroness Platt who first crystallised the idea. The
occasion was a seminar at the City University on the late
Sir Monty Finniston's report "Engineering our Future".
Little did I realise then that I should be asked to fill
the role of founding chairman to the Engineering Council
which resulted from Monty's work. When the time came to

recruit Council members, one of my first thoughts was of Beryl Platt, herself an aeronautical Engineer, who had made the stirring comment that engineering was the equal of any of the Liberal Arts as a basis for life's great experience. She became one of our best contributors.

My proposal that engineering should be accepted as a Liberal Art in no way detracts from the regard in which I hold the present body of that study. One of life's special pleasures is to converse with a well-read man or woman.

My dictionary describes the Liberal Arts as "the fine arts" and lists humanities, sociology, languages and literature, all of which are attributes that add to and round off the study and practice of engineering. My dictionary also describes a liberal education as one that aims to develop cultural interests and intellectual ability. To this I would add the preparation for the conduct of life itself in society.

What gives engineering these special claims and why are they not recognised in the great body of British society? The claims of engineering go back many centuries and are based on its vast contribution to human development and well-being without which intellectual progress would have been hindered just as it has been shown to be in isolated communities lacking access to the benefits of engineering. Water supplies, irrigation and waste disposal have, arguably, contributed more to man's welfare even than medical science, at least until the middle of this century. Mechanised transport by land, sea and air is exclusively a product of Engineering and, together with the printed word, is responsible for the greater parity of information and the promotion of understanding between peoples that constitutes a sea-change in international affairs of which the enormity is scarcely appreciated.

If war on the grand scale becomes extinct it will largely be the result of the spectacular growth of intercourse between peoples and the resulting diminution and diffusion of state power. In turn this will reflect the ubiquity of communications, started by mechanised transport, advanced by the printed word and cemented by the all pervasive telecommunications and broadcasting networks which again are the products of engineering.

Only some historic aberration seeded in tribalism and promulgated by perceptions of class and its distinctions, can have ascribed so much worth to the creation of ideas and philosophies for the spiritual guidance of mankind and so little to the creation of artefacts essential to the development of civilised society. Fortunately this uneven view of the relative contributions of philosophy and engineering shows signs of receding. It is by no means common to all developed countries.
Germany, Japan and the USA, Italy and France all lead the

U.K. in valuing Engineers and engineering in terms which the U.K. reserves for what it regards as the finer arts.

2 Current Perceptions of Engineers

There is a degree to which Engineers have ascribed to and propagated this notion and though it would be wrong to attempt to apportion any blame, it may be right to suggest a revision of some old and widely held ideas.

For example, it has often been put to me that the Engineer who leaves his calling for the perceived greener fields of management or investment or other commercial activities is tantamount to being a traitor to his profession. Nothing, to my mind, could be further from the truth. The more frequently Engineers are seen to number among the leaders of industry, commerce and finance, the more popular will be the calling and the study of engineering. One influential former captain of industry frequently extolled the virtues of an arts and humanities education as the basis for top management and leadership. I do not have to deny him his viewpoint to state without equivocation that the study of engineering is no less a virtue and even more a practical asset.

In its report 'Engineering at the Crossroads' the National Economic Development Committee in January this year (1992), claimed that too many managers, more than half those in engineering in Britain, lack the necessary skills, and that too much emphasis is given to business, financial and marketing training rather than to technical expertise. As to which of these should constitute the basic skill I would far rather take a professionally qualified Engineer and instruct him in business, financial and marketing disciplines than take a follower of one of those disciplines and put him or her through the rigours of an Engineering degree.

On the other hand the Engineer must face some awkward questions. Why do so few go on to absorb or at least familiarise themselves with the key principals of Finance, Business and Marketing? Once the jargon is mastered the subject matter will not prove an obstacle to those who have achieved a first degree in engineering. The need, not to discourage Engineers from moving to adjacent callings, but positively to encourage them would be a real step forward for the engineering Institutions.

There is another reason for this bold step to be embraced and I can only put it forward at the risk that one of its premises will seem to criticise the great body of Engineers whose cause I aim to champion. Sometimes a truth will hurt even though its subject is in no way to blame. The hurtful truth is that a chief executive looking for top level supporting personnel may place creativity, high IQ., inter-personal skills and that indefinable quality of leadership above any and every discipline whether arts or sciences. When this occurs it is often to the detriment of the well trained and qualified Engineer. Why? because people having these basic qualities are statistically less likely to enter engineering as a career. It's a chicken and egg situation: the perception of engineering as a career is not sufficiently high to attract the very brightest of the population.

It may also be that the cognitive element of engineering is so great as to bring with it an element of suppression of the creative talents, a possibly unproven but credible hypothesis which should be considered by all educationists.

For my part I believe that an element of election on the part of the student and an element of compulsion by reason of prejudice in the education and social systems combine to steer the best intellects in channels other than engineering. The education system by reason of its preference for research and pure science and the social system by reason of the well documented stigma still associated with engineering.

Of course there are notable exceptions to this generalisation as the annals of engineering make plain but, as long as the confusion exists over the meaning of the term 'Engineer' as stressed by Sir Graham Day in the Bridge Lecture to the Worshipful Company of Engineers in February of this year, so long will ignorance, confusion and prejudice dog the path of recognition for the Professional Engineer.

It has often been stated that many qualified Engineers are employed on work which is well below their capability and I have observed this phenomenon in practice. It stems from the wide church which comprises engineering and from the persistent uncertainty about the qualifications of Engineers and what exactly constitutes a Professional Engineer. As Sir Graham pointed out in his lecture:

"within the very broad church which encompasses engineering there may be a good and precise understanding of the various qualifications of Engineers, outside engineering there is not".

He goes on to say that one consequence of the confusion is that some Engineers are accorded by third parties a recognition which is less than their fair due, including by other professionals. The reverse is also true.

All this is contrasted with Engineers on the Continent and particularly with Germany and Italy. For those who would follow up this line, essential reading is 'German Engineers, the Anatomy of a Profession' by Stanley Hutton

and Peter Lawrence (1982). This study tells us as much
about British Engineers and how they are perceived as it
does about their German counterparts. The authors take
the view that the status of Industry has positive tangible
manifestations and that Engineers are a critical factor in
Industry's higher status in Germany.

It is usual in the UK to attribute the low status
enjoyed by the British Engineer to the low esteem in which
Industry is held thus suggesting that the quality of the
Engineer makes little impact on the situation. If, as
Hutton and Lawrence suggest, Engineers are a critical
factor in Industry's higher status in Germany, we should
be encouraged to believe that Engineers do have it within
their grasp to elevate the status of Industry by their own
endeavours, that perceived prejudices may be countered and
reversed by good example. Accepting this heartening
message what can we do to assist this and the next generation of Engineers to fulfil that promise and benefit not
only themselves as individuals but the whole profession
and the industries upon which we need depend for a major
part of our future wealth production?

3 Our Confused Education and Training Inheritance

It seems to me that there are two key areas in which major
changes are required, Training and Qualification. Training is haphazard and qualification is diffuse. Before
proposing changes let us look at the main weakness of the
present situation.

Having been closely involved, as founding chairman of
the Engineering Council, with the strongly held views of
the many Institutions of the profession, I am well aware
of the objections which proposals for change can engender.
Not only that but I do find myself in sympathy with a
number of the sincerely held views which will emerge:
notably the importance of training as compared with
education. Training or structured experience is essential
to the practical application of every discipline. In the
past, however, not only was too much expected of training
but the methodology by which it was valued and applied in
the process of awarding the qualifications of membership
and fellowship of the Institutions left the whole process
of qualification open to misuse: well-intentioned and
apparently innocuous misuse certainly, but misuse just the
same.

What person in some authority could, in the post war
era, maintain that there was no occasion on which he or
she placed a benevolent interpretation on the content of
the forms initialled in support of a decent hard-working
employee whose claims to be considered a qualified
professional engineer were, to say the least, tenuous.

After all, the educational requirements were decided and reviewed elsewhere and the wide differences in standards actually achieved in pursuance of diplomas and certificates dignified by the appellation 'National' were well known and rooted in regional differences in industries and populations. What would have been good grounds for the formation of technicians and mastercraftsmen became instead the body of knowledge on which were awarded the highest qualifications the Institutions had to offer. Their very existence was, in turn, dependent on achieving a sufficient level of membership to support no less than sixty sister Institutions.

That the real qualifications, on which major and technically orientated employers judged their engineer intake, included first and higher degrees and appropriate doctorates is not in doubt but many such judgements were peer judgements of managers actually prejudiced against university degrees - a situation for which the UK motor industry was well known in the days of its relative decline and which, I hasten to say, it appears to have put well behind it today.

With sixty qualifying Institutions, a hundred and more technical colleges, very few polytechnics and the economic pressures which made Universities the province of the few who could manage to fund both full time higher education and an equally full time practical training, it is small wonder that Sir Graham Day and others remain perplexed by the lack of clarity surrounding the status of the professional engineer. Advances have been made, C.Eng., F.Eng. and lately Eur.Ing. coupled with Institution letters signifying the area of specialisation and the general adoption of a degree requirement all assist in negotiating the minefield. But the minefield remains.

4 Formation of Engineers: the way ahead.

By way of contrast with the U.K. Sir Graham quotes the simplicity of the Canadian tag P.Eng for Professional Engineer which through rigorous requirements and government recognition conveys the unique meaning which might be reserved here for the barrister, the surgeon or the architect. Germany, however, provides an example more akin to my own proposals. There the Diploma Ingenieur (Dipl.Ing.) and the Ingenieur Graduiert (Ing.Grad.) are the two professional engineering qualifications awarded by the State and protected by Federal Law. The Diplom. Ingenieur qualification calls for one year of practical training and a minimum of 4 1/2 years at University, though in practice 5 1/2 years is the average time taken. One in ten go on to become Dr.Ing. Since they mostly start from the Abitur which is taken at age 19 they are

unlikely to achieve this high qualification before the age
of 26. The Ingenieur Graduiert may have left school at 16
but will have spent two or three years in a 'Fachober-
schule' before entering the 'Fachhochschule' which will
provide a three year full-time course leading to the
Ing.Grad. qualification at 21 or 22 depending on the
length of practical experience enjoined. In many cases
those qualified will not practice as engineers but will
broaden their studies and practical training to become
general managers, rather against the purist concept of the
profession, but very much in accord with my own wish to
see engineering as a Liberal Art.

You may have gathered from the foregoing that I
strongly favour the concept of full-time engineering
education in an establishment of higher education be it
called a university or polytechnic. I believe that it is
in such an establishment that the bona fides of the
student to become a professional engineer should be
established. Indeed I would not be opposed to qualifying
suitable such establishments to award the professional
qualification but, since the practical training necessary
may be beyond the capacity of the university or polytechnic to deliver, the need for a government-sponsored body
to set and uphold the standards and scrutinise the output
is one that I freely accept. On the other hand, the degree
which students require for their professional qualification would equally be open to students meeting the
necessary entry requirements but not having the profession
of engineering as their specific vocation. Such students
would, as do Arts and Humanities students, go on to
qualifications in Business Studies, International Law,
Accounting, Political Science, Economics or whatever other
specialisation would broaden their acceptability to a
world in which, increasingly, cross disciplines are essential. One of the best medical practitioners I have known
started his career as an engineer and never regretted it.
His understanding of engineering concepts not only helped
his understanding of biology and anatomy but also his
ability to explain them in engineering terms led him to
become a highly respected practitioner of environmental
medicine.

Of course engineering as a Liberal Arts subject would
have to avoid the narrow specialisations which have dogged
many such courses in the past. But this is already the
case in the best examples. In Germany, for instance, the
recommended allocation of time in an ideal engineering
course is roughly as follows:

> Humanities 10%; Foreign Languages 10%; Technical
> writing 7%; Industrial Administration, Economics
> and Social Science 12%; Design engineering 24%; and
> Basic and Speciality engineering combined 13%.

If two levels of professional engineering qualification were adopted, as in Germany, the first degree course would serve the somewhat lower level and a higher degree be required to achieve the upper qualification. This would demand the least change to the present system.

If the UK is to match the best of its competitors and thereby maintain its industrial trading base, we have to face the need for a serious increase in the quantity and quality of our education. Only about half the number of people are qualified by examination for the jobs they do in the UK as compared with Germany or Japan, say, 32% against 64%.

Rather than simply extend present courses I believe they should keep their present length of, say, 3 years for a first degree and two further years for a higher degree but that the higher degree should gradually become the norm. This would allow first degrees to include more generalised material and to take their place in a holistic education programme that would see a first degree in engineering become one of an increasing number of the Liberal Arts. Such a plan would have the effect of widening the scope of first degrees which would become rather more accessible and, looking into the next century, would comprise the rounded education of the majority of young men and women.

Graduates of this order, especially those who had fitted in a measure of practical training between terms, would have a sound basis for rewarding employment or an equally sound basis to go on to a higher degree and more advanced training to become the professionals of their generation and properly equipped to be the teachers and mentors of the next.

Extending general education in this way, rather than the unprofitable route of raising school leaving age still further, would so boost the quality of education in the UK that it could equal or better the industrial leaders behind which it has currently fallen.

The modular nature of this system enables each step to be deferred to meet the growth characteristics of the individual student. A school leaver might feel better equipped by a period of training in employment for the next academic step to a first degree. A first degree graduate might well chose practical experience in employment for a year or two before embarking on the higher degree by which his or her professional status will be confirmed. This practice would also enable graduates to go firm on the choice of specialised subjects in the final degree for which they may not have been confidently prepared at an earlier date.

5 In conclusion

The proposal that Engineering should be regarded and taught as one of the Liberal Arts is not revoluntionary but its inception as a key part of the evolution of a reformed education system, holoscopic in intent, may well be seen as such. The average increase of 2 years in full-time education will be challenged but this will do no more than meet the current standards of our best international competitors: what is more important is how many more students can advantage themselves in this way.

The associated proposal, that the wider familiarity with engineering in a population where it becomes one of the Liberal Arts should be accompanied by a distinctly higher standard for the qualification of professional engineer, by dint of the requirements for a higher degree, may be regarded as revolutionary; yet it does no more than emulate what is already achieved in best practice in the UK and elsewhere.

The proposal that the award of the ultimate qualification would be a national one fulfilling also joint E.C. standards and weighted strongly to academic success, will not find favour in most Institutions since the main reason for membership in the past has been the qualification which the appropriate classes of membership have afforded, and only afford, to those in continued membership. Because of the excellent role they play as learned societies, advocating, encouraging and publishing research and practice which might otherwise be neglected, I am at one with thousands of like minded people who wish to see the Institutions prosper. The leaders are clearly aware of the problems and are engaged in constructive dialogue aimed at solutions.

The Engineering Council through the initiative of its current chairman Sir John Fairclough has acted in a statesman-like way to raise the level of this dialogue to find a basis for the singular body which the U.K lacks.

The Fellowship of Engineering (or the Royal Academy of Engineering as it may become) also bids fair to be the U.K. focus of the Professional Engineer. Out of all these efforts one feels that a National leadership will emerge. My hope is that it will be forward looking, honour its roots without an overly conservative approach to their perpetuation and, above all, redefine the boundaries of practical training and academic study to ensure that future Engineers are increasingly drawn from the best intellects available and challenged by demanding levels of higher education.

Just ten years ago (Corfield 1982) I addressed the Imperial College Industrial Society on the future of the Engineer in Britain. I would like to close by repeating a few lines that seem to me to be equally applicable today:

"We have more material and energy resources than Japan or Germany. We have a fund of human talent better educated than most. We need some good management decisions backed by adequate financial resources to survive and grow."

"One of these decisions, and to me the key, concerns education. It is not enough to assure a reasonable supply of technologists and engineers, physicists and chemists, scientists and mathematicians. It is necessary for the population as a whole to have an appreciation of what is constituted by their Arts. A new renaissance in which the Arts and Sciences are blended to create a new Art for Living is required - nothing less."

References:

The National Economic Development Committee. (Jan.1992) Engineering at the Crossroads.
Fairclough J. (11992) Competitiveness for International Survival Conference on continuing professional development. Cambridge.
Corfield Sir Kenneth (1982) Imperial College Industrial Society.
Day Sir Graham (1992) Engineers and Other Selected Professionals in the British Aerospace Bridge Lecture given to the Worshipful Company of Engineers at The City University London.
Hutton S. and Lawrence P. (1982) German Engineers - The Anatomy of a Profession (Clarendon Press)

SECTION 2: INTERNATIONAL ISSUES

Engineering Education and Training in a Rapidly Industrialising Malaysia

A.A. Abang Abdullah

Faculty of Engineering, University Pertanian Malaysia

Abstract

Malaysia's aspiration to become a fully developed industrial nation by the year 2020, has brought about a shift from a predominantly agriculture based to a rapidly expanding manufacturing based economy. The role of the private sector in the economic development of the country has been simultaneously enhanced. Engineering education and training has to take cognizance of this changing trend in order to ensure relevance of the training provided to the country's current and future manpower needs.
Keywords : Engineering, Education, Industrial, Malaysia.

1 Introduction

Until recently, Malaysia with a relatively small population of some eighteen million people and abundant natural resources, has been a predominantly agriculture based country. Malaysia is a world leading producer of palm oil, rubber and pepper.

The Malaysian government is actively encouraging the intensification of agricultural activities and the use of new technologies to increase production. This increased agri cultural production, combined with being a major exporter of timber, tin, petroleum and natural gas enabled her economy to develop at a rapid pace during the last decade.

Uncertain commodity prices and the government's interest to see Malaysia enter the industrial world, have however, encouraged the development of resource based industries and the growth of production and manufacturing industries in the country. Today, Malaysia is a major exporter of semiconductors, electrical and electronic goods and has even gone to car production. There is now a very high demand for engineering manpower in the country, particularly from the manufacturing sector.

2 Changing work environment

Malaysia aspires to be an affluent industrialised nation by the year 2020. To achieve this vision, the government has given priority to human resource development, enlarging science and technology base and putting greater emphasis on research and development. The accelerated growth for the industrial sector demands a labour force of high quality; innovative and capable of absorbing and adapting technology. The upgrading of production technology from simple assembly and process-type operation to the more sophisticated automated processes will generate a demand for about 153000 engineering and engineering assistants during the 1991 - 2000 period (Anon,1991). This number is large compared to the country's relatively small population and her current small pool of trained engineering manpower.

There is thus a need to increase the supply of trained engineering manpower in the country, not only in terms of numbers, but also in the variety of disciplines to sustain the development of the new industries. The educational institutions must venture into these new areas to ensure technological advancements, industrial growth and competitiveness in the nation's industrial restructuring programme. While consolidating educational programmes in the traditional engineering disciplines and producing the required supply of graduates for the country's development programmes, there is a need to design new curricula to meet the demand for new skills arising from rapid technological developments.

The Malaysian government's look east and privatisation policies have to a certain extent changed the career prospects for graduates in the country. The look east policy paved the way for various training and technology transfer programmes in Malaysia's effort to emulate Japan and Korea. The privatisation policy means that there are now more jobs in the private sector. Graduates are then expected to absorb and adapt the new working styles and management practice brought about by the implementation of these policies.

3 Engineering education and training in Malaysia

At present, only six out of the eight institutions of higher learning in Malaysia offer degree or diploma level courses in engineering, producing some two thousand engineers and nine thousand engineering assistants annually. The engineering faculties of the six institutions are in various stages of development, with the majority being established within the last two decades. All the institutions offer the traditional disciplines of civil, mechanical and electrical engineering disciplines. There are plans to establish additional faculties of engineering in the other institutions as well as in the proposed new universities. The country's polytechnics which has been established in almost all of the country's thirteen states, offer certificate level courses for engineering technicians. A large number of Malaysian engineering students are trained overseas, with the bulk of the number in the U.K. and U.S.A.

The engineering degree courses in Malaysian institution of higher learning are accredited by the Institution of Engineers, Malaysia, Board of Engineers, Malaysia and the Public Service Department. Engineering degree courses are normally four years in duration with a 3 - 6 month practical training as partial requirement for graduation. The diploma in engineering courses takes three years to complete while the certificate courses are two years in duration.

Most local universities offer postgraduate programmes to local and foreign students. Unfortunately, there is a general lack of interest in pursuing advanced degrees among graduate engineers who are relatively comfortable in their professional career in the country. Despite the various financial support in form of research assistantships, graduate research fellowships or normal scholarships provided by the government, most postgraduate programmes in engineering are taken up by students only during periods of economic recession; even then, it is only for transitionary periods. There had been numerous examples of students leaving for employment even at the point of writing their theses without any effort to complete their degree. To a certain extent this phenomenon has affected the staff research programmes in the universities.

Despite the increasing importance of engineering in the economic progress of the country, the number of students choosing engineering as a career has not shown any increase in Malaysia. School leaving students generally believe engineering is a difficult course and it is not a very rewarding career compared to medicine, accountancy, law and business studies Thus there is a need to encourage more students to take up engineering as a career in the country if the lack of trained manpower, particularly in engineering and technical subjects is to be overcome.

The government has done its share in assisting the universities to overcome their problem in recruiting and retaining good academic staff members. In its New Remuneration Scheme announced recently, the government has improved the career prospects for university lecturers and has given special critical service allowances to lecturers in the professional fields in an effort to discourage them to leave for the private sector. During recent years the engineering and medical faculties in Malaysian universities have been hardest hit by an exodus of lecturers to the private sector due to dissatisfaction with the university lecturer service scheme. However there is a continuing dissatisfaction, particularly among the younger lecturers, and the effort by the government to improve the university lecturer service scheme is a serious reminder of academic staffing problem in Malaysian universities. As a result there is now an increasing number of expatriate lecturers serving in Malaysian universities.

4 The new engineering curricula

Recognising the increasing complexity of engineering today and the demand for new subjects such as management and finance to be covered in engineering degree courses, the academic and professional engineering community in Malaysia generally prefer four year engineering courses to the three year courses offered in the U.K. Even within the traditional engineering disciplines and with a four year course duration, there is a need to cut on the number of historical subjects to make way for the more modern and relevant subjects and to give students more time to develop analytical, design and computing skills.

While producing a sufficient number of graduates in the traditional disciplines, there is now an urgent need to increase the number of newer engineering disciplines offered such as manufacturing, computer, aeronautical and petrochemical engineering, to supply graduates to the growing number of new industries in the country. Here engineering faculties have to face numerous problems ranging from high cost of laboratory facilities to difficulty in recruiting qualified and experienced members of staff.

For both the traditional and newer engineering disciplines, there is a need to ensure the curricula are designed in such a way that the graduates are able to fit into their work place in industry. The number of jobs for graduates in the public sector which has been traditionally the biggest employer of graduates in the country, is diminishing due to the government's privatisation policy and the young graduates have to brace themselves for a career in the competitive world of the private companies. Engineering curricula has to be regularly updated to ensure relevance to the needs of the country.

5 External input

Most university degrees in Malaysia have external examiners appointed by the university senate in each university to maintain an acceptable international academic standard. External examiners, mostly from the U.K. or U.S., visit the relevant departments annually to examine the question papers and student answer scripts. Additionally, the examiner can be appointed as an external assessor and he would then be required to provide his assessment on the teaching and research facilities, curriculum, and give advise on future development of the department. Postgraduate theses are similarly examined by external examiners who are sent the theses for evaluation by post.

The British Council for several years had been maintaining a program to enable engineering departments in Malaysia to maintain an academic link with their counterparts in the U.K. The academic links facilitate interaction between staff members from the two departments through two-way visits, academic attachments and collaboration in research. A good number of engineering departments in Malaysian Universities have benefited from this programme.

Links between universities within Malaysia and between universities and industries have been actively encouraged but have not shown much progress. Cooperation between universities is sadly at the minimal level despite a realisation that much is to be gained from the sharing of limited teaching and research resources. Cooperation between universities and industries is slightly better, particularly in research and development and consultancy work where there are clear incentives for both parties to cooperate. Most local universities encourage university lecturers to take up consultancy projects and normally an Institute of Consultancy is established in the university to coordinate staff involvement in consultancy activities. The industries are not particularly enthusiastic about involvement in teaching despite efforts by the universities to appoint them to academic advisory boards and as academic associates of the various faculties. And there is no two-way mobility amongst staff members between universities and industries, except for the more of less permanent drift for lecturers to the private sector. Adjunct appointments although possible, have yet to take off in a big way.

Links between universities and the professional institutions such as the Institution of Engineers, Malaysia is good with a large number of university engineering lecturers becoming cooperate members of the institution. But although a regular accreditation exercise for each engineering degree programme is carried out by the institution, there has not been any direct effort in the past to assist the universities in curricula design and development.

6 Research and development

The Malaysian government has recognised the major role research and development plays in the progress of an industrialised country. The intensification of research in priority areas programme (IRPA) launched by the government during the fifth Malaysia plan has given impetus to research and development activities in the country and has enabled university researchers to plan for long-term research projects. This was not possible before the programme was introduced, because university researchers were limited to short term funds only. However the current emphasis on applied and development research as opposed to basic research presents a problem to university researchers. There is also the problem of managing these large and long term research programmes amongst relatively young university researchers who have only had some experience in independent research while studying for their Ph.D overseas.

The private sector although quite willing to participate in the research and development efforts of the country are at present too occupied in their manufacturing, production or construction activities to spare time for long term research and development efforts. Many industries are not prepared to invest in long term research; much less basic research. Quite often they employ university lecturers as consultants for problem-solving studies, even then only when they are not able to resolve the problem inhouse.

7 Continuing education

The Malaysian government is committed to providing sufficient opportunities for con tinuing education in the country. Various training institutes have been established in the country to provide training programmes for both the public and private sector employees and members of the public. There has been a number of special training schemes where students and serving officers are sent abroad to selected countries, particularly Japan and Korea to acquire special skills especially in engineering and technology. Training in specialist fields are augmented with enhancement of managerial, business and communication skills, appropriate to demands of the competitive world of the private sector. Civil servants are now given an option to retire at the age of forty, to enable them to start early in the new career in the business world. Opportunities for retraining to prepare them for a career in the private sector is also provided.

8 Conclusion

The relatively sudden shift from an agriculture to a manufacturing based economy has a great impact on engineering education and training in Malaysia. While maintaining and strengthening engineering curricula in the traditional engineering disciplines, much effort is needed to design and develop new curricula to satisfy the demand for trained engineering manpower for the growing industrial sector. There is a need for a more practical oriented education and training and direct involvement of the private sector in the design and development of the new engineering curricula. Malaysia's aspiration to become a fully industrialised nation in such a short period compared to the period taken by the other industrialised nations, means that the nation has to invest heavily on human resource development in the engineering field in the years ahead.

References

Abang Abdullah, A.A. (1985). Education and training of professional engineers. Institution of Engineers, Malaysia. Bulletin No. 7, July.

ANON (1990). Industrial technology development: a national plan for action. Ministry of Science, Technology and Environment, Malaysia, Kuala Lumpur.

ANON (1991). The second outline perspective plan (1991 - 2000). Prime Minister Department, Malaysia, Kuala Lumpur.

Jumaat, M.Z. et al (1989). Cooperation between Malaysian universities in engineering education. AEESEAP Journal of Engineering Education. Vol. 19, No. 2., September.

Jawahir, I.S. and Wong, W.C.K. (1985). An integrated approach to teaching of manufacturing engineering in less development countries. Proc. of Reg. Conf.

Matheson, L. (1988). Engineering the future. Keynote Address. Proc. of Int. Conf. on Civil engineering education, research and professional development.

The Mechanical Engineering Curriculum in Finland and the U.S.A.

T.E. Leinonen

Department of Mechanical Engineering, University of Oulu

Abstract

The author of this paper, who was visiting professor at the University of Florida, Gainesville, and Michigan Technological University, Houghton, in 1991, compares the requirements for degrees in mechanical engineering at both of these universities with the degree at the University of Oulu in Finland. The contents of the curriculum at each of these universities are evaluated and the lengths of the degree are compared in terms of credit hours. The contents of the curriculum are evaluated based on the university prospectuses and personal experiences.

Keywords: Education, Mechanical Engineering, Curriculum

1. Introduction

The engineering curriculum in Finland was originally based on the system in Germany. The first degree is that of MS (Eng), which takes 5 - 6 years to pass and corresponds to the master's level in the English system. The postgraduate degrees are those of Licentiate and Doctor of Technology.

The curriculum in the USA is derived from Britain. The original 3-year BS degree in engineering has now become a 4-year program. The postgraduate degrees are MS and Ph.D. The author of this paper was the visiting professor at the University of Florida, Gainesville, and Michigan Technological University, Houghton, in 1991.

The level of the mechanical engineering degree is the same at all four universities in Finland which offer it. Thus the University of Oulu is representative of the institutions awarding such a degree in this country.

In the USA the level of the degrees varies greatly from one university to another. The University of Florida and Michigan Technological University are very highly regarded in the field of mechanical engineering, but even they differ slightly in the basic arrangement of the BS program. Students at the University of Florida enter the mechanical engineering program after two years of studies in science and general subjects, which at the Michigan Technological University students are chosen directly for the mechanical engineering curriculum.

This paper evaluates the contents of the curriculum at each of these universities, and compares the lengths of their degree programs in terms of credit hours.

2. Credits and study weeks

There are two credit systems in use in the USA, one based on lecture and laboratory hours at the university during the semester (15 weeks) and the other during the quarter (10 weeks). The University of Florida observes the semester system and the Michigan Technological University the quarter system. The ratio between credits in the two systems is 15/10, the credits otherwise being determined in the same way. The lectures and demonstrations have the power 1 and laboratory classes the power 0.5.

In Finland lectures, laboratory classes and work done at home are counted together and 40 hours' work is taken to represent 1 study week (sw). A minimum grade of "good" (3/5) is required.

Direct comparison between these two systems is impossible, because it depends on the relation between the tuition given at the university and the work done at home. The tuition given in mechanical engineering in Oulu is theoretically 63% of the total work load, but the students are not present at all the classes and the actual relation depends heavily on the individual student.

Two numerical approaches can be taken to evaluating credit hours and study weeks from the prospectuses:

1. If credits for the courses required in Oulu are calculated in the American way, the study week is found to be 1.28 times greater than the credit hour.

2. Students in Florida obtain 34.5 credits during the 33 weeks that they study at the university per year, while students in Oulu receive 36.1 sw. for 36 weeks of work. In this sense a semester-based credit hour is more or less equal to a study week (sw.) in Finland.

Taking the average between these two evaluation methods, the study week in Finland may be said to be 1.14 times greater unit than the semester-based credit hour in the USA.

3. Curriculum

The courses as listed in Table 1 are classified to bigger groups into major units and assigned credits as indicated in the prospectuses (1, 2, 4, 5 and 6). Quarter-based credit hours are given for the Michigan Technological University (MTU) and then transformed to semester-based credits by dividing them by 1.5. The credits at the University of Florida (UF) are quoted directly from the prospectus (1 and 2) as are the study weeks (sw.) at the University of Oulu.

Table 1. Curricula in mechanical engineering.

Title	MTU credits (qr.)	credits (sem.)	UF credits	Oulu (sw.)
Hum. and soc. sc.	41	27.3	27	15.0
Mathematics	29	19.3	18	17
Sciences (chem. + phys.)	23	15.3	10	9.5
Machine Draw. and CAD	9	6.0	4	8
Engineering mechanics	12	8.0	14	22
Electr. eng. + control	19	12.7	10	8.5
Economics	3	2.0	-	5
Mechanical Engineering	38	25.3	34	33
- Machine Design	(9)	(6)	(2)	(11)
- Manufacturing	(4)	(2.7)	(3)	(9)
- Material sc.	(8)	(5.3)	(3)	(10)
- Thermofluids	(17)	(11.3)	(16)	(3)
Elective subjects	24	16	12	12
	198	132	138	
Professional elective subjects	30	20	24	20.0
Pract. training				10*
Thesis	15	10	6	20
MS (Dipl. Eng.)	243	162	168	180

* 3 weeks in industry = 1 study week

The BS programs in Florida and at MTU are quite similar in terms of credits, 138 and 132 respectively, the main difference between them being that there are more sciences, engineering mechanics and ME professional courses and less elective subjects in Florida. The material handled in the courses is also quite similar.

There is no BS degree in Oulu, but instead all the students are studying for the MS (Eng), which takes 5 - 6 years. If we compare the programs there are less humanities, social sciences and chemistry and more engineering mechanics in Oulu than in Florida. Also, the humanities component in Oulu consists mainly of foreign languages and that in the USA of reading and writing in English and of social sciences.

There are about 18.5 hours of teaching in an average week in the USA and 30 hours in Finland, the main difference lying in laboratory and design work. Where students in the USA have 16 hours of lectures and 2.5 hours of laboratory work in an average week, those in Finland have 15 hours of lectures and 15 hours of design and laboratory work.

4 Master of Science Degree

The MS degree in Florida comprises 168 credits and that at MTU 162 credits, including professional elective subjects and the thesis after the BS degree. The MS (Eng) in Finland is 180 study weeks, including about the same amount of coursework as in the USA together with practical training in industry (10 sw.) and a thesis, which takes 6 to 10 months to do and is evaluated as equivalent to 20 study weeks. At some other universities in Finland the practical training is 3 sw.

It is evident from this table that the MS degree in the USA is of about the same level as the MS (Eng) in Finland.

5 Doctoral programs

The doctoral program (Ph.D.) in the USA includes coursework and a doctoral dissertation: "A doctoral dissertation must demonstrate the ability of the author to conceive, design, conduct, and interpret independent, original and creative research. It must describe significant original contributions to the advancement of knowledge and must demonstrate the ability to organize, analyze, and interpret data (3 p. 54).

A student working for the Doctor of Technology degree in Finland is required "to develop a thorough familiarity with his own research field and its methods, an ability to produce new scientific knowledge independently, and a good grounding in the theory of science, the history of his own research area and the influence of science on society (6 p. 354).

There is also another intermediate degree known as "Licentiate in Technology" with course requirements similar to a doctorate but a more practically oriented thesis which must primarily demonstrate an ability to use scientific research methods, but is not required to generate new, independent knowledge in the same way as a doctoral thesis.

The credits required for these degrees are compared in the Table 2.

Table 2. Credits required for a doctor's degree after MS.

Title	MTU credits (qr.)	credits (sem.)	UF credits	Oulu (sw.)
Theory of sc. and history				6
Major studies	24	16	24	27
Supplementary studies	6	4	12	12
Licentiate thesis				35
Doctoral dissertation	<u>40</u>	<u>26.7</u>	<u>24</u>	<u>35-60*</u>
	70	46.7	60	80-105

*not mentioned in statutes

doctor's dissertation at MTU is somewhat more extensive, but there are other studies to be completed in Florida. The main difference between Florida and Oulu is that the doctoral dissertation is more extensive in Oulu, or at least takes more time. The theory of science and history of science are required in Finland, but are perhaps included in the curriculum at some other point in the USA.

6 Content of the Curriculum

The above chapters have concentrated on the work load and level of knowledge of the students, comparing these between the three universities. The theoretical level of the engineering studies is broadly the same, although there are some basic differences between the USA and Finland in this respect.

In addition to the theoretical level, practical skills are required also in Finland, where the diploma program includes 3 - 8 months of industrial experience. The curriculum in the USA is heavily analytical in orientation, whereas in Finland synthesis and design exercises are also required. A good example is the course entitled 'machine design', which in the USA includes only lectures and calculations, while in Finland it also contains an additional four individual design projects and laboratory work performed by the students.

The nature of the humanities is different in the USA and Finland. American students study reading and writing skills in English, whereas the main emphasis in Finland is on foreign languages. The average student in Finland studies 3 - 4 languages during his school and university career.

The main difference at the doctoral level is that the dissertation is more extensive in Finland, although the scientific requirements appear to be similar if we compare the wording of the statutes.

7 Summary

The mechanical engineering curricula in Finland and the USA are compared based on the situation at the University of Oulu, University of Florida and Michigan Technological University. Mechanical engineering at the University of Oulu is representative of the situation in Finland as a whole, as the requirements are similar at the other three universities. In the USA the level varies greatly from one university to another, the curricula at Florida and MTU representing the top level in the country.

Comparison of the credit systems in the two countries shows that 1 credit unit in the USA approximately equals 1 study week in Finland. 1 study week is a 1.16 times larger unit than the semester-based credit hour in the USA.

Comparison of the curricula in the three universities based on credits shows the MS (Eng) in Finland and MS degree in the USA to represent the same knowledge level, and the time spent in studying for these degrees is about the same, 5 - 6 years. The requirements for a doctor's degree are more extensive in Oulu than in the USA, the work done on the dissertation in particular being greater.

The main differences in the contents of the curriculum are in its orientation of the curriculum. In Finland the heavy theoretical background is complemented by synthetically oriented design exercises and industrial experience.

References

1. University of Florida; undergraduate catalog 1990 - 91.
2. University of Florida, graduate catalog 1991 - 92.
3. Martin, H. N. and Laurie, M. D. (1990) Graduate Student Handbook. University of Florida.
4. Michigan Technological University; undergraduate catalog 1990 - 91.
5. Michigan Technological University; Graduate School Bulletin 1989 - 91.
6. Opinto-opas, Oulun yliopisto, Teknillinen tiedekunta 1991 - 92.

New Trends, Old Schemes and Accreditation for Engineering Schools in Latin America

M.S. Navarro, R.P. Jetton

Department of Electrical & Computer Eng. & Tech., Bradley University

Abstract
When considering the direction that curricula might take relative to the questions of graduate school vs. industry, a lesson might be learned from a unique program in Ecuador. Historically third world countries have tried to emulate the engineering education endeavors of developed nations. it has become clear to some that this is not appropriate for their situation. The same questions that they are addressing are at the core of concerns in the developed countries relative to curriculum direction. While the exact mix of course work would be different it is nonetheless instructive to observe how others have worked to solve the same basic problem.
Keywords: Engineering Education, Latin America, Ecuador

I. Introduction

The creation of Universities in Latin America started soon after independence from Spain. Early models followed French standards and the regulatory office was the Secretary of Education. A century and a half later in most of Latin America the State still has the authority to "create" institutions of higher learning by decree. There are no local accreditation Boards and the standards are set by tradition and comparison with what it is done in the developed nations.

Most engineering curricula in South America require five years to complete and are very selective with respect to admission. They are very inefficient with respect to retention, due primarily to the high degree of complexity introduced in the early stages of the programs. Civil engineering is a special case, requiring six years. Chile also has a four-year program similar to the U.S., the so called Ingeniero de Ejecucion, which translates as "the engineer that executes". This can not be considered a parallel to the four year Technology programs of the U.S. which are of lower theoretical content and not aimed at creating a design engineer.

It is striking that countries with low industrial development still place emphasis on creating a five year engineering education and also dedicate resources to the creation of graduate programs in engineering. In several of these countries it would be more productive to create professionals for the lower ranks within the profession. The pyramidal distribution of professionals with a broad base of technicians, a middle group with four year degrees and a top with small highly trained (M.S. & Ph.D.)

professionals seems inverted in many countries where the possession of a five year degree is more a "social status" than an economic factor.

II. Comparisons

There is a tendency (at least in the U.S.A.) to think that Latin America is a group of countries with almost identical standards of development and afflicted with the same problems. This is not so. Literacy levels are very high in Chile, Argentina and Uruguay, approximating European standards. It is not hard to imagine that Universities have been stable with high standards and fulfilling the role given to them. Since the beginning of this century these countries were the models for many South American countries with less qualified institutions. Scholarships were granted to qualified students to pursue a college degree in the "South cone of South America".

But institutions of higher education must reflect the state of the industrial development of a country. Due to the uneven development caused mainly by the wealth (natural resources) of each country, its political instabilities and a range of socioeconomic problems, the state of harmony between industrial development and engineering education contains a "phase shift." The schools have advanced as have those in Europe and the U.S.A. but industry is stagnated and more and more engineers are produced without a job market at hand. It is estimated that more than 85% of the electrical engineers in Ecuador will not perform in a design position. Nevertheless they employ five years in obtaining a degree which is one year more than the bachelors degree in the U.S.A.

The universities prepare good engineers but these professionals have been unable to minimize technological dependance and what is worse, no alternate solutions for technological underdevelopment have been proposed within the ranks of the engineering schools.

It is obvious that an accreditation committee with representatives from industry can aid in the process of looking for "the right way" or the "right engineer". The engineer currently produced does not have a business mentality, does not know how to create industries, does not know about technology transfer. He or she is simply a good design engineer better suited for a developed nation.

III. Needs

In several South American countries the state has created institutions to promote scientific research. CONICYT (Comite Nacional de Investigacion Cientifica y Tecnologica) in Chile and Venezuela, to cite an example, have similar roles. The immediate benefit of the presence of these institutions is the availability of grants and scholarships. In our opinion, an additional mechanism is required. One not based on academic ideals but more related to industrial needs. Scholars are not best qualified to judge the need for technology transfer for the need of skills and techniques learned in industry. Obviously the graduates are in many cases overqualified for these jobs but the lack of other professionals in the technical arena made them the first choice. South America needs these new professionals. People without masters degrees but with skills typically acquired in industry. This is not to say that the traditional specialist with a Ph.D. degree is not required or not useful. On the contrary, traditional studies are still important, it is only suggested that a change on the "emphasis" be made. In years past, several programs supported studies abroad primarily for obtaining the masters and Ph.D. degrees. It is unlikely that a country without a good number of these degrees can really be independent technologically.

IV. Academic Development

The U.S.A. and some countries of Europe and the former Soviet Union, understood the crises of the developing nations and each with his own policies has been instrumental in the upgrading of University professors. LASPAU (Latin American Scholarships for American Universities) was established nearly 20 years ago, FullBright, AID and others generously have contributed to the academic development of Latin America.

Unfortunately, these professors acquired knowledge in a developed nation and in most cases with no direct application outside the academic environment of Latin America. Some of these individuals return to their host country (U.S.A. to a large extent) to work in academia or industry. As with any process involving humans, efficiency has not been 100%, but a good number of students, do return to their home countries to make contributions in science and technology. One question arises: Is it fair to support and emphasize studies with no direct application in a given nation? Research for the sake of research is sometimes the result of academic freedom. "I study what I like." When government funds are to be divided, a major portion should be allocated to certain projects or areas of development. In Latin America the grants are not abundant. The role of NSF in the U.S.A. is taken by CONICIT or by some estate University. Still the tendency is to match current world knowledge at the expense of a much less sophisticated industrial infrastructure. We believe that the support of the Latin American University has not been complete. Said in another way: The contact between a Latin American Professor and his institution in the developed nation must be a long term liaison. It is unfortunate that seldom are there follow up visits or simple "up to date" refreshing courses or seminars. Professors that return with a M.S. or a Ph.D. in the vast majority of the cases will not join industry and their full potential is rarely developed due to economic constraints in the Latin American University where research is a principle but not an established way of life. Summer work in industry in a developed nation would open the eyes of many professors of engineering in Latin America.

Perhaps ideas of technology transfer or technology development can arise with these contacts that now do not exist in a formal manner. Professors in a given discipline must learn the required skills. In the U.S. for example, studies are very academic with more scientific emphasis than technical detail. And this is so by design. It is not a flaw of the U.S. University. In simple terms, it is expected that those technically specialized skills and techniques are learned in industry. But what happens with the recent graduate, perhaps with an M.S. degree that returns to a country with little or no industry? The answer is simple, he cannot contribute efficiently because he doesn't know much outside the scientific world.

V. A Solution

With the background set in all these problems, a group of courageous individuals in Quito Ecuador, each one with a Ph.D. from prestigious U.S. Universities have decided to break old structures and create a university better suited to a developing nation like Ecuador. This is not in in opposition to the traditional five year programs it is simply an alternative that private sectors are supporting and it is a success story in

Ecuador. The fundamental idea is to form generalists, with an open mind to technology and to business ventures and with a down to earth philosophy about engineering and technology. A four year program for an underdeveloped nation is already an improvement!

VI. An Example

Universidad San Francisco de Quito in Quito Ecuador was chartered in 1985 and held its first classes in the fall of 1988. Originally it was housed in a residential mansion and 130 students enrolled initially. Its success can be at least partially measured by a growth in enrollment which is currently near one thousand. Another indicator is the support it has received which has allowed a growth and expansion of facilities.

It is a private university with a stated mission of providing both professional and general humanistic education. The general curriculum statement contains a clear and emphatic statement that all students will be broadly educated in the humanities. In addition no graduate is expected to become a specialist, that is, very narrowly educated.

Within this context the College of Applied Science has developed a curriculum that represents a very careful assessment of the potential for its graduates. It was realized that the level of technology in the basically third world economy of the country was relatively low. Therefore the students while being made aware of the latest trends need not be prepared to work in the high technology environment of the developed nations.

Additionally since many business institutions and government agencies are small by world standards, it was felt that the students should be exposed to a broad view of engineering, business and industry. Since many business ventures are in an embryonic stage a strong flavor of entrepreneurship is fostered. Relationships involving projects involving local industry are encouraged.

One must not conclude that the curriculum is superficial however. It is based on the very familiar preparation in basic science (physics & chemistry) and mathematics (calculus, differential equations, vector calculus). Typical Engineering Science components are included as well as such computer related topics as Data Structures and algorithms.

Some opportunity is provided for a limited amount of specialization with electives in areas such as Electronics, Systems Administration or Construction Management.

The level and depth of preparation in technical courses is such that should a student desire to, he or she would be able to enter graduate study in typical university in the United States. It must be emphasized however that that is not the primary intent of the program. The students are expected to take positions in their home country. They will have been educated in not only a technical area but will have an understanding of their country's culture and history and its relationships with other nations in the hemisphere and the world.

To insure an awareness of other cultures the university has been active in developing exchange programs for students and faculty from other nations. Several institutions in the United States are participating in a variety of experiences for students both in Ecuador and the U.S. Some institutions in the U.S., including The University of Illinois, The University of North Carolina at Wilmington and Rockford College have formal agreements with the University.

While the university has developed many interesting and novel programs the important point to consider here is not the details of each activity but the overall philosophy. Starting with a "clean sheet of paper" and avoiding the temptation to simply create a clone of developed nations engineering programs, curricula were developed to suit the needs of the students, industry and government. These needs were perceived to be less theoretically technical, more applied, broader in scope and containing a need for an awareness of business and social concerns.

VII. Conclusion

It is possible that educational institutions and systems in all environments can profit from a consideration of the example cited. Certainly the same set of factors will not be present. There are a great variety of situations, but the approach taken is the important issue. The output of the educational process must be a match for the requirements of society and educators have a responsibility to access those needs and respond to them rather than taking a narrow parochial view of "we know best what the content of the curriculum should be."

In addition, the programs will be much more effective if they are generated by the institutions on a proactive basis rather than waiting and responding to external influences, i.e. accreditation requirements, on a reactive basis.

Technology Transfer Through Postgraduate Training: Effectiveness of Current Practice and Feasibility of New Form of Cooperation

M.S.J. Hashmi

School of Mechanical and Manufacturing Engineering, Dublin City University

ABSTRACT

Traditionally, in engineering as well as other areas, the educational and economic planners in developing countries have sent some of their most able persons to obtain postgraduate training in developed countries. The primary objectives for seeking such postgraduate training are to expose these personnel to new technology, ideas and methodology and to transfer knowledge so gained to their own countries by training their own young engineers, and technological personnel.

In this paper the effectiveness of such postgraduate training modes is reviewed critically in relation to the cost effectiveness and real technology transfer. A new mode of cooperative postgraduate training scheme is discussed in the light of existing links between Dublin City University and a number of overseas Universities in the developing countries.

Keywords: Technology Transfer, Developing Countries, Postgraduate Training, Engineering, Cooperation.

1 INTRODUCTION

There are many good reasons for the transfer of technology from the developed to developing countries world wide. Apart from humanitarian, the most important other reason being the fact that unless the technology gap is narrowed down, the developing countries simply would not be in a position to provide a market for the developed world.

There are many ways in which technology transfer is facilitated such as training of technical personnel locally in developing countries, providing for equipment and facilities, training of technical and engineering educators at postgraduate level in developed countries etc. Some of these mechanisms have proved to be highly successful and beneficial for the developing countries concerned. Majority of these successful technology transfer mechanisms involve capital equipment in conjunction with man-power training support activities taking place locally in the developing countries, the manpower training being usually at lower to medium technical skills level.

Technology transfer at advanced level has been deemed to be relatively less successful, especially at postgraduate level. The principal reasons cited being the inappropriateness of the training received in the developed countries as well as the ineffectiveness of the current method in extracting meaningful benefit out of their trained personnel.

2 CURRENT PRACTICE

Traditionally, developed countries played a major role in training young and talented engineering graduates and technologists through postgraduate studies by course and research. In many cases the developed countries offered scholarships and other types of bursaries for these postgraduate students to be tenable in any University or other tertiary level educational establishment of the developed country concerned. Some of the scholarship schemes were dedicated towards maintaining and fostering closer links with the developing countries having previous colonial links with a particular developed country or a group of developed countries. For example the scholarship offered through the Association of Commonwealth Universities and Colombo Plan in relation to the UK. Similar scholarship schemes have been in existence for the former colonies and satellite countries of France and the erstwhile USSR.

In taking advantage of these fully financed postgraduate training opportunities, the economic and educational planners in developing countries have sent some of their most able graduates in engineering and other technological areas to obtain postgraduate training in developed countries. Bulk of these students were nominated by the authorities of technical and engineering educational establishments with the primary objective of exposing such personnel to new technology and research methodology, and to transfer the knowledge so gained to young graduates in their own countries. The idea being that such trained persons would take up teaching and research positions in university and research centres and perpetuate research and development activities through active participation as well as being catalysts for others.

The other group of engineering and technological personnel who availed of these scholarships belonged to the public sector industrial corporations. These scholars are usually nominated by high level government officials with the specific objective of appointing the returning scholars to management and high level administrative positions.

A very small number of bright but wealthy engineering personnel availed themselves of postgraduate education at their own expense or through support from a University in the developed country.

A substantial number of postgraduate engineering students who study at Universities in developed countries come from richer developing countries. These students are either state sponsored or privately supported. The state sponsored scholars are sent overseas as part of a planned manpower development strategy in an effort to acquire technological knowhow from the developed countries and to make use of these returning technocrats to develop the technological and industrial base of their own countries. Privately sponsored (mostly family supported) individuals had their own personal objective to upgrade their social status and standing in the job market in their own countries.

Whether these postgraduate students complete their studies with support from the host countries or from their own countries, the cost at present day value per student could be anything between UK£10,000 to UK£12,000 per annum. A typical postgraduate engineering student takes on average 4 years to complete his PhD studies. Thus, cost per PhD student to the sponsor could be in the region of UK£50,000. One reason why the developing countries are reappraising their strategy of achieving technology transfer through this mechanism.

A number of difficulties are proving to be seriously affecting the success of technology transfer through postgraduate training for higher degrees. Firstly, a fair proportion of the scholars never return and despite some stringent conditions, manage to settle in the host country or in another developed country[1]. Secondly, even if they return, a number of them find it extremely difficult to adjust to the prevailing social and economic conditions in their parent countries and decide to emigrate even after paying large sums of monies as penalty. Thirdly, those who return to their parent country find that the environment is not at all conducive for making the best use of the training they received and expertise they gained at their host countries. The most deciding negative factor being the fact that such newly trained engineering experts find themselves divorced from the research equipments and facilities they have been using and that there is no capital resource available for providing them with such equipments. These most enthusiastic engineering research personnel joining a University department or research organisation, make serious attempts to do something positive but in most cases their attempts get thwarted by obstacles of one kind or another.

The result is that within a very short period of time the majority of them become frustrated and eventually conditioned to doing very little by way of R & D activities. This situation is particularly true for most poorer developing countries. For richer developing countries the circumstances take a slightly different form. When the most technically able and promising engineers return after completing their postgraduate studies, they also find themselves divorced from their facilities and equipments with which they have been familiar and even though capital is available to provide them with new facilities, often it takes years before they are able to continue with their work. In some cases, the researcher's own requirements do not match with those of the establishment since the research carried out was not in line with what was intended by the sponsors or that the emphasis of the research area has changed from what was originally perceived.

In any event the net result is ineffective technology transfer and slowing down of the progress towards industrial development in the developing countries. In view of such inherent drawbacks of the current scheme, the educational and economic planners of many developing countries are re-thinking about their strategies so as to minimise the brain-drain and to enhance the effectiveness of technology transfer from the developed countries.

3 A NEW APPROACH

In this section a possible new programme of cooperation for training overseas postgraduate engineering graduates is presented which highlights the agreements between the Faculty of Engineering at Dublin City University and the corresponding Faculties in three overseas Universities and Research Institutions. These agreements have been developed based on previous links through sponsored research students and after holding extensive discussions and bilateral meetings.

The principal objectives of such a programme of cooperation are that it will:
  A. Minimise the brain drain from the developing country
  B. Permit more effective technology transfer and
  C. Make better economic sense (justification).

With these objectives in mind the backbone of the programme of cooperation was developed as follows:
1. Each scholar will spend at least one third of the total study period at either the host or the parent organisation.
2. Each scholar will have two supervisors, one from each organisation.
3. Where the scholar is supported by the host institution, he/she will be paid maintenance allowance only when the scholar is at the host institution. The course fees will be totally exempted by the host institution.
4. Where the scholar is supported by the parent institution, the maintenance allowance will be paid by the parent institution during the entire period of study (at different rates as appropriate). The parent institution will pay full or partial course fees to the host institution only for the period the scholar actually spends abroad and not for the entire period of study.
5. Where the parent institution is the sponsor, the research apparatus will be constructed at the premises of the sponsor and the research will be carried out mainly there.
6. Where the host institution is the sponsor, the research apparatus may be constructed at the premises of the parent institution if fully or substantially paid for by the parent institution.
7. The supervisors will meet frequently to exchange ideas, update knowledge base and gain supervision experience as appropriate.
8. On completion of their research the scholars will be examined at the

premises where the research took place (mainly at the parent institution).
9. The degree will be awarded jointly or solely by the host institution in the developed country.

The above clauses are typical of those one would find in a cooperation agreement between a University in a developed country and another in a developing country. Specific clauses may vary depending on the cooperating Partner Universities concerned. The following two specific cases are given as illustrative examples.

3.1 DCU - BUET Cooperation Agreement

This agreement was signed in 1990 between the Dublin City University and Bangladesh University of Engineering and Technology, and amended in 1991[2], the main clauses of the agreement being that:

(i) a number of junior staff of the Faculty of Engineering of BUET will undertake PhD studies under joint supervision.

(ii) about 50 percent of the students will be supported by BUET and be primarily located at BUET with short term visits to DCU during the formative and completion stages of the project. These scholars will be registered jointly between the two Universities.

(iii) the other 50 percent of the students will be supported by DCU and will be located there for the entire duration of their studies. These students will be registered with DCU.

(iv) the travelling expenses of BUET students coming to DCU will be paid for by the BUET, and the travelling expenses of the supervisors will be shared by the two Universities.

(vi) For the students located at BUET, the research apparatus will be purchased or constructed and paid for by BUET.

(vii) The degrees will be awarded by DCU solely or DCU/BUET jointly, the choice being made by the student concerned.

3.2 DCU - SSRC Cooperation Agreement

This agreement is between Dublin City University and the Scientific Studies and Research Centre of Syria[3]. The terms of this agreement has been agreed upon and is shortly to be put into practice. The main clauses of the agreement are that:

(i) a number of scholars from SSRC will undertake PhD studies and will be supported by SSRC.

(ii) each student will spend a period of 6 to 9 months of the first year and 3 to 6 months of the final year of study at DCU, and about 2 to 3 years at SSRC.

(iii) each student will be registered with DCU but supervised jointly by DCU and SSRC experts.

(iv) The research work will be carried out at the premises of SSRC who will provide for the facilities and apparatus. The supervisor from DCU will visit once a year.

(v) Each student will pay fees at full rate for one year and at reduced rate for the remaining years of study period.
(vi) Each student will be examined at SSRC by one external examiner and two internal examiners (normally the supervisors from DCU and SSRC) to be appointed by DCU and the degree will be awarded by DCU.

3.3 Benefits:

Thus, it can be seen from the two examples that the principal thrust is to create longer term research environment at the premises of the institution in the developing countries. The perceived benefits, specially where the research is undertaken in the developing country, are as follows:

(i) creation of research culture due to the involvement of local academic supervisors, technical and administrative staff.
(ii) appointment of local supervisors helps them to gain experience of supervising Doctoral research.
(iii) since research project is chosen at local level the topic area is likely to move relevant to the local need and hence changes of continued support is more likely.
(iv) the researcher will have the opportunity to continue to make use of his facilities and apparatus at post doctoral level.
(v) ongoing relationship with the Partner University in the developed country helps keep them up to date with any progress.
(vi) brings financial savings for the partner in the developing country at the same time provides reasonable return in cash and kind for the partner in the developed country.
(vii) the increased human mobility and interaction brings better understanding of the social and cultural background of peoples of partner countries.

4 CONCLUSIONS:

The inherent difficulties with the present system of postgraduate education as a means of technology transfer has been discussed.

Alternative schemes based on bilateral cooperation programmes are presented with examples and perceived benefits are discussed.

REFERENCES:

1. Commonwealth Scholarship and Fellowship Plan Tracer Study-Final Report, Commonwealth Secretariat, January 1989.
2. DCU-BUET Cooperation Agreement, Final Document, Dublin City University, 1990.
3. DCU-SSRC Cooperation Agreement, Draft Document, Dublin City University, 1991.

The Role of Engineering Technology Education in Technology Transfer between the United States and Developing Countries

W.R. Hager (*), J. Baguant (**)

(*) School of Engineering Technology and Commonwealth Campus Engineering, College of Engineering, The Pennsylvania State University

(**) School of Engineering, University of Mauritius

Abstract

This paper discusses the development and current state of the engineering technology discipline in the United States. In addition, the United States engineering occupational spectrum is discussed with emphasis on occupational specialty, discipline definition, degree, and job opportunities. A case is made for the suitability of faculty and graduates in engineering technology for participation in technology transfer from the United States to developing countries. More exchange and partnership opportunities for Engineering Technology faculty would be beneficial to both the faculty and the host institution.

Keywords: Engineering Technology, Occupational Spectrum, Technology Transfer

1 Introduction

In the United States the term engineering refers not only to a specific discipline but also to a spectrum of occupations and challenges that lie within a broader definition of engineering. This spectrum includes the craftsperson, the technician, the technologists, the engineer, and the scientist. Many individuals in all of these occupations may be in jobs carrying the title engineer. For example, a craftsperson may be a building and maintenance engineer, the technician may be a service engineer, the technologist may be a production engineer, the engineer may be a design engineer, and the scientist may be a research engineer. In interests, abilities, and job requirements this spectrum of occupations ranges from the practical and highly skilled to the esoteric and highly theoretical.

Thus, when someone is referred to as an engineer it is difficult to know the specific discipline or the degree attained. While job titles may cause confusion, the specific educational programs from which these individuals graduated are designed to prepare students with different interests, abilities, and career objectives with the necessary knowledge and skills to perform in specific jobs.

The distinctions and differences between the discipline designations, degrees, and job functions of these graduates are not well understood by career counselors, prospective students, or employers within the United States. As engineering education, in its broadest sense, moves towards a more international perspective it is necessary that the differences between these occupations, disciplines, United States degrees, and job functions be better understood by educators in other countries.

2 The Engineering Occupational Spectrum Degrees

The engineering occupational spectrum ranges from the craftsperson to the research scientist. Years of education range from minimal beyond high school to eight to ten years of post secondary education. Accordingly a broad range of degrees are offered for these programs in the United States.

The craftsperson is highly skilled in a specific area. These skills are learned through practice, experience, vocational education, or on-the-job training. They may also be graduates of a one to two year educational program and hold an associate of science degree (AS). This degree normally is not designed for continuing education.

The technician is usually a graduate of a two year associate degree program offered by a community college, technical institute, or university. The graduate normally works in support of engineers and technologists. He or she may possess a two year associate of applied science (AAS) or associate engineering technology (AET) degree. While most degrees are in the more general areas of civil, electrical, and mechanical engineering technology (Ellis, 1992), many more specific degree opportunities exist (Ellis, 1991).

The technologist is a graduate of a four year baccalaureate degree engineering technology program and holds a BS ET degree. These graduates are similar to engineering graduates, frequently occupy the same positions and job titles, but have specifically been educated to apply engineering principles for industrial production, construction, and operation. Typically these individuals are interested in the applications of engineering and technology. The BS ET degree usually builds upon the AET degree in a mode which often is referred to as 2+2. This format for engineering technology education has been reinforced in a national forum (Worthly, 1988).

The engineer is a graduate from a four year degree program with an emphasis on problem solving using the principles of science and technology to produce economically feasible designs. While the engineer's education is relatively broad and the engineer has an analytical, creative mind that is challenged by open-ended technical problems, the educational emphasis in the last two years is normally specific to a field such as chemical, electrical, or mechanical engineering.

Many engineering graduates continue their education to become more specialized in a specific engineering discipline. They may attain masters (MS) or doctorate (PhD) degrees in a specific engineering discipline. Alternatively, many BS degree recipients in engineering related fields, such as physics and chemistry, complete advanced degree programs in engineering.

From a discipline perspective AET and BS ET graduates fit into the engineering technology discipline, whereas, engineering graduates at the BS, MS and PhD level fit into the engineering discipline.

3 United States Historical Perspective on Engineering Technology

Within the United States the roots of the engineering technology discipline lie with the early development of engineering extension programs to supply technical education to locations remote from the parent educational institution. Penn State was an innovator in this area, formally extending its services in engineering education from its base in central Pennsylvania to the outlying communities in 1911 (Bezilla, 1981).

Penn State continued to be involved in the extension of engineering education and in the years following World War II worked to provide appropriate educational opportunities for returning GIs. As a result, associate degree programs in engineering technology were developed and many of today's engineering technology colleges were either established or extended at that time (Cheshire, 1992).

Following the launching of the USSR satellite Sputnik in 1957, a stronger emphasis was placed on science and technology education in the United States. In response, more science and mathematics courses were included in the engineering curriculum. Engineering courses and curricula were revised to include more theory, and laboratory components of many engineering courses were dropped. This resulted in a more theoretical and scientific engineering education curriculum. The art and practical elements of engineering education were lost in the curriculum transitions during the decade of the 60's.

These developing voids in the art, practice, and lab components of engineering education led to the expansion of the two year engineering technology curricula into four year baccalaureate engineering technology programs. These programs grew in both number and quality in the 1970's and 80's.

Simultaneously, with the development of associate and baccalaureate engineering technology degree programs, accreditation standards were developed by the appropriate professional engineering societies (ASCE, ASME, IEEE). The first Associate Engineering Technology (AET) programs were accredited in the early 1950's and the first BS ET degrees were accredited in the late 1960's by ECPD--now ABET (the Technology Accreditation Commission of the Accreditation Board for Engineering and Technology), the representative body for accreditation of the engineering professional societies.

ABET is composed of representatives of the professional engineering societies and has as its purpose the promotion and advancement of engineering education with a view to further the public welfare through the development of the better educated engineer, engineering technologist, and engineering technician. One of the basic objectives of accreditation is to identify for prospective students, student counselors, parents, potential employers, and officials the engineering and engineering technology programs which meet the minimal ABET criteria for accreditation. Accreditation is based upon satisfying minimum criteria. As a measure of quality, it assures only that a program satisfies a minimum standard.

Following accreditation of the first programs in 1967, baccalaureate engineering technology programs grew rapidly. In addition to an increase in quantity, the quality also increased in association with increasing accreditation standards.

Today there are 312 TAC/ABET accredited BS ET programs in 112 colleges and universities. This is a dramatic increase over the 221 programs that were accredited in 1980. At the associate degree level there are approximately 454 TAC/ABET accredited programs at 167 colleges and universities. In 1980 there were 416 (ABET, 1990). As these figures demonstrate, the most rapid growth in engineering technology has occurred at the baccalaureate level.

The most popular engineering technology programs are in electrical/electronics, followed by mechanical/manufacturing, and civil engineering technologies. These three categories account for 85 percent of the 22,793 TAC/ABET accredited graduates nationally (Ellis, 1992). These figures illustrate that engineering technology education has become a large enterprise in the United States.

Despite this fact many of the engineering societies and the engineering profession, in general, continue to debate the interface of engineering and engineering technology and many barriers still exist which either prevent or inhibit the professional registration of engineering technology graduates (Chapple, 1991). Among many engineering educators, engineering technology is viewed as a second class engineering discipline.

In addition, graduate engineering opportunities are limited for engineering technology graduates.

In recent years there has been an increased interest by BS ET graduates to pursue graduate education. In most engineering institutions the prevailing attitude is that if a technologist does not initially embark on an undergraduate engineering education, then he/she has no path to graduate education. This attitude has resulted in a recent increase in the number of institutions offering master level degrees in engineering technology.

4 A Brief International Perspective on Engineering Technology

Engineering technology or technology education has developed along side engineering educational systems in most industrially developed nations in the world. Regardless of the history, each country's economy has required and encouraged a spectrum of educational experiences from the highly scientific to the most applied. Some of the most developed structures outside the United States occur in the United Kingdom, Russia, Taiwan, and Japan.

As in the United States most of these programs train engineering professionals in a four year university environment and engineering technologists (associates) in either a two, three or occasionally a four year applied educational environment.

The educational systems in the European Community vary as much as the educational levels in engineering in the United States. In addition, there are many different titles, many schemes of education, and even some different schemes in several countries which lead to the same engineering degrees.

The British have evolved a model adopted by the Engineering Council which could be regarded as a rough model for the adoption of a European engineer. They see three different engineering qualifications as desirable: the engineering technician, the technician engineer, and the chartered engineer (Wald, 1987). This recognition of the duality of the engineering profession is encouraging. Also encouraging are the development of more recent models in Poland, Taiwan, and Australia. In these models the two systems are complementary and do not exclude or crucify each other (Grenquist, 1991). Hopefully this will develop into a trend and the two systems will be able to co-exist with equal merit with the objective to produce the best qualified people for both research and development and for general or applied engineering.

5 Characteristics of Engineering Technology Programs and Faculty in the United States

A definition recently adopted by the American Society of Engineering Education's (ASEE) Engineering Technology Council illustrates the educational emphasis of engineering technology programs and the most fitting placement for their graduates.

> "Engineering Technology is the profession in which a knowledge of mathematics and natural sciences gained by higher education, experience, and practice is devoted primarily to the implementation and extension of existing technology for the benefit of humanity. Engineering technology education focuses primarily on the applied aspects of science and engineering aimed at preparing graduates for practice in that portion of the technological spectrum closest to the product improvement, manufacturing, and engineering operational functions." (Engineering Technology Council, 1991)

In practice, engineering technology programs focus on both application and practice and devote equal time to classroom theory and laboratory experience. Typically faculty hold master's degrees in engineering, engineering technology or a closely related field and are required to have at least three years of relevant industrial experience. Faculty members are also expected to remain current by active participation in their professional societies, reading the literature, continuing education, applied research, consulting, and periodic return to industry. They are also encouraged to publish and to demonstrate other measures of scholarship (ABET, 1991). Within major research institutions many engineering technology faculty are experiencing difficulty in meeting university expectations for promotion and tenure. Opportunities in international education seem undeveloped and very appropriate for engineering technology faculty.

6 International Technology Transfer

Faculty in engineering technology are often denied international opportunities for lecturing, research, and curriculum development because they lack a PhD. For example, the United States Fulbright Scholar Program includes among its criteria for eligibility the requirement of a doctorate for a research award and a preference for a doctorate for a lecture award. While this criterion may be appropriate from a prestige viewpoint or for research and lecture awards in developed countries, it may not be appropriate for developing countries.

As a Fulbright Senior Scholar to the University of Mauritius from 1985 to 1987, I had the opportunity to work with faculty and local industry to develop appropriate research projects and to assist the University in establishing a research program. At that time Mauritius was rapidly moving from a third world to a second world classification. The country was experiencing rapid expansion of its tourism and textile industries. Growth in both of these areas was placing a strain on the country's energy resources and the development of indigenous resources of poor generation was desired. Bagasse, the fibrous portion of sugar cane, and solar energy for hot water and food processing were attractive alternatives, as well as hydro and wave energy. Projects were initiated or on-going in all of these areas. In all instances the projects were applied in nature and ideal for individuals with the appropriate engineering technology credentials. In addition to being better qualified than many PhDs for these projects, engineering technology faculty would experience a greater sense of satisfaction while gaining recognition and appropriate scholarly credit toward promotion and/or tenure in their home institutions.

As a result of these observations the University of Mauritius School of Engineering modified its criteria for awards to encourage individuals with master's degrees and industrial experience to apply. The 1992-93 Fulbright awards announcement for Mauritius includes the following (Council for International Exchange of Scholars, 1991):

> "**Computer Science**....MS degree with teaching experience acceptable. **Computer Science**....identify and initiate research ...for faculty; organize short courses and workshops for faculty and private organizations;...MS and 5 years industrial experience acceptable. **Industrial Engineering**....Initiate industrial projects and develop training modules for the private sector. Supervise staff research projects....Candidates with MS acceptable. **Surveying**....Assist in developing teaching and research programs...Candidates with MS and 5 years experience acceptable."

While it is still too early to assess the success of utilizing non-PhD faculty in these specific areas, it is encouraging to note that at least one developing country has recognized the engineering technology discipline and the appropriateness of engineering technology faculty for advancing and implementing technology in their local industries and for expanding the research base of their university faculty. More partnerships between **engineering technology** institutions in the United States and developed countries and **engineering institutions** in developing countries would be beneficial to both.

7 References

Ellis, E.A. (1992) "Degrees '91:," **ASEE Prism**, February.

Ellis, R.A. (1991) **Engineering and Technology Degrees 1991**, Engineering Manpower Studies, American Association of Engineering Societies, Washington, D.C.

Worthly, W.W. (1988) "NFET-A Mandate for Change in Engineering Technology Education", **Journal of Engineering Technology**.

Bezilla, M. (1981) **Engineering Education at Penn State**, The Pennsylvania State University Press, University Park and London.

Cheshier, S.R. (1992) "An Introduction to Engineering Technology", **1992 Directory of Engineering and Engineering Technology Undergraduate Programs**, ASEE, Washington, D.C.

Accreditation Board for Engineering and Technology (1990) **1990 Annual Report**, New York

Engineering Manpower Commission (1991) "Engineering Degree Trends, 1991," **Engineering Manpower Bulletin**, 115, December.

Engineering Technology Council (1991) "Eleven Principles that the ETC Endorses," American Society for Engineering Education.

Wald, M.S. (1987) "Developments in Engineering and Technology Education in Europe", **Proceedings National Forum for Engineering Technology**, Accreditation Board for Engineering and Technology, New York, NY, USA.

Chapple, A. (1991) "Engineers, Technologists Separated at Graduation by State Boards, **Engineering Times**, 13, 4.

Grenquist, S.A. (1991) "The Advent of Engineering Technology 'Down Under' in Australia", **Journal of Engineering Technology**.

The Accreditation Board for Engineering and Technology, (1991) **Criteria for Evaluating Programs in Engineering Technology**, New York, NY, USA.

Council for the International Exchange of Scholars (1991) **Fulbright Scholar Program: Research and Lecturing Awards 1992-93**, Washington, D.C., USA.

The Impact of the German Model Gesamthochschule on the Academic Education of Engineers

K. Detert

Inst. für Werkstofftechnik, Universität-GH-Siegen

Abstract
The model Gesamthochschule was intended to reform academic education in Germany. The historic system of advanced education of engineering and technology in Germany will be described. It is outlined how the Y model of integrated courses in the Gesamthochschule attains its objectives. Evaluation of its performance at present are discussed.
Keywords: academic education, university, curriculum, study-courses, objectives, advanced education, engineering science, engineering technology.

1 Introduction

After the second world war the old system of the university education before 1933 had to be restored, since Hitler had distorted this system by orienting the schools ideologically. But soon it became very clear that the old system of university-education had to be reformed to meet the requirements of a modern society. Georg Picht (1964) described the state of higher education in Germany in the early sixties as catastrophic, since only a small minority of young people, about 5 % of each age group were students in a German university. But this figure did not include all other kinds of tertiary education, which had been developed within the German system of higher education. For example he did not count the students at German polytechnics and at all sorts of colleges. When the so called student-movement at the end of the sixties tried to revolutionize the whole social system in Germany it was a natural consequence that a thorough change of the educational system in Germany was demanded. One of the objectives was the equal opportunity of entrance for everyone to highschools and universities as well, with no regard to the differences of educational background. As a consequence the new concept of Gesamtschule was born, which was supposed to integrate all kinds of secondary stage of education into one highschool. Furthermore a new concept of the tertiary system of higher education along this idea of equal opportunity was claimed by the Bundesassistentenkonferenz (1968), so that the University should be open for students of all classes of people. The institution which was claimed to be formed got the name Gesamthochschule (abbreviated GH).

2 The Objectives of the Concept Gesamthochschule

It was certainly the political pressure initiated by the German student movement of the late sixties which induced the German Wissenschaftsrat (1970) to recommend a change of the German university education through the formation of new institutions called "Gesamthochschule". The German Wissenschaftsrat acts as a national advisory council to the government of the Federal Republic of Germany. It consists of elected representatives from German universities and from industry together with delegates from federal and state governments. The recommendation to form GH was the result of thorough consideration over the years, on how to reform the German system of advanced education in order to meet the requirements of a modern society. This recommendation had phrased the objective of the new institution: *The Gesamthochschule represents a new educational system which differentiates in regard to its teaching courses and subjects, but it integrates its organisation and administration into one structure. This institution will meet in a more appropriate manner the expected requirement of advanced education of the future.* This recommendation met at its time with broad consensus among all political parties and groups and was introduced into the phrasing of a special law called "Hochschulrahmengesetz"(1976) (a law that lays the groundwork for the organization of the academic educational structure in Germany). Yet, efforts were taken only in some states to put those recommendations regarding the installation of Gesamthochschule into effect. Gesamthochschulen were founded only in the states of Nordrhein-Westfalen and in Hessen as new institutions to realize the new concept of advanced education. Those new schools teach in one place the same subjects as colleges and universities do in different institutions. The study-program of the GH is organised in the form of so called integrated study-courses, which offer as option two different final examinations. The one final examination was specified to provide the certificate that the qualification was obtained to apply the scientific background to solve the practical tasks in the professions related to the subjects offered. The other certificate, which requires one year longer study-period and the special choice of options during the course, shall certify the ability of participating in research and development on a scientific and theoretical base. It is also a prerequisite to be admitted to a doctorate. Therefore the objectives of the courses, which had so far to be studied in different schools remained, but are now combined in a differentiated course in one school with different kinds of final examinations. Those new schools however were not open for all classes of people, but the range of access to those new schools was wider than in universities, since the minimum requirement of admission was now defined by the requirements of the colleges.

It was a further objective to allow by this organisation the establishment of schools with advanced higher education in smaller cities in provincial areas apart from larger cities with the opportunity to pass the examination at university level, where one could not establish an university on its own. This was called regionalizing the offer of higher education. It was hoped to give those young people who did not live in large cities and did not want to move to a large city where living is more expensive and seducing, a chance to study at the highest level.

3 The Engineering Education in Germany

The application of the structure of GH on the academic education of engineering had a very special influence, which can only be understood by regarding its historic background. Education in engineering before this century was not regarded as having a level of university education. The German universities limited its education to science of humanities and natural science, which was directed to develop the human spiritual personality. Education in the field of technical application was regarded as kind of advanced professional training. So it was not a general agreement but a decision of the German emperor in 1900 that the Technische Hochschul emperor in 1900 that the status similar to the university with the same academic rights. Some of the so called Technische Hochschulen changed their names after the second world war into Technische Universität. However, up until now engineering sciences are in general not taught in genuine German universities, where humanities are the main subjects.

One furthermore should note that not all polytechnics which existed in Germany in the past had aimed to reach the university level. Therefore it came 1878 to a division among technical schools. Those schools which regarded themselves as being on a lower level in comparison to Technische Hochschulen emphasized education as a preparation for the different technical professions. The name of this type of engineering schools changed at first from technische Fachschule into Ingenieurschule. In 1968 the name was changed to technische Fachhochschule. But today the Fachhochschule aims very strongly to be regarded as equivalent to the university level, though their objectives emphasize the application of technical principles into practice more than technical universities. One has however realized, that the application of engineering today requires a thorough understanding of fundamentals and principles of science and engineering, to develop the flexibility to adjust to a rapidly changing technology and to technical progress. Furthermore, those schools claim that the organisation of their curricula is much more efficient than in German technical universities, where the organisation of the courses is more liberal and includes many options of different subjects, which extends the length of the studies substantially.

4 Engineering education in the Gesamthochschule

According to its objectives GH introduces two major changes in the traditional scheme of engineering education in Germany. Firstly, the academic education of a traditional university in humanities and science is combined with engineering education in the same institution. Secondly, both objectives of different types of engineering schools are integrated not only in one institution but into integrated curricula of various subjects of engineering science. The University of Siegen is an example of those GH with integrated engineering courses founded in 1972. Their objective states: *The academic curriculum should meet the demands of a changing professional world of engineering in providing the knowledge and skill in methods to act competent in science based projects, in critical evaluation of new technical concepts and revealing a responsible personal attitude.*

According to its regulation of curriculum the integrated course of mechanical engineering is divided in two parts as in the technical universities.

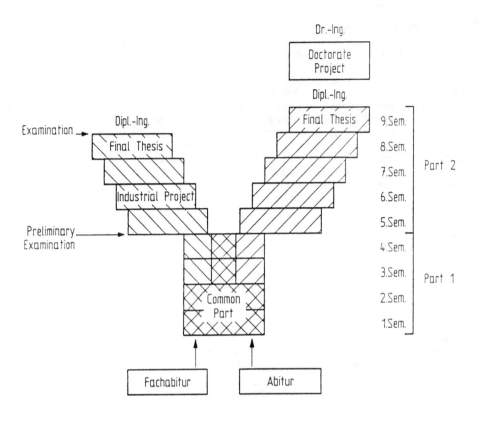

Fig. 1. Integrated Course Mechanical Engineering (Y-Model)

The figure gives a description of such an integrated course in mechanical engineering at the university-GH-Siegen, Germany. Part one, consisting of two year studies, puts emphasis on the fundamentals and principles of natural science and engineering, like physics, chemistry, mathematics, mechanics, material science, design-methods, production-technology, electrical engineering and some basic understanding of data processing and business of engineering. In this part a choice is offered to deepen the theoretical understanding of advanced mechanics, mathematics and thermodynamics. After studying this part of the curriculum the preliminary examination is to be taken which already differentiates to which optional branch of the second part of his study period the student will be admitted.

The second part of the study program of an integrated engineering course divides into two main branches, each containing an obligatory core and some options. This structure of the integrated course is called the Y-Model.
- The one branch is directed to teach the application of engineering science to design and production during the professional work as engineer. It consists of two year studies containing one semester to work in industry on an engineering

project and it includes a final thesis to obtain the Dipl.-Ing.
- The other branch strengthens the scientific background of the basic engineering principles in a two and a half year study programm. It contains the obligatory subjects to supplement the basic knowledge and to provide the students an overview of the different important engineering fields of the modern industrial world. It was an important issue, while developing this course, to point out that engineering solutions require the application of the fundamental concepts of systems-engineering and ergonomics as well. This study-programme also consists of a number of options of different subjects, so that the student can emphasize his studies in certain engineering areas including the choice of a more theoretical based knowledge of engineering science. This study-program will be completed by two minor engineering projects and a major thesis to obtain the Dipl.-Ing.

5. Efficiency of Engineering Education in the Gesamthochschule

Rimbach (1980) investigated the performance of engineering education of the GH Siegen very thoroughly and came to the conclusion that after eight years of operation the objectives were attained very successfully. Today, after twenty years performance, one can confirm his statements. In the meantime, other evaluations have ranked the academic education in the Universität-GH-Siegen as very effective. On the other hand, the concept of Gesamthochschule is not any more politically favored as a better structure of academic-education (BMBW 1984). The existing schools with this structure are now submitted to the competition between the different types of schools. The GH Siegen should not lose in its competition with other academic institutions for mainly two reasons. The students favor the smaller universities, for the large German universities with more than 30 000 students exert much more hardship and stress to the young students, who must learn to stay on their own in academic life. This hardship has caused an extraordinary number of suicides among students in the large universities. In spite of being only a small university with 12000 students, The Uni-GH-Siegen has earned a good reputation in the technical field as well. Students, who will not be admitted to an academic study on a university level, when their secondary school diploma only qualifies for admission to a college or Fachhochschule appreciate furthermore the advantage, that they can obtain admission to the integrated course of a GH. After taking some general subjects in order to condition their educational background they can get the Dipl.-Ing. of the university-level by taking the corresponding branch of the curriculum. It was at first a surprise that students with the lower level of high school diploma proved in general as qualified as those with the more highly regarded Abitur. Since most of the school with the higher level of final qualification (Abitur) stress the education in languages, the background in technical fields and physics is not better.
One must however admit, that two disadvantages exist in the new system. One is that faculties are divided in two classes of professors, one are more related to research and the others more to teaching. This division into two classes of professors with different qualifications causes some personal resentments, which burden the cooperation in the faculty. The other disadvantage is outlined by the fact, that the more liberal teaching system of universities has been adopted. Therefore the German Fachhochschule can claim that their system is more efficient than the academic education in the shorter branch of the

profession-oriented study programm of a GH. Indeed, the study period in this branch is normally likewise extended more than a year as in the longer branch. This might also be the reason, that only a minority of students chose the shorter branch of the Y-Model, though one had expected the opposite in regard of the demand of engineers with their specific educational background.

6 Future Outlook

Although the GH is at present not anymore in the focus of political attention, the existing GH have proved their worth. Of course their system can only survive when it proves the flexibility to adjust to the future-changes in the world and in the society. There is no reason to be afraid, that this school-system will lose its present attraction. The process to remove the barriers between the both classes of professors is at present already in progress. Probably the system Gesamthochschule with its integrated courses will only survive, when the character of learning can be changed from the overemphasized liberal style into a more rigorous schooling system at least during the first common part of the course, and perhaps also in the shorter branch of the second part. The political pressure to reduce the study-period will support the introduction of this more rigorous system, since it will prove as efficient and adequate to reduce the study-period. Therefore one can expect that this new school system will not only survive as a speciality of academic education, yet it can become a good example of a better structure for academic education in the future, since it enables the organisation of courses with smaller number of students to reach a high quality of academic education.

References

Picht, Georg (1964) Die Deutsche Bildungskatastrophe
Bundesassistentenkonferenz (1968) Kreuznacher Hochschulkonzept, Bonn.
Wissenschaftsrat (1970) Empfehlungen zur Struktur und zum Ausbau des Bildungswesens im Hochschulbereich nach 1970, Bonn.
Hochschulrahmengesetz, 26.Jan. (1976) (Bundesgesetzblatt I S.185)
Rimbach, Gerhard (1980) Studienreform als Prozeß - Zwischenbilanz des integrierten Studienganges Maschinenbau an der Universität-GH Siegen, FB 11, Siegen
BMBW (1984) Bericht der Expertenkommission zur Untersuchung der Auswirkungen des Hochschulrahmengesetzes, I B S 12 u.13, Bonn

Attributes for the Baccalaureate Engineer: What are the Desires of Industry?

D.L. Evans (*), D.L. Shunk (**)

(*) Mechanical & Aerospace Engineering, Arizona State University

(**) Industrial & Management Systems Engineering, Arizona State University

Abstract

This paper presents data on the perceptions of an informed group of industry representatives on the importance of and the preparation in a chosen set of attributes for BS level engineering graduates. This industry group's desires are compared with those of faculty and students. The most important attribute and the one of largest disparity between importance and preparation was *problem solving skills* (recognizing a problem, generating alternative solutions, executing a good solution). Preparation of new graduates was judged to exceed importance in the attribute of *computer skills*.

Keywords: Industry Desires, Customers, Attributes of New Graduates, Curriculum Goals, Preparation of New Engineers, Curriculum Outcomes

1 Introduction

The purpose of this paper is to discuss the "customers" and the desires of these "customers" for various attributes for the "product" known as the graduate engineer (Bachelor of Science [BS] level). Particular attention is given to the industry subset of these customers since this customer is often viewed as engineering education's most important one. It is also probably the most neglected customer, at least in efforts to systematically assess its desires. The data presented were collected as a part of a major study of undergraduate engineering curriculum needs that has just been completed in the College of Engineering and Applied Sciences at Arizona State University by a faculty Task Force (Beakley, et. al, [1991]). This study is somewhat unique in that it approached the curriculum design problem in a bottom-up fashion by first establishing a model for the future baccalaureate-level engineer, and then addressing the curricular components that might produce this model. The customer sampling described here was done to evaluate current performance and to determine desires for the future.

2 The Customers for the BS Level Engineering Graduate

As a part of the curriculum study, the 20 member, faculty Task Force spent several months evaluating appropriate attributes for the baccalaureate-level engineer that were suggested by its members. Although certainly not unique, ten somewhat distinct generic attributes were found to contain nearly all of the hundreds of individual

Table I.
The Chosen Set of 10 Important Attributes

A An ability to identify and define a problem, develop and evaluate alternative solutions, and effect one or more designs to solve the problem.
B A breadth and depth of technical background.
C A fundamental understanding of mathematics and the physical and life sciences.
D An effectiveness in communicating ideas.
E A high professional and ethical standard.
F A mature, responsible, and open mind, with a positive attitude toward life.
G An ability to use computers for communication, analysis, and design.
H A motivation and capability to continue the learning experience.
I A knowledge of business strategies and management practices.
J An appreciation and understanding of world affairs and cultures.

Table II.
The Customers of the University Enterprise

Customer	Service/Product Provided to the Customer
Students: Parents:	Marketable Skill Preparation for Life-long Learning Positive Learning Environment Preparation for Global Practice

Employers:	Technical Manpower Having
	  Competence
	  Communication/Interpersonal Skills
	  Professional and Ethical Attitude
	  Inquiring Mind
	  Economic Awareness
	  Sense of Responsibility
	  Motivation to Succeed
	  Technology

The Public: Society in General Taxpayers	Technical Manpower Having
	  Technical Competence
	  Social/Environmental/Political/-
	    Economic Awareness
	  Sense of Responsibility
	  Professional and Ethical Attitudes
	A Graduating Class that Mirrors the
	  Gender/Racial Mix of the Population
	Knowledge and Unbiased Assessment

Faculty:	Well-Prepared Students for
	  Graduate Study
	An Atmosphere to Pursue and Assess
	  Knowledge Freely
	Facilities and Rewards

suggestions. The ten chosen attributes are given in abbreviated form in Table I.

In establishing a curriculum based on these attributes, it is important to consider the opinions of a variety of "customers" on the relative merits of each attribute. In the presence of wide agreement on the relative merit of each, curriculum design can then proceed by insuring that each attribute receives the proper amount and type of attention. Following the Total Quality Management (TQM) revolution (Tribus [1987]), it is the customers who should determine the quality (that is, performance in pertinent attributes) of a product. Thus, it is timely and appropriate to consider who the customers are for the products and services generated by the enterprise called the University. Table II displays a rational customer set and lists some of the services and products that are provided to each member of the set.

3. Selecting and Sampling the Industry Customers

The second group listed in Table II, employers, was sampled by requesting chief executive officers (CEOs) representing the larger firms employing engineers in the Phoenix metropolitan area, to choose an appropriate person from their organization to participate in a one-day workshop on undergraduate education. Specific suggestions were included

Table III. The Requests to the CEOs

- No public relations or personnel employees.
- Appoint an engineering degree holder who:
  - Is in the management ranks of the organization,
  - Is considered a "rising star" in the company,
  - Is familiar with the capabilities and shortcomings of newly graduated engineers (from universities nationwide),
  - Is aware of the cultural changes that the industry must undergo to become/remain competitive,
  - Is capable and willing to articulate the company's specifications for engineering graduates of the future.

in the letter of request to the CEO's to help in their selection of the appropriate individual to send. These suggestions are listed in Table III.

This method was used to get an up-to-date picture of what is happening within American industry. As McMasters and Ford (1990) have pointed out, industry does not usually speak to academe in a common voice. Top management gives an engineering dean one impression, middle management gives a department chair another, dissimilar message, and faculty may get yet another from a variety of sources (alumni, etc.). This array of voices is particularly diverse in the current era in which much of American industry is trying hard to shift from its past position of world leader and/or government provider (e.g., aerospace and military equipment) to being competitive in a global economy.

In the short time that was available for bringing this industrial workshop together, representatives of 14 of the 23 companies contacted eagerly volunteered their services. The workshop, held in April of 1991, was a 5-hour long work session in which the participants were asked to verbalize their views as to specific abilities they thought were necessary of new engineers. Three Task Force members met with these participants and used a "nominal group" technique to keep the workshop on target.

The participants were an exceptional group. They came from a variety of disciplines and fields such as electronics (Motorola Inc.), mechanical (Garrett Divisions of Allied Signal Aerospace Company), aerospace (McDonnell Douglas Helicopter Company), communications (AG Communications Systems Corporation), computers (Digital Equipment Corporation, Intel, Honeywell, Hull HN Information Systems, Inc.), materials (ICI Composites Inc.), and consulting (Anderson Consulting, Inc.). However, because of the method of soliciting their participation, the participants represented only large firms. Government employers and other segments of the employment community were not represented.

The industry representatives worked together in harmony and reached consensus agreements with unexpected ease. Their individual views were amazingly similar and very much in line with the findings of the major (national) studies issued in the last two to three years. Although the faculty Task Force had already established the set of attributes listed in Table I, these attributes were not given to the workshop participants until near the end of the day. Instead, the participants were encouraged to discuss how their companies were changing and what they saw as needs for the new engineer in the decade ahead.

What follows is an abbreviated summary of the industrial participants' comments on the changes in their industry, their company, and/or in technology. Most of these comments were repeated by several of the participants: *Increasing emphasis on quality* ♦ *Foreign competition is growing stronger* ♦ *Seeking quality from a business standpoint* ♦ *Lifetimes of products are shortening* ♦ *The engineer is an integrator* ♦ *Cross-cultural relationships increasing; now working with both the French and Japanese* ♦ *Five years is a long time* ♦ *Moving away from military products toward commercialization* ♦ *Performance was once everything—now, maintainability, reliability, and cost are extremely important* ♦ *Technology* breakthroughs are a threat if they come from the competition ♦ Complexity is increasing—more emphasis on systems level problems ♦ Must reduce time to market ♦ Must learn to cope with change ♦ Beginning a continuous improvement process ♦ Beginning concurrent approaches, computer-integrated manufacturing (CIM) ♦ A cultural change is underway ♦ Shorter development times are necessary ♦ Emphasis on teamwork ♦ De-layering (reducing layers) of management is underway ♦ Trying to reward engineers by means other than moving performers into management ♦ Treating engineering as a business ♦ Automation is important ♦ Quality and metrics to measure quality are important ♦ Integration of different technologies (e.g., electrical and mechanical) is important ♦ Increasing emphasis on approach to the problem - scope it out ♦ Total customer satisfaction is necessary ♦ Engineers must now function in both engineering and social environments ♦ Must learn to manage people through changes.

Below is an abbreviated summary of industrial participants' comments on the need's for new talent. Many of these comments were repeated by several of the participants: *Need flexibility* ♦ *Need experience working with complexity—sort it out, condense, summarize* ♦ *Need ability to recognize how to do it better and how long it will take next time* ♦ *Ability to understand a process* ♦ *Need better problem solving skills— thinking, learning, independent, motivated* ♦ *Team skills* ♦ *Well-rounded—fewer "nerds"* ♦ *Business savvy* ♦ *Need loyalty toward a company (best students don't seem to have it)* ♦ *Ability to integrate* ♦ *Newer graduates are "software jockeys"* —need to understand the computer results ♦ *Have to understand entire systems in concept* ♦ *Need a "drive" for action—"drive" for a conclusion* ♦ *Fundamental core skills: mathematics (including statistics of the design of experiments type), physics, chemistry, computer skills - must keep and improve these* ♦ *More emphasis on communication skills: technical writing, listening, presentation skills, sketching skills* ♦ *Add thinking, hands-on skills (e.g., chip making), group dynamics, real-world product development cycle, scoping out the problem, defining requirements, continuous improvement philosophy (introduce early and nurture throughout), tolerancing* ♦ *Need to have some understanding of the business: intellectual property rights, proprietary data, international competition.*

4. Ranking of the Chosen Set of Attributes by the Industry Group

At the conclusion of the workshop, the attendees were asked to rank *the importance* of the Task Force's 10 attributes as well as to rank engineering education's *performance* in producing these attributes. Student and faculty customer groups had already been sampled as to their perceptions of the relative importance of and performance in each attribute of Table I. Individuals in each customer group were asked to give what they thought to be the most important attribute a relative rating of 1 and the least important (but not necessarily unimportant) attribute a rating of 5 (by marking the extremes of a five-part scale on a machine graded form). Respondents were then asked to establish relative ratings on this same 1 to 5 scale (integers only permitted) for the other 8 attributes according to their perceived importance. Individual results within each customer group were averaged to determine a *relative rating* number representative of that group. In a similar way, each respondent was asked to rank *the performance* of current engineering graduates in these attributes.

Figures 1 and 2 compare the relative rating of the *importance* of each attribute (Fig. 1) and the relative rating of *performance* of new graduates in the attribute (Fig. 2) as judged by each of the customer groups surveyed. Figure 3 displays the *disparity* defined as the numerical difference between the performance relative rating and the importance relative rating. All three figures show the relative ratings on each attribute as perceived by each of the four customer groups, ordered (from left to right) by increasing numerical place ranking (decreasing performance) as perceived by the

Engineering Education   127

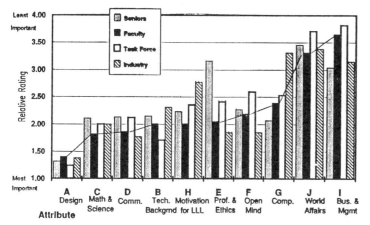

Fig. 1
The 10 Attributes Ranked by Importance
(Sorted by Place Ranking Determined by the Faculty)

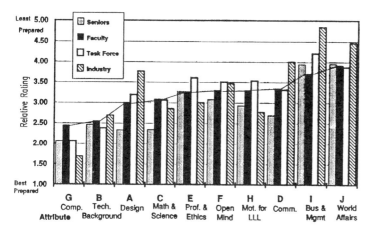

Fig. 2
The 10 Attributes Ranked by Preparation
(Sorted by Place Ranking Determined by the Faculty)

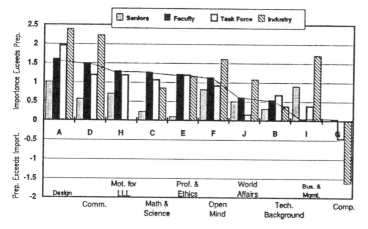

Fig. 3
The 10 Attributes Ranked by Disparity
(Sorted by Place Ranking Determined by the Faculty)

faculty group. These figures are intended to show the degree of agreement or disagreement among the four customer groups.

Figures 1, 2 and 3 lead to the following observations regarding this particular industry group's view of the chosen set of attributes:

- Attribute **A** was rated to be significantly more important than any of the other attributes by all 4 rating groups. This attribute combines several features that the industry workshop participants thought to be important. These include improved: intellectual ability at synthesis, analysis, and evaluation; thinking abilities; problem scoping abilities (i.e., how to define requirements, when to use the computer, the creation of data bases); and understanding of a real-world product development.
- Attribute **D**, communication skills, was rated by the industry group to be 2nd to only attribute **A**, problem solving, in importance but only 8th in preparation leading to nearly the same disparity as attribute **A**.
- The industry group gave large disparities to attributes **F**, open mind, and **I**, business knowledge. To this group, **I** seemed to signify an understanding of patents, intellectual property rights, proprietary data, and an annual report.
- Preparation in computers, attribute **G**, was considered by the industry group to be superficial. They viewed graduates as being unafraid, indeed eager, to use the computer but unable or unwilling to use it in a responsible manner by explaining results obtained. This group applied the term "software jockey" to new graduates.
- The industry group's discussion on skills within attributes **B** and **C** led to the conclusion that analysis skills (math, science, and engineering science) must be kept. In addition, skills in statistics (design-of-experiments type), critical and creative thinking, and tolerancing need improving. Improved thinking was a much-mentioned feature during the workshop. Interestingly, preparation in **B** seemed to be viewed as better than preparation in **C**, with less disparity.
- Surprisingly, the industry representatives showed less concern over current life-long learning capability, attribute **H**, than have many recent studies of engineering education.
- The industry group gave much importance to attribute **E**, ethics and professionalism. Their rating was higher than the faculty's rating and <u>much</u> higher than the students' rating of this attribute.
- Items of concern to the industry group that do not show up explicitly in the chosen set of attributes include teaming skills, hands-on skills (e.g., laboratory and project work), and the concept of continuous improvement (from TQM principles). However, these can easily be visualized as an implicit part of many of the chosen attributes.

This initial assessment of possible attributes now makes possible the definition of curriculum components that address the generation of these attributes. Beakley, et. al (1991) suggest such components.

References

Beakley, G., ch., (1991) *Engineering Education: Preparing for the Next Decade*, Arizona State University, Tempe, AZ, 85287-5506, USA.

McMasters, J., and S. Ford, (1990) "An Industry View of Enhancing Design Education," *Engineering Education*, July/August, 526-529.

Tribus, M., (1987) *Quality First: Selected Papers on Quality and Productivity Improvement*, Publication Number 1459, National Society of Professional Engineers, Washington D.C., 259 p.

Technology Assessment: An International Issue

M.A. Brown

Department of Communications, Humanities and Social Sciences, University of Cincinnati

Abstract

In his study of Finland's technology, Dr. Timo Myllyntaus observed, "Technology transfer. . . . is a complex process of transformation which necessarily takes place in a specific economic, social, and cultural context" (1991:294). In my comparative study of how the people of the United States and Finland gain insight into the impact of technology on their societies, I adopt Myllyntaus' "contextual filter" model: societies acquire new technology, sift it through their cultural value systems, then determine its result. Technological developments have resulted in similar benefits and detriments in both nations, from improved industrial production to environmental pollution; both nations also recognize the need for public participation in decision-making processes about technology and provide opportunities for that endeavor. Those discussions occur in useful forums from political and educational institutions to debating societies to public demonstrations; however, these forums often fall short of providing the broadest possible information base for assessing technology. As a partial remedy to that unfortunate situation, educators, by offering classes that address societal issues, by increasing interdisciplinary cooperation in the classroom, by relying more strongly on industry and by capitalizing on professional organizations, can take a strong leadership role in this crucial task. My study provides strategies for organizing curricula to reduce the gap between technology and the world society it serves.

Keywords: Technology - Social Aspects, Philosophical Aspects, Engineering Education.

1 Introduction

Assessing technology is everybody's business. While that statement is not revolutionary, its substance is problematic. As historian John Ziman observed in analyzing the philosophical and societal aspects of science and technology, the esoteric nature of those disciplines deters all but "the specialist" from participating in the application of their principles (1984:178). Unable to share in decisions about technology, many lay-people respond to these "profoundly revolutionary" developments with indifference or frustration, growing out of feelings of "uncertainty and. . . powerlessness" (Ziman: 118). Although other avenues for public discussion exist, Ziman, like many others, called upon educators to correct this inequity by creating curricula to broaden conceptual understanding for a larger population.

130 Engineering Education

Ziman's suggestions underscore the need for organizations such as ASEE, with a broad international organization of scientists, engineers and educators to capitalize on current world conditions to improve the flow of scientific and technological information. Currently, public curiosity about technology and its effects is at a high point: six million members of Greenpeace express widespread environmental concerns (Hällsten, 1991; Kuvaja, 1992); international committees, ranging from regional groups that discuss such issues as reindeer husbandry in Scandinavia's Lapland or acid rain in the United States and Canada, to super summits that address nuclear disarmament, genetic engineering and the economics of the world's workforce, signal a far-reaching commitment to our physical and societal environment. Unfortunately, the technological disasters that made Valdez, Chernobyl and Bhopal household words generated much of that attention. But the time is ripe for educators to channel that interest into positive results, in part by utilizing technology assessment methods already operating in such nations as Finland and the United States.

2 Curriculum Assessment

The ideal location for assessing technology is at institutions of higher learning: almost without exception, people world-wide hold education in high regard. But practical considerations, such as allocating resources and faculty energies, often create barriers to initiating new programs; however, technology assessment modules can be incorporated into existing courses.

In trying to expand the traditional parameters of most university courses in both the United States and Finland, I turned to the technology transfer module devised by Finnish historian, Dr. Timo Myllyntaus. If technology can transfer across national borders, why not use similar avenues to educate across national borders? As Myllyntaus concluded in examining Finland's rapid assimilation of electrical illumination, "technology transfer is not only a technical operation but also a societal procedure. . . .The technology itself and its transfer channels should be coordinated with the existing cultural and societal environment" (1991: 295). Today, improved communication has resulted in a more internationally uniform "cultural and societal environment," creating clearer channels to facilitate that transfer.

Myllyntaus identified seven factors that combined with the Finnish people's grassroots response to their perceived need for electrical technology: understanding cost effectiveness, implementing exchange programs, involving businessmen and engineers, recruiting foreign specialists, purchasing patents and licenses, importing equipment and attracting direct foreign investment. Many of these same factors, emanating from universities, can be exploited for furthering technology assessment programs. (See Fig.l). Particularly, by adjusting established curriculum, educators can have a strong influence in expanding awareness.

3 Curriculum Re-organization

In the United States, engineering educators must work within the confines of engineering accreditation board requirements; Finnish universities have less specific guidelines, and curriculum varies from university to university (Pietarinen, 1990). In both systems, transmitting the increasing body of knowledge in engineering and technology creates time constraints, resulting in limited space for related studies. I

Engineering Education 131

recommend three avenues to resolve this dilemma: first, by organizing studies in Course Clusters; second, by increasing interdisciplinary teaching; third, by assigning academic credit for intensified university-industry workshops and seminars. All three variations include built-in technology assessment modules, making critical analysis a habitual part of instruction.

Currently in the United States, academic departments distribute the required 24 semester hours for the study of communications, humanities and social sciences in various ways. For example, Purdue University specifies course hours in specific disciplines (Brown, 1990). The University of Cincinnati's College of Applied Sciences approaches the requirement by providing "minors" in subject areas: in the Technology and Society minor, for example, students choose from such selections as History of Technology, Literature of Science and Technology and Utopian and Science Fiction. In this program, students gain the perspective of scientists, dramatists, poets, practical mechanics, and politicians, among others. At Finland's Tampere University of Technology, courses such as History of Technology are scheduled according to need and time available (Riitahuhta, 1991). But both systems rely on traditional classroom formats.

To vary this system, I recommend that certain courses in both major disciplines and in liberal studies be designated "technology assessment" courses. Students would be required to earn a specific number of these credits by selecting from conveniently arranged technology assessment "course clusters." Faculty teams, working across disciplines, would provide students with information for critical analysis of engineering projects. For example, rather than teaching Technology and Society in isolation, I would present and discuss issues within the context of students' major academic work. Humanities and social science departments, in turn, would include discussions of science and technology in their classes. While this process would not work for every course, some are well-suited, such as studies of

Fig. 1. Mechanism for Technology Assessment Transfer. (Adapted from Dr. Timo Myllyntaus. 1990)

William Faulkner or Kurt Vonnegut, or History of American Culture. For example, I have invited engineers to discuss the mechanics of the automobile to help students comprehend its profound societal impact. This system has two significant advantages: students participating in technology assessment retain freedom to plan for breadth and depth of knowledge, and faculty can avoid a major overhaul of established curricula. Curriculum is our priority; bureaucracy cannot be allowed to complicate interdisciplinary cooperation.

A third element for curriculum variation is greater use of industrial expertise in the academic setting. Industry, faced with challenges presented by consumers and regulatory agencies, unravels the complications of technology from its practical application to its impact on society. Traditionally, many Research and Development (R&D) departments carried on their work in isolation, the results known and used by a limited audience. Of course, engineering faculty already tap this resource such as at Tampere's University of Technology, where legislative fiat stipulates that faculty "conduct technical research and product development" (Neuvo, 1988: 80). With such a strong basis for cooperation between academy and industry, institutionalizing channels for industrial contributions to traditional curriculum would enhance educational quality. Offering course credit for industrial experiences, beyond cooperative education, could increase on-site ability to assess technology and broaden the audience for corporate R&D offices.

Both the university and industry would benefit. Cost effectiveness is everybody's concern and broader shared knowledge is at the core of efficiency. The costs of cleaning up after errant technology are almost incalculable, not only on monumental projects such as the Chernobyls of the world, but to spills in rivers, in litigating biotechnology issues to mitigating workers' discontent. Moreover, management by re-action instead of pre-action creates negative images of technology. Where societies employ multiple resources to improve technology assessment, directed from many levels, the result is more thoughtful decisions about technological applications.

The workshop or seminar serves as an ideal model for institutionalizing industrial contributions to education. Industry, responding to consumer concerns, has a first-hand appreciation for the importance of technology assessment, and that assessment will emerge whether the workshops are grandiose or modest.

In Finland, for example, the R&D Department of Nordtrac Oy, manufacturer of snowmobiles, considers consumer needs across national boundaries. Their major customers, Scandinavia's reindeer herdsmen, have special needs. Working in proximity to nature, herdsmen strive to protect the natural environment so vital to preserving a healthy reindeer population and, at the same time, to improve their own efficiency. Snowmobiles, crucial to modern reindeer methods, must be reliable and rugged, but must not harm the wilderness. In developing a specialized vehicle, the LYNX GL 250 Syncro, Nordtrac engineers had to consider factors from international standards for noise emanations to protecting ground surfaces. Their R&D personnel, and indeed, all their employees, participated in this technology assessment, a resource for university-industry cooperation (Ohenoja 1992).

Similarly, Finland's Masa Yards, builder of seagoing vessels from tankers to icebreakers, expends enormous energies on R&D to meet customer needs, conforming to international naval regulations, and assisting the Classification Societies to formulate safety regulations. Much of this work, including technology assessment, evolves within the confines of the company. Masa Yards disseminates its technology primarily through educating customers about the complicated equipment they have purchased and by presenting public exhibitions that explain their technology (Juurmaa, 1992). Such resources should be incorporated into curriculum through specially-appointed industrial-university faculty whose main task would be to tap industrial sources for curriculum integration. Students would benefit greatly from a series of industry-based seminars that reflect a strong commitment to technology assessment. By assigning academic credit for these non-traditional programs, their importance will be underscored.

4 Curriculum Models

Many programs are already in progress in both nations that bring together the university and the community. The University of Cincinnati's College of Applied Science, for example, currently offers three models. The Chemistry Department provides a year-long workshop in the interaction of chemistry and the environment that attracts a national audience. The Construction Science Department has organized, "Safety in Construction," a workshop that links educational institutions, government agencies and professional associations. The Department of Communications, Humanities and Social Sciences, through its Center for Applied Business Communication, assesses communications technology within its main focus, "to provide clients with tangible and measurable results through practical application of proven communication principles" (OCAS Academic Newsletter).

Finland, too, has creative institutional approaches to technology assessment. The Arctic Studies Program at Lapland University interrelates science, technology and societal concerns. This flexible curriculum, directed to an international audience, ranges from a 15 credit program of two semesters to a 35 credit program of one year. In addition to technical developments and geographical circumpolar knowledge, the program includes international law, international environmental concerns, and a study of indigenous peoples (Heininen, 1991).

Another Finnish model is in the age-old occupation of reindeer husbandry. In 1898, the Finnish Senate mandated the Association of Reindeer Herding Cooperatives and charged the organization "to manage the relations between reindeer husbandry and the rest of society" (Huttu-Hiltunen, 1990:2). Traditionally, reindeer herders banded together to share their concerns about issues from food supplies to herd management. More recently, technology issues have occupied a larger portion of their time. While walkie-talkies, snowmobiles and helicopters have been a boon to the herders, they are well aware of the dangers involved (Katamajaa, 1991). Reindeer have extreme reactions to some technologies, from digestive upset and ulcers, to over 4,000 deaths each year in traffic (Huttu-Hiltunen, 1990:16). To find ways to balance technology's benefits and detriments, herders have joined forces with administrators of Finland's forest industry and with educators who provide programs in reindeer husbandry which include technology assessment (Kittilän Maatalousoppilaitos, 1990:14).

5 Conclusions

These models for incorporating technology assessment into established programs are available for international expansion. The time is ripe for overt and covert activity to invite a broader public to participate in making more informed decisions about technology through understanding its power to change our world. Public forums, the press, and private and public organizations contribute to raising world consciousness about technological applications; but educational institutions, with the infrastructure and the talent, must take the lead to make certain that technological innovation results from the most informed assessment the world community can muster.

References

Brown, Marion A. (1990) Instituting a Minor in Technology and Society at a College of Engineering Technology **Proceedings Seventh Canadian Conference on Engineering Education**, 98-102, University of Toronto, Ontario.
Hällsten, Annika (1991) There's Only One Planet Earth, **Finland**, 4, 5-7.
Heininen, Lassi, Coordinator of Arctic Studies, University of Lapland, Rovaniemi, Finland, September 19, 1991, Personal Interview.
Huttu-Hiltunen, Veikko, Koivuperä, Niilo, eds. (1990) **Reindeer Husbandry in Finland**, Pohjolan Sanomat Oy, Kemi, Finland.
Juurmaa, Kimmo, Project Manager, Kvaerner Masa-Yards, Marketing and Sales, Helsinki, Finland, February 18, 1992, Personal Interview.
Katajamaa, Timo, Reindeer Herdsman, Martinjarvi, Finland, November 26, 1991, Personal Interview.
Kittilän Maatalousoppilaitos (1990) 14.
Kuvaja, Sari, Campaign Leader, Greenpeace, Helsinki, Finland, February 18, 1992, Personal Interview.
Myllyntaus, Timo, Dr. (1991) **The Transfer of Electrical Technology to Finland, 1870-1930**, University of Chicago.
Neuvo, Yrjö, Mirja and Olavi Porri, eds. (1988) **Made in Tampere**, Foto-Porri, Tampere, Finland.
OCAS Academic Newsletter, 1, No. 2, (1992) University of Cincinnati, Ohio.
Ohenoja, Pekka, Chef de Ventes, Nordtrac Oy, Rovaniemi, Finland, March 24, 1992, Personal Interview.
Pietarinen, Anja, ed. (1990) **An Introduction to Higher Education in Finland**, Ministry of Education, Helsinki, Finland.
Riitahuhta, Asko, Prof., Tampere University of Technology, November 20, 1991, Personal Interview.
Ziman, John, FRS, (1984) **An Introduction to Science Studies: the Philosophical and Social Aspects of Science and Technology**, Cambridge University Press, Cambridge.

Engineering Education in Shantou University
T. Jiang
Shantou University

Abstract
This paper gives a brief introduction to the chief economic features of the Shantou Special Economic Zone and how the engineering education in Shantou University should be adapted to it.
To develop economy, science and technology should be taken as the guide, and education, the lead. Shantou can further develop, the key problem is to have large numbers of people with professional skills. Shantou University has been in engineering education carrying out the principle of taking the whole country into account and opening to the outside world, persisting in serving the local economic construction.
Keywords: Engineering, Education, University.

1 Introduction

Shantou University and the Shantou Special Economic Zone (SEZ) were born simultaneously a decade ago. They were marked as two milestones of both material civilization and socialist culture and ethics in the annals of Chaoshan (Shantou-ChaoZhou) region. As the founding of Shantou University adjusted to the demands of the nation's modernization programme and of the social, economic and cultural development of eastern Guangdong and the Shantou SEZ, it gained abundant support from people in eastern Guangdong, overseas Chinese and compatriots in Hong Kong & Macao, and Mr. Lika-shing, the noted Hong Kong patriotic personage, offered his generous financial aid (his donation has amounted to to HK$ 650 millions from 1980). It was a great event in the history of China's higher education and has produced a far-reaching influence at home and abroad.

2 Development and Reform of Engineering Education

Like other parts of the province, Shantou can further develop only if it depends on more input in science and technology, and man is the carrier of science and technology. Therefore, the key problem is to have large numbers of people with professional skills.

However, before the reform and opening to the outside world higher education was very backward, with only a normal school in Chaoshan region populated by ten million people, much less universities or colleges. Since the reform and opening to the outside world, by making use of foreign investment to introduce in a big way advanced technology and equipment from other countries, Shantou has registered rapid progress in industrial construction,and village and township enterprises have sprung up like mushrooms, As a result, an immense amount of labour force has shifted to the industrial production, and there is a great shortage of engineers and technicians, qualified technical personnel in particular. Therefore, it is an important task of engineering education in Shantou University to rapidly improve the quality of the staff and workers of all enterprises in respect of science and culture.

3 A Multidisciplinary and Comprehensive Pattern of Education

It is since its founding that Shantou University has conformed to this general trend and established departments and specialities in accordance with the needs of the society. In engineering education, efforts are made to support new and developing disciplines and establish such specialities as computer technology, electronic technology and architectural engineering. To promote enterprises of traditional Chaoshan arts and crafts the university established the speciality of industrial art design. To transform the traditional local mechanical and electronic industry, a few years ago it established the speciality of mechanical and electronic engineering by marrying mechanics and electronics, thus capturing the forward position of the integration of mechanics and electronics. In science education, it puts emphasis on the development of applied disciplines, the mutual infiltration of liberal arts and science, and the combination of sciences and engineering, actively and carefully developing such specialities as high polymer materials, fine chemical engineering, biological analysis and testing technology and new materials, coordinating with and participating in new and high technology enterprises in the SEZ. In brief, following the tracks of the general trend in the comprehensivization of modern science and technology and educational development, Shantou University has been turned into a comprehensive university which comprises liberal arts, science, engineering, medicines, laws, economics and arts, with a preliminary system characterized by close integration of social sciences and the humanities and the overlapping and mutual infiltration of natural science and technological science. Our practice has proved that this multidisciplinary and comprehensivized pattern of education if advantageous to our adaptation to the situation of new technical revolutions in the world and the academic environment of disciplinary comprehensivization,(advantageous) to the birth and growth of new and frontier disciplines and to the improvement of the quality structure of engineering and technical personnel.

4 Educational Pattern of Multiple-level and Form

Shantou University has in recent years managed to build an educational structure of multiple level, multiple standard and multiple form so as to train qualified personnel of different types who are much needed in the local economic construction. First, to do a good job of general higher education, by enrolling postgraduates in the departments where conditions are comparatively favorable, giving priority to undergraduate education, running sub-degree special training courses and enrolling students from factories, mines and other enterprises through contractual arrangements and directed building. Secondly, to develop adult higher education, by running evening colleges and other short-term traning courses, offering on-the-job training to the staff of the zone, thus extensively upgrading the quality of personnel of all levels in the SEZ. Thirdly, to attach great importance to professional technical education, for instance, to train sub-degree students who know not only electronic but also mechanics and cooperating teachers in accordance with the special demands which are made of skilled worker and teachers of technical schools. Fourthly, to set up new and high technology training centers in cooperation with high-tech companies abroad, for instance, to set up CAD, CAM soft ware training center, in cooperation with ATS company in Singapore.

5 Optimisation of the Process of Education

The most important aspects of their quality are moral character, cultural and scientific levels and capability. With regard to moral character, students must have patriotism and internationalism, advocate science, seek truth, value practicality, have noble character and self-cultivation and noble professional morals. With regard to cultural and scientific levels, they must keep abreast of the development of science. Just as science has gone through the three phases of chaos–division–comprehensivization, so the quality of engineering and technical personnel has undergone the change of encyclopaedic type –unitary structure– comprehensive structure. As would-be engineers, students of higher engineering education must acquire a systematic knowledge of basic theory, have a wide range of knowledge and some knowledge of relevant broad-fields courses and frontier courses. With regard to capability, students must undergo fundamental training in engineering and technology, including those in experiment, design and basic skills, so that they have the ability to synthesize, analyze and solve practical problem in engineering. In the past few years, in accordance with the new demands of future development on the quality of qualified personnel, we have adhered to the law of higher engineering education, carried out our reforms in the whole process of education and trained personnel of high quality needed by the SEZ.

5.1 Reform and Development of New Disciplinary Patterns
Since its founding, Shantou University has turned out over 3,800 graduates. With the deepening of reform in political and economic systems of the SEZ and the development of commodity economy, new features have emerged in respect of demands of industrial enterprises on qualified personnel. The original disciplines of engineering in our university, which were influenced by traditional education, laid undue stress on the special demands as set by different industrial on professionals

of different kinds. As a result, some disciplines were divided into smaller and smaller sections, and their fields were getting narrower and narrower, thus weakening the students' adaptability to the further shifting of their field of work and to the development of science and technology, and making their jobs unsuited to their special training. Accordingly, Shantou University has formulated the teaching principles of "laying a solid foundation, having a wide range of knowledge, broadening fields of study and attaching great importance to practices ", divided engineering disciplines according to their fields of study and fields of work, set up disciplinary patterns of "broad fields with divisions" and disciplinary readjustment mechanisms, in which disciplines with broader fields are divided into a number of specialities. For instance, architectural engineering divided into speciality of industrial and civil engineering and speciality of city construction, applied chemistry divided into speciality or high polymer materials and speciality of fine chemical engineering, industrial artistic design divided into speciality of decoration design, speciality of environmental artistic design and speciality of industrial modelling design. Enrollment is based on a departmental basic. at the earlier stage students are to receive a basic education, i. e. to take common core courses, and at the later stage to receive a professional education and to be trained according to their respective specialities. Specialities may be readjusted in a flexible way in accordance with social needs and information about preliminary job assignment. Students are permitted to choose specialities according to their aptitudes and desire.

5.2 Objectives of Training and Optimization of Educational Planning

The students of engineering in Shantou University must in the first place know how to take full advantage of the "two markets" (domestic and international), have a good knowledge of "two kinds of resources" (domestic and international) and acquire "two capabilities" (the capability of contributing to socialist construction at home and the capability of having dealings with other countries, doing business, learning advanced science and technology and making foreign things serve China). Secondly, we emphasize the comprehensiveness of the students' knowledge structure, require students to acquire in a systematic way basic theory of natural sciences and basic knowledge of technological science, have a good knowledge of his lines and of relevant engineering technology and keep abreast of the posture of science and technology st present, especially of new and high technology. In order to increase the cultural and artistic accomplishments of students of science and engineering, heighten their awareness of competition and economy and enrich their knowledge of public relations, we have in our engineering education planning strengthened education of the humanities and that of social science by offering such compulsory courses as political theory, Chinese, history, ethics, economics and management, and more than ten optional courses such as communications, fundamentals of aesthetics, knowledge of music and the ABC of painting. Thirdly, not only do we set strict demands on the basic training of engineering skills, but also emphasize the importance of cultivating the students' ability to synthesize, analyze and solve problem,their ability to work independently, their ability to manual operation, their ability to use computers, their organizing and administrative ability, their social skills, their creative ability and their ability to cope with new situations.

5.3 Curriculum Structure and Modes of Teaching

In the past few years we have taken the initiative in adapting ourselves to the change of social demands on qualified personnel, kept on optimizing the curriculum structure and updating the content of courses, tried hard to represent new achievements in science and technology, and attached great importance to the overlapping and mutual infiltration of various disciplines. With regard to undergraduate education, under normal conditions we intensify the teaching of common core courses at the earlier stage (lst year-3rd year), with core courses and technical core courses accounting for 85% of the total hours, and conduct branched training at the later stage (4th year) accounting to respective requirements. While intensifying the teaching of core courses, we make educate arrangements for such link of practicality as experiments, practice,designs and theses. Efforts are made to ensure the training of basic skills, especially computer language and computer operation. Laboratory course and field work are conducted in separate classes and students are examined separately. Students are encouraged to choose subjects for their graduation projects and theses directly from enterprises so as to promote the combination of teaching, scientific research and production. Besides, We have opened a second classroom, set up student radio clubs, amateur scientific and technological groups, organized a series of lectures on science and technology and art and literature so as to enrich the student' cultural life on the campus, broaden their field of vision and develop their aptitudes and hobbies. This serves as a supplement and an extension to the first classroom.

5.4 University and Industrial Enterprises

In the last few years the school has built up firm cooperative relations with enterprises. It has invited from enterprises experts, engineering and technical personnel to give lectures at school while teachers have gone to industrial enterprises to act as advisers and help solve problems in production. By strengthening our ties with the society, we have turned industrial enterprises into bases for teachers and students to carry on scientific research and production practice, update and substantiated the content of teaching, raised the teachers' level of scientific research, developed the students' ability to synthesize, analyze and solve practical problems and promoted the combination of teaching, scientific research and production.

References

Proceedinngs of International Symposium On Higher Engineering Education, April 17-21, 1990 Hangzhou, China

CADCAM in Developing Countries
M.A. Eason
Overseas Development Administration, U.K./ESPOL, Ecuador

Abstract
The paper gives a brief introduction into the reasons behind technological growth and the methodologies behind different implementation methods. The role of education in this process is highlighted and examples of applications of CADCAM in developing countries are given. A brief discussion of some of the recognised pitfalls is then followed by a possible model for implementation. The paper concludes with a brief description of the O.D.A./ESPOL project in CAE.
Keywords: Engineering, Education, CADCAM, Developing Countries

1. Introduction
In the early 1980s a report on the management of technology, ACARD (1983) stated "A manufacturing sector capable of competing in world markets must be a key ingredient of the U.K. economy". The implications of such a statement are far reaching in that to support such a sector requires not only finance and industry but trained personnel at all levels.

At that time there was a surge in available courses at higher education levels in Advanced Manufacturing Technologies.

The expansion was not limited to AMT nor was it limited to the U.K. All aspects of engineering utilising advances in computer technology saw major growth, particularly in the areas of CAD and CAM both in developed nations and newly industrialising countries NICs, as exampled by, Eason (1987).

2. Implementation
The case for the implementation of such technologies and for the expansion of available high level courses has brought many reports and papers, Nee and Hang (1989) considered CAD/CAM to be the most productive tools in modern industry widely used by large companies but lagging behind in smaller companies in NICs.

Sinha and Sinha (1989) noted the needs for technology transfer to third world countries and the difficulties of implementation into the industrial sector.

At this point it would be useful to define some of the requirements for the implementation of CADCAM or the wider term of CAE, with particular emphasis

142   Engineering Education

on the educational factors, though these cannot be divorced from the other forces of national policy, finance, and industry status.

3. Educational Requirements

Resources such as computers, machinery and work areas are obvious and well documented UNIDO (1990), as is the recognition of the human resource needs, Boon, Mercado (1990).

The specification of curriculum content is also well covered, Awaluddin (1989) for example and any university course planning document, but required training of personnel is often more difficult to define, FEU ().

The situation within a well developed Engineering department within a U.K. higher education establishment would be one of gradual development of resources, staff expertise and curriculum development with significant major steps being taken when sums of money became available for capital equipment or extra staff and training, often with the development of consultative work with Industry to fully utilise expensive state of the art equipment and integrate real case situations in the teaching environment.

Without this educational background, countries such as the NICs started further behind. The resulting methodology was often very large capital investments in equipment, foreign expertise and staff training which resulted in a decreasing need for external expertise, Nee, Hang (1989). This proved to be an effective route.

The situation for Developing Countries often falls short of this with lower resourcing, poorer localisation of staff and limited industrial transfer, Sinha K.K., Sinha A.K. (1989), Gosh (1989).

4. CADCAM Education in the Third World - Examples

Papua New Guinea, PNG, has a population of just over three million, a limited infrastructure and transport system because of the terrain and a relatively short history of technological development.

There is only one University level Mechanical Engineering course, which reflects in the level of development of the country in terms of new technologies and industry. The CAD facility was limited in terms of physical area, numbers and level of computers and available software.

Major developments in integrating CAD into coursework had not been undertaken until 1987 when basic courses aimed at students gaining some basic working knowledge of at least one CAD system were instigated. Limitations to this were the availability of staff expertise and curriculum time. The industrial sectors had at that time little need for CAD as manufacturing was limited and any large company used its overseas facilities for such work. CAD was therefore seen in educational terms to be an area of new implementation for graduates to take with them as a future development tool.

The CAM equipment did not reflect the machine limitations in that it consisted of two industrial size CNC machines with a DNC modification. The limitations in this sector were that little resources had been made available for staff training, the level of the technology was beyond existing industrial needs and the types of machinery were not readily available to student use. The short term contracts that most foreign staff had, left little time to develop this area.

Future needs were possibly being catered for but in a disorganised unbalanced framework.

Ecuador has a population of over ten million, a better infrastructure and transport system and a longer history of development than PNG, it is also closely placed to a number of larger trading partners.

There are a number of degree level Mechanical Engineering courses and all using computing to some extent. CAD is limited to older versions of well known packages but is often implemented on few computers. A central national development body (CEBCA) does have larger CAD and FEA facilities available for Industry to utilise and many companies use CAD in a range of different applications. The educational sector was to some extent lagging behind these initiatives.

ESPOL, the major technological polytechnic has a central facility running courses in CAD on more up to date versions but on a more occasional basis. The Faculty of Mechanical Engineering had a number of computers on which some work in CAD was done but no integrated facility existed. Therefore CADCAM had little educational based impact.

CAM was less well placed with a few industries having CNC machinery but relying on vendor training, no educational facility for CAM existed.

The need for CADCAM was recognised in order to feed into existing industrial organisations and also to act as a developing force by having experienced graduates in companies. Other major differences to PNG were that many of the students particularly by the final year had contacts or jobs within industry, as do many of the academic staff which is nearly completely Ecuadorean, whereas PNG was heavily foreign staff based.

5. Third World CADCAM Implementation Problems

Difficulties in technology transfer have been reported in a number of papers, such as lack of local knowledge on how to use and maintain equipment, often existing equipment is obsolete and cannot be integrated, a lack of coherent national technical policy, Sinha, Sinha (1989). The need for long term initial planning may not be recognised as it is in more developed nations, Nee, Hang (1989). Rapid expansion of programmes often produce resource shortages, and compromise quality when large student numbers are involved.

Conflicts of ideas and methodologies with overseas staff can occur and localised needs of the curriculum may not be seen, Gosh (1989). In some cases working practices differ and situations can arise where controlling a DNC machining centre may be seen as degrading manual labour. This may extend into a lack of recognition for professional practice in comparison to what is seen as academic pursuit. This situation does exist amongst staff and is therefore relayed into student teaching with the danger of isolating essential elements of an Engineering curriculum, Gnalalingam (1981).

Resource backup in all forms is a major problem and another is ensuring that what is initiated is appropriate in the widest sense. Often in 3rd world countries the students background in schooling and social life may have lacked any form of technology, whereas in developed nations high technology toys and computers are relatively commonplace.

6. An Implementation Model for the Third World

The first essential I would consider is that any CADCAM setup must be locally sustainable within a relatively short period of time.

The area should not be seen as a stand alone new technology but should be integrated into existing courses. Designing an alternative low technology water pump by CAD should be seen as an integrated use of appropriate technologies. Education in the area should not be seen as elitist but should form part of a continuing growth of new technology applications. In this way the students should become more self confident and self directed in furthering potential applications within the education sector and in later industrial work.

Most third world countries cannot implement in the same format as NICs and therefore appropriate small scale computer and machine set-ups that can readily be used by larger numbers of students at initial levels are preferable.

Low cost training and consultative links with local industries must be forged to facilitate not only the transfer of technology out, but real case situations in. The resistance to practical work and availability of real case studies has been highlighted in a number of the references quoted.

It is essential that local staff have the resources to gain confidence with the technology and that they then plan for future developments. Having new expensive equipment should not be seen as a goal, utilising that equipment to an ever increasing level necessitating expansion at a later stage must be seen as the goal.

Start small, understand and maintain it, be confident to apply it, utilise it to teach, train and thus disseminate, utilise the student to industry link to expand into centers of expertise and income.

The limited numbers of qualified people in third world countries makes extensive uptake of new technologies difficult on a wide scale.

The CADCAM must be appropriate.

7. O.D.A./ESPOL CAE Project

Overseas development aid from the British Government was used at ESPOL previously to establish improved laboratory teaching facilities for the Mechanical Engineering Faculty. The present project is aimed at providing an integrated laboratory to teach CAE, with CAD, CAM, FEA software facilities integrated with CNC machine tools.

One technical expert was provided for a period of two years, along with funding for computer equipment, machine tools, and books. Funding was also made available for the three counterpart staff to travel to the U.K. for periods of training in relevant areas.

With a limited budget an initial laboratory consisiting of two computers, and two Denford bench top CNC machine tools, enhanced by the faculties less powerfull computers was set up to run CAD, CAM, and FEA software, as well as machine tool simulation.

Ultimately the counterpart staff will be responsible for the centres development and to this end they are undertaking ongoing training.

It was felt that students should be able to use the area as soon as possible and to this end the facility has started to be integrated into existing courses as well as providing new specialised modules of study.

To help integrate a teaching methodology final year student projects are being undertaken where the students are producing didactic material from their own experiences of learning the technology as well as undertaking an assessment of local industry needs for training.

Initial work with local Industry in terms of training and joint development work is underway on a small scale as it is all too easy to swamp a limited facility. General awareness of the facility is growing and it is hoped to extend the training to other tertiary establishments thus providing a wider pool of experienced staff able to then pass on their experiences to students and also to utilise existing industrial establishments in joint development.

The initial goals of the project are a throughput of students who have used and can apply basic techniques in CAD, CAM with more specialist work being done on a project basis. Courses and seminars for industry form an important part of the dissemination process.

An important factor is the localisation of the technological expertise, which was already there but needed enhancing.

So far we have seen a great deal of interest and application from students, a pressing awareness from industrial sectors that they need a foothold in these technologies, and an educational institutional response linked to external bodies that such technology transfer forms an important part of the competitive growth for such a country.

When motivation awareness and application are engendered in a student or employee the development of applications are already well underway.

References

ACARD (1983) New opportunities in Manufacturing: The Management of Technology, **A report from the Advisory Council for Applied Research and Development,** HMSO ISBN 011 630 823 0.

Eason, M.A. (1987) A study of New Zealand Manufacturing and Education in relation to new technologies, **Dept. report Mechanical Engineering,** Manchester Polytechnic, England.

Nee A.Y.C. Hang.C.C. (1989) CAE/CAD/CAM Curricula implementation-Experience at the National University of Singapore, **WEEAT- World Conference on Engineering Education for Advancing Technology,** Sydney 13-17 February, 1989. The Institutuion of Engineers, Australia, Preprint of papers pp.145-149 (National Conference Publication No. 89/1)

Sinha K.K. Sinha A.K. (1989) Rationalisation of Engineering Education for

Advancing Technology for the 3rd World, **WEEAT-World Conference on Engineering Education for Advancing Technology,** Sydney 13-17 February, 1989. The Institution of Engineers, Australia, Preprint of papers pp.75-79 (National Conference Publication No. 89/1)

UNIDO (1990) Planning and programming the introduction of CAD/CAM systems, **A reference guide for developing countries, General Studies Series UNIDO Industrial Planning Branch, Industrial Institutions and Services Division,** Vienna 1990. UNIDO publication sales No.;E.89.III.E.7 ISBN 92-1-106236-5

Boon, Mercado (1990) Automatizacion Flexible en la Industria, Difusion y produccion de maquinas-herramienta de control numerico en America Latina. **The Technology Scientific Foundation /Noriega Editores,** ISBN 968-18-3320-1

Awaluddin M.S. Saleh Y.M. Evaluation Techniques in CAE Curriculum Development, **WEEAT-World Conference on Engineering Education for Advancing Technology,** Sydney 13-17 February, 1989. The Institution of Engineers, Australia, Preprint of papers pp. 53-58 (National Conference Publication No. 89/1)

FEU () CNC in FHE, **An occasional paper, Further Education Unit/Longman,** SBN 0 582 17354X

Gosh A.K. Engineering Education in a Developing Arabian Country- Problems: Iraq a case study, **WEEAT-World Conference on Engineering Education for Advancing Technology,** Sydney 13-17 February, 1989. The Institution of Engineers, Australia, Preprint of papers pp.270-273 (National Conference Publication No.89/1)

Gnalalingam S. The Education and Training of Engineers for Developing Countries, **Journal of the Institution of Engineers Sri Lanka,** Vol 1 1981/82 pp.3.

Developments in Engineering Education at Cambridge University

S.C. Palmer

Department of Engineering, University of Cambridge

Abstract

This paper reviews current developments in the teaching of engineering to undergraduates at Cambridge University, with particular reference to the curriculum and to maintaining the high quality of teaching. Regarding the curriculum, a new four-year engineering course is about to be introduced in October 1992, to replace the long-established three-year course. The paper summarises the structure and content of the new course and discusses the factors influencing this reform. Regarding quality, the paper outlines recent initiatives in the Engineering Department to monitor and to continue to improve the high standard of lecturing.

Keywords: Engineering, Education, Cambridge University, MEng, Curriculum, Quality, Lecturing.

1 Introduction

The Engineering Department has approximately 1000 undergraduates and 400 graduate students, and forms about 10% of Cambridge University. There are about 130 teaching staff and 70 post-doctoral workers, in addition to technicians and clerical staff. A hallmark of the Cambridge undergraduate engineering course, called the Engineering Tripos[*], is its tradition in providing a broad technical education covering the full range of engineering disciplines, followed by specialisation in a selected field. All students study the same subjects for the first year, and for about 70% of the second year. This broad approach helps to improve students' understanding of concepts which are fundamental to more than one engineering discipline. The aim of the course is to train flexible engineers who will ultimately be able to lead multi-disciplinary teams, and to apply new technologies in novel situations.

Curriculum development and other teaching matters are managed through a committee structure. Each subject is steered by a group of teaching staff, whose chairman represents the Subject Group on the Department's Teaching Committee. Since staff may teach a wide variety of subjects, staff are free to join any Subject Groups in which they have an interest. Teaching matters are coordinated centrally by two academic staff, a full-time administrator and a secretary in the Department's Teaching Office.

The last major reform of the course occurred in 1984. By July 1987 there were many indications that the time was approaching for the long-established and successful three-year course to evolve into a four-year course. Planning has gathered pace since that date.

---

[*] A Tripos is an honours examination - supposedly so-called after a three-legged stool on which candidates formerly sat during oral examination.

2 The Four-Year Course

The four-year course is about to be introduced in October 1992. The structure and content of the new course are summarised in Table 1. The reasons for change are outlined below.

2.1 Breadth and Depth
The aim of the Cambridge course remains that of providing a broad scientific education covering the full range of engineering disciplines in the first two years, followed by specialisation in considerable depth. To achieve this ambitious aim against a background of expanding technology and other pressures for change, recognised by the Engineering Professors' Conference (1991) and listed below, the addition of a fourth year was essential.

2.2 Changes in School Curriculum
There is evidence that the changes taking place in school sixth forms, following the changes in the National Curriculum, are leaving students less well prepared for university engineering courses, specifically in the areas of mathematics and physics. Whilst the Department continues to issue Preparatory Documents in Mathematics, Mechanics and Electricity to successful applicants well in advance of the course, additional lecture time is now needed in the first year to adapt to the lower levels of knowledge in these subjects.

2.3 The MEng Degree
The new course will enable students to achieve the degree of MEng, as recommended by the Engineering Council for the top 20% of university engineering students. The award of the MEng degree to the majority of Cambridge students may be justified by their exceptionally high entrance qualifications. The majority of students have three (or more) A level passes at grade A, with an average UCCA point score of 29.9 out of 30. Upon successful completion of the fourth year, students will be awarded simultaneously the degrees BA (with Honours) and MEng. (Cambridge has no BSc or BEng degrees). It will still be possible for students to leave the course after three years with a BA Honours degree, but the new course is designed around a four-year structure, and the student would be leaving his or her engineering training unfinished.

2.4 Participation in Europe
In general the longer course will help to prepare students for participation in Europe, where many courses are five years in length. More specifically a modular structure has been introduced in the final year of the course, to promote European exchanges. Opportunities for foreign language study are also being expanded, and include new third-year projects and fourth-year modules.

2.5 Projects, Design and Applications
The extra time will allow student learning to be enriched by the provision of more opportunities for project work and the development of communications skills.

The successful and popular first year Structural Design Project has been retained; pairs of students design, manufacture and test a metal structure to carry given loads at minimum cost. So too has the Product Design Project.

A new second year Integrated Design Project has been developed which demonstrates the need for teamwork and communication skills. Teams of up to six students design, build and test a robot vehicle in competition. Each team member is responsible for a particular sub-system as well as contributing to the overall system design.

In the third year, the written examinations have been brought forward to allow a full term to be devoted to projects. Students choose any two of four activities; a design project, from a wide range of topics; a computing-based project, related to engineering design and analysis; a language project in French or German; and a practical surveying field course.

In the final year students are expected to spend up to half of their time on an individual Major Project. Industrially-linked projects are encouraged.

Lecturers will also continue to place great emphasis on design and engineering applications throughout the course. The requirement for students to acquire at least 8 weeks of industrial experience has been retained. Approximately 65% of students are sponsored and can obtain this experience with their companies before starting the course.

2.6 Student Workload

Reduction of student workload has been a major objective, particularly in the first two years of the course, which had become severely congested. The aim of the reduction is to improve student motivation, and thereby stimulate understanding and enjoyment.

The average number of student contact hours for the new course will be approximately 19* hours per week of term in the first year, and 17* hours per week of term in the second year. The reduction of contact hours during term is accompanied by the removal of all vacation courses from the first two years, and a reduction in the rate at which the syllabus is covered in lectures.

Regarding the timing of lectures, all Cambridge courses have two unusual features; the duration of the teaching terms is only 8 weeks, and the timetable rarely involves lectures in the afternoons. A consequence has been the regular occurrence of lectures on Saturday mornings; this will no longer be necessary in the first two years of the new course.

3 Related Courses

The facility to transfer between subjects is another hallmark of Cambridge University. The first two years of the Engineering Tripos provide a basis for several other courses. For example, by specialising in the electrical and information engineering papers in their third and fourth years (Groups D & E in Table 1), students can choose to study the Electrical and Information Sciences Tripos. (Students may also read this Tripos after studying Natural Sciences or Mathematics for two years.) Other third and fourth year options available include the two-year Manufacturing Engineering Tripos, the two-year Chemical Engineering Tripos and the one-year Management Studies Tripos.

4 The Colleges

The Cambridge Colleges are autonomous bodies, separate from the Departments of the University. In addition to providing accommodation and social facilities, the Colleges have responsibility for interview, admission, academic progress and pastoral care of all undergraduates in Cambridge University. The Colleges provide individual tuition (known as supervision) for their engineering students, to supplement the lectures and coursework offered in the Engineering Department. Undergraduates spend one or two hours a week, normally in pairs, with one of the College Fellows (many of whom are also lecturers in the Engineering Department) to discuss queries arising from lectures, problem sheets or past examination questions. Supervision is an integral part of all Cambridge teaching, and is regarded as invaluable by staff and students alike.

---

* The estimates of contact hours include 2 hours per week of College supervision, and a total of $1^1/_2$ hours per week for Examples Classes and Applications lectures, not included in Table 1.

5 Quality of Teaching

Review of the curriculum and the quality of teaching is a continuous process. These matters are discussed at regular meetings of the Staff-Student Joint Committee, the Subject Groups, the Coursework Committee and the Teaching Committee. The decision many years ago to establish a Teaching Office to coordinate all teaching matters reflects the importance of teaching in the Department. The Teaching Office also ensures that course documentation is regularly reviewed and updated.

To maintain and improve the high quality of teaching and learning in the Department, there have been numerous recent developments; further training for laboratory demonstrators is being encouraged; student study skills sessions are being expanded; and the Department continues to help in the training of new College supervisors. However in view of the large size of the Department, with approximately 330 students attending most lecture courses in the first two years, lecturing remains the keystone of the Department's teaching. Hence in addition to the staff development activities arranged by the university-wide Committee for the Training and Development of University Teachers, there have been initiatives in the Engineering Department in recent years to monitor and to continue to improve the standards of lecturing, as discussed below.

5.1 Appointments procedure

The appointments procedure for engineering teaching staff now includes a requirement for all applicants to present a 30-minute lecture before interview. Applicants can choose a topic related to their research or other interests. The lecture is attended by the members of the Appointments Committee and other interested teaching staff. This procedure allows the applicants' lecturing ability to be assessed realistically before appointment.

5.2 Lecture Questionnaires

For many years student feedback on lectures has been obtained through termly meetings of the Staff-Student Joint Committee, and through the regular issue of various questionnaire forms. However the value of feedback from questionnaire forms issued to 330 students has often been limited by the low student response rate, typically only 25%. Consequently, an extensive computer-based questionnaire was designed and introduced in 1990 to obtain annual feedback from first and second year students on the quality of all lecture courses. During a two-week period at the end of February, each student is expected to log on to the Department's computer to enter numerical multiple choice responses (1-7) to 5 questions, concerning the pace, difficulty, presentation, handouts and problem sheets, for each lecture course attended. There is also a facility for entering text comments.

The questionnaire has been extremely successful in attracting a response rate of approximately 80%, thereby ensuring credibility of the numerical results, which are displayed in histogram form. These have been particularly useful in identifying those courses containing too much material, and have justified a reduction in syllabus content. They have also highlighted examples of lecturing of outstanding quality, and this has been valuable in promoting the spread of existing good practice throughout the Department. The number of text comments received on each lecture course was small, and hence the replies could not be considered representative. Since the demoralising effect on staff of isolated comments which may be unconstructive and possibly unrepresentative can be remarkably counter-productive, the value of the text comments is questionable.

To complement student opinion, corresponding feedback is also sought from College supervisors, who are well placed to assess students' understanding of the lecture material.

5.3 Video-Recording

Since 1989 the voluntary use of video-recording equipment in undergraduate lectures has been encouraged, to enable engineering lecturers to study their lecturing styles, privately or in groups. Throughout the lecture the video is operated by a technician, who hands over the cassette to the lecturer for retention at the end of the lecture. The majority of lecturers participating also volunteer to contribute to a half-day video-playback session, in which groups of 5-6 engineering staff discuss extracts from their recordings. Those involved are unanimous that the exercise is valuable. The success of these sessions relies upon the development of a constructive, cooperative and confidential atmosphere, which is facilitated by drawing all participants from the same Department, and by adherence to an agreed framework for giving feedback during the discussions.

5.4 Seminars on Lecturing Techniques

Although there are many organisations offering instruction on the 'art' of lecturing, very few address the specific problems associated with lecturing engineering (and other technical and mathematical subjects) to large classes of 330 students. Consequently since Easter 1991 the Department has offered in-house seminars on lecturing techniques, for both newly-appointed and experienced lecturers. The seminars concentrate on the presentation techniques commonly employed in the Department, including the use of blackboards, whiteboards and overhead projectors, the use of two projectors, pre-prepared overhead transparencies, complete handouts, partially-complete handouts, demonstrations, computers, videos and slides. These successful seminars also provide an ideal forum for disseminating current national developments on quality matters, such as those by Sparkes (1989), published through the Engineering Professors' Conference, and its Programme for the Improvement of the Quality of Engineering Education (PIQUEE).

5.5 Lecture Theatres and Equipment

The in-house seminars described above were of immediate benefit in prompting a major review of all lecture theatres and equipment in July 1991. In response to requests from lecturers, all large lecture theatres are now flexibly equipped to allow many types of presentation, including the simultaneous use of two overhead projectors (or slide projectors) and blackboard (or whiteboard).

6 Conclusions

The initiatives listed above reflect the growing national awareness of the importance of quality in engineering education, and suggest that there are still significant benefits to be gained from the dissemination of existing good practice in teaching within and between all institutions of higher education.

References

Cambridge University Engineering Department (1990) Undergraduate Engineering at Cambridge University, an introductory booklet, available from; the Secretary of the Department of Engineering, Trumpington Street, Cambridge, CB2 1PZ.

Engineering Professors' Conference (1991) The future pattern of 1st degree courses in engineering, **Engineering Professors' Conference**, Occasional Paper no. 3.

Sparkes, J. J. (1989) Quality in engineering education, **Engineering Professors' Conference**, Occasional Paper no. 1.

Table 1. Four-Year Course Summary

Year	1			2			3			4		
Term	Mich	Lent	Easter	Mich	Lent	Easter	Mich	Lent	Easter	Mich	Lent	Easter
Choice	CORE						OPTIONS					
Lectures	1.Mechanical engineering 52; 2.Structures & materials 44; 3.Electrical & info eng 50; 4.Mathematics 40; Lectures for coursework: Mechanical drawing 8; Principles of design 8; Computing 3; Dimensional analysis 8; Engineer in society 8			1.Mechanics 16; 2.Structures 20; 3.Materials 16; 4.Fluids & heat 26; 5.Electrical eng 26; 6.Info eng 22; 7.Mathematics 34; Coursework lectures: Company structure 8		8.Options; Select 2 from 6; 2 x 14 = 28	A1-4 Civil eng; B1-4 Mech eng; C1-4 Thermo/Fluids; D1-3 Elec eng; E1-3 Info eng; F1-3 Management & manufacturing; Select 5 papers, of which ≥3 from one Group (A - E); 5 x 32 = 160			A1-12 Civil eng; B1-12 Mech eng; C1-12 Thermo/Fluids; D1-12 Elec eng; E1-12 Info eng; F1-8 Management, manufacturing & languages; Select 8 modules, of which ≥4 from one Group (A - E); 4 modules per term; 8 x 16 = 128	Course-work for modules	
Contact hours	221			168								
Labs and other course-work	16 lab experiments 32; Mechanical drawing 27; Computing 16; Electronic instrumentation 8; Microprocessors 8; Dimensional analysis 4; Exposition 16; Engineer in society (essay) -			23 experiments 46; Computing & numerical analysis 16; Company structure (essay) -			12 experiments, of which ≥6 from selected Group; 12 x 4 = 48		Select 2 projects; • Design; • Computing; • Surveying; • Foreign language; 2 x 80 = 160	MAJOR PROJECT		
Contact hours	111			62						400		
Projects	Structural design project 30; Product design project 6			Integrated design project 40								
Total hours	36	⇒				⇒			⇒			⇒
Exams	4 x 3 - hour exams Classified			8 x 2 - hour exams Classified			5 x 3 - hour exams Classified			8 x 1½ - hour exams Unclassified		
Term	Mich	Lent	Easter	Mich	Lent	Easter	Mich	Lent	Easter	Mich	Lent	Easter
Year	1			2			3			4		

The Future of Engineering Education in Canada
A. Meisen (*), K.F. Williams (**)
(*) Faculty of Applied Science, The University of British Columbia
(**) Industrial Engines Ltd., Vancouver

Abstract
A recent report prepared under the auspices of the Canadian Council of Professional Engineers and the National Committee of Deans of Engineering and Applied Science examines the Canadian system for engineering education and makes recommendations for improvements in light of the changes occurring in engineering practice.
Keywords: Engineering Education, Canada.

Introduction
The future-oriented study (1992) of Canadian engineering education undertaken by the Canadian Council of Professional Engineers (CCPE) and the National Committee of Deans of Engineering and Applied Science (NCDEAS) reviews Canadian engineering education, identifies issues and problems, and recommends action to meet the country's need for increased numbers of engineers of calibre and expertise. The study was prompted by the rapid change in engineering practice resulting from the impact of the knowledge explosion, of technological development, of national and international competition, of globalization, of environmental concerns, and of pressing infrastructure renewal. The findings of this study are meant to provide a broad framework for action by universities, the engineering profession, industry, and government.
   That the Canadian engineering education system must change is no criticism of past accomplishments but rather an acknowledgement of changing engineering practice. The system has served the profession well to the present, but it will become inadequate if it does not evolve. Universities, the engineering profession, industry and government, all of which have contributed to the success of Canadian engineering, must concentrate their efforts and resources to ensure that Canadian engineering education meets the challenges of an increasingly competitive and complex world.
   Three principal challenges face the Canadian engineering education system: to provide an increased number of suitably prepared entrants into undergraduate programs and raise the interest in graduate engineering education; to improve the quality of undergraduate and graduate engineering education so that global competitiveness is achieved; to ensure the continuing high competency of practicing professional engineers.

Pre-university Education

In the area of pre-university education, the principal issues and problems are: elementary, junior and high school teachers' attitudes and competence regarding mathematics and science; insufficient familiarity with engineering work and engineering careers among teachers and counsellors at all levels of the school system; inadequate counselling of students regarding engineering; negative attitudes toward the perceived difficulties of engineering conveyed to students and, in particular, female students; grading practices of the schools; parental and student attitudes, especially among female students, towards perceived difficulties in mathematics and science; lack of knowledge of engineering career opportunities among members of minority groups including the handicapped; low awareness of the activities of engineers and the impact of their work on daily life. The principal recommendations are:

* Departments of Education and School Boards should require that a significant number of elementary school teachers have demonstrated competencies in science and mathematics;
* School Boards should promote the teaching of science and mathematics as a "prestige" career for school teachers;
* Industry, School Boards and Provincial Engineering Associations should enhance the awareness of science and mathematics teachers of engineering practice through workshops and strong links among industry, teachers, students and counsellors;
* Engineering Faculties should collaborate with Education Faculties so that teacher education programs develop a better understanding of engineering work and engineering careers;
* The understanding and appreciation of engineering by students in the school system should be developed by:
  - Industry, Universities and Provincial Engineering Associations identifying outstanding engineers who can act as role models, in particular, encouraging practising women engineers to show female students that engineering is an excellent career option;
  - Education Faculties, Departments of Education and School Boards striving to teach science as an "inquiry-oriented" activity rather than a "fact memorization" task to stimulate the natural curiosity of children;
  - Technical Societies and School Boards fostering engineering and science fairs in elementary schools to stimulate creativity and curiosity in technological activities;
  - Industry, Provincial Engineering Associations and Universities supporting engineering and science fairs through prizes and participation in their organizations;
* The image of engineering should be given better recognition by CCPE, the Provincial Engineering Associations and Engineering Faculties by collaborating in:
  - developing educational materials which display the activities of engineers and their positive impact on society and the environment;
  - providing summer camps for elementary school children which engagingly highlight the activities of engineers;
  - in arranging visits to industry for elementary school students to show them the work of engineers.

Undergraduate Engineering Education

In undergraduate engineering programs, the principal issues and problems are: the programs, in many instances, operating above full capacity; the present high, average student-to-staff ratios of 16:1 precluding individualized instruction; frequently inadequate teaching equipment and design tools to prepare for industrial practice; the threat to the broad base of Canadian engineering programs due to industry's desire for specialized expertise and government's predilection for programs which do not appear to duplicate existing programs; the high course load due to the required inclusion of increased subject matter relevant to engineering practice; the inability of many students to complete their studies in the prescribed period due to excessive demands imposed by programs; inadequate training in leadership and team-work skills; the lack of exposure to engineering practice in international settings; high school graduates without appropriate preparation in mathematics and science lacking opportunities to upgrade their skills for entrance into engineering programs. The report recommends that Engineering Faculties, Universities, and Government should collaborate to achieve the following objectives:

* The capacities of the undergraduate engineering programs should be expanded by approximately 20% (to 46,000 students) by the year 2000 to meet industrial demand for graduates and to provide the economy with young engineers who can create valuable jobs;
* The student-to-staff ratios in engineering should be lowered by at least 10% before the end of the decade to provide better instruction in design and unstructured problem solving; increased teaching assistantships for engineering graduate students could improve the student-to-staff ratios and provide valuable teaching experience as well as financial assistance for engineering students;
* Policies should be adopted and funds provided for the orderly replacing and updating of teaching equipment, design tools, and space commensurate with industrial standards. Since the effective lifetime of equipment varies widely, it should be set in consultation among engineering educators, university administrators, industry representatives and government officials.

Engineering Faculties should undertake: 1, the preservation of the broad base of Canadian engineering programs to ensure the adaptability of engineering graduates to technological and societal changes. Undue specialization should not occur at the undergraduate level; 2, a review and restructuring of engineering curricula to broaden the scope of programs. This broadening could increase the nominal program length or raise the instructional period beyond 26 weeks per calendar year; 3, an examination of the integration of mathematics and science into engineering courses on an as-needed basis rather than teaching this material primarily in separate courses.

Industry and Engineering Faculties should collaborate on: 1, the development of the students' competency in realistic design and problem solving by project work and modern computer-based design methods; 2, the development of leadership and team-work skills through lectures, case studies, and team work which includes students from other engineering and non-engineering disciplines; 3, providing increased opportunities for students to become familiar with engineering practice through improved work experience programs and mentoring by experienced engineers from industry.

Community Colleges and CEGEP's (i.e. Collèges d'Enseignement Général Et Professionnel) should expand transition programs which facilitate entry into engineering programs and which assist students deficient in science and mathematics.

Graduate Engineering Education
In graduate engineering programs, the principal issues and problems are: the excessively long time required to complete master's and doctoral programs; the limited range of master's programs and programs not sufficiently accessible to part-time students; lack of master's programs which combine graduate-level engineering studies with other disciplines such as management; difficulties of admission into doctoral programs from undergraduate programs; transfer from master's into doctoral programs resulting in no significant savings of time;; the participation of Canadians, and particularly Canadian women, in graduate programs likely to be insufficient to meet the needs of Canadian industry and academia; the low financial support for graduate students compared with starting industrial salaries. The report's principal recommendations are:

* Engineerings Faculties should undertake measures so that master's degrees with research can be completed readily within 16 months of full-time study. Similarly, measures should ensure that doctoral programs can be completed within three and four academic years after the master's and bachelor degree, respectively;
* The entry of outstanding undergraduate students into doctoral programs upon completion of the bachelor degree should be improved. Similarly, master's students who have demonstrated aptitudes for independent research should be able to transfer into doctoral programs before completing a master's thesis;
* Industry should be encouraged to provide engineers to teach in graduate programs on a part-time basis and to participate in research projects. Such measures would orient graduate work to industrial problems and facilitate technology transfer. Whenever possible, industry should make women engineers available since they could be important role models for women students;
* Government and Industry should be encouraged to establish, in conjunction with Engineering Faculties, a distance education network to facilitate access to specialized graduate programs, to link Engineering Faculties in Canada and to avoid unnecessary duplication;
* In conjunction with Industry and other faculties, the range of master's programs should be increased with particular emphasis on highly specialized programs which could focus on particular engineering topics or could combine engineering subjects with other areas such as management, business, finance and law. These programs should be accessible to part-time students and should be offered in the distance education mode;
* Strong representation should be made to Industry and Granting Agencies to raise the financial support for graduate students to approximately 75% of industrial salaries for new graduates. Without this effort, the participation by Canadians in graduate engineering studies is unlikely to increase sufficiently to meet demand.

Engineering Faculties and Universities
The issues and problems of Engineering Faculties and Universities are: the university reward system for tenure and promotion which appears to favour research over teaching, advanced design and contributions to engineering practice; lack of programs and structured initiatives to help young faculty members become proficient teachers; young engineering faculty members experiencing great pressures in obtaining research grants and contracts and in producing publications for archival journals; an impending shortage of new faculty members with industrial experience to replace retiring engineering professors; a shortage of professors with experience in engineering design; few women engineers choosing academic careers. The report recommends that Engineering Faculties and Universities should:
* broadly define the concept of scholarship to include excellence in teaching and research. Such definition should help to end the unproductive debates on the merits of teaching versus research;
* appropriately recognize the importance of undergraduate and graduate teaching in decisions regarding appointments, tenure, promotion and salary changes;
* introduce teaching programs, possibly on a shared basis among Canadian Engineering Faculties, to assist young faculty members to become effective teachers;
* assign faculty members, and particularly young faculty members, work loads which strike a balance between undergraduate teaching, graduate teaching, research, and innovative engineering practice;
* structure and administer the university reward system, which governs appointments, tenure, promotion and salary changes, so that engineering research and innovative engineering practice are recognized as similarly valuable scholarly endeavours;
* develop special incentive programs in conjunction with Industry and Government to encourage more young engineers and, in particular, women engineers to choose academic careers.

Engineering Professoriate
The issues and problems with the engineering professoriate are: many engineering professors lacking extensive design and current industrial experience; faculty members experiencing difficulties in becoming or remaining conversant with industrial practice due to heavy teaching loads and emphasis on research; the age distribution of engineering professors in Canada indicating that large numbers of replacements will be needed to compensate for retirements. The report recommends that Engineering Faculties and Universities should make increased efforts to develop the communication and teaching skills of new faculty members and make available the necessary resources to employ more new faculty members who have several years of appropriate industrial experience. Industry and Engineering Faculties should undertake to:
* provide, on a part-time or secondment basis, experienced engineers to collaborate with engineering professors in design and applied research;
* establish University-Industry programs so that engineering professors can have greater opportunities to acquire international experience;
* introduce policies and support initiatives to ensure that engineering

professors remain conversant with industrial practice through, for example, leaves in industry and joint industry-university projects.

Post-university Education
For post-university training and education the report principally recommends:
* Industry and Provincial Engineering Associations should enhance the quality and length of engineering experience before registration as a professional engineer through better training programs and greater exposure to challenging engineering activities;
* Employers of engineers should maintain a work environment which ensures high standards of professional practice and strict adherence to the code of ethics;
* Industry, CCPE, and Provincial Engineering Associations should encourage industry to provide professional engineers to act as mentors for engineers-in-training;
* Employers and organizations which represent large numbers of engineers should support, financially and through release time, the participation of their engineers in continuing education programs;
* CCPE, Provincial Engineering Associations, Engineering Faculties, training institutions and learned societies should collaborate in offering continuing education programs for engineers;
* Faculties of Engineering should be provided with resources so that they may become more flexible in the timing and format of continuing education courses;
* Television networks, Cable Companies and Engineering Faculties should encourage the creation of a Canada-wide communications network (such as "Télé-Enseigne" in Quebec) to enable continuing education offerings to be disseminated efficiently;
* CCPE should develop a Canada-wide data bank on continuing education offerings for engineers;
* The Government should permit the training of engineers and continuing engineering education programs to be partially funded through personal and corporate tax remission;
* Employment and Immigration Canada (EIC) should modify its policies so that continuing education programs for professional engineers are eligible for its financial support.

Finally, because of the complexity and long lead times inherent in engineering education and because inaction could seriously jeopardize Canada's future prosperity, the report's recommendations should be implemented immediately by a CCPE-NCDEAS Task Force.

Reference
A. Meisen and K.F. Williams (1992) "The Future of Engineering Education in Canada" Report of the CCPE and NCDEAS Task Force, Ottawa, Ont., Canada.

The Development of Postgraduate Studies at Xavier University, Cagayan de Oro, Philippines (A Link Programme Supported by the British Council)

M. Cui (*), M. Mendoza (*), T.J. Oliver (**)

(*) College of Engineering, Xavier University
(**) School of Systems Engineering, University of Portsmouth

Abstract
The Philippines are a partially industrialised nation with enormous differences in social and technological development. Engineering degrees are of five years duration normally starting from age sixteen. Universities may be funded by central government, by religious organisations, or by either non-profit or profit making bodies. Undergraduate engineering programmes are under-funded compared with Europe and North America but are popular with government, industry and students. In order to enhance the local economy students are encouraged to engage in design activities which will develop manufacturing using locally available or recycled materials.
Keywords: Philippines, Appropriate Technologies, Research, Academic Development, International Cooperation, Professional Development

1 Introduction

Xavier University in the city of Cagayan de Oro on the Island of Mindanao is one of the premier institutions in the Philippines and is one of the six Ateneos (high schools) of the Society of Jesus and was accorded University status in March 1958.

The University is thus a Catholic, Jesuit run, Filipino Institution and perceives its role as primarily that of teaching undergraduates. Research has a secondary role but at a level and quality that undergraduate courses would be enriched by a research environment.

The University's ideal is to develop men and women intellectually equipped and morally imbued to creatively respond to the pressing social, political, economic and spiritual demands of the developing Filipino national (Reference 1).

Section 6 of the Presidential Decree 223 specifies

that Review Boards shall have the responsibility for the enhancement of the engineering professions within the Philippines. The Review Boards have launched programmes for the continuing education of engineers which involve the attendance at postgraduate courses, seminars and lectures which are all credit rated; these credits are assessed during a three year cycle (Reference 2).

One of the accredited activities of the Review Board is to encourage the development of recognised innovations leading to patented inventions through creative research.

Research activity in the College of Engineering at Xavier University is at present minimal and this fact presents an enormous challenge to both academic staff and management.

Laboratories, whilst spacious, are devoid of advanced equipment or have equipment which is non-useable due to lack of spare parts. There is also a low level of technician support (2 technicians for the College of Engineering). Laboratories also are used extensively by undergraduates. All of the above factors limit the development of research.

This paper outlines a six year plan which will enable the University to come to grips with the challenge of developing its human and other resources in order to produce Christian engineers with the enhanced skills required for high level manufacturing research.

2   Areas of Research Opportunity in Mindanao

Xavier University is situated in the city of Cagayan de Oro in the province of Misamius Oriental on the Philippine Island of Mindanao.

The city had a population of 328,792 in 1990 and at that time the population was continuing to grow at the rate of approximately 5.9% per annum due to migration.

Improvements in employment prospects, access to better quality education and other factors associated with life in a city will lead to a doubling of the population by the year 2000.

Cagayan de Oro is also the regional seat of government and serves as a nucleus for light and heavy industries, is the hub of the banking, trade and commercial enterprises for both domestic and foreign markets and plays an extended role as the centre for educational, cultural, research and data processing services (Reference 3).

By 1989, 14% of the total businesses in Northern Mindanao were located in Cagayan de Oro. Of these 474 (9.0%) were in manufacturing, 1305 (24.86%) were service industries, 3462 (66.02%) were trading as shops and commercial enterprises and 3 (0.06%) were in the agricultural sector.

The most dominant areas of manufacturing were food-processing (Del Monte Philippines Inc, Nestle Philippines Inc and Lim Ket Kai Inc) mineral processing (Philippines Sinter Corp, Metro Alloy Corporation), chemical processing (Resins Inc, Sageco Inc, Pryce Gases Inc) and steels based manufacturing (Mindanao Steel Corporation, Philippines Iron Construction and Marine Works Inc) (Reference 3).

Manufacturing is often however based upon small urban and rural workshops and it is recognised by the authors that these businesses probably offer the areas where an indigenous Philippine industrial base could be created. The larger manufacturing companies are often transplants of multinational companies with the accompanying research and development activities occurring outside of the Philippines.

At present the areas of research identified for further study by the Engineering Faculty at Xavier University include lower limb prostheses, appropriate technology electronic instruments using locally available components, investigation of drainage systems suitable for a rapidly growing urban population, investigation of traffic control systems, investigations into the local environment with regard to the safe disposal of domestic and industrial wastes, the development of a clean air policy for the city area, further development of the Asian Utility Vehicle for local manufacture, and the fabrication of tools and components using appropriate expertise and available materials.

3   Research Development Plan

With no track record and little experience in research, it is of course very difficult to plan into the future. Project planning was obviously essential and as a basis for the plan the Teaching Company approach, developed in the UK, was used.

After extensive consultation, a six year plan was proposed which comprised the following main areas of activity:-

(i)   Market Survey

Xavier University produced 13 Chemical Engineering graduates, 46 Civil Engineering graduates, 21 Electrical Engineering graduates and 23 Mechanical Engineering graduates in March 1992. Most of these graduates should be successful in passing the Review Board examinations in November. It is a national requirement that they pass those examinations before they are allowed to practise as professional engineers. The vast majority of these engineers, based upon previous experience, will be absorbed into the industrial base of Northern Mindanao. These ex-students will be surveyed to gain information in order to develop a database containing first destination statistics and also in subsequent years to gather information of professional development.

Industrial leaders will be subjected to an intensive schedule of visits by academic staff and will be encouraged to visit and support the activities of final year undergraduates within the College of Engineering. Such activities will include project displays, laboratory demonstrations, explanation of the business plan of the College of Engineering as well as the more normal visits associated with industrial placements and graduate recruitment.

(ii) Definition of Undergraduate and Postgraduate Development

Academic staff have already identified areas within the College of Engineering as being suitable for the development of initially, undergraduate, and finally postgraduate course and research opportunities (see Section 2).

Laboratory equipment will be specified as a basis of 'must have' ie this equipment is vital to the courses offered. New equipment will also be identified during this phase and will be studied by undergraduates in the Feasibility Study project in the third year of their course.

A survey of local manufacturers, leading to a database of those capable of manufacturing laboratory standard equipment will also be developed as well as a computer model which will attempt to evaluate the impact of such changes upon the local environment.

(iii) Detail Design and Manufacture of Laboratory Equipment

Fourth and Final Year undergraduates will be assigned the detail design of laboratory equipment under the guidance of academic staff from Xavier University who will liaise with visiting research staff from Portsmouth Polytechnic. The linking of the two institutions through the British Council funded Link scheme enables Xavier academics to work with Portsmouth research groups on a three month exchange and also supports shorter visits to Xavier by Portsmouth academics. This link has encouraged the development of research at Xavier and the Polytechnic staff will assist in the evaluation of the detailed designs and the research programme plans of the academic groups at Xavier

(iv) Induction of First Cohort of Research Students

Year 3 of the Research Development plan targets the manufacture and inauguration of research groups. The initial cohort sponsored by industry and part time teaching contracts, will be supported by final year undergraduate projects and academic staff.

Additional technician staff, again sponsored by industry and further supported by income generating projects in the Centre for Industrial Technology, will be recruited so that laboratory and hence research activities will be enhanced.

(v) Implementation of New Technologies
Data capture and the development of intelligent controls are seen as vital areas in the plan of the Philippines to be a Newly Industrialised Nation by the year 2000 (Reference 4). The implementation of interfaces and computers to the previously designed equipment will assist in the collection of information and the development of experience required to formulate the specification for second generation research. This will provide the link with the other colleges within the University such as the College of Art and Science and the College of Medicine.

(vi) Continuing Research Programme
The objective of this Research Development Plan is to improve the undergraduate provision and to develop a postgraduate research based profile. At the end of the six year plan a major objective is the graduation of ten students with advanced engineering experience capable of leading research and design teams as outlined in the National Action Plan (Reference 4).

These Masters of Engineering will also assist in the expansion of the the programme and in defining the next stage in the re-appraisal of Philippine higher education required for the first ten years of the new millennium.

4   Discussion

There is little doubt that the Research Development Plan for the College of Engineering is adventurous in the light of the current economic and political climate of the Philippines. Academic staff regularly work in excess of 20 hours contact for a salary of US$200 per month and in some institutions individual lecturers work more than 40 hours contact and then teach additionally in Review Centres for the all important Professional Board Examination.

However the National Action Plan of the Philippine Metals and Engineering Industries published in 1991 recognises the requirements of the development of a Design and Engineering capability. This proposes the setting up of centres of excellence and has produced its own five year Research and Development Programme (Reference 5).

The six year programme of the College of Engineering is based upon realisable goals using locally available resources with funding plans reliant upon the existing close relationships of industry and academe and thus has a firm foundation based upon these factors.

5   Conclusion

There is a nationally recognised requirement for an increase in the number and quality of engineers and scientists in order that the Philippines should be able to support the independent design and development of products.

The enforcement of the requirements for laboratory equipment in engineering undergraduate courses is often approached in a pragmatic manner.

The cost of imported laboratory equipment is in the main prohibitive and thus new equipment must be designed, developed and manufactured locally.

The level of technology in Philippine industry is recognised as being five years behind Newly Industrialsed Countries and 15-20 years behind those of developed countries and this engenders a need for high calibre engineers.

The Research Development Plan accepts the premise that in the long term self help is more satisfactory than relying upon external grant aid which in the past has provided equipment which is not maintainable locally and which in the main has been unsuitable for both undergraduate and postgraduate provision.

References

1 Xavier University - A History
2 1987 Constitution of the Philippines, Presidential Decree 223
3 Cagayan de Oro - Northern Mindanao Investors Guide, Philippine Department of Trade and Industry: Region 10 1990.
4 Engineering and Technological Manpower Study Executive Management Group 1980.
5 National Action Plan - Philippine Metals and Engineering Industry 1991 1990-2000.

An Exploration of the Tendency for Development and Reform of Higher Engineering Education in China

X.M. Zhang

Higher Education Research Institute of Northeast University of Technology

ABSTRACT

The development of contemporary science and technology characterizes a tendency of high division as well as high incorporation, i. e. faster pace of development, more intensive and narrow division among disciplines, more indistinct border lines among disciplines and technology, more frequent overlapping between science, technology and humanities, social science, shorter period for the transformation of scientific achievements to productivity and closer links between S & T and economy. This fast paced tendency proposes more requirements on the personal quality of engineering personnel. They should have a solid foundation of knowledge in natural science, extensive knowledge of engineering, balanced knowledge structure which will include economic management, human and social sciences. They should have not only ability to solve engineering problems with their knowledge of various disciplines and with the communication with related disciplines, but also the ability to update their knowledges in practice. To answer this challenge, key Chinese universities of science and technology must be made into centers for education so as to offer society highly qualified engineering personnel. At the same time, they must be seen as centers for scientific research to enrich and improve, with their high—level achievement in research, curriculum, academic quality of faculty, teaching methods. This will help to lead students to the frontier of a respective discipline as well to the engineering practice. The world revolution of new technology and China's socialist modernization have proposed new requirements to higher engineering education. In order to explore these requirements smoothly it is necessary to abstract something of direction and characteristics from various practices and technology transfer between academy, industry and society. This paper will make preliminary research on the two important issues in the development and reform of HEE in China.

Keywords: Engineering, Education, University, Polytechnics, Reform, Quality.

1 Technology Transfer between Academy, Industry and Society

Modern science and technology have become powerful in force productivity in today's society in China. It certainly helped to bring the HEE closer to industry and society. HEE must now modify itself to the development of industry and society. Currently, the development and reform of HEE have achieved a remarkable progress in such areas as 1.) strengthening the cooperation between university and industry and society, 2.) the combination of teaching with scientific research and production, 3.) the realization of the lifelong educational system, and 4.) the development of the versatility of structures and patterns. But it also puts forward many questions worthy of further discussion.

2 Cooperation between Academy, Industry and Society

China is a developing country, compared to the developed ones, she underwent a tortuous development and is underdeveloped in education. Since her opening and reform in the 1980s, higher engineering education in China appears as a new aspect of opening to industry and society. But this development has just begun most of the cooperations are at lower level. Many conflicts and difficulties remain to be solved.

(a) We must fully understand that the development and reform of HEE should serve the development of technology transfer between academy and society, that the training of senior engineering personnel relies on the cooperation between university and industry and society, on the combination of teaching with research and production. HEE must respond to the requirement of industry and society. The separation of education from the practical needs of the socialist construction should be avoided.

(b) Objective laws and characters should be followed and favourite policies be made in the cooperation between university and industry and society, so as to encourage, support and promote the various cooperations. It is only in this way each part in the cooperation can make the best use of its strength, learn from each other and make up deficiency, help each other, share benefits and achieve mutual development. Cooperations like these will be stable, durable and developed further.

(c) The cooperation between university and industry and society should neither merely meet the present needs of industry, nor be involved too much in the service at lower level. With a strategic foresight, university should keep academic independence in order to give forethought to important issues that may be raised in the future development of science, technology and society. Science and technology, leading the industry, will promote the development of industry and society.

(d) Training of qualified personnel is most important in developing and combining the three—dimentional function (teaching, research and serving society). In dealing with the relations among these three functions, we should focus on coordination and promotion among them.

3 Structure of HEE to meet the needs of industrial and social developments

Modern industrial production and engineering require cooperation of the technical personnel from various disciplines and in different levels, therefore require structural variety in HEE. Responding to this requirement, new disciplines and subjects, not only emerged but also intersected, have become new research areas. It also creates a training method which works on different levels and in various forms. The learning process should exist not only before graduation but also after that. The traditional concept that higher education is a final education gives way to that of continued system. In short, variety in industry, society and the persons who receive education brings about the variety of disciplines, levels and patterns of HEE in different areas and universities. This is an important reform in HEE system.

(a) The preliminary stage of our socialism and the slow progress of our industry decided that we should focus on a lower level of HEE, that is, we should rectify the trend to pursue blindly to higher the level of education. In order to reserve necessary high level personnel for developing engineering, we can only set up graduate school in a few qualified universities.

(b) We should make adjustments among disciplines and specialities so as to further develop education. Special attentions should be paid to the newly emerging technical disciplines as well as to those with basic and wide engineering knowledge. Old specialities should be reformed and cross disciplines be established as well. To follow the latest development of the world's high technology, we should set up some key disciplines in universities with strong infrastructure.

(c) When paying atterntion to the education of different academic degrees, we should carry on the continued education program in engineering fields. Forms of programs can be flexible. Joint—Venture programs between university and enterprise, being able to take advantage of both sides, should have the priority in development. In addition to that, teleducation network such as broadcasting and TV educational programs should be further improved. In these ways, we can form a continued education system.

(d) Admitting that there exists an unbalanced development and a variety in HEE, we should make a strategic plan for its overall development, in order to optimize the structure, suit different measures to different universities and coordinate its development. Within HEE, there must be a reasonable division of focuses, to have universities for different levels, fields and forms take niches with their respective strengths, so that the system of HEE as a whole can pursue the highest efficiency.

4. Focusing on the improvement of quality in the future

As to the HEE in China which experienced a quantitative development in 1980's, we must shift our attention from increasing its numbers to improving its quality and efficiency. In this way the tendency to deepen reform and to improve quality has turned to be the common key point for the world in HEE. With the progress of education ideas, the deepening of educational research and educational technique, HEE has made some breakthrough in its methods and objectiveness. In recent years, HEE has experimented with teaching reforms. Of course these experiments have gained achievements and produced errors as well. We should analyse them seriously.

(a) To make an overall improvement on personal quality, with an emphasis on development of students' social responsibility and initiativeness. The improvement of personal quality of the future senior engineers relates directly with the effectiveness of HEE's service for industry and society. The potential senior engineers of China should adapt themselves to the requirements of modern development in engineering with their comprehensive basic knowledge, pratical ability in application and their personal development. They should think and study independently and able to help developing science and technology in engineering and be qualified to solve problems in their fields.

Future engineers should not only fit themselve for social and industrial requirements but also be inovative so as to push both industry and society forward. The development of students' creativeness has become a new aim in training qualified personnel in the world's HEE. We must get rid of the obsolete traditional educational ideas and pay attention to the creativeness of students. Education for developing creativeness must be based on the comprehensive knowledge acquired, taking continued education study as its centre and focusing on developing the ability of creative thinking and practice. This practice should go through each of the whole education procedures in order to explore new modes of training high quality technical personnel.

(b) To follow the objective law in training senior engineering personnel to optimize curriculum structure and to improve teaching process. The responsibility of senior engineering personnel is to make the best use of all natural resources to serve mankind and to solve engineering problems in a creative way with basic science and technical know—how, specific technical experience accumalated in the process of production and the most reliable and most economic plans.

World's HEE adopts the policy of strong basic curricula. Practical issues in modern engineering involve knowledges of various disciplines. Emphasis should be laid on the infiltration and combination of engineering curricula with those of natural science, technology, economic management, humanity and social science, so as to make the students's academic foundation more solid, their knowledge broader and have a relatively strong points area in one field or in one discipline. Thus, a more reasonable structure related to morality, knowledge, technical and personal ability can be formed and a more flexible adaptability achieved. So the optimization of curriculum structure is one of the important items in the development and reform of the current HEE.

(c) Modern engineering and the development of modern industry request that senior engineering personnel not only have wide theoretical knowledge in their fields but also be able to design and to manufacture. Scientific technology can be transferred into productivity when the above mentioned two elements are combined. The principle to pay equal attention to the teaching of basic theories and to the practical training is one prominent characteristics of modern HEE. There should be a coordination between them. It is necessary to combine university education with social practices, especially with engineering practices. According to schedule, students go to industrial practice in enterprises while studying in universities. These necessary basic practices (including experiments, internships, research and design) enable them to bear social responsibility, scientific spirit and engineering consciousness which enhance students' ability in research, design, manufacture and operation management, etc.

(d) We should provide the students with an active and animated educational environment. For that purpose, teaching schemes should be more flexible, changing from the uniform way to a more adaptable mode. To engage the enthusiasm of the students, teachers should be more devoted to their work, more strict and more objective in evaluating their teaching quality. The cramming method of teaching must be replaced by the heuristic method, so that students will be able to study independently, to experiment and to do research more enthusiastically. Students will have time and chance to think, to organize extra curricular activities, to make a self—education and to become a new generation which will be able to have an overall personal development and be more succesful in the future.

In general, by exploring the general trends of development and reform of engineering education, catching up with the world trends, combining world trends with the specific conditions of our country and giving full play to our potential, we are sure to educate tens of thousands of qualified future engineers to meet the challenges of the 21st century.

References:

Robert, M. Anderson. Jr. Industry/ University Continuing Education. Engineering Education V. 77. No. 7/8. 674—677. 1987.

Lin, A. X; Xie, B. Z. Issues on development and reform of higher engineering education

Liang, Y. N; Zhou, Y. Q; Yuan, D. N. Enhance the adaptability of higher engineering education to society (1990) P. 91—97; 111—119 poceedings of international symposium on Higher Engineering Education in China.

Massachusetts Institute of Technology Bulletin 1990—1991

"CIM in Orbit" - Masters Course in CIM (Computer Integrated Manufacturing). The Surrey Experience

P.G. Ranky

University of East London, Department of Manufacturing Systems Engineering (formerly with The University of Surrey)

Abstract

Integration is largely a process of *motivating people,* applying sound methodology and technology and, last but not least, trying to overcome internal problems which are often related to in-house politics. *CIM leads to the diffusion of information, decision making and authority, often creating major management and cultural changes in the enterprise. CIM projects inevitably fail, unless management recognizes that CIM is before anything else, an organisational or human problem rather than simply a series of technical problems.* In terms of the future, we have to recognise that we need to create system environments designed using sound methodology and technology in which able humans and powerful machines work in harmony. This can only be achieved by professional education at postgraduate level.

The main objective of the "CIM in Orbit" research project and post-graduate continuing education programme is to understand the importance and explore the requirements, the design and integration aspects of shipping CIM information via low orbit satellite linked Local Area Networks. Furthermore, to study EDI/CIM/OSI networking and integration strategy and architectures and to gain practical experience by using different CAD/CAM systems linked to computer networks and the University of Surrey Communications Satellite (designed, manufactured and operated by the University of Surrey itself) during the project. The project is sponsored by the European Comission and by several Universities and companies in the UK and Europe, including the main three university partners: The University of Surrey in Guildford (project leader), VIKING UETP and the Royal Institute of Technology (Stockholm, Sweden) and the University of Porto (Portugal). Sponsoring companies at the time this paper was written include large multinationals as well as small-to-medium sized manufacturing enterprises. (For a detailed list of contributing companies and sponsors, please refer to the *Acknowledgements* at the end of the paper.)

Keywords: CIM, Computer Integrated Manufacture, CIM Education, CIM Research, Satellite Communication, EDI (Engineering Data Interchange)

1 Introduction

Computer Integrated Manufacturing (CIM) is a multidisciplinary subject. According to our definition *CIM is a complex, multilayered system designed for the purpose of minimizing waste and creating wealth in the broadest sense.*

The winners of the rapidly changing and information technology driven manufacturing race will be those who can transform the research results quickly into innovative industrial applications. This means that increasingly more *knowledge intensive products* will appear in the marketplace created by the best educated designers, manufacturing engineers, managers and workforce. *For the reasons outlined above CIM education, manufacturing system modelling and design are strategically important for Europe's manufacturing industry.*

System modelling techniques have evolved because of the need to represent the operations and activities that occur within the growing number of increasingly complex flexible manufacturing and other types of systems that are being designed. This is primarily in order to facilitate the study of existing systems and to aid the design of new systems.

172 Engineering Education

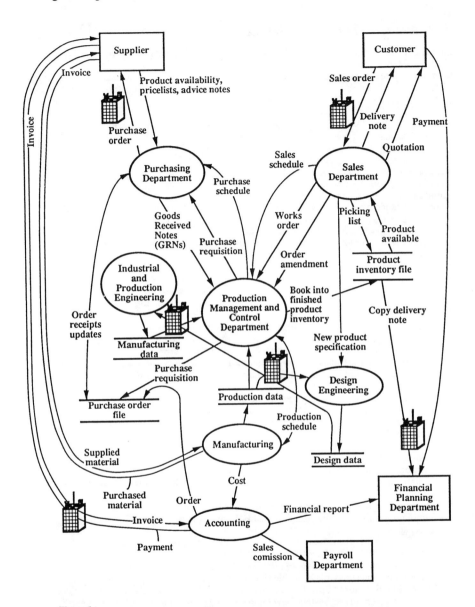

Figure 1
The data flow diagram (DFD), highlighting some of the most important data stores
(i.e. databases) of a typical medium-to-large size manufacturing enterprise applying
CIM and Concurrent /Simultaneous Engineering technology (via satellite).

 Note that as illustrated, the satellite link provides the networking
infrastructure for the Design/Manufacturing link, for the Total Quality
Information System, for purchasing via EDI, for sharing databases and
various other resources.

Engineering Education 173

The *design* is a particularly pertinent example of this need for system description by modelling. Fully operational CAD/CAM, MRP, FMS, etc. are systems of relatively high complexity - even more so if they are integrated into a CIM environment. The complexity of these system designs is further complicated by the emphasis that must be put on design for manufacture and simultaneous, or concurrent engineering. This is an area where a compromise must be found between what is best for the manufacturing system and what is best for the product. *It is increasingly acknowledged that designers and production engineers can no longer work separately and need to follow concurrent design for manufacturing rules.* Considering the above, the need to describe the product data, the manufacturing system and the other CIM components and processes in a concise, *multi-layered, structured and intelligible* way becomes obvious (**Figure 1**).

It must be realised that the concept of shipping meaningful data (i.e. information) versus piece parts is becoming an increasingly important field of CIME (Computer Integrated Manufacturing Enterprise, or the advanced, flexible manufacturing factory with a future) operating on a world-wide basis.

The main objective of the "CIM in Orbit" research project and post-graduate continuing education programme is to understand the importance and explore the requirements, the design and integration aspects of shipping CIM information via low orbit satellite linked Local Area Networks (Ranky (1990a), Ranky (1990b), Ranky (1991) and Gerhardt (1991)).

2 The "CIM in Orbit" research and continuing education programme

The research and continuing education programme (**Figure 2**) incorporates the following main CIME (Computer Integrated Manufacturing Enterprise) areas:

* To understand cellular manufacture (FMC) and FMS (interpreted as not just flexible machining, but assembly, electronic manufacture, test, inspection, painting, welding, packaging, etc.) within a satellite linked Computer Integrated Manufacturing (CIM) architecture.
* To establish a broad understanding of the underlying specification, design, modelling and simulation techniques, and their role in formulating the FMS plan. To provide a generic FMS design method, based on the structured analysis and design technique (SADT, IDEFx), discrete event modelling, solid modelling, simulation and other methods.
* To become familiar with some of the CAD/CAM, FMS, Production Management, Software Engineering, AI and Quality Control methods, standards, systems and available tools used in cellular manufacturing and FMS system design and operation control (Gerhardt (1991) and Ranky (1991)).
* To provide a comprehensive overview of the main components of Concurrent / Simultaneous Engineering (CE/SE) methods, tools and technologies as applied to manufacturing and design data interchange in the design office environment, on the shop floor and via the orbit by means of communication satellites (Gerhardt (1991), Douthwaite (1990) and Ranky (1991)).
* To become familiar with some of the most important CE/SE methods, tools, technologies and standards available (including: Quality Function Deployment (QFD), Simulation, Structured Specification, Analysis and Design, Database Management, Distributed/Networked System Development, CAD/CAM integration, fast prototyping and others) (Towill (1991)).
* To learn about Design for Manufacturing/Assembly (DFM/DFA) methods, tools and systems and gain some "hands-on" experience in redesigning electro-mechanical, electronic and mechanical products and components for Concurrent / Simultaneous Engineering and cellular manufacture (Anjanappa, et. al. (1991, Towill (1991)).
* To explore the application of such methods and tools as a means for achieving effective integration of information within the evolving, satellite linked CIME (Computer Integrated Manufacturing Enterprise) context.

Computer Integrated Manufacturing Courses
University of Surrey, Guildford, UK. Tel: (0483) 509296

Module F-02: "CIM in Orbit" Computer Networks for CIM (CIM/OSI. Network design & Implementation)

Duration: *24-28 February 1992* To be held: *Univ. of Surrey*

	9 am	10.30–11 am	11 am	12.30–2 pm	2 pm	3.30–4 pm	4 pm	5-6 pm	8 pm
Monday (Lecture Room 13 AC 20)	Introduction and Overview of EDI/CIM/OSI Networking (PGR)	Coffee Break	CIM Network Architectures & Design (PGR)	Lunch Break	Network Design Management & Multimedia [MG]	Tea Break	Network Design Management & Multimedia [MG]	Dinner Break	Network Integration Case Study (Novell) (JH)
Tuesday (Lecture Room & Labs. 13 AC 20)	Networking Lab I. (AK and JP)	Coffee Break	Networking Lab II. (AK and JP)	Lunch Break	Electronic Data Management via Comp. Networks I. (PF)	Tea Break	Electronic Data Management via Comp. Networks II. (PF)	Dinner Break	Networking Lab III. (AK and JP)
Wednesday (Lecture Room & Labs. 13 AC 20)	Satellite Engineering (MS)	Coffee Break	Satellite Data Transmission Laboratory (MS)	Lunch Break	Satellite Engineering & Data Comms. (BE)	Tea Break	Satellite Engineering & Data Comms. Lab I. (BE)	Dinner Break	Satellite Engineering & Data Comms. Lab II. (BE)
Thursday (Lecture Room 13 AC 20)	Network Design Case Studies (Integr. Shop floor Contr.) (RT)	Coffee Break	Network Design Case Studies (Networked Tool Management) (RT)	Lunch Break	Network Lab I. (RT)	Tea Break	Network Lab II. (RT)	Dinner Break	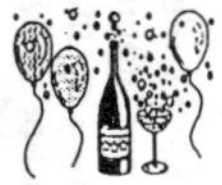 Course Dinner
Friday (Lecture Room 13 AC 20)	CIM-OSA, EDI and X.400 [PGR]	Coffee Break	CIM Network Design and Integration Case Studies (PGR)	Lunch Break	Course Summary and Team Presentations (PGR]	Tea Break	Course Summary and Team Presentations (PGR]	Dinner Break	Individual Study/ Departure

Lecturers and guest speakers in the order of appearance:
PGR = Dr. Paul G Ranky, U of S; Mr John Hunt (Netw. Consultant); Dr Terry Hinton (UofS); MG = Mr. Michael Grove (CIS Ltd.); AK = Dr. Anthony Kehoe, U of S; JP = Mr. John Pretlove, U of S; Mr Paul Francis (Director) ICC ; MS = Professor Martin Sweeting, U of S; Professor Berry Evans (UofS); Mr Roland Thomas (Director) ERT Eurotechnik;

Figure 2
The University of Surrey, Department of Mechanical Engineering, MSc in CIM Course Module example

3 The Modular Masters course in CIM at Surrey

The aim of this course is to provide a solid manufacturing science foundation and relevant practical experience to professional engineers and managers working in industry, universities and other institutions. The one week modules can be taken on an individual basis and/or can lead to the award of an MSc or Postgraduate Diploma in Computer Integrated manufacture, subject to satisfying the entry and course requirements.

A professional CIM training programme spans several levels. Top management has to understand what CIM is, how it can be used, and what its value is in order to endorse and support a corporate wide strategy to exploit the technology. Middle management has to learn how to select and manage an application of CIM technology, as well as to oversee the knowledge acquisition and system integration efforts. A technical group must be acquired or built that can design and build the desired systems. Three distinct but complementary types of module are offered. Study is normally part-time, in the form of one week modules, thus minimising disruption to work and carrier progression.

In addition to attending the approx. 55 hour week at the University, of which examples are shown in **Figure 2**, students registered for the MSc or the Postgraduate Diploma must successfully complete and submit *assignments* for each module and a *thesis* at the end. The MSc/PgD thesis can be carried out in the company or in one of the advanced manufacturing laboratories at the University of Surrey. The course structure is very flexible. There are a minimum number of precedence rules attached to each module, otherwise they can be considered as "interchangeable units" of theoretical and practical study and labwork. Modules are normally run at least once per year, so a missed module can be made up relatively easily on the next occasion. Students may apply for registration at any time during the academic year. (Other structures, such as *full-time* MSc courses are available too.)

Students gain a very comprehensive foundation and state of the art knowledge in applied manufacturing science, or more precisely "manufacturing art and science", taught by several leading academics and industrial managers. The course has a strong international dimension in both students and lecturers, who come from the UK, Europe, USA and other parts of the World.

Due to successful UK and European COMETT grants and industrial cooperation, many of our students are placed in various UK and European companies for the period of their thesis research and development. Sponsoring organisations immediately gain access to various new aspects of manufacturing systems engineering, computer science, design engineering and integration methods and technology which can form the basis of an ongoing link with the University. The thesis is of immediate relevance to the company, because it provides solutions to the sponsor's own problems, using methods and technologies learnt on the course.

4 Summary and Conclusions

To stay in business and prosper, manufacturing industry must learn to react with speed to market needs and product changes. Furthermore, the products that designers and manufacturing engineers create must be of high quality, achieved at low cost. What this means is that both design and manufacturing will become increasingly dependent on accurate information, often delivered by means of CIM satellite linked networks.

The sharing of design information with marketing, manufacturing, total quality management and the customers is the key to reducing the design and the "time-to-market" cycle. The main objective of the project is to understand the "CIM in Orbit" integration strategy, the methods, tools, technologies and architectures, to gain practical experience by using different systems and to analyse different product design solutions (including: electro-mechanical, electronic, computing, mechanical, automobile and others).

The research tools currently under development will establish a framework within which one can analyse, research and design satellite linked concurrent/simultaneous engineering, remote diagnostics, remote control of CNC and robotics equipment, cells and FMSs (i.e.

DNC - Distributed Numerical Control via the orbit) and other new areas of academic research, education and industrial interest. The Surrey Modular MSc in CIM Course provide a solid manufacturing science foundation and relevant practical experience to professional engineers and managers working in industry, universities and other institutions. The one week modules can be taken on an individual basis and/or can lead to the award of an MSc or Postgraduate Diploma in Computer Integrated manufacture, subject to satisfying the entry and course requirements.

Acknowledgements

The author would hereby express his sincere thanks to all individuals, sponsoring companies and institutions who have supported his *"CIM in Orbit"* research & continuing education programme, including: The University of Surrey, the Centre for Satellite Engineering Research at The University of Surrey, the European Commission, VIKING UETP in Sweden, Mr. Ivan Oberg, Mr. Hans Oberg, Professor Francisco Freitas, Professor Berry Evans, Professor Martin Sweeting, Mr. Cliff Douthwaite, Mr. David Montgomery, Professor A C Walker, Professor G A Parker, Mr. P Francis, Mr. M Grove, Dr. A. Kehoe, Mr. John Pretlove, Dr. B Becker, Mrs. M Huckerby, Professor Francis N-Nagy, Dr. T Hinton, Mr. Roland Thomas, Mr. Andrew Denford, Mr. Geoff Harrington, Professor Hans Skoog, The University College of Eskilstuna/Vasteras, ABB (Sweden), The Royal Technology Institute (Stockholm), IBM (Sweden), IBM (UK), ERT Ltd (Brighton) UK, CIS Ltd (UK), ISIS Informatics Ltd (Godalming) UK, The Fraunhofer Research Institute in Stuttgart (IPA), Germany, IDIT - Instituto de desenvolvimento e Inovacao Technologica, Portugal, Deckel Ltd., AUTOCAD, Denford Machine Tools (UK) and others.

References and Further Reading

Ranky, P G: Computer networks for world class CIM systems, CIMware Ltd., Guildford, 1990. 233 pp.

Ranky, P G: Total Quality control and JIT management in CIM, CIMware Ltd., Guildford, 1990. 256 pp.

C.Douthwaite: Satellite Hardware Manufacture using 3D CAD to 2D CAM interfaces.6 th International Conference Computer-Aided Production Engineering pp 329-334, 12 - 13 th November,1990 ,The Royal Society,London

C.Douthwaite and M.N.Sweeting: CAD/CAM in Orbit with University of Surrey Spacecraft, Effective CADCAM89 Conference ,IMechE , London, 6/7 December 1989

Ranky, P G: Flexible Manufacturing Cells and Systems in CIM, CIMware Ltd., Guildford, 1990. 262 pp.

Ranky, P G: CIM Information System Modelling Methods with Case Studies - Training: from CAD to CIE in Europe; IFIP-IFAC CIM Workshop, Alesund, Norway, April 1991.

Gerhardt, Lester: Strategic Issues and Technology Management Aspects of CIM, International CIM Conference Singapore, 1991.

Ranky, P G: Master Course in CIM. The Surrey Experience; International CIM Conference Singapore, 1991.

Anjanappa, M, Kirk, J A, Anand, D K and Nau, D S: Automated rapid prototyping with heuristics and intelligence: Part I - configuration, Int. J of CIM, 1991, Vol. 4.

Anjanappa, M, Kirk, J A, Anand, D K and Nau, D S: Automated rapid prototyping with heuristics and intelligence: Part II - implementation, Int. J of CIM, 1991, Vol. 4.

Towill, D R: Supply chain dynamics, Int. J of CIM, 1991, Vol. 4.

Planning, Development and Implementation of a Joint European Masters Degree: A Test Case for Cooperation under Erasmus Scheme

M.S.J. Hashmi

School of Mechanical and Manufacturing Engineering, Dublin City University

ABSTRACT

A group of academics from a number of University and Higher Education Institutions from different member states of the EEC joined a cooperation and exchange programme under ERASMUS scheme. Subsequently, a means of promoting further cooperation, a masters degree course in Manufacturing Engineering was planned and developed jointly by this group and the proposal received additional support through the ERASMUS Scheme.

This paper presents the outline of the scheme and discusses the difficulties which were faced during the planning, course development and implementation stages of this course. The course is currently being run at three different institutions despite some practical problems which are also discussed in the light of the experience of the group.

Keywords: European Masters Course, joint degree, course planning, ERASMUS Scheme.

1 INTRODUCTION

The third level general education systems in different countries of the world vary considerably. However, such variation appears to be even more contrasting within the European scene, especially for technical and engineering education. Such diversity exists in terms of entry level, duration of the course and the title of awards both for undergraduate and postgraduate studies. For example, the minimum age of school leavers at entry to an undergraduate degree course can vary from 16 to 19 years, and the duration of the first degree course can vary from 3 to 6 years[1]. The successful candidate may come out of the system with one of a variety of awards such as BSc, BEng, MEng, Dipl. Ing., Lauria, Diploma and so on. Such diversity of titles of the first degree does not in itself cause any problem within any given country. However, with the formation of EEC and resulting human mobility within different member states of the EEC, it is vital that the qualifications awarded by each country is seen to be and accepted as equivalent to those from the other countries. This is extremely important for firstly, the entry of graduates from one country into postgraduate education programmes of another country and secondly acceptance of such qualifications from one country by the employers of another country. With increasing emphasis on, and progress being achieved towards political and economic unity within the EEC, it is becoming imperative that the third level educational system of all the member states be looked at closely with a view to rationalisation of the schemes and harmonising the standard of first degree awards. Given the long history of diversity in tradition and attitude of each member states towards its existing system, it is clear that such a rationalisation and harmonisation task will face serious opposition and will require considerable diplomacy, tact and financial resource to achieve any success.

2 EXCHANGE AGREEMENT

Those who are familiar with ERASMUS scheme will know that the single most fundamental objective of this scheme is to facilitate interaction of academic staff and students from different member states through
(i) students spending part of their period of study in another country
(ii) staff undertaking teaching/research activities for a period at a university of another country
(iii) partners organising seminars, meetings on areas of common interest in different countries.

The idea being that such interaction will enhance the understanding of the educationalists of member states, of the educational system in each other's country. Since the launching of the scheme, it has proved to be one of the most subscribed and seemingly effective initiatives which facilitated wider interaction between large number of young and senior persons from different member states.

In 1988 a group was formed with the name "Euro-Net" comprising of the following tertiary level Institutions:
1. Dublin City University, Ireland
2. Staffordshire Polytechnic, U.K.
3. Instituto Superior Tecnico, Lisbon, Portugal
4. ENS de Cachan, Paris, France
5. Technical University of Athens, Greece
6. Instituto Polytecnico, Setubal, Portugal

The primary objective was to promote exchange of staff and students with support from the newly launched ERASMUS initiative. This pilot exchange programme proved very stimulating and effective. As a natural development of this programme the group decided to plan and develop a joint Masters degree course in Mechanical and Manufacturing Engineering[2].

3 COURSE DETAILS

It was agreed by all the members of the group that the format of the proposed Masters course will be similar to a typical one year duration MSc degree by course and dissertation offered by many Universities in the UK and Ireland. The taught course component will be completed in two terms from October to March and the dissertation project will be completed over the period from April to September.

It was also agreed that initially the proposed Masters degree will be run at two institutions, namely, at Staffordshire Polytechnic and ENS de Cachan[3]. After the initial two years, other institutions would be encouraged to run the taught part of the course should demand continue to grow.

The principal novelty of operation of this course is that:

1. Each participating institution would recruit 2 to 3 postgraduate students who will have the option of joining either of the two institutions running the course. There would, of course, be other students on the course outside the ERASMUS scheme and recruited from both national and international pool.
2. Academic staff from the participating institutions would give lectures on selected topics/subject areas and also take part in the examination process.
3. After successful completion of the taught component, each of these ERASMUS supported students will be required to carry out their dissertation projects in a partner institution other than their parent institutions or the institution where the taught part was completed. This regulation was adopted to encourage maximum possible exposure of students and staff to different systems and cultures of different member states.
4. Each student would submit his dissertation in either English or French language as well as a copy in his own language for those whose mother tongue is neither English nor French.
5. The examination Board would consist of one representative academic staff from each partner Institution in addition to the regular examiners and the supervisor of the dissertation.
6. The successful candidates will be awarded the degree of MSc jointly by the parent and host institutions (both for taught and project components).
7. These students will not be charged any fees and the travel and maintenance expenses will be shared by the student and the ERASMUS office.

Whilst it was a relatively easier task to formulate and agree the mode of operation of the course, there were a number of issues which could only be resolved after prolonged and sometimes heated discussions. A number of such issues are discussed in the following text.

3.1 SUBJECTS TAUGHT

The main discussions on this issue were directed towards the subject areas and the level at which these should be taught. The problem arose due to the fact that graduates from different countries would have studied different subjects at different levels in their undergraduate courses given the fact that the duration of these courses varies between 3 and 6 years. It was not possible to resolve this issue entirely satisfactorily. However, the group agreed to accept the subject areas and levels of a typical UK taught masters degree in Manufacturing Engineering with the provision of a certain number of core subjects (60 percent of the course) and a large number of optional subjects to cater for different levels of knowledge base of students, coming from different member states.

3.2 STUDENT INTAKE LEVEL

This was the single most contentious issue during the planning phase which required tactful addressing. Once again, the problem arose due to the fact that partners from the institutions offering longer duration courses (5 or 6 years) regarded their first degrees to be superior than those offered at other institutions. For example, the Italian partner was unwilling to send his graduates to do a Masters course alongside with a student from the UK or Spain having completed a 3 year duration degree course.

The added complications arose due to the fact that in Spanish system engineering qualifications can be gained by taking 3½ years as well as 4½ years duration courses in the same way as in the UK engineering degree courses can be of 3 or 4 years duration taking account of the full-time and sandwich modes. Eventually, it was agreed that any partner in whose country the first degree is of 5 or 6 years duration would be permitted to nominate students who have successfully completed the 4th year of the course, for admission into the proposed MSc course.

3.3 MECHANISM FOR JOINT AWARD

This issue needed less effort to be resolved and it was agreed that steps would be taken to form a single validation committee with representatives from each partner institutions and then revalidated through by taking it through the individual academic council of each partner institution.

3.4 LANGUAGE SUPPORT

The difficulties of running a joint MSc course in English or French with some of the students having very little exposure to either of these two languages are obvious. In order to ease the language problem, it was agreed that the prospective student will receive language training in two stages, once at his parent University prior to embarking onto the course and then at the host University concurrently with the course itself.

3.5 FINANCIAL SUPPORT FOR STUDENTS

Under the ERASMUS scheme no course fee is chargeable to the exchange students by the host institution and partial support is available from the ERASMUS office towards the travelling and maintenance expenses of the students. This meant that the nominated student had to arrange additional financial support from his own source thus making the studentship less attractive than those offered through other national bodies.

4 PRESENT STATUS

The proposal received support through the ERASMUS initiative and the joint MSc Course in running for the past 3 years but not without difficulties. The course is currently offered at three institutions, IST, Lisbon being the third one to offer the course[4]. However, out of the three, Staffordshire Polytechnic has proved to be the most popular venue for students from both Portuguese and Spanish partner institutions. No student joined the course from the Italian partner institution primarily due to the practical difficulties in taking a years break from their 6 year first degree course.

Other practical difficulties experienced are (i) the bureaucratic hurdle encountered in adopting an agreed method of awarding the degree jointly which resulted in the host institutions awarding the degree, and (ii) the lack of adequate financial support for the student necessitating, in most cases, students to return to their parent institutions for the project component of their course.

Despite these problems the scheme has proved to be very effective in terms of the participation by academic staff in teaching at other partner institutions. These exchange activities have, no doubt, made both the students and staff of various partner institutions much more familiar with each otner's system and standard of qualifications and their cultural attitudes.

Such familiarity brings better understanding and acceptance of the diversity which exists at the moment. This may lead to ideas about newer and harmonised first degree courses in different member states. However, the diversity of language and funding mechanism for postgraduate studies in each member state will need to be addressed to before any significant progress can be achieved.

5 CONCLUSIONS

A critical review is presented of an initiative to develop and run a joint Masters degree course in Mechanical/Manufacturing Engineering with support through ERASMUS scheme.

The difficulties encountered in developing and subsequently running the Course are discussed so that others may take note of these before planning any initiatives.

REFERENCES

1. Hashmi, M. S. J. (1988), "25 years of evolution of engineering education and training in western European countries", Proc. First World Congress on Engineering Education & Training, Kathmondn, Nepal.
2. Definitive Scheme Document (1988)-MSc in Computer Aided Engineering, Staffordshire Polytechnic, UK.
3. Course Document (1989)-DEA Production Automatise, ENS de Cachan, France.
4. Course Document (1991)-MSc. in Computer Aided Engineering, IST, Technical University of Lisbon, Portugal.

Towards Internationally Recognised Engineering First Degrees

M.S.J. Hashmi

*School of Mechanical and Manufacturing Engineering,
Dublin City University*

ABSTRACT

There is substantial diversity in the nature of first degree engineeirng courses worldwide. This diversity manifests in the duration, subject area and mode of teaching of the course. Some of these factors determine the acceptability of the equivalence of an engineering first degree from one country to another and have some degree of validity of the assessment. However, there are some other reasons which are purely subjective in nature and arise due to the ignorance, prejudice, protectionism and political and economic considerations which are unacceptable.

This paper attempts to discuss the issues which may be playing a major role in promoting and hindering international understanding and cooperation towards mutually recognised first degrees in Engineering.

Keywords: Engineering Degrees, International Recognition, Equivalence and Accreditation.

INTRODUCTION

Engineering education is very capital intensive due to its professional and vocational nature and the human resource planners of different countries of the world attempt to provide adequate facilities for educating their engineering personnel. Due to the diversity of requirements of each country there are wide ranging disciplines in which first degree engineering courses are available. Such diversity in discipline is understandable and reflects the very nature of the engineering profession. However, in the international scene, considerable diversity also exists in the duration, mode of operation, entry requirements and the range and level of subjects taught in each discipline. Such diversity is the major obstacle in promoting mutual recognition and acceptance of equivalence of first degree engineering courses from different countries especially since there is no formal mechanism fo ascertaining the standing of an engineering degree course from one country by anothe country. In addition, there are some other factors which are purely subjective in nature an arise due to the ignorance, prejudice, protectionism and political & economic consideratior which often play a major role in hindering the recognition by one country of engineering fir degrees awarded in another country.

2 REASONS FOR DIVERSITY

2.1 DISCIPLINES

In most developing countries first degree engineering courses are continued to be offered the more traditional disciplines of Civil, Mechanical, Electrical, Chemical, Metallurgic Mining/Mineral and in some cases Agricultural Engineering. For many developing countrie offering degrees in Manufacturing/Production or Electronics Engineering disciplines eve only as a major part of Mechanical or Electrical Engineering course has proved to t adventurous. The principal reason for sticking with such traditional courses is the fact th the job market in these countries will simply not accommodate anything but. By contra there has been a proliferation of new as well as hybrid types of degree courses in develope countries during the past two decades. In addition to the more familiar aeronautical, HVA(automotive and textile engineering, one can see some new titles such as Computer Aide Software, Mechatronics, Aerospace, Ceramic, Polymer and Biomedical engineering mention a few from amongst about fifty offered in the UK alone[1]. Similar diversity disciplines is also evident in other Western European countries and to a lesser extent in Nor American countries and in Australia.

2.2 COURSE DURATION

In general the international norm for the duration of a first degree engineering course seer to be <u>four years</u> with the exception of European and a very small number of other no European countries. Outside Europe, most countries in North America, Africa, Middle Ea Indian Sub-continent, South East Asia, Far East and Australasia offer engineering degr courses which are of four years duration[2]. Within Europe, however, the system is ve complex and hence confusing to the external observers. The duration of first degr engineering qualifications from different European countries can vary from three to six yea of full time study[3]. Most UK Universities offer first degree engineering courses of thr years duration. Universities in Italy generally award their engineering first degrees after s years of full time study. Scottish and Irish Universities offer four year degrees whi Universities in other mainland European countries offer first degree engineering cours which last either four, five or six years. To make the matter more complex one can fi some engineering degree courses in U.K. are of four year duration full-time and/or sandwi mode.

2.3 COURSE STRUCTURE/SUBJECTS AND LEVELS

Internationally, the four year duration engineering degree courses are run on annual, semester or three terms mode of full time study. Courses of sandwiched or cooperative (with industry) modes are also predominantly four year duration. Five to six years duration first degree courses can be full-time as well as with built-in training element (different from part time study). Irrespective of the course structure, the idea is that the range of subjects pertinent to a discipline will be taught at certain acceptable level. As long as the graduates attain certain expected level of proficiency in the normally accepted/recommended range of subjects within a discipline, it should not matter whether the course is full-time or sandwiched.

2.4 ENTRY REQUIREMENT/STUDENT QUALITY

The international norm for entry to a first degree engineering degree course is the possession of School/College Leaving Certificate or Higher Secondary Certificate obtained following 12 years of schooling. Generally, students with good results in three or four science subjects including mathematics prove to have the necessary grounding to follow a four year duration first degree engineering course.

In Europe, Scotland and Ireland follow this system. Other European countries including rest of the U.K. have differing entry requirements e.g. 'A' levels in the U.K. for entry into either a three year or four year duration degree course whilst HSC for entry into a 5 year (Spain or Portugal) and 6 year (Italy) duration first degree course. However many Universities in the U.K. are currently accepting students with 'S' level or 'HSC' School Leaving Certificate as alternative qualifications for entry into a degree course.

In most developing countries the demand far exceed the number of places available (due to resource implications) in engineering disciplines which means that only the very high calibre students can study engineering. In most developed countries the number of places is evenly matched with the demand and is closely related to the employment prospect. The students need to have better than average grades in their HSC or other entry qualifications in order to get a place and follow the course.

The minimum age of students at entry to a first degree engineering course in most countries is 17. However, in some of the European countries the minimum age at entry can be 18 or even 19.

QUESTION OF EQUIVALENCE

It is desirable to establish the international equivalence of engineering first degrees from different countries whether developed or developing. Ordinarily, it would seem to be not an important issue since only a small percentage of engineering professionals actually leave their own country. Nevertheless certain level of mobility takes place mostly from developing to developed countries for postgraduate training and education and the question of equivalence becomes very important. Within Europe, however, it is becoming essential to not only establish the equivalence but also to rationalise and harmonise engineering first degrees. This is even more important for the member states of the European Economic Communities as unrestricted human mobility is one of the main features of the treaty. The diverse nature of engineering first degrees within the member states of the EEC will make it a very difficult, sensitive and long drawn process. No doubt various organisations and authoritative bodies at European level are taking steps towards bringing some kind of uniformity for the existing systems.

Whilst such steps are being considered it would perhaps be prudent to take note of the systems accepted by the majority of the international communities (countries). It appears that the post HSC, four year duration engineering first degrees are by far the commonest system prevailing internationally and hence it would make sense to adopt a common system within the EEC which would be similar to those operated by countries outside of the EEC and other European countries. In the absence of such a worldwide accepted common system the question of equivalence and recognition of engineering first degrees from different countries would continue to create controversy and discontent within and external to the EEC.

There is no assurance that the adoption of such a common system would lead to instant solution as acceptability of the equivalence of an engineering first degree from one country by another. Presently, there are other reasons beside the technical ones such as length of course, range & level of subjects, course structure, entry level of student etc. which influence the international equivalence of first degree engineering courses. These reasons are purely subjective in nature but very often used to ascertain the standing of an engineering degree from another country.

4 AN INTERNATIONAL ACCREDITATION BODY

The technical factors which determine the equivalence of engineering first degrees may, in principle, be dealt with relatively easily through the cooperation of international community of engineering educators, professionals and their professional institutions. Perhaps, an international accreditation body for engineering first degrees should be set up. Such a body would have the responsibility and authority to accredit degree courses either directly or indirectly through various acceptable national bodies e.g. Professional Institutions. The formulation of the precise mechanism of how such a body can be formed and made to work to the satisfaction of the world community of engineering educators and employers will require tremendous effort and support from all concerned. An international task force could take on the responsibility to oversee the process of setting up such an Accreditation body.

One of the main objectives and responsibilities of an International Accreditation Body would be to remove the undesirable factors (ignorance, prejudice, protectionism, political & economic consideration) which very often influence the equivalence determination process. At the same time, it would provide guidance and advice for use by the Community of engineering educators worldwide as to the subject norms and standards which would be expected to be incorporated in a first degree course in a given discipline.

There are, however, a number of hurdles which appear to be formidable and unsurmountable at least in the European scene. Within Europe, countries currently offering three year duration first degree engineering course are reluctant to offer four year course since amongst other things resource is a serious factor. Others who currently offer five or six year duration first degrees are not willing to offer four year duration courses primarily due to desire to preserve their traditional value and concept of their own engineering degree qualifications. Resource issue is of secondary importance in these cases. One probable solution for such cases could be to award a first degree after four years of full time education and then award a second degree (MSc, MEng etc) for the remaining one or two years of further study. This will require some re-structuring of the course curriculum and in the interest of long term international uniformity this is a small price to pay.

The prospect of establishing a truly independent international accreditation body for engineering degree courses and its effective functioning will be very much dependent on close cooperation between different national bodies and most importantly on the desire of these national bodies and educational establishments of different countries. Once a rational solution can be found for the European communities the principal hurdle towards the realisation of international solution will be that much easier to cross. Unfortunately, such an international accreditation body will only resolve the question of academic standing of a given first degree engineering award; the subjective factors such as prejudice, political and economic considerations and protectionism are likely to be unresolved and probably will always be there unless the gap between the developing and developed countries is narrowed down significantly. In any event, a better system is needed to be developed for realistic assessment of international equivalence of engineering first degrees.

The alternative is to rely on systems currently operated by individual developed countries, e.g. GRE (USA), British Council guide lines (UK), European guide lines (Netherlands). The GRE test puts additional financial burden on individual candidates and often is inconvenient to take. Other guidelines offer some help but are not free from biased and subjective assessment of first degree engineering courses especially if these are from developing countries. A typical example is the assertion according to the British Council that four year duration engineering first degrees from a number of Asian countries are not regarded as equivalent to the three year duration engineering degree from the UK. The amazing thing about this is the fact that such assertion is based on the course content and student quality at entry level but on the basis that the local entry qualification of HSC is not equivalent to 'A' levels in the U.K. yet the four year duration engineering degrees from Irish and Scottish universities are accepted as equivalent even though the Scottish and Irish school "leaving certificates" are also not equivalent to 'A' levels. Clearly such assertion is not based on sound criteria. Similarly, the US administered GRE system, although fair, can only provide shallow assessment of first degree engineering courses from other countries.

5 CONCLUSIONS

The factors which influence the standing of an engineering first degree have been discussed in relation to determining the equivalence within Europe and internationally. Idea of and international accreditating body is put forward and major issues which may promote or hinder international cooperation towards worldwide recognition of first degree engineering course have been discussed.

REFERENCES

1. Entry Handbood (1992), The UCCA, U.K..
2. Internation Guide to Qualifications and Education (1991), The British Council, Mansell Publishing ltd.
3. Hashmi, M. S. J. (1988), "25 years of evolution of engineering education and training in Western European countries", Proc. First World Congress on Engineering Eudcation & Training, Kathmondu, Nepal.

International Cooperation in Engineering Technology Education: An Interim Evaluation

G.J. van Woudenberg, J.P. Rey

Department of Electrical Engineering, Leeuwarden Institute of Technology

Abstract

In the last decade new impulses toward a further unification of Europe have been triggered by the policy-makers of the different member countries of the European Community.

One result is a set of directives and the corresponding financial funds, stimulating collaboration amongst the educational institutes in the various European countries.

Since 1989 our institute has been involved in a number of such cooperative activities.

This paper presents some background information, experiences and an evaluation concerning these joint undertakings.

Keywords: International Cooperation, Education, Evaluation, Strategy

1 Introduction

With the establishment of the European Community for Coal and Steel (1951), the first attempt towards unification became a reality.

Subsequent developments were Euratom, the European Economic Community and finally, the fusion of the three units in 1967.

In the 1980's the political-psychological climate changed, the economy recovered from a recession, but competition with the US and Japan forced Europe for further unifying steps, which ultimately has lead to what nowadays is known as Europe '92.

In addition, fundamental changes have taken place in East-Europe. As a consequence, cooperations in various fields and amongst the different countries have been proposed in order to stimulate the unification.

One attempt to work jointly is, for example, in certain areas of education.

Since 1989 the Leeuwarden Institute of Technology -the Netherlands- has been actively participating in a number of international cooperations.

The basic aim of this paper is to share some of our experiences with colleagues facing the same challenges of how to organise and to improve these joint undertakings.

The remainder of the paper is organised as follows: section 2 explains in concise form the structure of Dutch technical education for those readers who are not familiar with the system, section 3 discusses programs and regulations, section 4 deals with initial and current activities, section 5 is devoted to the evaluation of achievements and it ends with a number of recommendations/conclusions.

The final part of the paper is, as usually, a listing of references.

2 The System Of Technical Education

Basically, the present Dutch system of technical education can be divided into three levels [1].

Level 1: The lower Technical School (LTS), which is a 4-year vocational school; supported by theoretical subjects, the emphasis of the training is on manual skills.
  Pre-requisite: 8-year basic school and the nominal age on entrance is 12 years.

Level 2: The Middle Technical School (MTS), an intermediate level vocational school consisting of 3 years in-school- and 1 year industrial training.
  Pre-requisite: LTS or MAVO (a 4-year course of a general nature after the basic school).

Level 3: There are two types on this level:
  a) The Institute of Technology (IT) or Technical College, the official name given by the Ministry of Education.
  An IT is roughly the equivalent of an English Polytechnic or a German Fachhochschule.
  Pre-requisite: 3-year MTS or HAVO (a 5-year course of a general nature after the basic school).
  b) The Technical University (TU).
  Pre-requisite: a succesful 1-year attendace of the IT, or a VWO (a 6-year course of a general nature after the basic school).

Our Institute, the Leeuwarden Institute of Technology, belongs to catagory a). It is located in the north eastern province of Friesland and is about 100 km from the border with Germany.

An IT is a 4-year course with three distinct parts:
- The first two years are basic with emphasis on mathematics and electrical engineering fundamentals (e.g. networks, electronics, control, computers & programming, power engineering, telecommunications).
- Both, theoretical and laboratory courses are offered.
- The third year is spent in industry under the supervision of the academic staff; this term is considered of great importance.
- In the fourth year the student returns to school; an area of specialisation is compulsory and the year is concluded with a qualifying examination and a final project.

After succesful completion of an IT, the student is granted the "ing"-degree (about the equivalent of a B.E. or a B.Tech.).

3 Authorities, Programs, Regulations

In the spirit of European Unification, Dutch and European authorities, universities and colleges believe that tertiary education must take an international position to maintain or to improve the level of education and to give the students the opportunity to prepare themselves for an international labour market [2].

For the benefit of this need the authorities set up a variety of international cooperation programs to stimulate internationalising activities with financial support for costs of development and extra costs due to movements (staff-travel, student and teacher mobility).

Different programs with specific objectives, but with respect to application, all characterized by the need of profound preparation, much paperwork and a long term application time (about one year ahead), and meant as a temporarily contribution to financial needs. The latter implicates that ultimately the involved institutes must finance their international activities themselves.

Since at the moment there are still many stimulating grants available, it is now the time to get things done, i.e. a policy and infrastructure favourable for internationalising activities.

Both authorities, the Dutch Ministry of Education/Sciences as well the European Task Force (Human Resources, Education, Training and Youth of the Commision of the European Communities), prefer a system of boundary exceeding networks. It is within those networks of higher educational institutes (in European connection all indicated as "universities") that the exchange of students, lecturers and curricula must be accomplished.

Such networks are regarded as the most effective instruments to shape and to promote international integration in education. They give intensive cooperations a clear structure and continuity. Networks make the system of contracts surveyable and the costs controllable [2].

Payments of grants for student mobility are effected through the National Grant Awarding Authorities (NGAAs, e.g. the Dutch Nuffic). Other payments of the EC-programs are effected through the coordinating institute; those of the Dutch funds through the so called HBO-raad (Council of Higher Professional Education).

Besides the network schemes there have always been and will be individual initiatives of students and lecturers. For those initiatives one can also apply for support of extra costs of traveling and housing.

Our department makes use of the following existing programs [3]:

* The Erasmus-program (European Community Action Scheme for the Mobility of University Students) of the EC is mainly focussed on enhancing the mobility of students within the European Community. Final mobility target is 10% of all higher education students (1991: about 3%) [4]. Erasmus grants are available since 1987-'88 and given for a study at a high school or university in another EC-member state or in one of the EFA (European Free-trade Association) countries. The program comprises four actions: Inter-university Cooperation Program (ICP), Student Mobility (SM), Teacher Mobility (TM) and Curriculum Development (CD).
* The LINGUA-program of the EC started in January 1990 and is exclusively aimed on the promotion of linguistic education. In this sense the program is often combined with Erasmus.
* In COMETT-cooperation programs (Community Program for Education and Training for Technology) of the EC, industry and education institutes jointly set up curricula and the program creates possibilities for industrial placements of students and detachments of staff.
* The TEMPUS-program (Trans European Mobility Scheme for University Students) of the EC offers financial support to institutes both education and industry, who want to co-operate with countries in the Middle and Eastern Europe to improve the higher education there.
* The STIR-program (Stimulating program for Internationalising of the higher education) of the Dutch Ministry of Education and Sciences is intended to speed up the Dutch internationalising activities and initiatives. The program started in 1988-'89 and comprises four actions: projects, staff-travel, student grants and guest lectures.

4 Initial And Current Activities

ERASMUS

Our first application for the Erasmus program has been in 1989 for the course 1990-'91 with Béthune, France on the subject: Power Electronics and Power Engineering. A modest network with rather old fashioned sounding subjects, which have probably been the reasons why the application was rejected.

In 1990 the program has been enlarged:
1. The network: with Kiel, Düsseldorf, Brussels, Charleroi and London.
2. The spectrum of the subject: "Ecological and economical handling of electrical energy"
3. The period of application: '91 through '94.

4. The number of actions of the program: intensive courses of one to three weeks, student mobility, teacher mobility, curriculum development and linguistic preparation.

This proposal was accepted, although not entirely. The most attractive part, the intensive courses, has, unfortunately, not been accepted.

Since January 1992 we have the first exchanges under this scheme:
- one English student from London to Leeuwarden for industrial work for six months.
- three Belgian students from Brussels and Charleroi to Leeuwarden for three and five months, respectively, following regular courses both theory and practice. (in Dutch!)
- one German student from Kiel to Leeuwarden for his final project for six months.
- one French student from Béthune to Leeuwarden for his final project for 2½ months.
- two Dutch students from Leeuwarden to Kiel for their final project for five months.

Every exchange requires two or three visits from at least one lecturer of the sending institute to the receiving institute: a preparing visit, an interim and/or an evaluating visit.

TEMPUS

Last year, 12th of june 1991, prof. Wilhelm Riesner from the Fachhochschule Zittau, Germany and member of the World Energy Council, has visited Leeuwarden on our request for a guest lecture on: "Energy and environment". He combined his guest lecture with a Tempus-grant for staff travel to investigate the possibilities for cooperation between Zittau and Dutch institutes.

This year (1992) several contacts resulted in two applications:
1. Cooperation amongst the institutes of Leeuwarden, Zittau, München, Gliwice (P), Liberece (P) and Prag (CS) and two partners of industry: IBM-Germany and Siemens A.G. with the objective of joint development of laboratories in the field of automation technology in the Euroregion "Dreiländereck" (PL, CS, G).
2. Cooperation of the institutes of Leeuwarden, Wilhelmshaven (G) and Budapest (H) with the objective of joint development of lectures and laboratories in the field of electrical measurements, in order to raise the level of education in this field to an European standard at the Kandó Kálmán Technical College in Budapest.

COMETT

In November 1992 we will deliver a contribution to a post-tertiary educational seminar in Wilhelmshaven (G) on variable speed drives for technicians/engineers in industry.

Through this invitation we have the opportunity to get acquainted with the conditions, regulations and organisation of a Comett-program.

STIR-HBO

Preliminary to the exchanges under the Erasmus scheme, two Dutch students went with STIR grants to Béthune for their final project of five months. Mainly due to misunderstanding they came back much earlier than planned with no result.

Fortunately, they have been able to accomplish the compulsory final project in a nearby Dutch hospital.

A few other students went by individual initiatives with STIR-grants, to Luxemburg and to England for industrial placement and final project, respectively.

5 Evaluation And Recommendations

The first period of exchanges has not yet terminated, hence to give a definite evaluation is premature.

For example, we still have to discuss standard of grading and/or how to judge the achievements of students.

However, it is possible to give a global review of expierences and effects; they are:
* a more perceivable set of international activities and an increasing interest of our students/lecturers within this context, due to the presence of foreign students;
* a refreshing look on our system and situation, caused by closer contacts with other educational systems, cultures and institutes;
* better communication amongst the people involved, students and lecturers, provided the basic problem of language can be adequately solved;
* favourable circumstances for entering into European networks caused by the establishment of a minimum requirement of infrastructure such as:
  - a central office for foreign affairs,
  - a course "survival Dutch Language",
  - a host student project; in this project a student of our institute is house holder of a split-level appartment and lives together with three foreign students; he/she helps the students from abroad to get familiar with the new environment.

Although a number of positive effects and results in the context of international cooperations can already be observed, we believe that some recommendations are still necessary:
* to avoid problems in communication and adjustments, it is preferable to give courses like Survival Dutch at the home institutes followed by a more advanced one in the receiving institute;
* we consider intensive courses as the most attractive part of the Erasmus scheme; in case this component will be rejected in the application for '92-'93, we will have to reconsider this specific partnership in Erasmus. The principal reason for reconsideration is the fact that the remaining parts can be realised in other programs for which higher budgets are available;
* while institutes and authorities are encouraging more students to go abroad, the grants per student are decreasing too fast (1992: 20% compared to 1991). Therefore it is necesary to find a way to stabilze the budget/student on an acceptable level;
* In general, is is advisable to be alert on the financial aspects in the (near) future; to illustrate our point of view we will take investments for international-like infrastructure, such as (foreign) student housing and the office for foreign affairs, as examples; presently, these infrastructures are financed by external funds, but in case of termination of these funds it is not yet clear which parts of the institute will be responsible for preservation; all departments or only those involved with international cooperation.

We believe that the continuation of international cooperations will still have many stimulating impulses of different nature on our curriculum, though external financial inputs are already diminishing.

As a final remark, we have to realize that within measurable time internationalising will form a substantial part of the total budget of our institute.

References

[1] Ministry of Education, "The Institutes of Technology in the Netherlands", Information Booklet, The Hague, 1984
[2] Mrs. A.S. Birkhoff, "Policy outlines on internationalization" (in Dutch), General Office of Education, Research, Innovation and Planning, Leeuwarden IT, March 1992
[3] Nuffic, Dutch organisation for international cooperation in higher education, "Nuffic Bulletin 1992, nr.1", The Hague, January 1992
[4] Erasmus bureau, "Erasmus and Lingua Action II Directory 1991/92", prepared for the European Commission of the European Communities, Task Force: Human Resources, Education, Training and Youth, Luxembourg, 1992

Masters Degrees for Professional Practice and Management

S.K. Al Naib

Department of Civil Engineering, University of East London

Abstract This paper discusses the new concept of Masters courses for professional development of engineers, particularly on a part-time basis, which are significantly different from the conventional specialized postgraduate awards. Such an approach has taken root in some universities in the United States but not in the UK Universities/Polytechnics. The article identifies the two major challenges of practicing engineers, namely design and management, corresponding to the two professional areas designated by Chartered Institutions for their membership examinations. Such broad courses provide the skills, techniques and approaches which may be applied to every professional challenge encountered. The discussion is illustrated with details of the MSc programme at the University of East London (UEL).

<u>Keywords</u>: Professional Practice and Management; International MSc Programmes; Curriculum Development; Design and Construction; Professional Examinations.

1 Importance of Graduate Study

The increasing complexity of many phases of engineering and construction and the rapid technical developments have created a strong demand for engineers with training beyond that included in undergraduate courses. Among the fields of work for which graduate study is desirable and for which it prepares the engineer for professional practice are : advanced planning, analysis and design, consulting practice, management and public administration, project organisation and contract procedures, research and development work in various specialised fields. Formal postgraduate work and participation in creative research enable the engineer with postgraduate training to go beyond the limitation of present practices and to contribute to the progress of the profession.

Internationally Masters degrees are becoming more and more the norm. This system of education allows the institutions to broaden their undergraduate courses and increase the depth of subjects with postgraduate studies. In Europe, four to six years of education has always been the norm. It is felt that British engineers with an MSc working in EEC countries from this year will be looked upon more favourably than to those with a degree only.

2 Masters Course Rationale

The concept of professional development degrees is not new. Such approaches have taken firm root in the United States of America. The idea is apparently not widely popular in the U.K., although a number of institutions have developed four year MEng courses with technical specialisation. Professional development degrees differ significantly from conventional post-graduate awards. Masters' programmes in engineering have proliferated in recent years, and these are usually in the form of advanced analytical study of a specialised branch of the discipline which the student has already studied to bachelor degree level. Their aim is to produce research workers and highly specialised practitioners who then fulfil specific roles in the engineering industry. It appears that only a small number of engineers deem it advisable to embark on a Master's level programme.

The concept of a professional development degree is different. The starting point is not so much the specialised discipline of the conventional approach, but rather the professional needs of the young practitioner. The professional context in which the engineer works provides the framework for the development of the degree programme. Hence the subject areas should be broader and more numerous than for a conventional programme, and the topics included should be those which the engineers meet daily in the context of their professional work.

For the civil engineer, the broad areas of the profession have been defined by the Institution of Civil Engineers (ICE) as TECHNICAL, examined at PE1, and PROFESSIONAL examined at PE2. Working within this broad framework a Masters Course in Engineering should identify these two major challenges facing civil engineers as DESIGN and MANAGEMENT, corresponding to the two areas designated by the I.C.E. Thus some course elements should cover the technical aspects of the design process, while others should develop the civil engineer as a manager.

3 Fields for Advanced Study

A Masters programme should complement undergraduate training in technical subjects and offer relevant areas of study in management and business to help engineers with their professional career. The following paragraphs suggest some possibilities in the non-technical field.

3.1 Accounting & Finance
Study in this subject should enable the engineer to examine the role of accounting in the administration and management of varied phases of business operations. Integrated with mathematical concepts, it increases the understanding of business decision.

3.2 Project Organisation and Management
Work in this area should involve the interrelationship of the technical, human and financial factors which affect the organisation and management of construction projects. It should discuss the choice of management structure, decision-making policy, team building, design and contractual relationships.

3.3 Construction Law and Contract Procedures
Study in this area should emphasize the legal aspects and institutions relevant to construction practice. It should deal with contract administration, performance of engineer-client and owner-contractor relationships, surety bonds and insurance.

3.4 Managerial Economics
This subject should synthesize those principles and findings of economics which are necessary to improved performance by the engineer. At the company level, the emphasis should be placed on economic analysis for decision making in the functional areas of planning, design and construction.

3.5 Human Resources and Industrial Relations
The focus should be on the factors in the successful management of the firm's human resources. The forces in the economic and political environment which have resulted in the organisation of labour and worker-employer relations including labour legislations and union powers should be explored.

3.6 Public Administration
Since the public sector determines the shape of the economy and the operation environment of private firms, general familiarity with public administration with special reference to local and central government is desirable for the construction professional.

3.7 Systems Modelling Techniques
This field entails the study of a broad range of mathematical model building and its application to the computer analysis and solution of engineering problems.

4 The UEL MSc Programme

4.1 Aims and Objectives
The University of East London Master's programme is designed to develop in students those qualities which reflect the need of modern industry for broadly educated engineers who possess a generalized point of view, adaptability to new situations and high degrees of analytical and management skills. The approach to professional education for engineers is made possible through the identification of concepts and techniques that are used in various areas of engineering practice. They constitute a framework of technical and decision-making tools, as useful in design and construction as they are in management and financial control.

4.2 Attendance Pattern

Arising from both external enquiries and requests from students, there was clear evidence of demand for the individual course modules to be offered on a semester basis, i.e. four hours of attendance per week for each module over half a session.

4.3 Course Structure and Organisation

The course is a linked Postgraduate Diploma/MSc and is offered on a full-time/part-time basis in two stages:

i Students will be required to attend the Polytechnic for four modules to complete stage 1, ie the Postgraduate Diploma course (PGD).

ii Subject to the recommendation of the Assessment Board, a student may, after completion of the stage 1 studies proceed to the stage 2 of the MSc course. Students will be required to attend the University and complete two further modules and submit a dissertation on a research based project undertaken.

4.4 Course Modules

A wide range of self-contained optional modules is currently available to allow the engineer to acquire advanced technical knowledge and management skills relevant to professional practice. The structure of the course permits the range of modules to be altered according to the needs of industry. The range has been gradually developing over a period of 18 years and caters for the wishes of engineers engaged in design and construction. The modules are: Applied Hydrology & Irrigation Engineering; Fluid Mechanics & Hydraulic Engineering; Public Health Engineering; Water Power & Dam Engineering; Engineers in Management & the Community; Design of Civil Engineering Works; Soil Mechanics & Foundation Engineering; Analysis & Design of Structures; Highway & Transportation Engineering; Computing - Technology & Applications.

4.5 Course Curriculum

The following table gives the number of hours for the part-time mode:-

Semester	Stage and Modules	Average Teaching (hours/ week)	Lecture (hours/ week)	Sem/Tut/ Pract (hours/ week)	Total (hours/ year)
	Stage 1				
1	Module A	4	2	2	96
	Module B	4	2	2	96
2	Module C	4	2	2	96
	Module D	4	2	2	96
	Total for Four Modules.(PGD)	16	8	8	384
	Stage 2				
3	Module E	4	2	2	96
	Module F	4	2	2	96
4	Project (MSc)	-	-	10	260

A to F are any of the modules listed in (4.4).

4.6 Management Training for Professional Practice
A module on the "Engineer in Management and the Community" is aimed to improve the communication skills of the engineer, to study management skills, finance and contracts.

4.7 Foreign Language for Personal Development
A number of languages e.g. (French/German/Spanish) are offered depending on available resources and student choice.

4.8 Research Dissertation
To qualify for the award of an MSc students are required to complete a satisfactory research dissertation of around 15000 words. Most of these research projects are industrially-based.

4.9 Awards
Students are normally required to satisfactorily complete four modules for the award of the Postgraduate Diploma, and six modules plus a research project to qualify for the award of the degree of the Master of Science Degree. Students who achieve a suitably high standard (normally an average of over 70%) would be

awarded the MSc with Distinction.

5 Professional Recognition and Benefits

The course and the module syllabuses have been assessed and approved by the Institution of Civil Engineers as a continuing education course at the two levels of Professional Examinations PE1 and PE2 following postgraduate training and practical experience.

6 Conclusion

Education for practice management recognizes the need for future professionals to develop their capacity to adjust to rapidly changing circumstances, and to have a generalized rather than specialist viewpoint. The skills required in managing complex construction projects demand those with ability to react flexibly and creatively. Not only must they understand the separate activities which make up the building process but they must be able to cope with the wide range of responsibilities which construction imposes on them. Education for a professional career in construction therefore focuses on the needs of industry in constant change and development. Thus it provides the techniques and approaches necessary for continuous development by the engineer, which are applied to every future challenge.

7 References

C R A C Graduate Studies (1991). Hobsons Publishers, London.
Higher Education in the United Kingdon (1990). Longman Press, London.
Higher Education in the European Community (1989). Denthscher Akademisher Anstranchoheimst, Bonn.
Comparative Guide to American Colleges (1979). Harper & Row, New York.

Internationalising the Technical Curriculum

S.M. Kazem, E.L. Widener

*Department of Mechanical Engineering Technology,
Purdue University West Lafayette*

Abstract
With freedom "breaking out all over," surely today's buzz-word in college circles is INTERNATIONALIZATION. This meld of altruism and self-preservation has been known variously as League of Nations, One World, United Nations, World-Wide Vision, Global Village and Planet Earth. What is new is an accelerating demand for firm programs and effective performance, endorsed by the university faculty and endowed with a sense of urgency. At Purdue University, the Technology Advisory Council on International Programs (TACIP) has active departmental committees (8) in the School of Technology. Actions since 1989 are described.

Keywords: Global Education, Communication, Graphics, Pedagogy.

1 Issues

Excellence in education clearly includes reducing illiteracy and innumeracy[1] everywhere, but to what ends? To industrialize the Third-World and revitalize the Super-Powers? To redistribute wealth and reduce conflict? To promote spirituality and suppress materialism? To control drugs and disease, crime and pollution? To solve ozone-depletion and polar-warming? To avert species-extinction and resource-exhaustion? Appropriately, the 1990 ASEE International Conference theme at Toronto was "Technology Advancement Through Global Interchange." Their 1990 Frontiers-In-Education Conference at Vienna selected a bilingual format for "Engineering Education 2000." The 1991 ASEE Conference theme at New Orleans was "Challenge of a Changing World." And their 1992 theme is "Creativity by World-Class Engineers."

2 History

Members of ASEE (American Society for Engineering Education) are aware of chronic debates between "fine-tuners" versus "over-haulers" of our primary schools, now called "America's 200 Year Old Conundrum."[2] Horace Mann (mid 1800's) pushed for statewide systems, and Charles Eliot (late 1800's) promoted a free nation-wide system, but only 10% graduated from high-school. John Dewey (early 1900's) espoused "progressive" curricula with vocational emphasis, along with immigration laws and child-labor legislation. The "Sputnik" era (1950's) then hastened the evolution of "bifurcated" engineering education into separate engineering and technology curricula. Thereafter, civil-rights reforms (1960's) increased the pool of eligible students, but social excesses (1970's and 80's) escalated the dropouts.

Since World War II, our programs to INTERNATIONALIZE technical education were sometimes seen as paternal gestures by the "Ugly American," flaunting affluence. Nevertheless, an "open-door" policy attracted diverse and gifted visitors, yearning to be free and planning to remain. The "melting-pot" process also maintained English as our common language. Ironically, as current events bring the "decline of dictators" abroad, there are concurrent calls for "domestic czars and federal plans" in the United States. Therefore, our foreign graduates go home in ever-increasing numbers, while our domestic students are slow to enroll and to graduate. Fear of foreign competition now signals a return to academic rigor for the U.S. in the 1990's. The present climate of competition is best illustrated with a "truestory": A buyer solicited bids for finished parts, specifying "a maximum of 3-defects" per 10-thousand parts. A successful supplier from overseas punctuated the low bid with a shipment of 9,997 parts labelled "zero defects." So much for "normal confidence" of 99.73% (plus or minus 3-sigma) or 27-defects in 10-thousand. That's history.

3 Tactics

Worldwide technology in the 1990's brings challenges of growth, diversity, and method to engineering education[3]. Schools should be alert to adopt new courses and to adjust existing contents, correlating with advances in other fields. A variety of choices characterizes any given science or technical discipline: a) fundamental or corollary; b) essential or incidental; c) important or informative; d) complex or elementary . However, a decreasing time-frame has forced faculty to be more selective. This involves personal experience, independent judgment, textbook reviews, and alternative curricula as never before.

Imperial China, a world-class society with a great-wall philosophy, is the classic case of defensive isolation and subsequent stagnation. Germany and Japan are converse classics, where intellectual vigor and technical excellence bring renaissance from ruin. Just how do we deal with their conspicuous lack of atomic weaponry? As international exigencies escalate, nationalism and isolationism seem less acceptable. Recent divisions of the U.S.S.R. and deficits of the U.S.A. are reflections of misdirected technology. Meanwhile the U.S.A. is asleep in metrication procrastination; sickly with litigation paralysis; suffering from technical innumeracy and ethical evaporation.[4]

Nonetheless, disturbing cries now arise in the U.S. to buy American and stiffen tariffs; to boycott unfair traders and ban foreign investors; to restrict immigration and deport aliens; to cap enrollments of foreign students and eradicate accents from teaching assistants. Wiser choices involve more domestic recruiting and increased technical enrollments; less attention to accents and more appreciation of differences; improved retention of students as well as more graduates.

4 Technical Development

After World War II, President Truman's U.S. program, "Point Four", established vocational schools and engineering colleges overseas. Renamed "International Cooperation Administration (ICA)," and then "U.S. Agency for International Development (US/AID)," this program is still active. Similar activities by British, French, and German administrations have accelerated technical training in emerging nations.

Purdue has long maintained an Office of International Programs, headed by a Dean, to facilitate agriculture engineering projects in other lands. Also, there is an Office of International Student Services, plus an International Center, to assist visiting students. However, in November 1989, a different need was recognized. TACIP, the "Technology Advisory Council on International Programs," was initiated.

With sub-committees now active in all eight departments of the School of Technology, their emphasis is now upon preparing our graduates to serve the world-wide marketplace; i.e., to compete for jobs in a global economy. Faculty members, drawing on industry and college contacts, suggested new coursework and improved procedures. These lead to a "Global Initiatives Faculty Grants Program," which have produced summer placements in Germany for two professors. Next, a 3-hour course was developed as a 1993 elective (Technology from A Global Perspective). A third professor was invited to visit Japan, as fluid-power consultant. Recently, a "Japanese-Interest Group" has been organized, to utilize local employees of Subaru-Isuzu (SIA plant) in Lafayette.

Modern pedagogy in the European Economic Community (EEC) embraces some 15-thousand students from twelve member-nations in nine schools of Western Europe. Multi-lingual fluency, cultural exchange and muted nationalism are hallmarks.[5] Programs for our exchange students and coop employees should expand into institutions and enterprises beyond U.S. boundaries. With Hispanics becoming our largest minority, technology students can cultivate cultural and commercial understanding with Central and South American programs. The progress toward consistent degrees, international accreditation, universal standards and global technology should accelerate.[6]

In a lighter vein, high-schools and colleges support popular initiatives for "Technology Olympics," with international contests to come: Rube Goldberg (silly complexity); Go-carts (engine performance); Pedal-cars (vehicle drag force); Bridge trusses (balsa wood, cardboard); Calculator skills (reverse polish, abacus); Simple engines (mouse trap, rubber band).

5 Communication

If French is the discourse of diplomacy, English is becoming the international language of commerce. However, leading educators and industrialists now find non-native languages to be necessities, not luxuries, to develop international markets. Technologists probably need to study a second language and develop fluency through continuing education; to discover media for accelerated study of new tongues as needed; to have reading-writing-speaking skills in their native tongue; to encourage neighbors and colleagues to retain and share native fluency; to learn idioms and colloquial words. Therefore, it seems necessary to add a foreign language to technical curricula, especially for graduate degrees; to expect proficiency in several foreign languages from the faculty; to evaluate such proficiency in hiring and promoting staff.

Suggested ways to a second language are:
a) Rejuvenate the language department in each high-school.
b) Restore the language requirements for a high-school diploma.
c) Include foreign terms and phrases in college-entrance tests.
d) Require other language competency tests for college graduation.
e) Include foreign-language questions in professional license (PE) exams.
f) Encourage students to learn the language and culture of a country likely to provide job opportunities.
g) Exchange ideas and information via computer-networks and translator-software.
h) Provide studio facilities for satellite transmission/reception of overseas education.
i) Promote international exchange programs for students and staff.
j) Teach songs and poems, directed toward thinking and dreaming in a second language.
k) Seek "total immersion" in that chosen language to enhance pronunciation and improve retention.

Each department should have a computer facility for independent study via language software, readily available. Language scholarships and industrial subsidies could swell the ranks of exchange-students and summer-scholars. Concentrated study of a second language should appeal to summer students and faculty. Using mnemonics, rhythm, melody, and drama, courses can turn letters and words into pictures, for easier retention of Japanese, Arabic, or Russian.

6 Graphics

Technical graphics already is an international language for "blueprint-readers" in design, production, and marketing. More attention to mechanical drawing is needed in high school, to reduce remedial coursework in colleges and to reduce technical illiteracy in general. Although letters and numbers are taught early in grade-school, simple legibility tends to be neglected in college. Worse yet, mistakes and waste arise from confusing symbols and careless lettering:
a. "ϕ" may be Phi (Greek), Phase (electric) or Zero (number).
b. "1" may be One (number), El or Eye (letters).
c. "6" may be Six or Zero (numbers); Oh or Bee (letters).
d. "W in lb" may be "weight in pounds" or "work, inch-pounds."
e. "10 lbs" may look like 10,165. Is the comma a decimal point?
f. Easily confused are 4 and 9, D and P, 1 and 7, 2 and Q or Z.

The myriad of fonts and serifs poses a serious problem. Europeans tend to use extra strokes: 1 (one), 7 (seven), Z (zee). Telegraph operators, draftsmen, and computer operators have disagreed on conventions. Word-processing equipment has various styles of type, print, and display; some are faster to scribe, some artistic but illegible.

We need universal technical standards, where single strokes (0,1) only represent <u>numbers</u>, most used by <u>quantitative</u> problem-solvers. Then, Greek letters (hand-scribed and printed) could be standardized (both upper and lower cases) to extend the selection of "key symbols" beyond Roman letters for technical formulas, units, and properties. Countries disagree on the meanings of million-billion-trillion, by factors of thousands. Is "micron" a millionth of a metre, or of an inch? Is "mil" a thousandth of a metre, of an inch, or wire area? What is a pound or a ton? Is weight a mass or a force? Yes, technology is difficult, but too many find it impossible!

7 Books

Give special attention to basic mathematics and science, introducing just enough theory to elucidate practical applications and develop general solutions. However, textbooks must not over-simplify concepts and sacrifice clarity or exactness. The expository style of J. Willard Gibbs (1839-1903 mathematician and physicist) is a balance of efficiency and precision. General equations should precede specialized forms. Unfortunately, the recent trend in academia is toward too many courses and topics, too little time, larger enrollment, and textbooks. Excessive length with extreme rigor is debatable; abridged and incomplete coverage is excusable; but excessive length plus generality is unacceptable. International consensus on model textbooks and consistent curricula is not yet here. International transfer of college credits and equivalent degrees is improving.

Let us revive, revise, and translate our classic and concise textbooks. Comparison of curricula from various countries is also desirable: teacher-student ratios; size of classes; laboratory emphasis; semester-trimester-quarter systems; years and credits to graduate; high-school preparation; graduate-school enrollment; typical technology job-titles. More Faculty can compare textbooks from other nations, comparing difficulty and scope. This includes computer software, videotapes, and industrial standards. Neglected areas (quality control, waste disposal, space flight) benefit by guest lecturers from abroad.

8 Conclusion

Ethnic humor has for years been popular in all cultures. Comedians exploit jokes to denigrate and devastate opponents and strangers, politics and religions. Civil rights advocates use legislation and litigation, to control contempt with contention. Ted Turner (Cable Network News) banned the word "foreign," implying estrangement and misunderstanding; "global" and "international" are preferred, as implying unity and friendship. Offending reporters incur fines ($100) which are sent to UNICEF, the United Nations Children's Fund.[7]

However, anyone can decide today to replace petulance with patience; to covet peace and prosperity, not petty grudges nor festering feuds. While hatreds separate Israel and Palestine, Iran and Iraq, Serbia and Croatia, even Dublin and Belfast, the efforts of academicians to "fine-tune" their profession seem puny and futile.[8,9] Nonetheless, INTERNATIONALIZING is inevitable for technology graduates to be competitive.[10] It is most desirable for pooling global knowledge, talent, and resources. Tolerance can bring understanding, then greater respect and fuller appreciation, in Century 21.

References

1. Paulos, J. (1988) "Innumeracy: Mathematical Illiteracy," Hill and Wang, NY.
2. Toch, T. and Cooper, M. (1990) "Lessons from the Trenches," U.S.News and World Report, v 108, 8, p 50.
3. Wakeland, H. (1987) "Internationalization of Engineering Education," ASEE Annual Proceedings, p 1156.
4. Threlkeld, D. (1989) "Metric Standardization and Global Market," MAPI Economic Report (ER-144) 1200-18th ST. NW., Wash., DC.
5. Mapes, G. (1990) "Polyglot Students Weaned Early Off Mother Tongues," N.Y. Wall Street Journal, Mar. 6.
6. Sissom, L. (1990) "U.K. Engineering Accreditation," ASME News, (International Windows), v 9, 10.
7. Turner, T. (1990) "Let's Not Alienate," Chicago Tribune, Mar. 16.
8. Kazem, S. and Widener, E.(1990) "Barriers to Technical Education," F.I.E. Proceedings, Vienna, Austria.
9. Kazem, S. and Widener, E. (1987) "Standardizing the U.S. Curriculum," ASEE Annual Proceedings, Reno, NV.
10. Widener, E. (1991) "Manufacturing Designs: Competition or Chaos", SME Autofacts Proceedings, Chicago, IL.

Experiences in Providing Educational Support for a Developing Country

W.A. Barraclough (*), M.D. Bramhall (**),
Z.A. Famokun (***), R.G. Harris (*)
(*) School of Engineering Information Technology, Sheffield City Polytechnic
(**) School of Engineering, Sheffield City Polytechnic
(***) Department of Electrical Engineering, Federal Polytechnic, Ado-Ekiti

Abstract
This paper describes the contribution made to the World Bank Technical Education Project (Nigeria) by staff of Sheffield City Polytechnic. This involved both development of local staff and commissioning of much of the equipment provided by the Project. The effectiveness of the contribution is appraised and recommendations made for future work.

Keywords: Developing Countries, Technical Education, Educational Support

1 Introduction

During 1991 Sheffield City Polytechnic (SCP) was involved in providing educational support as part of the Nigerian Technical Education Project funded by the World Bank. The project was to upgrade the technical teaching facilities in four selected Federal Polytechnics. This was to be done by the provision of a large quantity of modern equipment; of fellowships for the training of Nigerian staff in the U.K.; of consultants from the U.K. to work at Nigerian Polytechnics. The project was multi-disciplinary, SCP providing support in the areas of electronics, telecommunications, materials and mechanical engineering. A particular aim was to increase and improve the practical elements of the relevant programmes and their relevance to the needs of industry in Nigeria.

The project was financed by a loan to the Nigerian Government from the World Bank. The equipment was procured by the International Labour Office based on lists of requirements initially drawn up by the Nigerian Polytechnics involved. These requirements were modified in the light of financial and practical constraints.

The British Council had tendered for the project in partnership with a consortium of U.K. Polytechnics. They provided the overall management of the educational aspects of the project. The individual polytechnics provided the specialist expertise through consultants from their teaching staff.

Fig 1 shows the relationships between the various bodies involved.

2 Role of SCP

The fundamental role of SCP was to provide the support necessary to ensure that the large volume equipment arriving at each Nigerian Polytechnic could be efficiently used. The major element of this was staff training. This was to be undertaken at SCP prior to the equipment arrival, and in Nigeria subsequently. The educational consultants had the task of preparing for the arrival of the equipment; commissioning the equipment; providing the on-site staff training; and developing laboratory work based on the new equipment.

Various support activities were also undertaken, such as curriculum development, production of laboratory exercises and teaching notes, investigation of local sources of supply, provision of safety advice, development of laboratory procedures etc.

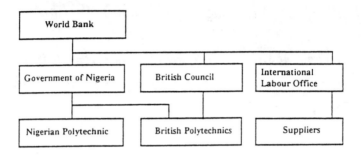

Figure 1

3 Installation and Commissioning of Equipment

The equipment arrived at the Polytechnics in containers. Unloading and subsequent unpacking of equipment was a major time-consuming task involving technicians, academics and consultants. Timing of this would have been better prior to the arrival of the consultants so that more time was available for installation and commissioning of equipment.

Difficulties were encountered in unpacking due to local resources. Only one fork-lift truck for example at most of the Polytechnics.

Installation and commissioning was achieved on only a small proportion of the new equipment for various reasons:-

- Unfinished laboratories.
- Inadequate power supply to laboratories.
- Lack of water supply.
- Missing parts of equipment/instruction manuals.
- Lack of consumables.
- Incorrect voltage of new equipment.
- Inadequate time for the volume to be dealt with.

It would therefore be recommended that for future projects of this kind that some of the resources for the project should be spent on upgrading laboratories to the required standard ready for arrival of equipment. Equipment suppliers should take care to supply the correct specification equipment and ensure there are no missing parts. Instruction manuals should be comprehensive and perhaps specially written for self-installation. In a European country a specialist engineer from the supplier of the equipment would install and commission. Consultants on this project were working from manuals supplied as if the equipment was already installed by a company engineer!

4 Development of Nigerian Technician and Academic Staff

It was recognised from the outset of the project that long term success would only be possible if training was available for both academic and technician staff. It is useful to examine the knowledge of the staff prior to their training. Most have good academic qualifications, the academic staff commonly having degrees and the technologist staff National Diplomas or Higher Diplomas. However staff needed more training in practical skills. The practical standards are likely to be low due to the lack of equipment previously available, and the highly academic nature of Nigerian courses. This led to the need to adopt training of these staff to a more practical rather than theoretical nature.

Training was given to both groups of staff. It was to provide the technologists with the skills to:

- Understand the modern, and in some cases, sophisticated equipment being supplied.
- Allow the technical staff to operate and demonstrate the equipment and to generally support academic staff.
- Monitor the condition of the equipment and to service and maintain it.

Additionally, the training had the task of helping the technicians to develop suitable procedures for maintenance, laboratory organisation, and the construction of demonstration equipment.

For the academic staff the aims were similar except that a more detailed knowledge of the underlying principles of operation were needed.

Prior to the arrival of the technicians in the U.K., meetings were held at SCP to determine the optimum strategy for providing the training. This exercise involved the consultants who would eventually go to Nigeria, as well as other academic staff members with a particularly relevant expertise. It was decided that the actual plan of work for a particular individual would be decided by discussion between them and the consultants. This would allow flexibility in meeting the individual's wishes and use their knowledge of the equipment that they anticipated receiving. The general techniques that were used were:

- Project work, which allows the freedom to explore a chosen particular topic in detail.
- Working with SCP technicians to gain an appreciation of the methods and skills used.
- Exploratory tasks, for example fault finding on equipment and characterisation of systems.

In practice a significant level of support was needed, although it had been initially envisaged that much of the work could be self managed. The support and supervision was provided by both academic and technician staff at SCP as appropriate. SCP staff thus gained a great deal of expertise in training staff from a developing country.

The training at SCP in general met the original aims although the depth that was achieved was less than that originally envisaged. The reasons for this are not easy to determine but some points emerge.

The pace and hours of work in the UK were somewhat higher than is typical in a Nigerian Polytechnic.

- The environment was initially unfamiliar to the Fellows and a substantial "settling in" period was needed.
- The UK weather during the period (winter) led to many minor ailments such as coughs and colds.
- In a few cases, a fundamental understanding of the underlying principles was lacking.
- Some areas of the work (particularly computing) required new concepts. These tended to be only slowly formed and required careful development.

More specific training was provided on-site in Nigeria. This had the primary aim of providing a structured introduction to the new equipment that had been commissioned by the consultants. In practice this was hampered by the late arrival of the equipment. During this time investigations of existing facilities and discussions with the staff proved valuable in forming a clear picture of what was required. Consequently, time was spent developing work based upon existing facilities and equipment as these were not being used to their full potential. Short courses, seminars, demonstrations and practical workshop sessions were used to provide training and academic updating. These were designed to put the skills acquired in the U.K. into the context of the local environment. It must be pointed out that attendance on the short courses was often erratic because of other pressures on the local staff. A practical solution to this was to design the courses so that they could be used in a self-learning manner. The staff could then work at their own rate, and when they had the opportunity. The consultants then acted as a resource, providing advice and information where needed. Experience shows that it would not have been possible to prepare the material prior to arrival in Nigeria. Local needs and facilities are highly variable, the requirement is for the consultants to be flexible and resourceful and to devise the work as necessary.

All staff showed a great enthusiasm for the new work and an eagerness to acquire as much knowledge as possible. In many cases there was a marked reluctance to actually use the new equipment and to experiment with its facilities. This is understandable as the equipment is new, expensive and unfamiliar, resulting in the fear of damaging it and then being blamed. Overcoming this is not difficult. Demonstrations can be given to show how easy the equipment is to use and how hard it is to damage it. Laboratory sheets that have been written primarily for student use can be worked through as well. These methods allow access in a controlled way and gradually build up confidence in the ability to use a particular item.

5   Curriculum

One of the intended aims of the project was to examine and where necessary develop the curriculum. Upon examining the curriculum as defined by the Nigerian Board for Technical Education (NBTE), there was obviously little room for change. The curriculum is prescriptive and defines the laboratory work that should be done and the equipment that should be available. The equipment supplied as part of the project should have ensured that all the NBTE required equipment was available. In practice budget cuts within the project meant that this was not the case. However, the shortfalls are unlikely to be significant.

The only area in which it is possible to have an impact is in the delivery of the curriculum. A more practically based approach, aided by the new equipment, should now be possible.

Accreditation for running HND courses was a major concern of the individual institutions. The provision of the new equipment and the associated staff training seems likely to lead to successful accreditation. Whilst the NBTE documentation gave reasonable guidance on the basic requirements for accreditation in terms of equipment and facilities, there seemed to be a lack of "case law" on these and other requirements.

6   Industrial Links and Local Resources

Industrial links were actively encouraged, through numerous visits to local manufacturing companies. These links are a source of local specialist knowledge and can be a useful resource to both staff and students. In addition, valuable training placements were obtained for students through such links.

The use of local resources for consumables was also encouraged, rather than import from European countries. For example, the Foundry at the Polytechnic at Afikpo needed sand for laboratory castings. Local sands were therefore analysed for clay content and particle size. Local foundries were visited to determine their sources of sand. In many cases free samples were obtained as well as continued offers of support for the Polytechnic.

Links with other educational institutions were also encouraged.

7   Long Term Support

Long-term support is necessary if the project is to be a continued success. Staff require further specialist training on the new equipment through further short-term consultancies or through training from equipment suppliers. Local training in some instances could be possible through collaboration with other academic institutions that have similar equipment.

It is hoped that continued links between SCP and some of the Polytechnics in Nigeria could be maintained with staff and student exchanges, provided that financial backing can be found.

8   Conclusions

The project was successful in that the Nigerian Polytechnics involved have been able to develop. The provision of new equipment and expertise have been gained in a relatively short space of time. The four polytechnics are now in a position to build on this success.

SCP has gained a great deal of expertise in supporting the training of staff through this project. For the long term further support is necessary for these Nigerian Polytechnics. SCP is therefore talking to the British Council about long term links. Ongoing assistance, even at a low-key level, will greatly help in maintaining the momentum and ensuring that final success is achieved.

9   Acknowledgements

The authors would like to record their appreciation of the support received from the World Bank; the British Council both in the U.K. and in Nigeria; the Nigerian Federal Ministry of Education via its Project Implementation Unit; Sheffield City Polytechnic; and the staffs of the Federal Polytechnics of Ado-Ekiti and Afikpo.

10   References

NBTE (1987) National Diploma in Mechanical Engineering syllabus, NBTE, Kaduna, Nigeria.

NBTE (1988) National Diploma in Electrical and Electronic Engineering syllabus, NBTE, Kaduna Nigeria.

Some Experiences of Curriculum Development and Technology Transfer between the U.K. and Developing Countries

T.I. Pritchard (*), J.M. Bullingham (*), D.G. Rivers (**)
(*) School of Engineering, Polytechnic of Huddersfield
(**) International Office, Polytechnic of Huddersfield

Abstract
Experiences in co-operation over many years between a Polytechnic in the UK and institutions in three developing countries, Egypt, Indonesia and Malawi are described. The co-operation covers aspects of staff development, curriculum development, assistance with equipment as well as aspects of validation and quality assurance procedures. The paper also describes procedures and some necessary steps for the transfer of technology to developing countries to be successful.
Keywords: Technology Transfer, Curriculum Development, Developing Countries, UK, Polytechnics.

1 Introduction

The Polytechnic of Huddersfield has a long history of involvement with the development of technical and vocational education and training systems in other countries.
The early focus of these activities was human resource development both in the United Kingdom and overseas. This emphasis was mainly due to the role of the School of Education as one of the four national centres for the training of technical and vocational teachers in England. From this well established foundation Huddersfield became the main United Kingdom resource in the establishment of the Colombo Plan Staff College in Singapore and then in the early 1970's the four national Technical Teacher Training Institutes in India at Madras, Bhopal, Calcutta and Chandigarh. The School of Education provided a series of staff on long term secondments to assist in these developments.
As the Polytechnic grew in size and diversity there were more opportunities for its staff to contribute to development programmes. This expansion has involved a wider range of countries and a more diverse spread of project types including engineering.
The International Office, which was formerly called the Centre for International Technical Education, has during the last fifteen years conducted courses and consultancies in over forty countries.

Currently we are involved in projects in ten countries, these initiatives being funded by national Governments, the British Government Overseas Development Administration, UNESCO, the Asian Development Bank, The World Bank and the European Development Fund.

In brief the current projects include an in-service training programme for science laboratory technicians in Jordan; the training of technical teachers in Ethiopia which has extended over the last nine years; the institutional development of the Gambia Technical Training Institute; staff and curriculum development for four University Faculties of Technical Education in Turkey and the installation and commissioning of equipment with related curriculum and staff development at four Polytechnics in Nigeria. In the following sections a more detailed description is given of three other typical initiatives involving engineering where the Polytechnic of Huddersfield has played a leading role.

2 Activities in Egypt

The Polytechnic of Huddersfield has had links with higher education in Egypt for many years. In recent years, however, our closest links have been with Benha Higher Institute of Technology (BHIT). Benha is a town with a population of about 500,000, situated on the banks of the Nile, approximately 45 kilometres north of Cairo.

In 1979 the UK Government set up a Committee of Enquiry into the Engineering Profession (The Finniston Committee) to review amongst other things "the requirements of British industry for professional and technician engineers". Similarly the Egyptian Ministry of Higher Education (EMHE) started a debate in the mid-80's in an attempt to produce graduate engineers better able to breach the gap between applied science (academic engineering) and real engineering (industrial engineering). Just as Finniston concluded that the traditional analytical type of BSc should give way to BEng/MEng courses, "orientated to the synthesis, application and practice of engineering in the environment and incorporate elements of practical training and project work (EA1 and EA2)", so the EMHE concluded that their traditional methods of training engineers were also in need of similar modification. However, instead of asking the existing and well established Egyptian Universities to address these perceived deficiencies a more radical approach was adopted and a number (3) of new 'Institutes of Technology' were established. Hence the Benha Higher Institute of Technology was formally opened in 1988.

The new institutions were expected:

1   to be complementary to the Universities,

2   to be equal to and different from the Universities,

3   to have parity of esteem with the Universities,

4   to be comprehensive Higher Education Institutions providing both full-time and part-time higher education,

5   to provide degree and higher degree courses equal in quality and standard to those provided in the Universities.

The above specification will be very familiar to those who know UK Polytechnics and is no doubt the main reason why a UK Polytechnic was selected to participate in the development of the new institutions.

The courses offered at BHIT are 5-year programmes leading to BSc(Eng) in Technology degrees in 3 general areas:

1   Civil Engineering Technology
2   Electronic Engineering Technology
3   Mechanical Engineering Technology

Approximately 300 students per year are recruited on to a joint first year (first intake September 1988) but not all students stay for the full 5 years. There is an opportunity after the 3rd year for some students to terminate their studies and be awarded a Higher Technology Diploma. Because we do not have Civil Engineering at Huddersfield we have confined our advice to the Electronic and Mechanical Engineering routes. To date our contributions have included:

1   The Polytechnic Rector is a member of the Institute of Board of Trustees and has given advice on the setting up and organisation of the Institution.

2   Visits from 4 senior academic staff from the Polytechnic to BHIT, on 3 separate occasions, to contribute to the discussions on course structures and content.

3   Advice on assessment, course management, peer review and quality control, in line with traditional CNAA methods.

4   Advice on workshop, laboratory and computer facilities.

5   Visits to Egyptian companies to advise on industrial training requirements and the benefits of strong academic-industrial links.

6   Provision of staff development for BHIT staff through postgraduate degree programmes, industrial visits, and direct observation of our teaching programmes, in the UK.

7   General advise to BHIT staff on-going UK developments in engineering education.

All of these activities have been pursued with the financial support and the collaboration of The British Council.

3 Activities in Indonesia

Indonesia is a large country consisting of over 3000 islands and has the fifth largest population in the world, 155m and is growing at an annual rate of 2.3%. Current estimates of its population is over 215m by the end of the century. The country by its position on the Pacific rim is determined to become a major power and is currently going through a rapid period of industrial expansion and a major problem limiting this is the lack of trained people. In order to overcome this over the last ten years there have been a series of large projects funded by the World Bank amongst others. One area of our involvement has been in the establishment of around 20 Polytechnics throughout the country starting with the Polytechnic Mechanic Swiss established with help from the Swiss Government in 1976. Our involvement with bodies such as The British Council has included:

1. a major project over 3 years to train 60 engineering students to degree level in Engineering (Mechanical and Electrical) who then returned to become Polytechnic staff

2. curriculum development with the centralised Polytechnic Educational Development Centre (PEDC) at Bandung which develops the curriculum, equipment procurement and staff training

3. carrying out a series of short consultancies (typically involving 3 staff for a period of 4 weeks) on aspects of the development of the polytechnics such as management, quality and future direction.

Overall this has benefitted Indonesia in that it has been possible to transfer technology to a developing country. It has also been useful from our point-of-view to be able to take a small part in the development of technical education in a large country whose total student base, including Universities both Public and Private, and Polytechnics now exceeds 1.8 million students.

4 Activities in Malawi

The Polytechnic of Huddersfield has had links with the University of Malawi - The Polytechnic at Blantyre for around 16 years. The University is the only Higher Education Institution in Malawi which teaches Engineering up to degree level. Initially the Polytechnic provided staff development and training in teaching methods in Malawi for Malawians newly appointed to the staff. Selected staff were then trained in Huddersfield to degree level with a few subsequently pursuing Masters qualifications.

The link then moved naturally on to senior engineering staff from Huddersfield visiting Malawi to advise on course content and carry out examining activities on an informal basis and senior staff from Malawi visiting Huddersfield to look at the way we teach Engineering in the UK and to be involved in curriculum development. More recently, since 1990, a formal link exists between us, initially for a period of 5 years, sponsored by the Committee for International Co-operation in Higher Education of The British Council. The general aims of this link are to assist in the development of a national cadre of high level human resources qualified in the three main branches of Engineering (Civil, Electrical and Mechanical) and capable of contributing specialist skills to the implementation of the Malawi Government policy in engineering education.

At a more detailed level the following wider objectives have been set:

1 to use UK expertise in curriculum, syllabus, laboratory and staff development and monitoring academic standards.

2 to arrange visits to Malawi by UK staff. During such visits, the staff:-

 hold meetings with Polytechnic staff to discuss proposed changes in curricula and associated laboratory and staff development;

 give staff and student seminars in their own fields of expertise;

 give academic guidance to local staff seeking to acquire higher degrees by research and, where applicable, to act as (or identify) supervisors for such candidates;

 discuss collaborative research projects between staff in Malawi and the UK.

3 to arrange visits to appropriate institutions in the UK by selected Malawian academic staff to gain first-hand experience relevant to the short and long term objectives in curriculum development in Malawi.

4 to assist in identifying and purchasing minor items of teaching equipment and reference materials from such funds as are available in the link arrangement.

The Link has now been in place for nearly two years and to date four staff visits (each of around two weeks duration) have been made from the UK and five staff visits (each of about four weeks duration) have been made by Malawi Polytechnic staff. It is the view of all concerned that the Link is generally working well and is achieving its objectives. Amongst specific objectives which are producing concrete results in Malawi are the conversion of the existing BEng courses in Malawi to a modular-structure and, at the same time, the addition of an Honours stream. The Link is also helpful in overcoming the difficulty in Malawi in obtaining a local supply of spare parts and components such as electronic components. The facility under the Link for us to purchase some such resources in the

UK and ship them out to Malawi is seen as very useful by colleagues in Malawi to support laboratory and project activities. On the negative side the rate of progress is obviously limited by the resources allocated to the Link resulting in a limited number of visits by staff each way and the level of support in terms of purchases etc.

The activities under the Link are reviewed annually and changes made as necessary, subject to British Council approval.

5 Conclusions

This paper has attempted to show how curriculum development and technology transfer has been carried out over many years between the Polytechnic of Huddersfield, and quite different higher education institutions in three other countries. It is the view of all those concerned here at the Polytechnic and at the British Council, and overseas that the process has been very successful and worthwhile and it is our intention to carry on and develop more such Links in the future.

6 Acknowledgements

We would like to acknowledge the help of colleagues at the Polytechnic of Huddersfield, the British Council and overseas without whose assistance and co-operation over many years this work would not have been possible.

Working Together to Update the Engineering Profession - The East-West Distance Education Project

M. Markkula (*), A. Hagström (**)
(*) *Centre for Continuing Education, Helsinki University of Technology*
(**) *International Association for Continuing Engineering Education, IACEE*

Abstract
This paper describes a pilot project initiated by the International Association for Continuing Engineering Education, IACEE, to provide video-based courses and other distance education material to serve the educational needs in management, technology, and environment in the former USSR. The project utilizes existing western programs and delivers them through a network of distance education centers to be established throughout the Commonwealth of Independent States. Course selection procedures, delivery mechanisms, learner support, and project organization are described.
Keywords: Professional Updating, Distance Education, Video-based Training, East–West Cooperation.

Introduction

Bringing the economy of the former Soviet republics up to western standards of quality, productivity, free market management, and environmental protection will require a substantial amount of training and retraining of scientists, engineers, managers and technical workers. Serving the millions of engineers and scientists in the former USSR with a western level of basic management training and continuing professional education will require a major effort within the continuing education system of the new countries.

The International Association for Continuing Engineering Education, IACEE, has started a project called the "East–West Distance Education Project" to develop an economic and effective way of delivering the quantity of training and education required in the Commonwealth of Independent States (CIS). The project extends current western video-based instruction networks into the CIS, initially using videotapes. At a later stage, satellite delivery and other media, such as audiographics, will be considered.

The objectives of the Project have been summarized as follows:
• to help the countries of the CIS adapt to the free market system
• to develop an equivalent of NTU or EuroPACE in the CIS
• to provide continuing education for thousands of engineers and managers
• to help the recipient countries and organizations deal with environmental issues, product quality, and prepare for western joint ventures
• to develop the local distance education production capability.

The project will also be used as a model for future IACEE projects in Asia and Latin America.

Project Concept

This project will enable professional engineers, scientists, and managers in the former USSR to participate in courses offered by satellite consortia and individual western universities. By providing access to these courses and a mechanism for selecting, administering and distributing them, this project will produce a foundation of courses which can be utilized directly by the recipient universities and also incorporated into other, more specialized training and assistance programs.

The project goal is to provide video based courses to serve most economically the education needs of the CIS in management, technology and environment and to assist in integrating these courses into local curricula or structuring new joint programs and curricula as needed.

Western courses are utilized within the project in two ways: as a part of the regular programs of undergraduate, graduate and continuing education at the receiving universities; and as a basis for new programs of study tailored to the needs in the CIS of emerging businesses, international joint ventures, and government agencies with changing roles and responsibilities.

As the project develops, it will not only provide technical assistance, but will develop working relationships between the consortia and the participating CIS universities. These universities will thus become producers of video-based courses as well as consumers and eventually full participants in multinational educational consortia.

Integration of the video-based courses delivered by the participating consortia with courses delivered by other media and by other organizations is also being considered. This could include locally produced classroom courses as well as courses delivered by other means, such as print, audio, audiographics or computer conferencing.

The project will also serve to build the necessary infrastructure to support local continuing education and develop the video production capabilities which will allow locally produced courses to be fed back to the consortia for international distribution.This will mean developing the necessary organizational and administrative mechanisms and telecommunications infrastructure throughout the CIS. Organizational development, staff training and hardware installation are necessary to create the required administrative mechanism to provide for course selection and learner support as well as overall coordination, budgeting and evaluation.

Project Organization

The project originated with the IACEE Council in June 1990 and was endorsed and expanded at a meeting of the major organizations involved later that month. Since September 1990, IACEE through the project management of Helsinki University of Technology has worked closely with, at first, the Soviet, and later the Russian counterparts to elaborate further the project and to establish it within the overall system of continuing engineering education in the USSR/CIS.

The project has received endorsements from the project partners and from their respective institutions at the highest levels—both in the former USSR and later in the individual republics.

The entire project is responsible to a governing board that represents the major organizations involved. The principal western sponsor organizations and course providers are represented on this board, as well as the coordinating institutions in the CIS.

The Russian Ministry of Education has appointed a Supervisory Council, with representatives of the institutions involved. This body helps coordinate the activities in the Russian Federation with those of the international partners.

IACCE is the project organizer. An umbrella organization representing continuing engineering education societies and agencies all over the world, it provides an appropriate base for initiating and marshalling support for projects like this. IACEE will closely evaluate the experiences during the progress of the project to be able to develop other projects of this type and and replicate it in other regions of the world.

On the western side, the project is managed by *Helsinki University of Technology (HUT)*., who conducts the feasibility study and pilot programs and support the operational phase of the project. Being the administrative location for IACEE and located in Finland, a close neighbor to the Russia, HUT is an ideal base for the coordination of the project.

The *University of Wisconsin–Madison*, a member of NTU and a leading university in distance education in the USA, provides project coordination in the USA and is a major contributor in project design and support.

Program Providers

The major program providers for western course material are currently the *National Technological University (NTU), EuroPACE,* and *EUROSTEP*, and their member universities.

In addition, several individual institutions in North America and Europe have also shown an interest in the project. *Stanford Instructional Television Network* and *Oklahoma State University* have provided two of the pilot video courses.

In addition to being course providers, these consortia also provide technical and organizational assistance as necessary and are involved in the management of all phases of the project.

Industrial Partners

Private industries, which can offer a special benefit to content understanding, delivery of courses, or other kinds of support are the other important western group of project partners. The can provide an independent source of evaluation to ensure that the project is meeting the needs of emerging industries in the CIS and their international business partners.

Potential industrial sponsors include companies active in the CIS market, joint-ventures, and other businesses with an intense training need for their partners in the CIS, but also providers of educational hardware and software.

IBM, who is also sponsoring other major educational programs in the former USSR, supplying the educational system from kindergartens to business schools with computer equipment, and *Barco International*, manufacturer of video monitors, have already committed themselves as sponsors.

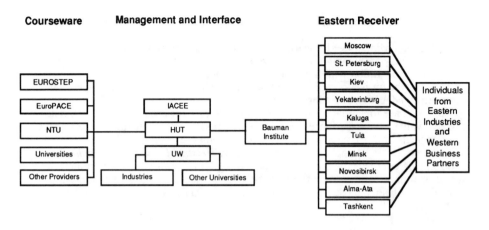

FIG. 2. Structure of the pilot phase of the East–West Distance Education Project.

CIS Partners

On the CIS side, the project manager is at the *Moscow Bauman State Technical University*. Their Inter-Branch Institute for Advanced Engineering Training is responsible for conducting the needs assessment, developing the administrative organization, conducting the pilot projects and managing the operational phase of the project. They receive guidance and assistance from the western partners through a close working relationship with HUT.

To date, thirteen regional centers representing universities and industry have been established in five republics of the CIS:
Russia:
• Moscow Bauman State Technical University
• Leningrad Institute of Precise Mechanics and Optics, St. Petersburg
• Urals Polytechnic Institute, Yekaterinburg
• Kaluga Regional Center "Tsentroekt"
• Tula Regional Centre "Postgraduate"
• Novosibirsk Institute of Electrical Engineering
• Moscow Aviation Institute
• Moscow Institute of Electronic Machine Building
• Moscow Institute of Engineering Physics
Ukraine:
• Kiev Polytechnic Institute
Byelorussia:
• Byelorussian Center for Educational Technology, Minsk
Kazakhstan:
• Kazakhstan Ministry of Education, Alma-Ata
Uzbekistan:
• Tashkent Polytechnic Institute

FIG. 1. By the end of the first pilot year, thirteen distance education centers were established in five republics of the CIS.

There has been a lot of interest on the part of industries, universities, and regions to participate in this program and far more requests have been received than could be attended to. The concept has been to work from the thirteen original sites and consider them as regional so that satellite sites will be managed from each of those regional locations. Because of the enthusiasm for having sites in local areas, the next years hold the potential of expanding to 150–200 regional centers.

Finance

On the Western side all partners have to date covered their own costs through the feasibility stage. The funding, the major portion of which still remains to be raised, will be expected to come from sponsoring industries, foundations, and governments. The receiving costs will be covered by the receiving organizations except for the hardware which includes telecommunications equipment, video cassette recorders, and possible subtitling equipment.

The CIS side has covered their internal costs except for equipment for the first pilot sites with local government support. The local commitment in roubles is estimated at more than US $1 million per year. This includes project management, deliveries, translation, training, subtitling and copying, but not reconstruction of classrooms in receiver organizations.

Pilot Experiences

The pilot phase, which ends in the spring 1992, includes a comprehensive needs assessment, feasibility study and five pilot courses during 1991. The result of the first phase will be a detailed design for the project, a report on conditions necessary for success, and a summary of current and planned technical assistance projects which relate to this one.

Specific items that are being addressed during the first phase include needs analysis by topic, job function and industrial sector, an examination of technical delivery options, a plan for administration of receive sites and overall program administration.

Training in course administration, printing and publishing of course notes, as well as evaluation of student acceptance of and performance with video-based courses, are also being addressed.

The following courses were delivered in July 1991 for subtitling, duplication and distribution:
- Marketing of Technology-Based Projects (EuroPACE)
- Lessons from Japan for Manufacturing Industry (EuroPACE)
- Quality Management (NTU)
- Engineering Risk Analysis (Stanford)

To test the logistics of delivery a few initial courses were offered in English. It has, however, been recognized that effective delivery to large audiences requires translation into Russian. Subtitling of the videotapes was expressed as the preferential means of translation by the target audiences as it would offer the opportunity to learn technical English while watching the tapes. As subtitling would require quite substantial investments in hardware, dubbing was, however, chosen as a more affordable way of translation. External cassette recorders has proven a very convenient and cost effective dubbing technology.

Six curriculum committees have been established which have identified courses of interest from the providers. The curriculum areas are: project management, business management, quality and standards, industrial ecology and environment, flexible automation, and business English.

The receiving organizations have expressed a strong interest in extending the program into credit programs. However, in the pilot phase it has been thought best not to enter the difficult area of international accreditation.

During phase two, the start-up phase, the necessary administrative organization will be put into operation.

During the third, operational phase sites will be added continually, but the technical and administrative details will have been worked out, so that system growth will be smooth and new sites will operate successfully.

Trans-European Cooperation Through International Student Seminars

A.J. Miller (*), H.W. Holz (**)
(*) Department of Building, Brighton Polytechnic
(**) Fachbereich Bauingeneurwesen, Fachhochschule Hannover

Abstract
This paper describes the co-operation of members of staff at four higher education establishments in four European Member States. It describes the organisation of a series of international student seminars and discusses the benefits experienced by those who have taken part in them.
Keywords: Environmental, International, Seminars.

1 Introduction

Graduate engineers will be required to work throughout Europe either with home based companies or working abroad with companies in their own native country. Successful integration into such an international working environment relies on prior knowledge of the relevant culture as well as practice and procedures. Graduate engineers need to be at ease with the opportunities and constraints of different working environments.

This paper describes the work of a group that has brought together staff and students of four higher education establishments in four separate EC-member states. The group has been keen to develop the interchange of knowledge and experience between different countries and to develop the potential for their students to succeed in the broad arena of the European working environment.

The group had been formed through contacts made during short educational study tours in different countries. Over a number of years the staff involved had informed themselves of the nature and content of the courses in different establishments, but perhaps more importantly had established individual links with staff in other countries. They were therefore able to organise inter-country activities on a personal level which simplified matters especially considering the language differences.

The established network has facilitated the organisation of short term work experience for a few students in different countries. It has also led to one or two students following lecture courses in other institutions but without any evaluation or specific credit within their own courses.

In this paper we will describe one particular activity of the group which has been successful in developing young engineers with a broader outlook towards career prospects within the European Community.

2 Cooperating Organisations

The four educational establishments cooperating in this work are the IUT Cergy in France, the Hogeschool Midden-Brabant in the Netherlands, the Fachhochschule Hannover in Germany and Brighton Polytechnic in the UK. Within these establishments it has been the departments of Building and Civil Engineering that have developed the cooperation.

The success of the group is such that it has recently been joined by a fifth member, Thesolonica University in Greece.

3 Rationale

The long term goal of the group is to be able to set up a system of credit transfer so that individual students could include a three to six month study period in another country as an integral part of their own award from their own institution. However alongside this goal the group decided to develop an opportunity for small groups of students to experience a broader, European outlook to their studies.

It was easy to see that travel to different countries, visiting their educational establishments, discussing experiences with similar students and visiting appropriate construction projects would be of benefit to all students who had the opportunity to do so. Further meeting and working with students in other countries would bring the advantage of establishing contacts which might be of benefit to future professional careers.

It was therefore decided to develop the opportunity for students from different member states to meet as a group and undertake a small part of their studies under common conditions. It was intended that the students should be able to study together and undertake projects in multi-national groups to encourage the interchange of ideas and expectations.

4 Educational Aim and Objectives

The aim of the seminar organisers was to provide a week of study that would be exciting to undertake and be relevant to the individual student's courses without duplicating them. In so doing it would:-
1. Present new material representing current thinking in different countries.
2. Develop knowledge of different regulations and standards.
3. Broaden perspectives from different national

viewpoints.
4. Provide the opportunity to exchange ideas and develop friendships and future professional contacts.

5 General Format

A common theme of interest between the members of staff involved in the group was that of the environment and the damage to it caused by the construction industry. It was therefore decided to take this theme and to develop a one week programme of lectures, workshops and other activities that could be supported by all four educational establishments.

The first seminar was organised and run by the Hogeschool Midden-Brabant in Tilburg. This seminar was intended to be a trial for future seminars and was undertaken on a very low budget. However its success gave encouragement for future developments and it is the second seminar which was hosted by the Fachhochschule Hannover that is being described in this paper.

It was decided that the seminar should take the form of a one week programme with each participating establishment sending 6 student delegates. This gave a total of twenty four students who could be catered for in most standard teaching rooms. It gave the students the opportunity to get to know all of the rest of the group but also made it possible to have small, multi-national groups for detailed discussion in workshops.

Whenever possible the seminar was attended by at least two members of staff from each establishment which gave considerable flexibility for the organisation of small group work.

The seminar was broadly developed around four main subjects, Groundwater pollution, Re-cycling of Materials, Energy in Buildings and Architecture. On each day one of the subjects was addressed through lectures and discussion groups. There were generally lectures from at least two of the participating establishments or from local industry involved on each day.

Each establishment took responsibility for the organisation of one day of the seminar. It was therefore responsible for the programme of lectures and workshops on that day and coordinated the appropriate input from the other establishments.

The fifth day of the seminar was devoted to study visits to organisations and projects of particular relevance to environmental issues in the locality.

6 Hannover 1991

The seminar hosted by the Hannover Fachhochschule took place in Nienburg, Germany between 11th - 15th November 1991. It was partially funded by the ERASMUS Bureau under their Inter-university Cooperation Programme which supports intensive programmes of this nature.

During the week we were treated in grand style both by our host institution and by the local community. We were welcomed at Nienburg Town hall and addressed by leading statesmen including the Environment Minister for the region. This kind of hospitality not only adds to the enjoyment of the week but also broadens the international experience for all participants.

Each of the major topics within the seminar were presented through lectures from at least two of the participating countries. In the case of Groundwater Pollution this included a presentation from a German company active in the field, who also hosted a study visit and tour of their head office and laboratories.

The workshops operated in different ways, some were based on small discussion groups considering the same subject and then meeting back as a full group to present their findings, while others were based entirely on full group discussion. There was also one occasion where the students were given a choice of activities with independent outcomes.

7 Discussion

A major consideration in the organisation of any seminar such as this must be that of language. The common language used was that of English, thus leaving the Dutch, French and German students with the immediate disadvantage of operating in a second language. It is to the great credit of these students and their associated staff that no major problems of communication were experienced. It should also be realised that there are currently very few undergraduate engineers in this country who would be able to participate in such a seminar using a second language.

The opportunity to communicate in English on subjects that will affect their professional lives was seen as a benefit to the Dutch, French and German students. There can be little doubt that their confidence and ability to communicate in the English language was greatly improved by the seminar. It has also fired the incentive in some of the UK students, and staff to become competent in another European language.

The students attending the seminar were from different backgrounds and so it was necessary to prepare them for the course content before attending the seminar. Lecturers making presentations, or leading workshops at the seminar were asked to prepare written papers and to circulate them together with a maximum of 30 pages of additional background material. Where possible these were given to the participating students three weeks before attending the seminar and therefore gave some common ground on which to base the work during the seminar.

8 Benefits

The benefits of this international cooperation have been many and have been experienced by both students and staff who had the opportunity to attend. Moreover, some of these benefits have been passed on to the main body of students in the individual universities even though they did not attend.
 Students attending the seminar have made presentations to their fellow students when they have returned home. This has disseminated the information gained to a broader group of students. It also develops further links to the network of information flow between establishments in different countries.
 The benefits to staff attending the seminar has been the immediate contact with experts in other countries and the opportunity to discuss differences in practice, procedure and regulations.

9 Conclusion

The seminars organised to date have been both enjoyable and educationally beneficial. They have improved the knowledge of both staff and students who have participated. They have also developed relationships between individual members of staff in the different establishments such that future developments are more easily organised through correspondence with 'known faces' at the other end of a telephone.
 The group intends to continue these seminars on an annual basis and is already looking forward to the next one which will be held in Cergy, Paris later this year.

A New Journal to Assist in Teaching and Research in Engineering for International Development

P.H. Oosthuizen, J. Jeswiet

Department of Mechanical Engineering, Queen's University

Abstract

A new journal entitled "JOURNAL OF ENGINEERING FOR INTERNATIONAL DEVELOPMENT" has recently been launched. This journal will publish papers dealing with all aspects of engineering problems that have an application in "developing" countries. Papers of an applied and of a fundamental nature and those dealing with socio-economic aspects of engineering problems will be published in the journal. The purpose and aims of this new journal are discussed in the present paper. Various ways in which the journal can be used in teaching and the potential significance of the journal to engineering research in international development are discussed.

<u>Keywords</u> Publications, Teaching, Research, Case-studies

1 Introduction

There is increasing interest in undertaking research on engineering related problems in "developing" countries. If this is to be done effectively, it is necessary to ensure that the mechanism exists to report on such work in a widely circulated form and to ensure that the publication of the results of such research will be in a form that is widely available and that is accepted as indicating that the research is of a high standard. Further, universities in "developed" countries are increasingly attempting to give engineering students some experience in working on engineering problems in "developing" countries. To do this effectively, it is necessary to have available full documentation on a wide range of problems that are typical of those that arise when engineers from "developed" countries undertake work in "developing" countries. In an attempt to cater to both of these needs, a new journal entitled "JOURNAL OF ENGINEERING FOR INTERNATIONAL DEVELOPMENT" has recently been launched with financial assistance from the International Development Research Centre (IDRC) of Canada and Queen's University at Kingston, Canada. In the present paper, some of the aims of this journal are presented and ways in which the journal can be used in teaching will be discussed as will the potential significance of the journal to engineering research in international development.

2 Purpose and Scope

The purpose of the journal is to provide a forum in which papers dealing with all aspects of engineering problems that have an application in "developing" countries can be published. For example, papers on various aspects of agricultural processing, water supply and control, energy supply and use, transportation, manufacturing and environmental problems will be published. Papers of both an applied and a fundamental nature and those dealing with socio-economic aspects of engineering problems will be published in the journal. An attempt will be made to publish approximately equal numbers of papers by authors from "developing" countries and from "developed" countries. To this end, if economic conditions in an author's institution prevent them from preparing the manuscript and/or the figures to the standard required for submission to the journal, an attempt will be made to have them prepared to this standard by the journal staff.

The success of the journal in covering a wide range of subject areas from a wide range of countries is partly indicated by the papers published in the first issue of the journal, these being as follows, only partial titles being given:

- Production of Nodular Iron for Developing Countries
- Modelling of an Indirect Natural Convective Solar Rice Dryer
- On the Mechanization of Palm Crown Operations
- Performance of Abrasive Disk Dehullers
- Design and Performance of a High-Temperature Pre-Dryer
- Biogas Potential in Nepal

These papers originate from work done in The Netherlands, Canada, Saudi Arabia, Senegal, The Philippines and Nepal.

The journal is published three times a year. Papers for inclusion in the journal may be written in either English or French. Letters to the Editor that pose significant questions or comment on any technical paper that appears in the journal will be considered for publication.

3 Use In Research

Those who have undertaken research or design and development work involving engineering aspects of international development will be well aware of how badly work in this area has often been documented in the past. This is particularly true of research work undertaken by workers from "undeveloped" countries with the result that such work is often not appreciated by a wide audience · This has sometimes caused mistakes to be repeated and led to frequent "re-inventions of the wheel". A number of data-bases have been established in recent times to deal in part with this problem. These data-bases, however, require that the work be documented in a form that will be accessible to the compilers of the data-base and this is often not the case. It is believed that the new journal will play an important role in documenting engineering research and development work involving "developing" countries in a widely available form and thus help to avoid the repetition of past work and mistakes. Research undertaken by workers from "undeveloped" countries has often been the most poorly documented with the result that such work has often remained unknown by and, therefore, not appreciated by a wide audience. It is hoped that the new journal will help to overcome this problem and in this way encourage research efforts in such countries.

It also has to be accepted that progress at many academic institutions is to a major extent based on the number of papers published by a faculty member in so-called "high-class" publications and the available forums for the publication of engineering work in "developing" countries have often not been well regarded in this light. The "JOURNAL OF ENGINEERING FOR INTERNATIONAL DEVELOPMENT" will try to be accepted as a "high-class" publication by fully and widely reviewing all papers submitted to it, thereby ensuring that only papers of a high standard are published in the journal. In this way it is hoped that it will help in establishing research in engineering aspects of international development as a valid, high-class academic endeavour.

The new journal will also publish announcements and calls-for-papers for conferences that are concerned, in some way, with the subject matter dealt with by the journal. This will help ensure that research workers in this area have information in a single place on almost all conferences that are of interest to them.

4 Use In Teaching

As already mentioned, it seems that there is increasing interest in giving at least some undergraduate engineering students exposure to the particular considerations that are involved in undertaking engineering work in "developing" countries. It also seems to be quite widely accepted that a case-study type approach should be used in providing training in this area. However, it is often difficult for an instructor to find suitable material on which to base the cases. In is hoped that the new journal will help in providing such material. For example, the papers published in the first issue of the journal, in conjunction with the references they include, provide most of the information required to develop case studies of:

- Dehulling of cereal grains
- Biogas generators
- Drying of rice
- Foundry operation in "developing" countries

To make the journal particularly useful for the purpose of developing case-studies for teaching purposes, an attempt will made to ensure that, when appropriate, papers include a thorough discussion of the background to the problem being considered.

Beside providing information that can be used in the development of case studies, many of the papers in the new journal should be useful as background study material for class discussions of engineering in international development.

5 The Editorial Board

The Editorial Board of the JOURNAL OF ENGINEERING FOR INTERNATIONAL DEVELOPMENT is believed to have a more important and proactive role than the boards of journals that deal with more conventional research areas. The range of topics covered by the new journal is very broad and the input of experts from a wide range of fields and from a wide range of countries is believed to be vitally important in ensuring that the journal meets the needs of its target audience. The initial editorial board has members from the following countries:

- Belgium
- Canada
- Indonesia
- Senegal
- Tanzania
- Thailand
- United States of America

Additional board members are being recruited from various countries particularly in Asia, South America and Europe.

The board members are actively involved in the reviewing of papers. for the journal. They are particularly asked to review papers that deal with material that would not normally be published in the journal and papers that have received very mixed reviews. They are also asked to considered the types of papers that are being published in the journal and to have the Editors encourage the submission of papers in areas they feel are important but that are being ignored. The board members are also asked to assist the Editors in developing general editorial policy such as the amount of assistance that can be given to authors in preparing their papers for the journal.

6 Reviewing of Papers

An attempt is made to ensure that papers submitted to the journal are reviewed by workers and Editorial Board members from both "developed" and "undeveloped" countries. While the review process is conducted in a conventional manner, the decisions about accepting or rejecting a paper are often not as cut and dried as in a more conventional journal. In many cases, if the changes are of a non-technical nature and can be made by the Editors, this will be done because of the long time it frequently takes to communicate with authors or because the journal office has facilities that are not available to the authors. If a paper has to be returned to an author, an effort is made to explain what changes and extra work would required to make the the paper acceptable rather than simply rejecting the paper.

7 Distribution of Journal

In order to ensure that the new journal is widely available, an attempt has been made to keep the subscription rates low, these presently being US$75 for multi-reader institutions and US$35 for individuals. In addition, attempts are being made to procure additional support from government institutions and private foundations that will allow free subscriptions to be given to libraries in many "developing" countries hopefully thus assisting the research and development activities in these countries.

8   Page Charges

Many technical and scientific journals find it necessary to levy page charges on the papers they publish. While, in most cases, publication of a paper is not dependent on the page charges being paid, invoices for these charges are usually sent several times and the impression certainly given that payment is expected. In the initial planning for the new "JOURNAL OF ENGINEERING FOR INTERNATIONAL DEVELOPMENT" it had been assumed that page charges would be leveyed on papers from "developed" countries but not on those from "undeveloped" countries, it being assumed that even if authors from such countries were told that payment was not mandatory, the receipt of an invoice might deter then from submitting another paper to the journal. However, it was later decided that such a policy would be undesirable and unworkable. The impression that papers from the two "types" of country would in any way be treated differently was to be avoided and the country of origin of the paper was too difficult to define e.g. was it the country where the work was undertaken or or was it the country where the lead author was working when the paper was submitted. It was decided, therefore, that page charges would not be levied. This has meant that the journal has to be produced at a much lower cost than most other journals and that additional funding from public and private agencies has had to be sought.

9   Conclusions

The new "JOURNAL OF ENGINEERING FOR INTERNATIONAL DEVELOPMENT" appears to fulfil a need for a high quality journal for the publication of the results of research and development work of an engineering nature that has an application in "international development". The journal should also provide much needed background material for use in the teaching of the nature of engineering in international development.

Establishment of an MSc Programme in Indonesia

J.S. Younger (*), A.D. May (**), T. Soegondo (***)
(*) O'Sullivan & Graham, U.K. and Indonesia
(**) Institute for Transport Studies, Leeds University
(***) Institute of Technology Bandung

Abstract

In 1981 a decision was taken to set up a Master's degree programme in Highway Engineering and Development in Indonesia, in order to provide an indigenous high level education programme to serve the urgent and increasing demands relating to development of the highway network in this very large developing country. The Institute of Technology Bandung (ITB), the leading Indonesian academic establishment in this field was chosen as the location for this project, and initial funds were provided through a World Bank loan to the Indonesian Department of Highways, Bina Marga. The programme commenced in 1982.

The history of the programme over the past ten years is described, along with the various elements which are needed in order to establish a high level programme of this type in a developing country and ensure its sustainability. Issues discussed concern the further training of the local staff, their gradual and increasing involvement in teaching duties in line with the decrease in assistance from foreign academic expertise, the importance of research in the programme and its management and longer term sustainability. The possibility of expanding the programme to serve wider regional aims is now under consideration, in which case a commitment will have to be made to offer the total programme in English.

1. Introduction

There has been growing awareness of the need to provide education and training at higher levels within developing countries, rather than sending students overseas. Larger numbers can be trained at a much lower unit cost and centres of excellence can be fostered. The major world lending agencies have begun to play a significant role in the provision of assistance with the establishment of training programmes.

Expert input from the developed world is essential in the early stages of establishing such programmes, but in order for the programme to continue in the long run, the impact of this initial outside expert input will need to be sustained. The following is a review of a particular case in which an MSc programme in Highway Engineering and Development has been established in Indonesia.

2. Factors to be Considered

These involve identifying the need, choosing an appropriate centre, developing the programme itself, ensuring suitable facilities, having a properly structured local staff development programme, establishing and continuing to improve individual courses, developing the research potential with provision of useful contributions towards national development, and ensuring sustainability.

Identifying the Need

As a country's economy expands, the need for skills, and therefore for training at all levels increases very significantly. However, while this need is generally recognised, it is essential that aid is focused on major areas if it is to be used effectively.

It is often the public sector which experiences the most severe skill shortage, since government cannot compete on salary with the private sector. The highways sector in developing countries falls into this category of problem.
The need therefore is for education programmes which are of sufficient quality to attract the better students, and continuing education programmes which can train staff to fill posts vacated by others.

Indonesia is one of the largest and most rapidly developing countries in South East Asia, with a population of 180 million. Growing motorisation is leading to an increase in congestion and a greater accident toll in major cities. Against this background, it was clear that the skills base of the Public Works Department and other government agencies urgently needed enhancement.

One response to this was the decision, in 1981, to establish this MSc programme in Highway Engineering and Development at the Institute of Technology Bandung. The programme was designed primarily to train staff of Bina Marga, the highways directorate of the Public Works Department, and World Bank funding was received for this purpose.

Choice of Centre

ITB, Bandung was an obvious choice as the centre for the programme. Of the premier universities in Indonesia, it has the strongest engineering base, is adjacent to the Indonesian Road Engineering research centre, and is close enough to Jakarta, the headquarters of Bina Marga, to permit easy access. Moreover, ITB could provide suitable office and lecturing accommodation and laboratory facilities and the nucleus of a local teaching staff within the Department of Civil Engineering.

Development of the Programme

The development of a new educational programme will require the definition of the objectives, content, standard and means of assessment of the courses to be provided. All of these should be able to be determined from the initial assessment of need. Having specified content, it will be possible to judge the extent to which local staff have the necessary knowledge and teaching skills; similarly specification of means of assessment and standards will help identify shortfalls in academic qualifications and in experience in student assessment.

A clear understanding is needed of the time which will be taken to train local staff, establish the source and its assessment procedures, and transfer teaching and assessment skills to the local academic team. In our experience, the time required is invariably grossly underestimated.

The programme initially involved a total of eight expatriate academic staff, some working full time and others on part time assignments, between them covering the full range of subjects from geotechnics to transport planning. In the first years, these staff developed the courses and provided much of the teaching, while local staff received MSc and PhD training overseas. The original intention was that the course would have become self-sustaining, based solely on local staff input, by the end of the contract in 1986.

In practice this did not prove feasible. It became clear that expatriate staff input would continue to be needed for teaching in certain subjects, for support to provide experience in teaching methods, assessment and course development and, in particular, for guidance in the establishment of research programmes on which students' theses could be based. Support for this continued programme, in which expatriate staff inputs have been gradually reduced and become more focused, has been provided from 1986 to 1987 under the World Bank, from 1987 to 1989 by the UK Overseas Development Administration and, from 1989 to 1992/3 again through the World Bank. Current outside assistance has been provided by a joint venture of The University of Leeds and Frank Graham International Ltd (now O'Sullivan & Graham), and has recently been recognised with the 1991 IBM award for Sustainable Development.

From the outset, it was decided to teach most of the courses in English.

Facilities

A new programme will usually require its own premises, teaching facilities, library, computing equipment and software, and laboratories. While these can generally be made available as part of the initial capital costs of the programme, they will all need maintaining. It is essential that the host institution is committed to continuing provision of library books and periodicals, replacement of computers, purchase of new software, and procurement of improved laboratory equipment.

While adequate accommodation was provided, and initial laboratory computing and library facilities, the latter had to be substantially enhanced by the initial grant, and have been boosted since. However, it has been difficult to maintain a regular supply of periodicals and to ensure that software, computers and laboratory equipment are updated. These represent recurrent costs which must, in the end, be met by the host institution from fees received.

Staff Development

A programme of staff training will be an essential element in the establishment of any successful programme. In the early days much of this must be provided overseas, and can often most successfully be provided by an academic organisation which is providing the expatriate staff. Such two-way exchanges of staff can help cement a continuing relationship between the academic institutions. In the later stages, much of the training, particularly in teaching, assessment and research methods, can be provided on site, using local conditions as the basis for coursework, examples and research projects. However, it will be essential that the expatriate staff stand back at this stage and let local staff develop their own capabilities and confidence.

For this programme at ITB, three senior local staff provided the management of the programme, and were assisted in supportive duties by staff from Bina Marga, while junior ITB staff were sent for training overseas. Since commencement, some

12 ITB staff have received training to MSc and PhD level in five different countries, and five are currently overseas for PhD's.

Programme Development
Any initial programme outline will generally be formulated by the expatriate team working together with a nucleus of local staff. It is obviously sensible if not essential to involve potential employers of the students and trainees, to ensure that their needs are to be met.

For the ITB staff, in addition to academic training, the programme has provided experience in academic and research development, by financing attachments for senior local staff within the institutions providing expatriate support. This has proved invaluable in strengthening links and encouraging further development of the programme. An initial review of the programme in 1986, conducted by expatriate staff, led to some restructuring and updating of coursework material, with the particular aim of focusing it more directly on Indonesian needs. Since then, one member of expatriate staff has acted as an informal external examiner to ensure that standards match those in other countries, while local staff have taken over the responsibility for examinations and student assessment. Further course reviews are in hand, although it has taken longer than anticipated to develop the confidence among local staff to update their own teaching programmes.

Research Development
It is generally accepted that the best academic institutions are those which combine teaching and research, and ensure that their teaching programmes are kept up to date by remaining in contact with advances in research. Such contacts are even more important in developing countries, where the limited number of experts can easily become isolated from advances in their subject area.

Most of the last six months of the ITB programme are devoted to the undertaking of an appropriate research project, which is proposed, written up and presented in English. In the early days the expatriate staff identified all the subjects and provided most of the supervision and assessment, with local staff providing a supporting role. Gradually, these roles have reversed, with the local staff now managing most of the assessment. However, it has taken much longer for local staff to take over supervision and even longer for them to gain confidence in proposing thesis topics. To remedy this, the current stage of the programme is encouraging the development of research programmes and projects, in geotechnics, pavements and traffic engineering. It is hoped that this will provide the experience on which to base specification of topics of appropriate complexity and rigour for student training.

Sustainability
Superimposed on all these requirements is that of sustainability. Course structures, facilities, staff skills and research capabilities are only worthwhile if they can be sustained during the period in which expatriate support is withdrawn. There are, sadly, too many examples of programmes which have been developed to the point where they appear to be self-supporting, where financial aid has ceased, and where, perhaps through changes in key local staff, the momentum of the course has been lost. Once standards start to fall, they are very difficult to resurrect, and the benefits of several years of professional and financial support can easily be lost. Continued contact with academic institutions in other countries is essential if programmes are to

be sustained.

In retrospect, the initial four year programme at ITB was too short to achieve the establishment of a self-sustaining two year MSc programme. A six year programme would seem to be the absolute minimum, given a full complement of academically qualified staff at the outset. It is not surprising, therefore, that the ITB programme, in which the majority of the staff needed initial training, has taken much longer. Nevertheless, the programme has been cost-effective. Over the ten year period, some 190 Masters students will have been educated at a cost of about $ 9m, which compares reasonably with the costs that would have been incurred had the same number of students been sent overseas for a similar 2 year programme, but with the additional benefit of having established an international standard programme in the home country. Future per capita costs for training at this standard will thus be very much reduced.

Some limited support will continue to be needed well after the programme is run fully by local staff. The main areas for this will be in academic assessment, research links, and maintenance of library and software materials. Without these, it will be difficult in an institution remote from other international academic centres to sustain the high standards achieved.

3. Conclusions

This paper has reviewed a case study of the establishment of an education programme in the highway sector in a developing country. Such a locally based programme has the potential advantage of providing training at lower cost, more immediately focused on the country's needs, and in a form which establishes a centre of excellence in the country.

However, if such programmes are to be successful, a number of requirements need to be met in their planning and execution, the main ones being identified as :
* identifying the need
* selecting a centre
* planning the programme
* providing and maintaining facilities
* developing staff skills
* continued development of the programme
* developing a research capability
* ensuring sustainability

There has been a number of successes and shortcomings, these being:
(1) failure to maintain the supply of library materials and software, and the performance of computers and laboratory equipment;
(2) competing pressures on staff time, induced both by low salaries and desire for consultancy expertise, leading to failure to develop courses and research programmes;
(3) lack of availability of staff to be trained in the necessary academic skills;
(4) lack of confidence to develop research capabilities and research related teaching.

All of these can be overcome, in part, by a clearer commitment from the host institution, and higher levels of financial support. However, the continued support from other academic institutions is also needed if the standards are to be sustained. This is harder to provide where the number of suitable academic institutions is limited. External support will continue to be needed to provide this expatriate input, but the costs are small relative to the benefits, and the benefits accrue to both participating institutions.

This emphasis on sustainability is perhaps the most important message from the experience gained to date. The limited expenditure on sustainability can be compared with that for infrastructure maintenance. If it be incurred on a regular basis, the benefits of the initial investment in establishing the teaching programme will be maintained and enhanced. If ignored, those benefits will be rapidly dissipated.

4. Acknowledgements

This paper has been made possible by the support for the programme described from the World Bank, the Overseas Development Administration and the British Council. Many of the ideas have been developed in discussion with colleagues. However, the authors take sole responsibility for the views expressed.

KEYWORDS

Engineering, tertiary, education, developing country, sustainability.

International Technical Education Program
R. Massengale
John Fluke Mfg. Co., Inc.

Abstract

Over the past 24 months, some historic changes have occurred through the world's political and social arenas. Whether you are discussing the breakup of the Soviet Union; the Gulf War; formation of the European Economic Community; North American free trade; or the quest for democratic reform in the Peoples Republic of China; the reality of a global community has become more apparent than ever. This rate of rapid change will require workers to have skill sets that will allow them to compete in the global economy. However, they are the same people that have not had the benefit of economic and educational tools employed by the countries in the developed western world for many years. This is further complicated by the need for these countries to build an infrastructure that will allow them to compete.

This transformation will have a particularly strong impact in the area of technical education. Colleges and universities in economically developing countries must play an increasing role in preparing their scientists and engineers to develop skills necessary to produce quality goods and services throughout the world. To facilitate the process, foreign universities must turn to their counterparts in the West for assistance. New international partnerships and alliances must be formed if the developing countries institutions are to achieve comparable status to western schools.

Keywords: Technical Education, International Partnerships, Global Economic Development, Industry/Education Cooperation.

Background

John Fluke Mfg. Co., Inc., located in the Northwest corner of the United States in Everett, Washington, is a world leader in compact professional electronic test tools. The company boasts sales of $239 million in fiscal year 1991, 35% of which came from outside the US with products available in 85 countries throughout the world.

Over the past decade, the company supported in excess of 500 colleges and universities with a wide variety of programs ranging from millions of dollars in equipment grants, to faculty sabbaticals, to extensive summer internships and co-ops. In the mid-80's, Fluke embarked on a program that strengthened the ties between U.S. institutions and various technical universities in the Peoples Republic of China (PRC).

More recently the company has become interested in Mexico and other Latin

American Countries as well as Eastern Europe. With this as a backdrop, in 1990 the company established an International Technical Education Program (ITEP) as part of its College/University Relations efforts.

Mission

The purpose of the ITEP is to demonstrate how domestic schools (lead institutions) can "partner" with John Fluke Mfg. Co., Inc., to develop new engineering curricula and better laboratory facilities on U.S. campuses and corresponding institutions around the world. The program is designed to provide incentives for greater cooperation between U.S. technical faculty and investigators and their overseas counterparts, while exposing numerous students, who will eventually become Fluke customers, in their respective countries.

The purpose of this paper is to examine new initiatives of the program and explore the possibilities of opening up to other schools in other countries. We will examine the workings of the program and the roles and responsibilities of each participating partner. Finally, we will look at the benefits derived from the ITEP of all involved.

A Tradition of International Cooperation

Seattle University, in 1985, had a relationship with Beijing Poly Technical University. Both schools were interested in developing a Data Acquisition Laboratory. Fluke was and is very interested in establishing itself as the major player in the potentially overwhelming PRC marketplace. In order to facilitate all parties desires, a partnership was formed between the three entities.

Fluke had been doing business in PRC for the previous decade and had some of the academic contacts to entice Beijing Poly Tech's participation. That went hand in hand with an already productive relationship between Seattle University and Fluke which allowed for a smooth transfer of information between all parties. Fluke acted as a catalyst for the exchange and arranged for the Department Heads at Seattle U and BPTU to discuss details.

Coming to America

Two faculty members from BPTU were selected to come to Seattle to start phase one of the project. The faculty were selected on the basis of their interest in Data Acquisition and their ability to read and write English (this was very important, since the course materials, including many technical terms, had to be translated into Mandarin Chinese).

BPTU faculty spent a total of three months learning the technical aspects of the Fluke products. They also attended laboratory sessions to get a feel for the instruction techniques in the U.S. The next step was to translate manuals, lesson plans and workbooks to their native language. Finally, the equipment was shipped to the BPTU campus for installation. Fluke assisted in securing export licensing since some equipment had to be cleared by the State Department prior to being sent to PRC.

After the experience in the PRC, the company turned its sight on the two

American continents to look for similar initiatives that may be able to use this existing model. They became founding members of Ibero-American Science and Technology Education Consortium (ISTEC).

ISTEC Profile

"The Ibero-American Science and Technology Education Consortium (ISTEC) is a non-profit organization comprised of educational, research and industrial institutions through the Americas. The Office of the Secretariat of the Consortium is currently located in the Department of Electrical and Computer Engineering (EECE) of the University of New Mexico, Albuquerque, New Mexico. The Consortium has been established to accelerate economic development in Latin America by fostering scientific and engineering education, joint international research and development efforts among universities and scientific centers, and to provide a cost-effective vehicle for technology transfer.

The above mentioned goals are to be accomplished via the creation of new and enhancement of existing undergraduate and graduate programs of study, the exchange of personnel, the creation of state-of-the-art laboratories to provide a hands-on approach to education, the development of up-to-date educational materials and techniques, and the use of direct telecommunication links for the delivery of courses, seminars, workshops, and other pertinent activities. An additional, major activity which is emphasized by ISTEC is transnational cooperation in joint research and development projects, efforts carried out by participating members in two or more countries. These research and development efforts will in turn encourage additional technical activities through new business opportunities for local and international business." (1)

ISTEC membership now contains over 25 colleges and universities representing 14 countries in the Americas and Spain. Fluke participates as one of 10 industrial members including; IBM (Argentina, Brazil), Motorola, Cray Research and Texas Instruments.

UTM/CSU Profile

Although ISTEC represents a significant coordination effort between institutions and corporations in many countries and may lead to sweeping changes in the participants' organization, the best interaction still takes place school to school. One case in particular is the relationship between Colorado State University (CSU), Department of Electrical Engineering, Fluke and Universidad Technologica de la Mixteca (UTM) located in Oaxaca province in Mexico.

UTM is located in Huajuapan de Leon, a city approximately 200 miles south of Mexico City. The university was created in June of 1990 by the Governor of the State to improve the quality of life of the inhabitants of the province of Mixteca, to improve the economy of one of the poorest regions in Mexico, and to increase the retention of the college age students who typically emigrate to Mexico City or the United States. As of January, 1992, the University has established bachelor's degrees in electronics, computer engineering and engineering design.

Colorado State University located in Fort Collins is a school of 20,000 students. The Electrical Engineering Department (currently under the chairmanship of Dr. Jorge Aunon) has long enjoyed a prosperous relationship with John Fluke

Company. Equipment donations and discounts totaling over a quarter of a million dollars has made the department one of the strongest in the country.

In 1991 Dr. Aunon approached Fluke for assistance under ITEP in helping UTM establish their newly formed engineering laboratory. The company's college relations committee agreed to provide some state-of-the-art data acquisition equipment as well as high end digital multimeters to assist UTM's efforts. The intent was to duplicate the equipment set up that CSU has so that there could be an exchange of ideas between the two schools. The result was Dr. Aunon was able to deliver the equipment in January, 1992 and help with setting up a laboratory that will be viable for several years.

The following section can be used as a model of how an ITEP can be structured to work effectively for all participants:

Lead Institutions should develop the proposal for the use of Fluke equipment, preferably in a teaching laboratory. The request should list what equipment is desired to be donated; how it will be used; which and how many students in each location will benefit; how it will be incorporated into the curricula; and how it will be maintained overseas.

Overseas Institution sends one or more faculty members to be resident at the lead institution for training purposes, for a period of time. The training should include a thorough combination of technical skills development and particular applications along with any Fluke sponsored training courses intended to give the participant a greater understanding of specific applications of Fluke equipment.

Transportation and lodging costs for the overseas facility are usually covered by one or both of the institutions.

After training has been completed, the overseas faculty then return to their home institutions. It is recommended that a faculty member from the Lead Institution also accompany them to assist in the start-up phase of the new laboratory. The Fluke equipment from the U.S. institution is placed on long term loan to the overseas institution.

The Lead Institution will be granted new equipment directly from Fluke upon completion of the required shipping arrangements (including all required export and import licenses).

Benefits

There are numerous benefits to both institutions and Fluke to participate in ITEP.

Lead Institution
1. Increased interactions with international peer institutions.
2. Provides contacts for overseas study for engineering students.
3. Helps develop state-of-the-art lab facilities on campus.
4. Increases interaction with a leading instrument supplier that leads to lab equipment for the school.

Overseas Institution
1. Assistance from leading U.S. universities in developing new labs.
2. Low cost state-of-the-art labs.
3. Increased interactions with leading multi-national equipment supplier that may lead to employment for their graduates.

Industry
1. Company's equipment will be seen by more students, leading to easier access to export markets.
2. Applications information that may be tied closely to a particular industry in a particular country (i.e., petroleum in Mexico) or localizing manual or software instruction.
3. Education can be a non-threatening way to enter new markets by using the schools as an intermediary.

Environment for Success

Sponsoring Company
1. Management commitment.
2. Focal point - individual responsible for assuring resources are available.
3. Long term outlook regarding developing markets.
4. Technical champion to act as an advisor for equipment configurations.
5. Clear understanding of the deliverables - what to expect as a result.

Lead Institution
1. Administration commitment
2. Release time for faculty
3. Clear understanding of the deliverables
4. Internal sensitivity to possible cultural difference
5. Access to language personnel

Overseas Institution
1. Commitment for government to change technical standing
2. Commitment from institutional leaders
3. Faculty member should read and write English, especially technical terms.

Conclusion

In conclusion, the ITEP can be a method for colleges and corporations to work together to increase interaction between overseas institutions who are attempting to make a quantum leap into the twenty-first century. It is an excellent way for universities and industries to work cooperatively on a project of great mutual interest and important to the development of the technical communities in the United States and abroad.

There is also an opportunity to include other interested parties such as US Aid for International Development, National Science Foundation and Department of Education and their corresponding agencies in other countries in the joint sponsorship of laboratories around the world. As the globe shrinks and we all become closer, many types of creative programs such as ITEP will not only be interesting to the few, they will become a necessity if we are to ensure the positive spread of technology continues throughout the world; thus allowing new market

248   Engineering Education

opportunities and an enhanced standard of living for many.

Programs that support better understanding between peoples, who in some cases were mortal enemies, can serve to strengthen the cause for greater harmony on the planet. By ensuring that all people have access to quality educational facilities, we can teach more self reliance while providing the tools to solve today's most complex medical, environmental and economic problems.

(1) From Ibero-American Science and Technology Education Consortium Charter

Tempus Projects "Computer Aided Learning and Simulation Technologies" and "New Curricula and Courses in Theoretical Engineering Education" at the Czech Technical University

K. Květoň, M. Starý

International Centre for Scientific Computing, The Czech Technical University in Prague

Abstract
The CTU takes part in TEMPUS Joint European Projects (JEP) oriented at modernization of engineering education with the aid of European Communities. Within the CTU, an International Centre for Scientific Computing has been established to coordinate, beside other tasks, the CTU participation in two projects: "Computer Aided Learning and Simulation Technologies" and "New Curricula and Courses in Theoretical Engineering Education". The paper presents briefly our activities concerning these Joint European Projects - above all our efforts to develop a joint educational software FAMULUS -, our future intentions and explains our experience with the East-West cooperation.
Keywords: Engineering, Education, Universities, TEMPUS, Joint European Projects, East-West Cooperation.

1 Introduction

The Czech Technical University (CTU) was founded at the beginning of the 18th century and belongs to the most ancient institutions of higher technical education on the Continent.

Nowadays, it consists of five faculties: Faculty of Civil Engineering, Faculty of Mechanical Engineering, Faculty of Electrical Engineering, Faculty of Nuclear and Physical Engineering and Faculty of Architecture.

The curricula and study programmes consist of two or two and half years of basic theoretical study and three years of flexible specialization with great impact on interdisciplinary education. Students are encouraged to develop self-reliance, responsibility and entrepreneurship.

The CTU is participating in 24 TEMPUS Joint European Projects oriented at modernization of engineering education with the aid of the European Community. International Centre for Scientific Computing (ICSC) coordinates the CTU's participation in two projects: "Computer Aided Learning and Simulation Technologies" and "New Curricula and Courses in Theoretical Engineering Education".

The first JEP enables cooperation of 24 European universities and other institutions from 14 countries. The project is coordinated by Prof. M. Wald, HAP, Hamburg. The CTU is developing within this project software for computer illustrated textbooks for mathematics, physics and mechanics.

The JEP "New Curricula and Courses" is coordinated by Dr. K. Květoň at the CTU and 13 universities from 8 countries are cooperating. The CTU working groups are developing new curricula for applied and numerical mathematics, physics, mechanics of fluids, applied mechanics and fundamental electrotechnical disciplines. One of the working groups is preparing new methods of teaching using multimedia systems.

In June 1992 we organized an international conference "Trans-European Cooperation in

250 Engineering Education

Engineering Education" in Prague. The participants presented about 70 papers on problems of computer aided education, on curricula design and development, on utilization of computing technology in laboratories and for information processing and transfer. Experiences with both TEMPUS projects were evaluated.

2 Project "Computer Aided Learning and Simulation Technologies"

The contribution of the CTU is aimed at forming Computer Illustrated Texts for Mechanical Engineering. The term Computer Illustrated Texts refers above all to classical textbook and associated PC programmes for undergraduate and postgraduate courses, not fully electronic textbook. This textbook specializes in three key subjects: mathematics, physics and mechanics. Within these subjects, the following topics are treated:
Mathematics: numerical methods, constructive geometry, statistical methods, symbolic algebra
Physics: simulation of physical phenomena, computer aided data measurement and processing, statistical methods in physics
Mechanics: kinematics and dynamics of multibody systems, symbolic algebra systems, finite elements method, computer aided design in mechanics, fluid mechanics and thermodynamics
Special subjects: interactive teaching programmes for multimedia systems, teleinformatics.

The textbook will enable students to understand fundamental theory of various topics, to learn different approaches to the subject, and to do practical coursework. Application to technical problems and numerical solution of technical problems are also dealt with. The database of program packages will contain exercises, with data from industrial applications, as well as simulated data (so that accuracy and reliability of presented methods can be investigated).

The software developed for this purpose will have a structure which attempts to lead a student through a series of teaching and testing steps. The teaching steps usually consist of text, sometimes with simple diagrams or animated graphics. Students' wrong answers to test questions will provide a chance to loop back to the relevant teaching material.

3 Project "New Curricula and Courses in Theoretical Engineering Education"

The targets to be met are:
- to create new curricula and courses in theoretical engineering education to enable eligible countries to reach the level of education in the EC member countries
- to update computing facilities to enable development of cooperation between eligible countries and EC countries in the area of theoretical engineering education
- to develop multimedia educational/training packages
- to encourage long-term cooperation of eligible countries with partners in the EC through joint activities and relevant mobilities.

Creating of new curricula and courses in theoretical engineering education
is including:
- restructuring of complete curricula of postgraduate higher education, first of all at the mechanical, electrical and civil engineering faculties
- large scale development of teaching material
- development of software and multimedia educational packages
- development of open and distance learning
- short intensive courses, seminars and language courses, bringing together postgraduate students and teachers from partner organizations.

The CTU working groups are developing new curricula for applied and numerical mathematics, physics, mechanics of fluids, applied mechanics and fundamental electrotechnical disciplines.

One of the working groups is preparing new methods of teaching using multimedia systems. This working group projects to provide an intelligent multimedia authoring system, which would enable authors to define a high-level structure of a course graphically, in terms of a set of concepts which are linked to multimedia frame modules. Students will be offered audiovisual data representations, treated digitally at least at some point in the course of processing.

The contribution of the ICSC is aimed, beside other tasks, at the development of software educational packages. It is very important to use a common software educational environment to support cooperation between TEMPUS partners. The availability of a suitable authoring program allows the teacher to develop particular instructions, with the requirement that training should be as highly individualized as possible. An interaction between the material and the learner is required.

The result of the efforts of all TEMPUS JEP partners should be development of a system of self-paced learning, based on an appropriate engineering courseware.

New curricula and courses do not mean only new structure of lessons but above all special programs permitting the implementation of lessons involving examples, exercises, tutorials, laboratory experiments, etc. into a new teaching system.

4. Professional educational software FAMULUS

The Czechoslovak made educational software package FAMULUS (see the References) was found to be a very convenient environment for developing new teaching packages. That is the reason why it has been selected to be used by all JEP partners as a unified tool.

FAMULUS is an integrated system for numerical computing and for synoptical presentation of its course and results in the form of tables and graphs. The user is given the possibility to influence simply, and also to change, not only format of the presentation of the results, but also the algorithm of the computing itself. It is possible to use FAMULUS to figure both functions and measured data, to model many phenomena and processes from various spheres of application, to process data and to compare results with theoretical models, for quick interactive jobs in scientific and technical practice and in other fields of engineering activity.

FAMULUS does not force the user to learn complicated procedures, it enables him to do without programming knowledge. It directs the user by means of offers and suggestions.

In the case of more sophisticated tasks, the user can select requested jobs, insert necessary data and the system produces quick solutions. The user can make use of ready programmes (models) and of a subprogrammes library (procedures and functions).

It is possible to look at the results from various points of view in form of various graphics or to change quickly the form of figures. Later the user becomes able to modify flexibly the calculation method itself and to produce shorter models.

FAMULUS offers its own language the use of which enables to programme in the field of numerical calculations practically all what the languages Fortran, Basic or Pascal enable. The programmes created are often more compact or nearer to standard mathematical records. The language is very similar to standard programming languages and partisans of Basic, Pascal or Module-2 or even of Fortran 77 and of language C adapt themselves quickly to its use.

FAMULUS will serve also as an environment for the development and perfection of algorithms. It is possible to figure partial results during the computation or to stop the computation process and to look for the value of variables. A debugger enables to perform the calculation step by step and to check values of expressions and variables.

It can be used with advantage at all levels of the education system. In mathematics, physics and engineering it can replace a long range of teaching programmes of both demonstration and simulation types. It can be used even in chemistry, biology and in economics.

FAMULUS is a universal tool. Its task is not to replace teachers but to be a helpful means stimulating the teaching process.

The environment of the FAMULUS system saves a lot of work related to the programming of inputs and outputs, etc. A possibility to modify quickly the developed models according to the needs is very important.

FAMULUS will provide students with possibility to look at various aspects of the phenomena studied, to solve more interesting problems, to understand function of processes they will try to model. Last but not least, it provides them with enormous computing possibilities.

By the end of 1992, an improved version of the FAMULUS program, not only in Czech but also in English language, will be available. It will enable us a multilateral exchange of software products, developed on the basis of FAMULUS, among all JEP partners. A low price of the FAMULUS Software together with other advantages will enable its widespread distribution to eligible universities which usually do not dispose of higher amounts of money.

5 Activities planned to start in the academic year 1992/93

Three new JEPs have been proposed by the ICSC:
"New Curricula and Courses in Postgraduate Engineering Education", "Development of Selected Courses of Engineering Education Presented in English" and JEP "Introduction of Engineering Education Ending with Bachelor's Degree".

In the time of sending this manuscript to the Programme Committee of this Conference we do not yet know, if our new projects will be accepted. The financial means, the EC TEMPUS Office disposes of, are a welcome support for the East-European universities. Nevertheless, TEMPUS grants are relatively low in respect to the rather obsolete school systems in post-communist countries and enable development of a relatively small number of projects.
It is supposed that from about 500 submitted Czechoslovak new TEMPUS projects, only not more than 30 projects could be accepted.

6 Brief evaluation of our experiences gained up to now in cooperation within the TEMPUS Joint European Projects

As we already remarked, one of the obstacles of a more effective cooperation of European countries in the field of higher education within the TEMPUS program is lack of financial means. However not even projects awarded by relatively high grants must in all cases yield satisfactory results. There are more reasons for that and we do not claim to mention all of them due to our limited experience. Some of them were subject of discussions at the TEMPUS Workshop held at Luneburg, Germany in November 1991. These reasons are doubtless known to the representatives of the EC TEMPUS Office in Brussels and of the national TEMPUS Offices. We will mention here therefore only some of the reasons that are usually not presented in public documents.

Every cooperation within a TEMPUS project should bring advantages both to East-European partners and to EC countries partners. Most of the granted money being allocated to East-European universities (e.g. financial means for equipment purchase may be used only in East-European countries), the roles of Coordinator and of Contractor are very important. It depends on them how they will make the cooperation interesting also for the West-European partners. We can mention a case where no financial means for mobility of West-European partners were allocated. The richer and more helpful West-European partners covered once or twice their travel costs from their own sources, it is true, but later they cancelled their cooperation in the project because their representatives could not take part even in the Management Board meetings. On the other hand, some West-European partners try to shift almost all financial means intended for mobility to support long-term stays of outstanding postgraduate students from East-European countries in their institutions where these students are used as cheap employees to develop e.g. sophisticated software packages. In spite of an important contribution for eligible East-European universities, this type of mobility becomes, if it prevails and suppresses other forms of mobility, rather a form of help of the poor East to the rich West and a manifestation of inadequate cooperation concept.

The problems of East-European countries are then to a large extent determined by their lack of experience in international cooperation with West-European universities, this cooperation having been obstructed by communist regimes. That is the reason why the posts of Coordinators and also of Contractors should be occupied (at least in the second and third years of the project) by good organizers from eligible universities. They have in comparison with their western colleagues two main advantages: they are personally much more interested in the success of the JEP and their personnel costs are about ten times lower as those of their western colleagues. This would enable to reduce considerably the personnel costs of the project. On the other hand, not only Chief Coordinators should be remunerated but also local national coordinators in eligible countries and the key specialists working on the projects. Informal and successful coordination of international cooperation and particular contributions to the project require a lot of time and even the most willing people work better if they are motivated by an appropriate remuneration, even if their financial awards would not represent in most cases the main reason why they participate in the cooperation.

References

Valášek M. (Edit.) (1991) : Proceedings of The International Workshop on Computer Aided Learning and Simulation Technologies, Prague

Květoň K. (Edit.) (1991) : Working Plan on TEMPUS JEP "Computer Aided Learning and Simulation Technologies", (JEP 1087-90) in Academic Year 1991/92, Prague

Dvořák L. at al. (1991) : Famulus 3.1, Manual, Famulus Etc., Prague

Collaborative Railway Roller Rig Project

S.D. Iwnicki (*), Z.Y. Shen (**)

(*) Department of Mechanical Engineering, Design and Manufacture, Manchester Polytechnic

(**) Railway Vehicle Institute, Southwest Jiaotong University

ABSTRACT
This paper describes a collaborative project between Manchester Polytechnic in the U.K. and Southwest Jiaotong University in China. Both institutions are currently constructing railway roller rigs to help analyze and improve vehicle behaviour. During this project the 1/5 scale rig in Manchester will be used to help set up the data capture instrumentation and software for the rig in China. The scale roller rig is being designed and built by a small team including both the authors, postgraduate students from Southwest Jiaotong University and Undergraduate students from Manchester Polytechnic.
KEYWORDS: Vehicle dynamics, Simulation, UK-China cooperation

1. INTRODUCTION

In a collaborative project partly funded by the British Council a 1/5 scale railway roller rig is being built at Manchester Polytechnic to assist with the development of a full size roller rig in China. The full size roller rig is part of a laboratory in the Southwest Jiaotong University which will be used to improve the performance of railway vehicles in China. The scale roller rig is being designed and built in Manchester by a team which is led by the first author with the various aspects of the design being carried out by postgraduate students from the Chinese University and Undergraduate students from Manchester Polytechnic.

The aim of this project is to establish a suitable control system and a data capture and analysis system on the scale roller rig in the UK before transfer to the full size roller rig in China. The various parts of the roller rig and the data capture system are considered by individual members of the team under the guidance of the authors.

1. THE FULL SIZE ROLLER RIG

The Institute of Railway Vehicle Research at Southwest Jiaotong University in the People's Republic of China is in the process of setting up a traction power laboratory. The laboratory will provide a major testing facility for Chinese Railways and will help to achieve four main aims:

1. Raising the freight train weight from 3000 t/train to 6000t/train and increasing passenger trains from 15 cars/train to 25 cars/train.

2. Increasing the speed of freight trains from 80 km/hr to 120 km/hr.

3. The introduction of special high speed trains capable of speeds up to 250 km/hr.

4. Prevention of derailment.

The main part of the laboratory is a full size roller rig which will allow a vehicle to be run at high speeds in controlled laboratory conditions. A bogie vehicle with four axles will be accommodated and excitation in the lateral and vertical direction will simulate the alignment of real track. The excitation is provided by 8 actuators in the vertical direction and 8 in the lateral direction. The total excitation power is 1100 kW and the control program runs on a VAX 4000 computer.

The vehicle will sit on top of the rollers which will rotate to give the required linear speed at the wheels. The actuators will be controlled to force the rollers to move according to stored data representing real track parameters. This motion will simulate the conditions encountered when running at speed on various qualities of track.

The vehicle will be fitted with instrumentation to allow its performance to be monitored. The data will be compared with that from a mathematical model of the vehicle and this model will be used to suggest improvements to the vehicle design. The effect of these changes on the full size vehicle on the roller rig can then be evaluated.

2. THE 1/5 SCALE ROLLER RIG

A 1/5 scale roller rig has been designed and is being manufactured in the Department of Mechanical Engineering, Design and Manufacture at Manchester Polytechnic. This rig will be used as a test bed for the control system which will then be installed on the full size rig in China.

Although the sizes of the two rigs are different the same principles of excitation and data acquisition are being adhered to. The 1/5 scale roller rig is shown in figure 1.

Scale roller rigs have been used by many researchers to investigate the dynamic behaviour of railway vehicles. Probably the earliest was at British Rail Wickens (1965) but more recent work has shown the value of data form roller rig tests when used correctly Jaschinski (1990) and Heliot (1985).

The errors inherent in using a roller rig with rollers which have a finite radius have been analyzed at Manchester Polytechnic. These differences have been simulated with computer models and excitation techniques developed which tend to minimise these errors and provide realistic excitation that matches the situation of a railway vehicle running along a track. Track alignment data has been collected using equipment designed and manufactured for this work described by Iwnicki and Brickle (1992).

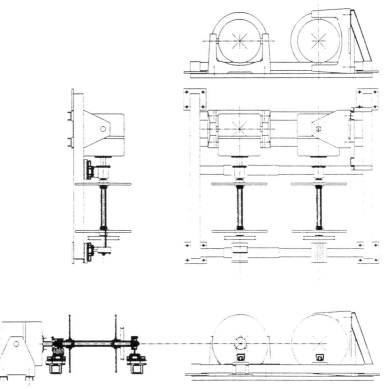

Figure 1. The 1/5 scale roller rig at Manchester Polytechnic

A computer program follows the algorithm that has been developed to control two shakers which are required to excite the rollers in the lateral direction with forces up to 500 Newtons. These motions then accurately simulate the effect of the rail misalignments in the lateral direction. Vertical excitation on the full size roller rig can be controlled in a similar way utilising the vertically mounted actuators.

3. METHOD OF DATA CAPTURE

For acquisition of the data required from the roller rig a microcomputer based system has been set up. This is initially designed to capture data from the 1/5 scale rig and therefore uses four channels of analogue input. If successful a similar system with more inputs will be set up for the full size rig. The data is required to compare the behaviour of the vehicle with mathematical simulations that have been carried out and to assess vehicle performance against safety and passenger comfort criteria. Changes to the suspension of the vehicle can be assessed at a range of speeds and track conditions.

The Computer models that have been set up produce results in the form of a frequency spectrum of vibrations or as a time history of the selected parameter. The roller rig is required to validate this data and therefore accelerometers are mounted in suitable locations on the body and the wheelsets of the roller rig. The signals from these accelerometers are conditioned and, if necessary, integrated by charge amplifiers and integrators and then digitised before being sent to the data acquisition package.

The computer package Spiders is used to monitor four channels simultaneously. The signals are combined in the required way and can then be stored for presentation and comparison with the simulation results.

The signal conditioning flow chart is shown in figure 2.

4. PROJECT MANAGEMENT

Several people were involved in the design process of the 1/5 scale roller rig at Manchester Polytechnic. Each member of this small design team had specific responsibilities and regular meetings were held to coordinate progress. The objectives for the roller rig performance were set out by the two authors to allow the process of technology transfer to the main roller rig in China. The specification was then set out by the team in

Manchester and one undergraduate member designed the mechanical parts of the rig. The method of excitation was agreed by the group taking into account the work of one of the visiting Chinese scholars who had investigated the inherent errors in the motion of the rollers. Component selection and overall design was coordinated at the group meetings.

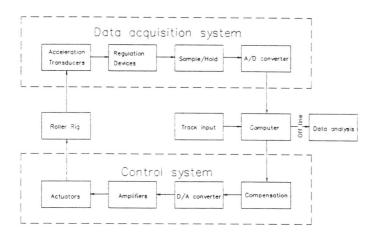

Data Flow Chart

Figure 2. Flow chart for data acquisition and control systems

The British Council ALCS funding made it possible for scholars from China to come to Manchester to work on various parts of the project and for the two authors to meet regularly to coordinate the developing design of both roller rigs.

The expertise developed in setting up the 1/5 scale roller rig, particularly with regard to the control and data acquisition methods can be easily transferred to the system for the rig in China.

5. CONCLUSIONS

A 1/5 scale roller rig for simulating the behaviour of a railway vehicle has been designed and is being constructed. A control system has been set up for this roller rig to allow realistic track alignment data to be used as an exciting input to the rollers. A data acquisition system has also been designed to collect exactly the data that is required to validate vehicle models and analyze behaviour.

These designs have been carried out by a group with members from both institutions and the systems developed will be transferred to a full size roller rig which is currently being constructed in China. The work on the scale rig has helped the parallel development of the systems for the full size rig.

REFERENCES

Jaschinski,A On the application of similarity laws to a scaled railway bogie model Doctoral Thesis Delft University 1990

Wickens, A.H. Correspondence Proc. I.Mech.E. vol 180 1965 part 3F p149-150

Heliot,C Small scale test method for railway dynamics Proc. 9th IAVSD Symposium Linkoping 1985 Swets & Zeitlinger

Iwnicki, S.D. and Brickle, B.V. (1991) Application of dynamic simulation methods to rail vehicles used in British Coal mines Proc. Railway Dynamics for Today and Tomorrow I.Mech.E 1992

Simultaneous Degree Programs in Engineering and a Second Language for American Students

J.S. DiGregorio (*), T. Krauthammer (*),
K.M. Grossman (**), W.R. Hager (*),
M.E. Keune (***), H.J. Sommer III (*)
(*) College of Engineering, Penn State University
(**) Department of French, Penn State University
(***) Department of German, Penn State University

Abstract

Two simultaneous degree programs are currently being developed at the Pennsylvania State University (USA) in a cooperative effort between the College of Engineering and the Departments of French and German. In addition to the credit requirements of the selected engineering major, students will complete a major in French or German language and culture, and participate in an engineering internship, which includes the study of engineering at a French or German school and a work experience with an engineering firm in either France or Germany. Students who complete this program will both broaden their knowledge and understanding of engineering culture and design practices in France and Germany, and gain an appreciation of French and German language, history, culture, and tradition.
Keywords: Engineering, Education, International Curriculum, French, German, Simultaneous Degrees, Internships.

1 Introduction

It is clear that today's engineers must have an international perspective if they are to compete successfully. Modern technological advances are made globally, and their impacts ripple throughout the economies of nations worldwide. Furthermore, the relationship between a country's history and culture and the evolution of its engineering activity must be understood to gain a comprehensive perception of engineering practices.

In such an intensely interdependent world, a leadership position in engineering and technology is required to maintain a leadership position in the international marketplace. Thus, the global technical environment is directly related to a nation's industrial competitiveness and places demands on the graduating engineer beyond learning basic technical skills.

Many engineering schools worldwide have recognized the need to adopt an international engineering component in their undergraduate programs. Concomitantly, numerous organizations in the United States, such as the National Research Council (1991), the American Society of Civil Engineers (Wright, 1990), and the American Society of Mechanical Engineers (1991), have issued reports calling for the introduction of such activities into the U.S. engineering educational system.

Penn State's College of Engineering recognizes the need to produce highly competent graduates who possess the intercultural skills necessary to help the U.S. maintain its position of worldwide leadership and innovation in engineering. In addition to study abroad and student exchange opportunities, the College is re-examining its curriculum to determine how to internationalize the educational experience. One such innovation is the development of simultaneous degree programs in "Engineering and a Second Language." The educational experience provided by these programs will increase students' competence in engineering as well as in intercultural relations, enabling them to work more effectively in the international environment of modern business and industry.

2 Simultaneous Degree Programs at Penn State

A "simultaneous degree program" is defined at Penn State as "a combination of two major programs of study which combine two distinct fields where an unusual combination of background and breadth is desired." Such is the case in the internationalization of the engineering curriculum, where proficiency in a second language, and in the history, tradition, and culture of the corresponding country, are as important as the technical skills acquired in the study of engineering.

Grandin (1989, 1991) at the University of Rhode Island was the first to report on a simultaneous degree program in Engineering and German in the U.S. A similar program was introduced recently at the University of Maryland (Berman, 1991). Challenger and Feehan (1992) have described a curriculum for engineering students at the University of Cincinnati involving German or Japanese language proficiency, a sequence of courses on the history, politics, economics, culture, and geography of Germany or Japan, and a work experience in one of those two countries.

Although a period of study and/or work experience in Germany is included in the Rhode Island, Maryland, Cincinnati, and Penn State programs, these programs differ significantly from typical "exchange programs" and "internships," offered by a large number of engineering institutions in the U.S. These latter programs offer students the opportunity to study or work abroad for an extended period of time, usually up to one year. The remainder of the engineering educational experience is spent at the home campus, with little or no exposure to the international aspects of engineering. Simultaneous degree programs, on the other hand, continuously expose engineering students to engineering, language, and culture throughout the entire period of study.

The development of simultaneous degree programs in "Engineering and a Second Language" at Penn State is a cooperative effort between the College of Engineering and the College of the Liberal Arts.. In late 1988, faculty in the Department of German were contacted by several companies in North Rhine Westfalia interested in offering work experiences to Penn State engineering students. Both the Department of German and the College of Engineering recognized the opportunity to provide our engineering students not only with exposure to international business practices, but also with a unique insight into German engineering practices.

It was obvious from the start, however, that for our students to gain the maximum benefit from this opportunity, a pre-immersion in the language and culture of Germany would be required. The College of Engineering and the Department of German agreed to establish a simultaneous degree program in Engineering and German and to infuse in it the requisite academic quality. Soon afterward, the Department of French indicated similar interest in developing a simultaneous degree program in Engineering and French. A coordination committee, consisting of faculty from Engineering, French, and German, was formed in mid-1991 to oversee the efforts.

To determine the level of language skills that our engineering students bring to the University, the committee conducted surveys of the "second language" courses our entering first-year students had taken in high school prior to enrolling at Penn State.

Table 1. Percentage of Penn State's Entering First-Year Engineering Students Who Had Taken "Second Language" Courses in High School*

| | ------Years of "Second Language" Taken------ | | | | | |
	1	2	3	4	5	Total
Spanish	1%	12%	21%	10%	3%	47%
French	1%	6%	9%	5%	2%	23%
German	1%	6%	8	2%	2%	19%
Latin	2%	3%	1%	1%	0%	7%
Russian	0%	1%	0%	0%	0%	1%
No Language	3%					3%
					Total	100%

* combined survey data for academic years beginning Fall 1989, 1990, 1991; University Park Campus.

As Table 1 shows, 97% of surveyed students entering the first year of engineering study at Penn State's University Park campus had studied a second language in high school. The largest number (47%) had studied Spanish, while smaller numbers had studied French (23%) or German (19%). Only a few students had studied Latin (7%) or Russian (1%).

A separate study revealed that many Penn State engineering graduates take at least one "second language" course while at the University. Of this group, 34% take courses in German, 22% in Spanish, and 17% in French. The remaining 27% study Chinese, Greek, Hebrew, Japanese, Korean, Latin, Russian, Polish, and/or Portuguese.

Encouraged by these facts, the coordination committee consulted with the Colleges of Engineering at the Universities of Maryland and Rhode Island to learn about their simultaneous degree programs in Engineering and German. Proposals were then prepared by the committee to obtain internal administrative approvals for simultaneous degree programs in Engineering and French, and in Engineering and German. The committee has begun to identify engineering schools and companies in France and Germany willing to become our partners by accepting our students during the internship phase of the program.

We plan to accept the first group of students into the two programs during academic year 1992/93. Students who complete these simultaneous degree programs will receive the Bachelor of Science (B.S.) degree in the appropriate engineering major, as well as the B.S. degree in either French or German.

The primary aim of the program is to allow qualified engineering students to add an international dimension to their studies at Penn State. This will be accomplished by intensive study of language and culture at Penn State and abroad, along with an international work experience. At some future date, we plan to expand the idea to include several other languages. Presently, we envision the development of simultaneous degree programs in Engineering and Spanish, Russian, Japanese, or Italian.

3 Content of the Curricula for Simultaneous Degree Programs

The minimum number of academic credits required to graduate in each of the eleven engineering majors available to undergraduate students at Penn State varies from 132 in Civil and Environmental Engineering to 168 in Architectural Engineering.

Undergraduate engineering curricula at Penn State are designed to be completed over a four-year period, except for Architectural Engineering, which is designed as a five-year program. The French or German degree portion of the program, discussed below, as well as a requisite one- or two-semester study and work experience in France or Germany, will add another year to the period of study.

Before acceptance into either simultaneous degree program, students must demonstrate proficiency in the corresponding language, as evidenced by high school transcripts, advanced placement tests, and/or examinations administered by the Departments of French or German. Students who show such proficiency will be awarded a certain number of credits - normally 8 to 12 - toward the French or German degree.

Engineering-French:

To receive B.S. degrees in both Engineering and French, students must complete 37-40 credits of French language and culture beyond the intermediate level in addition to the appropriate number of credits in the engineering major. Table 2 shows the distribution of credits for the French requirement.

Table 2. Courses in French Language and Culture Required in the Engineering-French Simultaneous Degree Program

---

FR 201 (4 credits) Oral Communication and Reading Comprehension
FR 202 (3) Grammar and Composition
FR 203 (3) Applied Phonetics
FR 308 (3) Business Writing in French
FR 330 (3) French Culture and Civilization
FR 350 (3) Introduction to French Literature
FR 401 (3*) Advanced Oral Communication
FR 402 (3*) Advanced Grammar and Writing
FR 409 (3-6*) Commercial and Technical Translation
FR 430 (3*) Contemporary France
plus 3* additional credits in French at the FR 4XX level
plus 3 additional credits in French literature at the 4XX level

* credit requirements for these courses may be satisfied by the engineering internship in France, at the discretion of the Department of French.

Engineering students can apply up to six of the additional credits in French language and culture toward the "General Education" credits required of all Engineering majors. To receive simultaneous degrees in Engineering and French, the total number of credits required will vary by Engineering major, from as few as 163 in Civil & Environmental Engineering, to as many as 208 in Architectural Engineering.

Engineering-German:
To receive B.S. degrees in both Engineering and German, students must complete 36 credits of German language and culture beyond the intermediate level in addition to the appropriate number of credits in the engineering major. Table 3 shows the distribution of credits for the German requirement.

Table 3. Courses in German Language and Culture Required in the Engineering-German Simultaneous Degree Program

GER 100 (3 credits) German Culture and Civilization
GER 200 (3) Culture in East and West Germany
GER 201 (4) Conversation and Composition
GER 301 (4) Intermediate Conversation and Composition
GER 308 (3) German Business Communications
GER 330 (3) History of German Literature I
GER 331 (3) History of German Literature II
GER 399* (3) Foreign Study - German
GER 401* (4) Advanced Conversation and Composition
GER 408* (3) Advanced German Business Communications
GER 499* (3) Foreign Study - German

* credit requirements for these courses may be satisfied by the engineering internship in Germany, at the discretion of the Department of German.

Engineering students can apply up to six of the additional credits in German language and culture toward the "General Education" credits required of all Engineering majors. To receive simultaneous degrees in Engineering and German, the total number of credits required will vary by Engineering major, from 162 for students in Civil & Environmental Engineering to as many as 204 in Architectural Engineering students.

4 Funding

Support for this program has been provided in part by the National Science Foundation (NSF) through the Engineering Coalition of Schools for Excellence in Education and Leadership (ECSEL), a coalition of seven U.S. schools and colleges of engineering: Penn State, the University of Maryland, the University of Washington, City College of New York, Morgan State University, Howard University, and the Massachusetts Institute of Technology. Established in 1990, ECSEL is engaged in a five-year effort to improve dramatically the effectiveness of undergraduate engineering education in preparing students for lifelong, productive careers. Among ECSEL's goals is the internationalization of the engineering curriculum. ECSEL is responding to a call for action by NSF, which is providing $3 million per year to the coalition.

5 Summary

Technological decisions today are greatly influenced by global politics and economics. To practice their profession effectively, engineers must be familiar with the non-technical conditions under which they make technological decisions. This presents a formidable challenge both to engineers and to the engineering academic community.

To meet this challenge, the Pennsylvania State University, in a cooperative effort between the Colleges of Engineering and the Liberal Arts has begun to internationalize its engineering curriculum. One innovation is the introduction of two simultaneous degree programs in Engineering and a second language. These programs are designed to increase students' competence in international engineering, as well as in intercultural relations, and to enable them to work more effectively in the modern business and industrial environment.

Such training can produce not only better engineers, but also synthesize the different approaches of technology development and application into a Global Science of Engineering.

References

National Research Council (1991), **Improving Engineering Design - Designing for Competitive Advantage**.

Wright, R.N. (1990), **Prospects for International Engineering Practice**, American Society of Civil Engineers.

ASME NEWS (1991), ASME Must Lead Efforts to Improve US Education, Vol. 10, No. 3, p. 1.

Grandin, J.M. (1989), German and Engineering: An Overdue Alliance, **Die Unterrichtspraxis**, 22 (No. 2), 146-152.

Grandin, J.M. (1991), Developing Internships in Germany for International Engineering Students, **Die Unterrichtspraxis**, 24 (No. 2), 209-214.

Berman, Marilyn R. (1991), University of Maryland, private communication.

Challenger, Kenneth D., Feehan, M. Therese (1992), University of Cincinnati International Engineering Program, **Proceedings of the Engineering Foundation Conference**, Santa Barbara, California.

Foreign Language Learning for Engineers

D.J. Croome (*), G.K. Cook (*), S.E. Poole (**)
(*) Department of Construction Management & Engineering, University of Reading
(**) Department of French Studies, University of Reading

Abstract
The recent changes in the internationalisation of engineering have brought with them a need to strengthen the communication links within it. Communication is required for the successful completion of any task. This paper describes the innovative language element of the Building Services Engineering Design and Management (BSEDM) course at the University of Reading which provides engineering undergraduates with skills in a modern European language (French, German or Italian). An industrial placement period is seen as an essential part of this course since it marks the point at which its content is re-orientated from a general to a strongly technical engineering vocabulary besides offering personal development of language skills within a foreign country. Although specific in its engineering objectives there is a need to share and understand how other countries express themselves in the spoken word, the written word, as well as in more general cultural ways. This course aims to prepare the student to become acquainted with technical aspects of a foreign language and also to reach a stage which would enable the student to become fluent reasonably quickly.
Keywords: Engineering, Languages, Education, Europe.

1 The Need for Language - The Cultural Link

> "Thought development is determined by language"
> Vygotsky.

English has now become the principal commercial and business language of the world. The development of the Commonwealth undoubtedly saw English become a major international language but today it is the second language for Japan, China and Eastern Europe as well as most countries in Western Europe. Those of the "All the French/Austrians/Germans (etc) speak English" school of thought might therefore question the necessity of studying a foreign language.

They may indeed raise the issue while enjoying a glass of Bordeaux or a Beethoven symphony - while appreciating, that is to say, the best of a foreign culture. Those cultures whose lifestyles, art or other forms of expression so many appreciate have their root, their essence, in the language they use to express themselves: The sharing of that foreign language thus provides the most intimate link with different countries and cultures.

From a more practical angle, a people's language gives valuable insight into its attitudes, approach and priorities. The delicate formality of French business correspondence, for example, provides clues as to French work practice and initiates the British or American businessman, used to meetings where policy is decided, to the notion of the meeting as *debate*, wherein ideas are thrashed out...but do not necessarily have to lead to instant *decisions*.

2 The Language of the Poor Communicators

The need for a language dimension to vocational courses is now being recognised, Lloyd (1992), Saunders (1992). This was not based on the academic merits of stretching science and engineering students into new areas where debate and interpretation were recognised as useful tools, but rather as a grudging recognition that some of the rest of the world either did not, or did not want to talk in English. National identity is a potent force. So too of course is commercial pressure linked to international markets and it is probably this which has focused minds onto the teaching of languages to engineers. This has been given new stimulus recently with the single European Act and the Maastrict summit. Although these points are generally true for all engineering disciplines they are particularly relevant for the building services engineering industry since its business is truly international.

There is a strong belief that engineers are very poor communicators. Although this may be due to the fact that use the language of numbers, an area of low knowledge for the population at large, it was recognised as sufficiently important to call for the introduction of communication skills/structure of the industry/role of engineer in society courses to be included in undergraduate engineering courses, Ellison (1991).

3 Planning the Choices

Through the SARTOR guidelines the content of engineering courses is firmly controlled. A failure to meet the requirements will result in the course being designated as non-engineering, it may be a very interesting course but not an engineering one. The guidelines are not specific about detailed content although the 60 to 70% engineering requirement means that there is little place for what is seen as peripheral subjects.

The opportunity to add a language element into an engineering course would always involve some reappraisal of the traditional course structure, if realistic student hours and course content are to be balanced. However courses are at present undergoing many changes to try and attract school leavers but more importantly introducing a language course can achieve enhanced education, Saunders (1991).

In practice the time squeeze on all engineering courses makes them intensive due to their vocational nature and the increasing rate of technological change. This is more marked in the BSEDM course at Reading since it contains a core of engineering interwoven with the two distinct strands of management and a modern European language.

The Engineering Council also specifies the responsibilities of a Chartered Engineer and this includes having an international outlook.

4 Entry Conditions

The course was tailored to the probable student intake common for engineering courses. These candidates do not generally have a GCE "A" level language. Although the language departments can normally accept students without a language at GCSE "O" level, a preference during application processing is given to candidates with a good GCSE "O" level performance.

5 Course Plan

The course calender is shown on Diagram 1. At Reading we offer a choice of French, German and Italian which is studied during part II (three terms) and part III (three terms) for a total of c170 hours. The part II course is designed to consolidate and improve conversational, reading and basic writing skills, aiming to provide a solid foundation on which to develop a grasp of the more technical vocabulary relevant to the main degree discipline. In the summer following part II the student spends six weeks in the relevant country working for a company in the building services engineering industry. This differs from those courses which involve study periods in Europe, Lloyd and Harvey (1991), but offers many of the educational advantages.

The part III course aims to introduce a more technical vocabulary, building on the students work experience and using manuals designed for the relevant country's HNC/D - equivalent students. Two technical journals in the relevant language, chosen after consultation with The Building Services Research and Information Association are subscribed to, and multilingual specialist dictionaries are available in the library.

6 Industrial Placement

During the summer of 1991 all of the Part III students were placed with European building services engineering companies. Five were placed in Germany and six in France. Sometimes the sponsoring company arranges the placement. Details of the placements are given in Table 1. Some German companies were selected through an established link between the Reading and The University of Kassel which has a very well organised department that can place students into the workplace both during courses and following graduation. Although Kassel does not run a building services engineering course it has links with many engineering companies. One of the placement companies arranged for the Reading students to spend two three week periods in what was the East and West of Germany. Some companies were known from personal contacts in Germany and France.

Most of the French companies were selected with the help of U.K. consultants and engineers.

Accommodation was generally provided by a family from the placement company who were requested to speak the foreign language only, except in a case of emergency. This provided the student with a considerable amount of time to develop the foreign conversational skills. During working hours it was expected that some English would be spoken. This is due to the emerging technical vocabulary of the students and the need for a more commercial pace. The students were expected to keep a detailed diary and produce a 2000 word technical report in English and a smaller 500 word report in the foreign language. In this way duplication of the contents of the reports is minimised whilst the challenge of communicating technical aspects in a foreign language is not.

Table 1. Details of European Industrial Placements Summer 1991.

Country	Town	Company	Activity
Germany	Kassel	Rudolf Otto Meyer	Building Services Engineering Design and Construction (one student)
Germany	Betzdorf	Heinrich Nickel GmbH	Building Services Engineering Research and Development (one student)
Germany	Thuringia	Energietechnik GmbH	Building Services Engineering Design and Construction (two students)
Germany	Rodermark	JEWA Industrieplan GmbH	Building Services Engineering and Construction (one student)
France	Paris (Chatillon)	CGC Entreprise	Building Services Engineering Design and Maintenance (two students)
France	Courbevoie	Sulzer	Building Services Engineering Equipment Design/Installation (one student)
France	Nanterre	CEGELEC	Building Services Engineering Plant/System Design (one student)
France	St.Jean de la Ruelle	Herve Thermique	Building Services Engineering Site Installation (one student)
France	Cergy Pontoise	Spie Batignolles	Building Services and Electrical Engineers (one student)

7   Assessment Methods

The assessment of the Part II course is through continual assessment. An oral test is part of this assessment and this together with written work is considered by the Part II examination board. All students must pass this course to proceed to Part III.

The assessment of the industrial placement period is through the two reports described above.

The Part III assessment is carried out in two parts. The first part is an Oral Examination of c30 minutes. The second part is a 3 hour final examination. The weighting of each is Oral 50 and Examination 100. The examination of the language at finals level makes a clear statement to the students of its importance and ensures their appreciation of the language component as an on-going and integral part of their training for the internationalised *métier* engineering has become.

8   Conclusions

The language content of the BSEDM course at Reading has:
* Produced strong links with ten European building services engineering companies.
* Strengthened the existing links with continental university departments.
* Been received enthusiastically by the students
* Improved the quality of project and dissertation work where evidence of reading and understanding foreign papers and technical articles has added another dimension to their engineering studies.
* Highlighted the problems of differing language ability of the students. This has created a need to review the admissions procedures and course tutoring.

References
Vygotsky, L.S. (1962) **Thoughts & Language**, Cambridge M.I.T. Press.

Ellison, G. (1991) The Future Pattern of First Degree Courses in Engineering, **Report of the Engineering Professors Conference Working Party**, Prof. G.Ellison (Chairman), EPC Secretariat, University of Surrey.

Lloyd, D., Harvey, S.J. (1991) An Engineering Student Exchange Programme between Britain and Germany, **Proceedings of a National Conference on Innovative Teaching in Engineering**, (ed. R.A. Smith), Ellis Horwood, London, 47-52.

Lloyd, D. (1992) **Personal Communication**, Department of Combined Engineering, Coventry Polytechnic.

Saunders, R.J. (1991) The European Dimension in Engineering Education, **Proceedings of a National Conference on Innovative Teaching in Engineering**, (ed. R.A. Smith), Ellis Horwood, London, 65-69.

Saunders, R.J. (1992) **Personal Communication**, Department of Mechanical and Process Engineering, University of Sheffield.

272 Engineering Education

BSEDM Course

Year	Autumn Term (Weeks 1–10)	Lent Term (Weeks 1–10)	Summer Term (Weeks 0–10)	Vacation (Weeks 1–12)
1	Lectures (10 Weeks); Lectures (10 Weeks) + Languages course; Lectures (10 Weeks) + Languages course + Integrated Design Project & Diss (1.5 day/week)	Lectures (10 Weeks); Lectures (10 Weeks) + Languages course; Lectures (10 Weeks) + Languages course + Integrated Design Project & Diss (1.5 day/week)	Exams / Fieldwork; Lectures (9 Weeks) + Languages course; Proj Assessment	EA 1; Industrial Placement (10 Weeks)
2			Exams / AROUSAL; Options + Languages course + Integrated Project (8 weeks); Examiners Meeting	Industry/Europe (6 weeks); Industrial Placement (6 weeks)
3			Proj Assessment / Lectures; Lectures; Graduation	Integrated Design Project

Diagram 1

Languages for Engineers at Nottingham Polytechnic

E. Stewart, A. Jones

Department of Languages, Nottingham Polytechnic

Abstract
Curriculum development in language for engineers must aim to reflect the range of their language needs. In practice this involves a balance between subject-specific input and the essential functions required by all language users. Close collaboration between linguists and engineers is indispensable as are high quality tuition and resource development.
Keywords: Development, Languages, Engineers, Needs, Subject-specific collaboration.

Introduction

The foreign language skills of the British are legendary. Students entering Higher Education in the 1990s, however, seem determined to subvert our centuries-old reputation for linguistic arrogance, ineptitude and insularity. Nottingham Polytechnic's Engineering students find every encouragement to prepare for professional life in the European and larger international context: not only in the extensive placement and exchange programme and related initiatives, but also in the range of supporting language provision.

1989 saw the introduction of the Polytechnic's institution-wide programme offering students of any discipline an opportunity to learn a language. The Engineering Faculty was one of the chief motivating forces behind this initiative: the first French, German, Spanish and Italian classes were piloted there, as was Mandarin Chinese. In response to demand, Japanese will join them in September.

I should like to begin by acknowledging the help of BP, who gave a grant of £10,000 to enable a pilot programme to start in the Engineering faculty. I would also like to thank colleagues from the faculty without whose co-operation and goodwill the Language Programme in its current form would not have been possible.

In 1990 the Language Programme doubled in size, to include 500 applications from Engineering students and staff alone. For some of them language skills are a preparation for foreign exchanges and work placements, or professional life abroad: for others, with no immediate plans to travel, a language qualification is a more general enhancement of their cv and job opportunities. Survival skills; a foundation for development here or abroad; an opportunity to perfect knowledge acquired elsewhere: the Programme has to design and deliver syllabuses to meet all of these needs.

Language study in context

Language study in the Engineering Faculty involves 3 hours of class contact per week. For a significant proportion of students, languages form an accredited element of their main study. Since research underlines the importance of accreditation for sustained motivation we are working with colleagues from Engineering towards making languages an integral part of all courses in the faculty.

Students who are successful in their language study, whether or not it is accredited, receive an official Polytechnic Award (or Advanced Award).

The language element is examined entirely by coursework assessment. Assignments are designed centrally and based closely on syllabus topics. All marks and samples of work are subject to External Examination as part of the Programme's monitoring process, the final stage of which involves the completion of course evaluation forms by students and staff. This feedback is an important factor in pedagogic and administrative revisions to the course.

Student numbers 1991-2

Numbers of engineering students studying languages during 1991-2 was as follows:

French	65
German	114
Spanish	16
Italian	8

Philosophy

There are now over 20 institution-wide Language Programmes in the British Polytechnic sector, and their approach to languages for technologists has differed significantly from university provision. Universities have traditionally taught grammar, writing, reading and technical translation skills to small groups. Polytechnics have tended to reflect a more recent move away from this grammar-based approach by teaching their much larger numbers of students active situational language, speaking and aural comprehension skills. At Nottingham we feel that neither approach satisfactorily equips students for the range of demands made of them.

As interest in Europe grows, so too does the number of language learners, and the small-group approach is increasingly unworkable. The administratively convenient alternative - the same syllabuses for all students, whatever their specialism - is equally unacceptable, particularly at more advanced language levels. All learners need a general framework if they are to progress through the language learning stages. But there must be scope from the earliest stages for subject specialism. With this in mind the Language Programme team is re-designing its syllabuses for engineers to reflect:

  the need to enhance grammatical awareness, as well as reading, writing, speaking and listening skills, so that students will retain and can develop what they have learnt;

the importance of reserving more space within the broad
framework for broad Engineering-specific topics, so as to
optimize relevance and student motivation;

our concern to address the needs of the distinct learner groups,
that is, to offer essential cultural background to all learners; "get
by" skills to students about to go on work placements abroad;
academically-oriented language for exchange students; and
vocabulary extension and fluency practice to people with a good
level of language which they propose to use professionally;

All of this may seem rather obvious to European (and many non-European) colleagues. But as is often the case, Britain's linguistic pendulum swung from very traditional to quite novel methods in the 1970s and 80s. The old grammar-led approach produced students who could read and write fluently but not speak a word of their chosen foreign language. The alternative had students speaking and understanding the language from day one, but unable to retain or develop what they had learnt. The current notion of "best practice" involves taking the best from both approaches.

Course design; student needs

Our emphasis on topics of broad interest to engineers is quite deliberate. We can sometimes be misled into thinking that all our students necessarily have a common goal, and that they will all have exactly the same linguistic needs at the end of their course. In fact our students come with very different perspectives on the future, with differing motivations and in many cases quite ill defined career objectives. Moreover, in this institution, many different Engineering courses come together for language classes, so the most we can assume is a common interest in the field of Engineering.

It follows, therefore, that we are not so much teaching languages for specific purposes as preparing our students linguistically for the choice and variety of work which they will confront, and for the career changes they will almost inevitably undergo. The challenge, therefore, at least in the initial stages, is not so much an attempt to link up language and vocational input at all costs, but rather a search for the linguistic common denominators which will be useful in themselves and which will build a solid foundation for more specialist language-learning later.

Levels of Language Learning

At ab initio level, there is clearly less divergence. Most students (although not all) may not wish to do a placement in the country; they may have designs on visiting for short periods, or increasing career mobility later. At this stage, therefore, the reconciliation of a variety of needs is less problematic than at higher levels. The emphasis is communicative, the primary aim being practical proficiency in oral and listening skills, but with a good grammar base to complement this. Syllabuses are clearly structured, allowing time for revision as well as assimilation of new material. They involve elements of survival language while allowing students the opportunity to express themselves more freely in a variety of social settings. Socio-linguistic aspects and cultural awareness are also important, as language divorced from its context not only provokes boredom, but can lead to gaucherie and misunderstandings later on. At this stage some subject-relevant language is included, but it is related to Engineering, industry and business as a whole and remains fairly generic.

Subject-specific input

As we progress though the levels, the amount of subject-specific language gradually increases, but the syllabuses, rather than being too prescriptive, are sufficiently flexible to allow for some negotiation within individual groups. There is also an increasing balance between oral/aural skills and reading/writing skills. Not only does this allow for more vocational material to be used, but it equips students intending to do a placement abroad with the language and the skills they need for work and study. The essential vocabulary for survival is still there, but more structures are introduced which allow students to manipulate the language in a variety of situations, both specialist and non-specialist, depending on the vocabulary used. The gradual introduction of more vocationally-specific materials thus appears initially as reading material which at the same time develops students' gist comprehension skills. Prior knowledge of the subject covered may also help in this respect.

Non-linguistic skills

This brings us to the skills we are trying to develop when teaching languages to engineers. All the skills contained within a 'communication skills' course can in fact be incorporated into a language programme. We are not only interested in developing linguistic knowledge or cultural awareness, but also in incorporating the skills which our students will need in the work-place: negotiating skills; presentation skills; interviewing; chairing discussions, convincing, and so on. We want to develop interpersonal skills so that our graduates will use their newly-acquired language skills to the full, articulately and with self-assurance. For too long languages have been taught as a 'pure' subject, but they provide an ideal context for many transferable skills to be developed.

Utilizing staff expertise

Over the years, traditional methods have alienated large numbers of language learners. Some of these people are now in a position to make decisions about the value of language training for others. The old tendency to ask engineers with some knowledge of a foreign language to teach it has made many people's introduction to languages a rather negative experience. But paradoxically it has made those same learners rather suspicious of linguist teachers. And rightly so in some cases. If we are to maximize progress in all skills areas we need a *combination* of highly-skilled and technologically-aware linguists, and native-speaker technical input. Technologists are increasingly coming to see language learning as a serious (and enjoyable) activity, both interesting and professionally invaluable. But if we don't deliver this mix of professional linguistic and technical input, student demand alone will not produce a positive attitude towards languages and language learning.

Internal links - syllabus development

In Nottingham we have piloted a scheme whereby one fluent German speaker from the Engineering faculty has taught alongside a trained German teacher. Each has had a complementary input into the class despite very different starting points. Student feedback has been very positive, as they have appreciated the more 'adult' approach which contrasts with their secondary school experience. They also see language study as less of a 'bolt-on' addition and more as an integrated part of their whole programme. This co-operation between linguists and engineers is set to continue, as both see the benefits for students. Language teachers need to increase their knowledge base in order to be perceived as more than a necessary evil, and the contact between staff will gradually change the attitudes of any staff remaining who may regard language studies as peripheral to the real business of engineering.

Collaborative links- syllabus development

The time has gone when linguists could make lofty claims for the teaching of languages as an end in itself. We must now increase consultation and collaboration with engineers outside (as well as inside) the Polytechnic.

We are increasingly forging links with European-based companies in order to keep in touch with what they actually want and expect of our graduates. This feedback can then be reflected in syllabus and materials development. Most of the syllabus modification undertaken so far has been based on information from the world of employees. Good interpersonal skills, a basic knowledge of commercial language, and the ability to mix confidently in social situations with foreign colleagues are all required by employer contacts, as well as an ability to discuss in general terms the relevant industrial sector.

Feedback

No less important are the views of students recently returned from placements abroad, and feedback from young engineers sent abroad by employers as part of their training. They have emphasised the need for a sound grammar base in order to develop what has already been learnt. "Get by" skills are invaluable in the short term, but they will not prepare students for career mobility in the new Europe. For this we need to lay the foundations for further study in later life, whether in the form of independent study based on open learning, or through further in-company training.

Future development

As demand for special-purposes materials grows, collaboration has to extend between educational institutions. Informal consortia are consequently emerging in order to avoid duplication of effort. Since language study remains primarily interactive, we are developing oral skills by increasing native speaker input. We intend to use more foreign language assistants, foreign visitors to our own institution, and placement students from abroad.

As far as new technology is concerned, the increasing use of satellite broadcasts has already brought more authenticity into the classroom, but we look forward to a greater use of software, e.g. for self-correction in grammatical exercises. The more language learning is integrated into the whole curriculum, the better it can reflect the modes of learning to which students are accustomed. Thus the use of technology is appreciated by engineering students, as it adds a new dimension to existing skills.

Links abroad

We anticipate that links which already exist with our European counterparts will be further strengthened in the near future, giving further impetus to the Language Programme. Staff from Nottingham have visited ESTE (Ecole Supérieure de Technologie Electrique) in Paris and the Uni-Gesamthochschule in Paderborn, Germany as visiting lecturers: staff from Paris and Warsaw have lectured in Nottingham. Some students from ESTE complete a six-month project in Nottingham, while some students from Nottingham do an eight-week project with a French student in Paris. More recently, students from Nottingham have done their entire industrial training year in French industry in placements found by ESTE. Students from Paderborn have also come to Nottingham, while some Nottingham students have done their industrial training year in Germany. Students from Greece, Italy, Spain and Poland have also studied with us. We also have a collaborative project in power electronics in Romania.

Nowhere has the collaboration of Languages and Engineering been closer than in the introduction of Mandarin Chinese tuition. A visiting lecturer from Shanghai has been adapting the syllabuses while setting up a Chinese cultural centre in Nottingham. From September 1992 Mandarin Chinese will be a fully-fledged option on the Language Programme. With the advent of Japanese, we are now broadening our horizons to embrace the Far East as well as Europe. Staff have visited China with a view to creating links for the future, and several Chinese research students are currently studying in Nottingham. Staff as well as students have been attending Chinese classes. In fact the impetus for the development of Chinese has come from the Engineering faculty and represents a model for future co-operation between linguists and engineers. Paradoxically, the greatest degree of co-operation has been needed while introducing a language about which we initially knew very little. However, it has brought staff together in a common appreciation of the potential benefits of high quality language training for engineers. We are working hard to consolidate in the future.

Engineering Degrees and Foreign Languages

B.E. Mulhall

Department of Electronic and Electrical Engineering, University of Surrey

Abstract
A description is given of the main elements of courses combining the study of an engineering discipline with attaining mastery of a foreign (European) language. Some Final Year material is to be taught in the language. The adverse effects of studying in a foreign language are also discussed. It is questioned whether course structures which make efficient use of existing material are educationally the best.
Keywords Foreign Language, Learning, Engineering Education

1 Introduction

The origins of this paper are to be found in the development of Degree courses in engineering with foreign languages at the University of Surrey, and in a paper which has been published recently by Jochems (1991) on the effects on learning and teaching in a foreign language. From the latter the idea arises that what is effective from an institutional point of view when resources are limited may well be exactly the opposite of what is desirable on educational grounds. Such questions will be posed here, as a basis for discussion and further debate.

In the courses considered the study of engineering is combined with attaining fluency in a further language. These are timely because there is an ever increasing awareness of other countries as places in which one not merely passes holidays, but where one may carry on normal business activities or even live. Forty years ago the business horizons of most people in Great Britain ended at national boundaries; ten years ago it was the European Common Market; now even the EC is perhaps too restricted a region. Within these expanding horizons communication is essential, and the fact is that through much of the world the common language of business is likely to be English. On the one hand this is very satisfactory for the native speaker of the language - why spend time learning a foreign language, when there are so many other interesting things to do? On the other hand, the chances of the native English speaker being put in the situation of having to use a different language, and thereby have the opportunity to improve his skills, are, sadly, quite small.

Despite this, there is an increasing effort on the part of UK Establishments of Higher Education to address this problem, by offering courses which include a "European" component. The number of such courses, as judged by checking through the handbook listing all University courses, has increased by roughly 40% from 1991 to 1992. The precise growth is difficult to analyse, for some Universities which have previously had a low level activity, typically encouraging a few students to take a sandwich placement abroad, or a few exceptional individuals to study for a year abroad, are now making such arrangements an integral part of a distinct Degree, while simultaneously expanding the planned intake.

The final point which needs to be noted is that these developments are taking place against a background of considerable pressure on the curriculum and on resources. The curriculum has to cater both for the growth in technology itself, and from the recognition over the past decade that there is a considerable body of non-technical content - communication skills, project planning, business studies, team working, etc. - to be included in engineering courses if the graduate is to be fitted for survival in the modern industrial world. Resources, both staff effort and funds for materials and equipment, are becoming ever less freely available, so any new developments should ideally be less demanding in their requirements than present courses.

2   Who Wants to Study Engineering and Language?

The questions of which engineering students would also wish to study language, and the course structures which enable this, have been discussed elsewhere by the author, Mulhall (1992). The level of linguistic attainment sought by students can vary from a fairly low level right up to that which will allow living and effective working in a foreign country. It is anticipated that the lower level can be catered for by optional extra courses, typically an hour or two a week on top of the standard engineering studies, such as have been available for many years.

This paper is concerned with the more ambitious, where language is important enough to be included in the Degree title, and so must be, and be seen to be, an essential part of the course. In the case of some students this will form part of a quite definite career plan, though as often as not the aim is to keep open options, or even just to satisfy a personal ambition to become more broadly educated. From a national point of view the goodwill generated by showing a willingness to meet others in their own language and, by implication, cultural terms also, should not be undervalued.

The motivation for engineering will also be by no means uniform; there are always those who have less enthusiasm for the analysis and "hard" technology, and they may be expected particularly amongst those who seek a language enhanced course. Thus, the career pattern envisaged may be in sales or contract management, where the usual analytically based course may not be the most appropriate.

3   Course Structures

The courses available are inevitably heavily based on standard engineering courses, with appropriate parts added or substituted. Thus, for the first two years - the foundation stage -, language teaching will occupy typically four hours per week, and will displace the communication skills/business studies component. It is difficult to see what parts of the engineering content can be displaced, though in one such course it seems that production processes has been eliminated. In addition, at Surrey we wish to include an intensive course of at least four weeks duration in an appropriate country, in order to enhance greatly the students' competence and confidence in using the language. Although it might seem more logical for this to precede immediately any extended period abroad, experience over more than ten years suggests that it should be during the first summer vacation, as a result of which stronger motivation is maintained during the second year.

After the foundation stage it is usual to include a year abroad, though in Establishments offering "thin" sandwich courses this may be two separate six month periods. A common pattern is for the third year to be spent doing practical work, either in industry or a research project in a University laboratory, but in at least one case a fairly standard academic course is followed in the host country. At Surrey an industrial year abroad is the most natural pattern, arising from long and successful experience with "thick" sandwich placements in the UK and elsewhere. The final year is then spent at the home University, though the plans at Surrey envisage the possibility of some exceptional students undertaking the final year of study abroad.

Where the final year is taken in the home University it is clearly important to encourage the student to retain the competence achieved in the foreign language. Given the reasons for including the language in the first place, and the level which should have been reached by this stage, it is natural and logical for teaching to be in the language. However, the numbers studying any particular combination of language and engineering discipline are likely to be so small that the option of teaching technological subjects in the foreign language is hardly practicable. This is perhaps a little unfortunate, for the development of many engineering subjects is customarily given a different, often more abstract, flavour than in the UK. The alternative, being adopted at Surrey, is to give the business studies material, which has already been displaced from the earlier years, completely in the foreign language, and as a common course to all engineering disciplines. This could constitute up to 35% of the Final Year, giving an overall balance of engineering to language of around 3:1.

4 Learning in a Foreign Language

4.1 Anecdotal Evidence

Most organisers of courses in Higher Education have experience of foreign students, and form a view of how their performance is influenced by language problems. In general there are no concessions made to such students in the course material, although a language qualification will be required as a condition of admission, and there is a usually language support available. Since it is so readily accepted that these students should be taught in English, is it not illogical to be concerned about the problems English students may encounter when studying abroad?

Of course, as already suggested in the introduction, the use of English is so widespread that the student coming from, say, Norway or the Netherlands, or Singapore will have had considerable exposure to English even apart from formal schooling. In contrast, in the English-speaking world one has to make considerable efforts to get any worthwhile exposure to a foreign language. Moreover, the Norwegians, Dutch and Germans have the advantage of a native language which has a close relationship to English. Nevertheless, our experience has been that many of these students need several months before they can learn really effectively in English, though problems may not always be apparent. Factors tending to mask the difficulties are, first, that the previous technical education often overlaps significantly with the course being followed in the UK, and, second, that students who elect to go abroad are usually the better motivated and so able to work harder to overcome any problems. Students from southern Europe seem to have much greater difficulties.

Another aspect worth considering is the question of culture. By this is meant the way academic material is treated and examined in different countries. Put simply, the Anglo-Saxon tradition is to define rather precisely the time frame within which study takes place, and also to emphasise the practical and the pragmatic, rather than the theoretical. One result is that students have less freedom to study exactly what they wish at any particular time, and must sit examinations which are written, and relatively short. Another is that in practice, however much deplored by academics, students have to study strategically and learn only as much and as far as is necessary to pass examinations.

4.2 Analytical Studies

Jochems (1991) reviewed the results of a number of quite independent studies on the performance of students in North America whose native language was not English. Because the disciplines, the backgrounds and the ability levels of the students all varied widely - certainly between studies, and in some cases within a study - it is difficult to draw firm conclusions, though the evidence does at least seem consistent with what might be expected from experience. Thus, it appears that in engineering academic success is much more strongly correlated with mathematical ability than with linguistic ability, whereas the opposite appears to be true for humanities and social sciences. However, once the linguistic attainment becomes sufficiently high the correlation becomes much weaker, particularly in technological subjects. There is also a lower level, below which learning is seriously impaired, and an intermediate range in which the student can compensate for linguistic deficiencies by working harder.

In all the studies reported the measure of academic success used was the Grade Point Average (GPA), while the measure of linguistic ability was the Test of English as a Foreign Language (TOEFL). Both are widely used, though any single parameter used as a measure of ability and attainment is open to criticism. In the case of admission to University, for example, a test such as that run by the British Council, which yields a profile of abilities in different aspects of linguistic competence, is probably more useful, though more difficult to administer and to evaluate. It should be noted that the level (550) above which the correlation became weaker was that normally considered the minimum for entry to higher education. The overall impression is that above a certain threshold of linguistic attainment it becomes immaterial whether the student is learning in his native or in a foreign language.

5 Discussion

In the Introduction it was suggested that organisers of courses combining engineering with a foreign language face a dilemma in weighing educational values against efficient use of resources. The issue arises in three ways. First is that of using standard engineering course material, without modification, to make up some 70 - 80% of the total, even though the strong orientation towards educating for research, development and design may be inappropriate. But to create a special set of lecture courses in each discipline for perhaps 10% of the students is hardly realistic. The solution, yet to be explored, may be to adopt a much more flexible, student centred, way of presenting the courses.

Second, not mentioned above, is the cost of courses and placements abroad, and the cost of the liaison needed to generate and keep them. In principle much of this may be realised through reciprocal arrangements with Universities abroad, but the time and effort needed to maintain these is still considerable. In addition, welcome though incoming exchange students may be, they have to be allowed for in any distribution of resources.

Finally, to have reasonable class sizes in the Final Year requires the material to be taught in the foreign language to be that which is relevant to all engineers, yet this may be the subject matter where problems of learning in the language are most likely to occur. Moreover, any problems may be accentuated by the somewhat artificial situation of living in one linguistic and cultural environment, yet being educated for a minor fraction of the time in another.

How it will work out remains to be seen, but there are grounds for optimism. There are already examples of this type of course running successfully on a small scale, though probably with rather exceptional students. And, in the final analysis, our students are, fortunately, remarkably resilient!

References

Jochems, W. (1991) Effect of Learning and Teaching in a Foreign Language, **European Journal of Engineering Education, 16,** 309

Mulhall, B.E. (1992) The Place of Foreign Language Teaching in Engineering Degree Courses, **Engineering Science and Education Journal,** 1, 99-104

Language Teaching as a Key Element in Engineering Education or Why Should Languages be Mandatory in All Engineering Courses?

R. Meillier

Ecole Nationale d'Ingénieurs de Saint-Etienne

Abstract
The aim of this paper is to show that languages should be part of any well-balanced curriculum leading to the award of an engineering degree.
In spite of long-standing traditions which have often opposed languages to engineering sciences, there are very good reasons for recognizing sufficient compatibility between them and acknowledging the necessity for tomorrow's engineers to learn at least one foreign language during their initial training. When one considers the linguistic, communicative and cultural aspects of language learning as well as the professional assets derived from it, one is bound to admit that language teaching is also the right way to prepare efficient engineers for the economic and industrial environment in which they will operate.
Keywords : Engineering, Languages, Education, Competition, Communication, International Environment.

1 Introduction

Not very long ago, learning foreign languages was restricted to specialists such as: interpreters, translators, language teachers, commerce, business and marketing people, diplomats and those working in tourism essentially... With companies now looking towards global markets, the advent of the Single European Market, the greater mobility of people and the development of international cooperation, attitudes to language learning have changed significantly. This change in attitudes could be illustrated by an anecdote and a few figures.

The anecdote, and this will come as no surprise, is an epitome of British humour. It was related to me by the late Dr. H. Law, former President of Portsmouth Polytechnic. This is more or less what he said:

"Portsmouth Polytechnic used to be the butt of a joke because of its coastal situation. Drawing a circle with Portsmouth at the centre, our competitors would say: 'Look at your catchment area. Half your potential students are fish !' But the joke turned in our favour several years ago. As the Polytechnic developed, it became clear that we are in a privileged position to attract students and forge links, not just nationally, but internationally. Today our south coast location makes us a natural focus for developments in a European environment."

The figures concern a French Institution of Higher Education : the Ecole Nationale d'Ingénieurs de Saint-Etienne, E. N. I. S. E. (Saint-Etienne National Engineering College). When it was first established in 1961, its prime objective was to produce engineers who met the demands of local industry in terms of engineering staff. And it is true that, during the first decade of its existence, 70% of ENISE graduates found jobs on a local basis. But the times are changing and today only 10% find local jobs, the majority are employed all over the country and a growing percentage are working abroad. It must be added that, todate, 25% of the graduates spent at least one semester out of their ten-semester course studying in a foreign institution and approximately 8% of the final year students can really bear the title of "European Engineers" since they are awarded a double engineering degree after spending three semesters in a partner institution. Their field of action will be Europe and the world whereas the first graduates' target was Saint-Etienne and its area. This has been made possible only because languages have always been an integral part of their engineering syllabus, even if that principle was not easy to put into effect on account of internal pockets of resistance.

2   The various degrees of compatibility of languages with engineering

Arts versus Sciences : very often this is the way the problem is posed, but is it not the wrong way of putting it?
For reasons deeply ingrained in people's minds, languages should be reserved for linguists and sciences for scientists. One often hears about the impossibility to bridge the gap between "soft" sciences and "hard" sciences. From a mere linguistic point of view, the use of those two adjectives is far from innocent and speaks volumes about the way their related subjects are ranked !
Instead of bringing languages into conflict with sciences, it is more interesting to identify their common points, complementary aspects and degrees of compatibility.
   Most students are led to believe that if they study sciences, their brain configuration is such that they are unable to tackle the learning of a foreign language and vice-versa. But when one considers the problem in-depth, it is obvious that there are no grounds for that argument. All academics know that learning sciences requires sound reasoning, rigorous analysis, the ability to synthesize and quick thinking, among other skills. But the same applies to language learning: how could one understand the grammatical system of any language without the above qualities ? Would it be possible to reconstruct grammatical patterns in order to make sense if one lacked the capacity of analyzing and inferring ? Could one produce anything but broken, hesitant speech if one was not endowed with the gift of quick thinking ? And the list of common skills could be developed much further, since the process of LEARNING does not change, no matter what the subject of learning is.
   Rather than indulge in sterile opposition, it is much better to focus on the complementary aspects of languages and engineering. A good deal of language teachers hold the view that engineering students are very often better-equipped than many others for learning languages. As a matter of fact, many people tend to think that

learning a language is only a matter of memorizing words, structures, idioms, stock-phrases, sounds, etc... It is true that memory plays an important part, but understanding the internal logic of a language, reasoning on language patterns, analyzing grammatical rules or complex sentences, reading into the principles of a linguistic system, as all scientific minds can do so well, certainly help and facilitate the acquisition of linguistic knowledge and know-how. And if one points out to engineering students that they possess all the ingredients needed to learn a language quickly, they usually do so with efficiency and great pleasure when realizing that it works !

3   The communicative and cultural aspects of language learning

In addition to the linguistic aspect of language learning, students are enabled to develop their skills in communication and knowledge of foreign cultures.

All engineers and specialists in engineering education acknowledge the fact that, in today's industrial and economic context, a professional engineer spends 70% of his time communicating in the broad sense and the remaining 30% using his/her technical or engineering skills. No one denies that if a company wants to be highly competitive, it has to develop a service-driven philosophy and, as a result, extend its relationships with its customers, which means that it has to understand and communicate with those customers.

In this respect, the language classroom is certainly one of the best places where communication activities can be organized, whether it is orally with pair work, group dynamics, problem-solving, oral presentations, simulation of interactive situations such as phone calls, role-playing, chairing meetings, taking part in meetings, case studies, etc... or in writing all types of letters, fax messages, memos, reports, technical descriptions or translations, etc...

Students' interest can even be kindled - or rekindled - if they are given the opportunity to talk and write about what they know best, i.e. themselves and their specialist subjects and thus, sometimes, gain self-confidence which they lacked or regain the one they had lost. What language teachers often describe with delight are sessions in which their students report to their counterparts on their industrial experience abroad and expatiate on how they tried and managed to solve technical, administrative, human problems on the work site, asking their friends what they would have done had they been in their place. There, one sees real communication at work !

And it is in the language lessons too, that they are gradually made aware of cultural differences and therefore made to raise questions about what they had taken for granted before.

At this point in the paper, it seems appropriate to simply quote three students who did part of their courses abroad within the framework of the Tripartite Cooperation Programme operating between the University of Portsmouth (U.K.), the Universität Gesamthochschule Siegen (G.) and the Ecole Nationale d'Ingénieurs de Saint-Etienne (F.).

The first statement illustrates the point about the questioning of prejudices and was written by Colette Haseldine, a Portsmouth Mechanical Engineering student who spent her third year in Saint-Etienne in 1988:

"Although I really enjoyed my trip to France, I was initially worried at how I would find the money for travel expenses, and whether there would be any complications on the journey, what the accommodation and the company would be like, etc... Natural enough to worry, but I soon settled in and found that the townsfolk of Saint-Etienne were very friendly and helpful to strangers (not as I had been led to believe...!)"

The second statement was written by Rolf Kotte, a student from Siegen University who spent one semester at the Ecole Nationale d'Ingénieurs de Saint-Etienne in 1987: he draws the following conclusion from his experience:

"Exchanging ideas, living a different culture, another way of looking at things, having to accept different values will certainly improve tolerance in people of different nationalities and help strengthen the links between different countries."

The last one was produced by Philippe Ammirati, one of two students from the Ecole Nationale d'Ingénieurs de Saint-Etienne who followed an equivalent study programme at Portsmouth Polytechnic and were awarded a double degree in 1986-1987 :

"The friendly relationship with the lecturers and the administrative staff is what first struck me when I came over from France to Portsmouth Polytechnic...... I make myself practice some of the wide range of sports offered and meet students from various courses and countries in pubs and discos. My final year at Portsmouth Polytechnic: Great!"

4 The international dimension of engineering

With the reinforcement of their linguistic, communicative and cultural awareness, the new graduates are certainly in a better position to meet the challenge of international competition.

As was evidenced in the introduction to this paper, an increasing number of our engineering graduates will have various contacts with their peers from different countries in the course of their professional careers. The time of the domestic-only market is well behind and we now live in the global village. This goes for engineers as well. And if they want to be present on the world stage, they certainly have to be able to speak several languages. From design to manufacturing and even marketing, their sectors of activity are large and varied all over the world. It would be a pity if they were unable to grab tomorrow's and even today's opportunities just because they do not speak the same language as their customers, for instance.

That is the reason why more and more institutions of higher education have recently laid the emphasis on the international dimension of their engineering courses by implementing staff and student exchanges. Earlier on in this presentation, mention was made of the Tripartite Cooperation Programme organized by the University of Portsmouth, the University of Siegen and the Ecole Nationale d'Ingénieurs de Saint-Etienne. It is currently admitted that all those who participated have greatly benefited from it. It is equally admitted that none of this would have been possible if languages had not been taught in the three institutions.

An interesting comment which has been made several times by exchanged students is that languages should be given more clout in the educational system and their first reaction when they get back to their home institutions is to encourage the younger students to address languages seriously. All agree that the extra time spent learning languages was more than worthwhile !

It should also be pointed out that many a lecturer from the three institutions have felt the need to brush up their knowledge of foreign languages and a fair number registered on and regularly attended language courses prior to their periods abroad.

5 Concluding remarks

The interdependence of national economies worldwide, the job market in industrialised countries, the legitimate aspirations of people to greater mobility and the present trends in engineering education have already given ample demonstration that languages cannot but be included in any individual, well-balanced educational scheme. One aspect, which is too often overlooked, is that languages can also be a stimulating, challenging, gratifying and rewarding item in the training programmes of tomorrow's engineers who will also be citizens of Europe and citizens of the world. That is why it is the role and duty of all those involved in their initial and continuing training to supply them with all the tools required to be efficient and competitive in the environment of the next millenium. And if languages are one of those tools, then it will be possible to give their education one more touch of humanism which will provide them with a better understanding of the world outside and the people living in it.

Foreign Language Learning for Students of Engineering: Some Theoretical and Practical Considerations

P. Hand

Institution-Wide Language Programme, University of Portsmouth

Abstract

This paper examines some factors to be considered during the present rapid expanssion of language tuition for students of engineering in British Higher education. These include:

Why learn a language, when English is virtually a world-wide lingua franca in many professional fields?

Which languages are offered? Which should be offered? What are the limitations?

How is language learned? Linguistic, practical and psychological factors that affect language learning; comparison between first and second language learning.

What should a language course for engineering students consist of? Consideration of the content of a language syllabus, eg. how much specialisation?

Practical issues: support from the institution; timetabling, materials, staffing, accommodation.

Keywords: Language; First Language (L1); Second Language (L2); Ab Initio (A.I.); Languages for Special Purposes (LSP).

1 Introduction

The provision of foreign language courses in British higher education is undergoing rapid and unprecedented expansion. A small increase in undergraduates in language degrees is far outstripped by a huge increase in language courses for students of other disciplines, mostly Engineering and Business Studies (Rigby and Burgess, 1991). In this institution alone there has been, from 1990-91 to 1991-92, an increase of over 100% in language courses for engineering students - from 8 to 17 classes. Most higher education institutions either have, or are planning, a policy of foreign languages for all, both students and staff, are are introducing degree courses combining languages with subjects not previously considered partners for a language.

2 Background: Changing Economic, Political and Educational Factors

Behind this lie the changes in Europe. For the single European market, employers demand language competence. Ambitious mainland Europeans have long had to be polylingual. There is also the effect of the student mobility programmes - Erasmus, Lingua and Comett. By highlighting gaps in language proficiency, these have led both students and staff to realize the need for language tuition. Although the position of English as a world language is secure, the importance of language learning for speakers of English is now recognised. As stated by Girard (1988) for GRIPIL, a research group set up by the French "grandes écoles" of engineering to promote language study by engineers: "The monolingual European of the end of the 20th century will be the equivalent of the illiterate of the last century".

Another factor is the National Curriculum's requirement for language study from age 11-16; soon, every 18 year old entrant to H.E. will have had recent language experience. The interactive and communicative activities essential in GCSE language have already led to more A-level language entries; this popularity will also increase demand for language courses from non-language students.

3 Which languages?

For UK students, European languages seem the obvious choice; French, German, Spanish and Italian. This is in order of popularity and does not imply importance or desibibility. French outstrips other languages by more than 2 to 1, for mainly historical reasons. There is a large pool of teachers of French, which perpetuates it in schools, and a wealth of material. Both these factors are decisive. There is no practical reason why French should predominate; unified Germany is a larger market, the eastern European countries and the former USSR will soon be open for business; countries of the Pacific rim will become increasingly significant. But without financial support for training teachers and developing material for other languages, the situation will not change. Language teaching skills must be learned; one cannot simply put a native speaker, for example, in front of a class; work on materials for languages other than French is needed. It is likely, then, that many UK engineering students will continue to study French. There is also student preference. Engineering students often lack confidence in their language skills, and prefer to stay with what they may know, ie. French, rather than try something new. For students it is also relevant that France is the nearest European country, making visits easier and cheaper.

4 How is a language learned?

Research has yet to answer this vital question; but certain points should be borne in mind. Language learning is very different to students' experience in other parts of their engineering course. It involves cracking a code, instead of using a code already known (their own language) to learn about other things. For those whose strengths are in mathematical and practical skills, this can seem daunting. In addition, language learning with modern, interactive methods is very much a "hands-on" activity. If we distinguish between competence and performance, "knowing that" and "knowing how", it is the latter that ensures language survival, and it can only be acquired through practice. If students miss a lecture on frictional losses in fluids in a pipe, they can consult various texts to make up the gap; if they miss a language session in which one of the past tenses is practised, then certainly they can read up how it is formed, and used, but, without internalising the sentence patterns through practice, the knowlege will remain largely academic in situation when one urgently needs to say, for example, "My car's broken down". Language evolved in face-to-face interaction with other people, and is best learned the same way. For this reason language classes need to be small - 15 is a working maximum.

The process of second language (L2) learning is illuminated by comparison with first language (L1) acquisition. We are all good language learners; we learned our L1 with 100% success. We perform complex linguistic operations without thinking - verbs agree with subjects, word order is perfect, we manipulate tenses and subordinate clauses, we automatically address an employer in a different linguistic register to that used for friends. But for adults, it is difficult to gain this level of competence in a L2. Bley-Vroman (1989) concludes that adult L2 learning is characterised by lack of success and by fossilisation - inability to go beyond a certain level despite practice.

A child has certain advantages over adults in language learning. First, there is no other language to cause interference. Adults speak the L1 as automatically as breathing and are unaware of how thinking is permeated and conditioned by it; hence, with an attempt to learn an L2, there is conflict between the two sets of data.

Second, exposure time. If a child is exposed to the L1 for 6 waking hours a day over 2 years this gives 4,380 hours. Engineering students, at 2 hours a week, 25 weeks a year, for 2 years, have 100 possible hours of exposure. Despite good intentions, it is rare for classes to be conducted entirely in the L2 - certain things have to be explained in the L1, there is the student fatigue caused by straining to understand an L2 - subtracting 15 minutes from each session, this leaves 87.5 hours. As Scullard (1989) states, for Ab Initio students the minimum time needed for basic survival and social skills is 50 hours tuition. Hence in 2 years A.I. students will acquire slightly more than a basic language level; and even this is unlikely without work outside class time.

Thirdly, the environment. A child is constantly exposed to L1 patterns; even asleep, there is subliminal reinforcement of the sounds, the syllabic stress, the intonation of surrounding speakers. Later, reading skills are helped by street signs, the media, etc. For an adult in a language class, L2 input ceases on leaving the room. One can read L2 newspapers, listen to L2 radio, watch L2 television on satellite, but anyone who had tried to do so knows the effort it takes. A short news article may need an hour's work. Engineering students have a heavy timetable, and language tutors find it is rare for them to do preparation or revision outside classes; they are therefore dependent on tutor-contact time.

Fourthly, there appears to be a language acquisition device in children which operates for only a limited time. It is striking that all normal children acquire the L1 perfectly, even in conditions unfavorable for later learning. In childhood, language operates through both brain hemispheres; in adults, it is a left-brain activity. One hypothesis is that language learning ability diminishes at the time of the final "wiring-in" to the left brain, coinciding with Piaget's stage of <u>formal operations</u> (the development of abstract thought at adolescence).

Lastly, as well as being a means of communication, language has a powerful integrative function. It marks membership of a social group - usually national, sometimes racial. Much adult identity is tied up in the group - hence the fierce defence of minority languages. Children have no attachment to social identity. Learning an L2 implies accepting to some extent the values and identity of the L2 group. Negative attitudes towards this group will hinder L2 learning.

These factors show the importance for adult learners of strong motivation. It is this that enables one to make the time to struggle with the L2 newspaper, to practise the sentence patterns, to learn the vocabulary, and makes one decide to appreciate the L2 group's habits and assumptions. Along with the assets possessed by adults and not by children - general problem-solving ability and a high level of concentration - it should, given enough time, lead to adequate L2 acquisition.

The task is a wide one. In Klein's description (1986) it involves:

1: <u>Phonological knowlege</u> - the sound system. Foreign languages make sounds and distinctions between sounds that do not exist in one's L1. The French learner of English must learn to distinguish short from long vowels - the difference between "ship" and "sheep". The British learner of French must distinguish "rue" (street) from "roue" (wheel).

2: <u>Lexical knowlege</u> - much more than knowing which word is the equivalent of which (not simple, as so many words have multiple meanings). It includes knowing how words join - that English can say "language student", but French says "étudiant de langues", and German "Sprachstudent". There is also the vast range of idioms, and collocations - which word can go next to which other.

3: <u>Morphological knowlege</u> - the forms a word can take. In English this is rudimentary; but in German there are 8 forms of definite article. This field includes most grammar.

Each of these is a large field, yet alone they are not enough. To quote Klein again, if I am locked in a room with loudspeakers playing Chinese, I will never learn the language. For that, much additional information is needed; who is speaking, to whom, about what, and what the body language is. It is this living situation that classroom learners lack. One must also learn what is appropriate as well as what is grammatically correct, a point we shall return to later.

5 What should a language course for engineering students consist of?

It may seem obvious that, to prepare students for future careers, they should learn technical and engineering language. In practice tutors find this not possible, and there is also a good case for saying that it is not what students most need.

First, the question of level. We have already seen what is reasonable to expect from A.I. students; they therefore concentrate on the four basic language skills; listening, speaking, reading and writing. Oral work is predominant, based on survival situations - obtaining food, information, assistance, money; travelling. At this level, technical material cannot be tackled.

With students who are already fairly proficient one can introduce Languages for Special Purposes - LSP. In the UK this is normally post-GCSE grade C - that is, after 4-5 years' study. But few engineering students have grade C or above; students with good language ability tend to go into other fields. Engineering students often feel they are "language failures". In addition, for first-year students it is at least 2 years since they have studied language; longer for mature students. Their language skills have lapsed and they need much revision, which takes time away from that available for LSP.

Secondly, experience suggests that what engineering students need more than LSP is to improve general language skills. A few weeks into the course, post-GCSE students are given technical material in French. Their first reaction is enthusiasm that engineering matters can be expressed in French. After working on the material, they point out that some of the technical terminology is similar; what is not they can find in a dictionary. It is the non-technical language linking the specialist terms that is problematic. This is unsurprising, as certain construction much used in scientific language, such as the passive, are not easy. In other words, the students themselves realise that LSP depends on a good general language level. They also say that, despite their difficulties, they enjoy language because it is such a contrast to the rest of the course, with different subject matter and teaching style. They are confident of absorbing technical terminology fairly easily once they are abroad.

This is borne out by a study of Chinese-speaking student engineers in Singapore (Koh, 1988). These students, who had passed internationally recognised English language exams, were offered English for Special Purposes; results were poor. Students saw no need to practise engineering skills in the L2 - they felt they knew them well enough; and their grammatical errors were tolerated by English speakers, who "read meaning" into imperfect English. Thus there was no motivation to improve. Yet students used grammatically correct English in quite inappropriate ways - in a note to a tutor, "Be in your room at 1pm.", or in a job application, "I am willing to offer my help in your organisation". These examples show the vital importance of the right linguistic register. British students make a similar mistake by wrong use of the familiar form of you (tu in French - English now has no equivalent) instead of the formal vous. The Singapore study also investigated the relative importance at work of the various language skills - reading and writing, compared to speaking. Despite estimates that 70-80% of time was spent in oral communication, the Chinese students received no training for this at work. On the other hand, most had training in report writing, and yet reports were generally of a routine style, easily learned. The study concludes that these students benefitted most from practice in oral and communication skills; the importance of the latter for engineers is generally agreed.

In practice, then, engineering students are best served by language sessions consisting mainly of oral work consolidated with reading and listening comprehension - which is essential for those wishing to study abroad - with some input of technical material and practice of additional skills such as interpreting L2 visual data, oral presentation of research, etc.

6 Practical issues

Language learning is very time consuming. Both total tuition time, and frequency of sessions, are important. Because repetition and memory work are unavoidable in L2 learning, the more often one works on material, the better one's recall. Most mainland European professionals learning English have 2x2-hour sessions a week;

here, one 2-hour session is usual. Progress is thus slower. Timetabling should reasonable. If other lectures end before lunch and the language session is late in the afternoon, it needs exceptional motivation to attend. Such situations have led to the rapid collapse of classes. Accreditation is also an issue. If students receive no credit for language attainment, again motivation is strained.

Staffing and materials development need institutional support. At present, good technical language material is sparse (the planned Nuffield Foundation French course for engineers will be very welcome); there must be support for in-house materials development. As for staffing, because this type of language provision is new, it tends to be in a Cinderella-like situation. There is a case for saying that, due to the high number of A.I. courses and the lack of language confidence of engineering students, they need tutors even more skilled than those who teach specialist language students. Yet tutors in this area are usually part-time or temporary, which can bring problems of availability, continuity and comittment.

Rooming: because specialist language accommodation is fully used, engineers' classes often take place in the engineering faculty; but accommodation designed to teach engineering skills is not suitable for language activities. Audio and video equipment needs to be readily available; good accoustics are vital; even the arr- angement of furniture helps or hinders - language learning works best in small groups with a round-table atmosphere. Clearly more language accommodation is needed.

These difficulties can be solved with institutional support. The time has surely come when language provision for engineering students warrants permanent staff, ad- equate timetabling and accommodation, and material support for this important and expanding activity.

References

Koh, M.Y. (1988) ESP for Engineers: a reassessment. **ELT Journal** 42/2, 102-8.

Bley-Vroman, R. (1989) in **Linguistic Perspectives on Second Language Acquisition** Ed. Gass and Schachter, Cambridge University Press, 44-47.

Klein, W. (1986) **Second Language Acquisition**, C.U.P., 47-48; 44.

Girard, D. (1988) in **Actes de la Rencontre Inter-Langues**, Conférence des Grandes Ecoles (GRIPIL). Publications ENISE, CNAM, Paris, 118. Original text reads: "L'Européen monolingue de la fin du XXe siècle sera l'équivalent de l'analphabète du siècle dernier." Translation in text above is by P. Hand.

Rigby, G., Burgess, R. (1991) **Language Teaching in Higher Education: A Discussion Document**, University of Warwick, 10-11; 13.

Scullard, S. (1989) **The Provision of Foreign Language Training to Industry for the FHE Provider**, D.E.S. Further Education Unit, 23.

Case Studies for Use in Teaching Engineering for International Development

P.H. Oosthuizen

Department of Mechanical Engineering, Queen's University

Abstract
There appear to be good academic reasons to expose at least some engineering students to the special considerations that have to be taken into account when working in or with "developing" countries. A number of possible ways of providing students with this experience exist, all relying to some extent on the use of case studies. In preparing students for work with "developing" countries it is necessary, therefore, to have available a series of case studies that illustrate the particular difficulties and constraints involved in engineering related problems in "developing" countries. The present paper discusses some of the considerations that enter into the selection of suitable cases and briefly describes a few typical such cases. Some discussion of how these cases can be presented to undergraduate engineering students is also presented.
Keywords Case-studies, International Development, Teaching

1 Introduction

At some time during their professional careers, a significant number of engineers from so-called "developed" countries will be involved in some way with a project that entails working in or with a so-called "developing" country. If an engineer is to be successful in such work it is important that it be clearly understood that the constraints and techniques required in this type of work can be very different from those that apply in the typical situation encountered in the "developed" home country. For example, it is important to understand the constraints placed on a design by the local availability of materials, by local atmospheric conditions, by the availability of spare parts and, often, by the culture of the country involved. To be successful in this type of work, it is also important for the engineer from a "developed" country to realise that while modern design techniques can be bought to the project, the inhabitants of the country involved have a wealth of practical experience under the conditions actually existing in the area involved. A successful project, therefore, involves, wherever possible, cooperation between the workers from "developed" countries and workers from the area where the device is to be used or where the work is to be undertaken. Each group should clearly understand and respect the strengths and experiences that the other brings to the project. In undertaking work in "developing" countries, it is also important for an engineer to be aware of the difficulties that can sometimes arise in living and working in such countries. With these considerations in mind, it seems important that at least some engineering students should be given exposure to this type of work as part of the undergraduate program. The exposure could be obtained in such a program either by introducing suitable examples of this type of work into existing courses or by introducing a separate, probably elective, course on this subject into the program e.g. see Oosthuizen (1990). In both approaches, it is necessary to have available a series of case studies that illustrate the particular difficulties and constraints that are involved in engineering related problems in "developing" countries. The present paper discusses a few typical such cases and methods of incorporating these into an undergraduate engineering program.

2 Selecting the Cases

Cases to be used in teaching "Engineering for Development" should be selected with the following aims in mind:

1. To try to develop in the student a thorough understanding of the true meaning of "appropriate technology" i.e. to develop an understanding that the devices must be designed with the real needs of the potential "customers" in mind, with a thorough appreciation of the constraints imposed by local conditions whether they be environmental, social or political, and with a thorough knowledge of the level of maintenance that can be expected. The student must learn to appreciate that "appropriate technology" does not mean "low technology" and, indeed, that the design of a device for use in a "developing" country may require the application of highly sophisticated methods because of the very severe restraints placed on the design by local conditions. Large amounts of money have been squandered in the past by workers from so-called "developed" countries thinking that no real engineering knowledge was required to design devices for "developing" countries which is seldom true.
2. To try to develop in the student an appreciation of the vast experience and knowledge that the inhabitants of the country involved can bring to the project and to develop in them the realisation that a successful project involves, wherever possible, cooperation between the workers from the "developed" country and workers from the area where the device is to be used or where the work is to be undertaken. Since much of the experience in certain areas of work rests with the village women in a "developing" country, an attempt should be made to develop in the student an appreciation of the particular problems faced by this group.
3. To foster the understanding that the constraints and techniques that are required in this type of work can be very different from those that apply in the typical situation in a "developed" country.

4. To try to develop in the student an awareness of the difficulties that can sometimes arise in living and working in "developing" countries.
5. To try develop in the student an appreciation of the problems faced by the "developing" countries and their inhabitants.
6. To try to develop in the student an appreciation of the environmental consequences that can result from engineering decisions.
7. To try to develop in the student an awareness that the solution to most problems in "developing" countries involves such complex and interrelated issues that it will seldom be possible for the engineers alone to solve the problem. Instead they have to interact with experts from many other fields, developing multi-disciplinary teams appropriate to each problem.

In selecting cases it seems important to choose those that deal in some way with the following:

- Production and processing of agricultural crops
- Provision of an adequate water supply
- Provision of an adequate energy supply
- The interaction between development and environmental considerations
- The development of a manufacturing base

since so much of the engineering work undertaken in "developing" countries is concerned with these areas.

3 Presenting the Cases

It is extremely important to ensure that the person teaching the course, as well as being extremely competent technically, should have experience in working, in some way, in a "developing" country. It is preferable that the instructor be currently involved in development work. It is also important that this person feel comfortable in leading the unstructured class discussion that is involved in teaching using the case-study method.

Prior to considering the cases, it seems necessary to present some of the following material using a lecture type approach:

- An introduction to the "developing" world and to the social, political, educational and technological conditions existing in various "developing" countries in Africa, Asia and South America.
- A broad discussion of some of the problems commonly faced by "developing" countries.
- A discussion of some of the special considerations that must be taken into account when performing work for or in "developing" countries.
- A discussion of some of the difficulties that can be encountered when working in "developing" countries.

Following the presentation of the introductory material, the cases are considered one at a time. The procedure used in studying the cases is roughly as follows. First, each case is briefly discussed and some background material is provided to the students. After this material has been read, several relatively informal class discussions of the problem, the constraints, of past work, of socio-economic and environmental aspects, and of possible solutions are held. The class is then split into smaller groups to work on various different aspects of the case. Finally the groups are brought together to discuss their findings and, where possible, to discuss solutions. Each case typically, would take about three weeks.

4 Typical Cases

As mentioned previously, it appears that the best way of developing awareness in engineering students of the special problems that occur while working in or with a "developing" country is to examine a series of cases. These cases should be concerned with the development of a device or process, with which engineers trained in a "developed" country have been or could have been involved, for use in a "developing" country. The following cases are of the type that is believed suitable for this purpose. Space does not permit a detailed discussion of these cases and the selection of the cases discussed may be very much a result of bias by the personal experience of the author. Some background material for these cases is provided by Schiller and Souare (1989), Pokharel et al (1991), Bassey and Schmidt (1989) and Oosthuizen (1986).

Case (1) - Pumping of Water from Deep Wells in West Africa. Much of the water available to villages in certain West African countries is now in wells that are much deeper than those that have traditionally been used. Obtaining the water from such wells takes much more time and physical effort than required with the older more shallow wells. The traditional method of drawing water from a well is to throw a bucket tied to the end of a rope down the well and then to haul the filled bucket out of the well by hand. There is a need to replace this procedure with one that is more appropriate to deep wells. Many solutions have been proposed, ranging from simple wooden winch systems to photovoltaic cell powered pumping systems. In presenting the case, the basic problem will be introduced and the reasons that have lead to the problem will be discussed. The social and environmental conditions that exist in a typical village in this part of the world will be described and the implications that these conditions have for the solution to the problem will also be discussed. Various possible solutions will be presented and the advantages and disadvantages of each will be discussed. A solution will be selected and a detailed design of the device will be undertaken, the effect of local conditions and the availability of local materials on the design being strongly emphasised.

Case (2) - A Biogas Generator For The Himalayan Region. In many parts of the Himalayan region, significant numbers of livestock are kept in relatively small areas. The waste from this livestock can be used in a biogas generator to produce a gas containing roughly 60% methane. This gas can supply much of the cooking energy needs of such areas. While biogas generators have been extensively studied in many parts of the world, the particular conditions existing in the Himalayan region pose unique difficulties. For example, very wide ambient temperature variations occur and biogas generators do no operate well at low temperatures. The presentation of the case will begin with a description of a "typical" Himalayan village . The operation of a biogas generator will then be reviewed and the efficiency of the device will be discussed. The amount of animal waste being produced and the cooking energy requirements will be reviewed. Separate groups of students will then look at the modelling of the effects of changes in ambient conditions on the operation of the generator, at the construction of such generators, at the economics of the generator and at the ways in such generators could be used in the region. The groups would then work together in order to develop a proposal for a generator matching the conditions existing in the region.

Case(3) A Village Scale Dehuller. Cereal grains such as Sorgum and millet are important food-sources in many parts of Africa. Dehulling, i.e. removing the outer envelope, is an important part of the preparations of the grains for consumption. Traditionally, this dehulling has been done by the village women by pounding the wet grain in a mortar with a pestle. The resultant product is wet and cannot be stored for more than about one day. This method of dehulling is very time consuming and not suitable for producing a marketable form of the grain. Various types of mechanical dehullers have been suggested and produced but they have received only limited acceptance. In this case-study, the need for dehulling is discussed. The traditional method is then examined and the socio-economic constraints on a replacement of this method are discussed. Groups of students then examine various possible mechanical dehullers in light of the conditions existing in the villages where they are intended to be used. Particular emphasis will be placed on trying to decide on an optimum size of dehuller.

Case(4) A Solar Rice Dryer. Drying is the major method of preserving food products in many "developing" countries. In many of these countries crops such as rice and corn are extensively dried by spreading them on a hard surface in the sun i.e. by sun-drying. While a good quality product is usually obtained by using this procedure, quite high losses can be incurred during the process for a number of reasons. In order to reduce these losses, a solar dryer in which the crop is contained in some form of cabinet can be used. In this case, the drying of rice will be considered. The changes in the moisture content during the drying of rice and the constraints on the drying process such as maximum allowable crop temperature will be reviewed. The size of crop to be dried and possible local economic and social constraints will then be considered. A discussion of the various possible types of dryer will follow. The class will then be spilt into three groups, each of which will study a separate type of dryer, e.g. indirect cabinet type, in detail. Then, a class discussion of the findings of each group will be held.

As mentioned previously, these are meant only as examples of suitable cases. Many other cases dealing with, for example, the harvesting of crops, irrigation, improvements in wood stoves and desalination can be developed from available material.

5 Conclusions

Case-studies are important tools for preparing engineering students for work in international development. To be effective the cases must be carefully selected. Some of the criteria for selecting cases have been reviewed and some typical cases have been described.

References

Bassey, M.W. and Schmidt, O.G. (1989) Abrasive-Disk Dehullers in Africa: From Research to Dissemination, International Development Research Centre, Ottawa, Canada.

Oosthuizen, P.H. (1986) A Numerical Study of the Performance of Natural Convection Solar Rice Dryers, Drying '86, Proceedings of the Fifth International Symposium on Drying (ed. A.S. Mujumdar), 2, Hemisphere Publishing Corp., 670-677.

Oosthuizen, P.H. (1990) Teaching Engineering for Development - A Proposal, Proceedings of the 7th Canadian Conference on Engineering Education, University of Toronto, Toronto, 344-350.

Pokharel, S., Chandrashekar, M. and Robinson, J.B. (1991) Biogas Potential and Implementation Issues in Nepal, Journal of Engineering For International Development, 1, 45-56.

Schiller, E.J. and Souare, M. (1989) Solar Pumping in the Sahel: The Case of Senegal and Mali, Proceedings of the 15th Annual Conference of the Solar Energy Society of Canada, 454-459.

Engineering Education in the Global Context. Working Together
M.L. Watkins
WPI London Project Centre

Abstract
Engineers have to broaden their education and be made aware of the significance and relevance of their work to society. If young engineers, male and female, of different nationalities, work together in a mutually supportive role in a foreign environment, on an exciting assignment, insight and understanding are gained.
Keywords: International, Internships, Projects, Cooperation.

1. Introduction

If engineers are to persuade the public that their profession contributes greatly to the quality of life, it will be necessary for their attitudes to change and their outlook to be broadened. Enhanced status will follow the recognition of the value of their work. The engineers contribution to the raising of standards of life, health, comfort and enjoyment, need to be explained vividly and convincingly; therefore communication skills are essential. Good written and verbal communication with clients, the public and the media are required in order to show that what engineers do brings significant benefits to humans, animals and the environment. Of course this requires that engineers themselves are made much more aware of the effects which their work has on society. Not only in their immediate vicinity and time, but also the global impact and the consequences for the future.

One of the attempts to broaden an engineers outlook across frontiers is the organisation of international groups which can work closely together on projects of interest to different countries and different disciplines. This can be done in a variety of ways. My experience is with undergraduate and graduate students from the USA and Europe working in London for periods between two months to a year. Unfortunately there have, so far, been no British students on the scheme. I am still trying to interest our Universities to recognise this as a worthwhile learning experience which deserves academic credit. Progress is slow. However there is enthusiastic response from Universities in the USA, Belgium, France and Germany. Even a lone Basque student is participating.

2. Participants

An un-anticipated bonus is the number of women participating in the scheme. As we all know the number of women studying engineering is still small, although it is up to 25% in some universities. However the number of women participants in exchange and overseas placement schemes is much greater than their number at the relevant campuses. This seems to be due partly to the independence and the spirit of adventure of women taking up engineering and also because, according to some of them, it is a way of escaping parental supervision without hurting the parents feelings. Moreover when working in mixed groups it is often the woman that acts as leader and surprisingly the men do not resent it. This too is a broadening experience for all concerned.

3. Scenarios

There are a number of different scenarios which fulfil the aims of international cooperation and the broadening of experience and hence also of the outlook. The two which I like best are the 'internships across the frontiers' and the international projects. Both schemes are supported by COMETT money for EEC and EFTA members. This confers a considerable financial advantage to the participants as it can offset 50% of the costs.

4. The across the frontier internship

Here the candidate is placed with a suitable industry or business for periods of between 3 months to 1 year. The work program is mapped out in detail, allowing for possible language difficulties, to suit the requirements of the firm and of the intern. Previous experience and knowledge of the prospective intern as well as their personality are taken into account so that he or she should not take too long to settle in. It is also necessary for the intern to get to know the structure of the firm where he/she is placed as well as to get to know the staff and to get known by them. Only then can the internship produce worth while results for the intern and for the firm which has accepted the placement.

The benefits for the firm are the acquisition of an able and enthusiastic worker who can help in a number of areas, such as design or development, marketing or analysis and can bring an international outlook to the problems. Indeed any position where assistance is of benefit can be used for such an internship; provided it offers an environment where learning and the enhancement of skills can take place. The benefit to the intern is that there is a real work situation, account of the pressures of business has to be taken, with its constraints on time and quality. In international placements this is enhanced by giving an insight into the way businesses function in another country as well as the chance to work with peoples of other nationalities and different cultures. A lot of friendships develop from such placements, some lasting for many years and resulting in mutual visits (including family) in each other's country. Language learning is rapid of necessity. A slight disadvantage is that it is only practicable to place one intern at a time at a firm, so there may be a feeling of isolation. Close monitoring by an academic supervisor is indicated as this enhances the quality of work given to the intern as well as his/hers performance.

The COMETT program may help to defray travelling expenses, special language tuition and some of the extra costs of accommodation. The firm accepting intern may pay a small salary or offer an unpaid situation, depending on individual cases. There is also a sum available from COMETT for administrative costs of the organiser of the internship.

5. Projects

Another way of working together in a mixed group is to undertake a project with a firm in the host country, the project team consisting of different nationalities. The advantage of this is that close cooperation between individual members of the project group as well as between the group and the firm is required and a good rapport will develop if the project is well lead. The disadvantage is the effort required to originate suitable projects and the close supervision such mixed project groups have to have, if the outcome is to be of value to the firm where the project is carried out.

There are three main requirements for a project to succeed. First and foremost the project investigation results have to be of importance to the sponsor. If this is not the case it is difficult to motivate either the sponsor's liaison person or the project team. Secondly the topic has to be suitable, for the investigation to be completed and written up in the time allocated to the project, but must be sufficiently challenging to require substantial effort, knowledge and imagination from the project team. The third requirement is the sponsors critical assessment of the results; both the written project report and the oral presentation.

6. Examples

An example of an internship is that of Tom Knight from the USA who is working in a London public affairs consultancy collating manufacturing data. He is seen here attending a presentation.

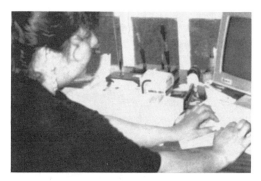

Another intern is Keren Rachum from Israel working at a cosmetic factory in London producing environmentally friendly products.

My experience of projects is mainly with Worcester Polytechnic Institute. Projects are initiated by various industrial, government and institutional sponsors. Normally three students form a project team. There will be students from European Universities in Germany, France or Belgium in some of the teams.

306 Engineering Education

The WPI projects have a 7 week preparatory period at their home University where the team learns and researches the background information required for the investigation suggested by the sponsor; a preliminary scheme outline and a plan of action is then submitted to the sponsor, who makes alterations as necessary. The team then meets in London, lives under the same roof, and works full time on the project at the sponsor's business. A professor from WPI supervises the day to day progress and the writing of the report as well as the preparation of the oral presentation. A supervisor from the sponsor meets the team at intervals to give advice and to steer the investigation in the right direction. The result is a typed report, which can have anything between 100 and 400 pages. It must include an executive summary as well as the usual literature review etc. and an oral presentation to interested parties.

COMETT money is also available for this scheme giving assistance to EEC and EFTA participants who receive help with fares, accommodation costs and language tuition.

Nineteen projects were completed in London in 1991. One of them won the Presidents award for best project.

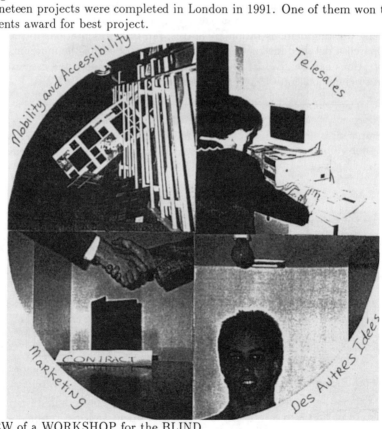

A VIEW of a WORKSHOP for the BLIND

This interesting project was the reorganisation of the PVC workshop for 'Action for the Blind' to make it more user friendly for the blind workforce and more efficient in producing and selling the workshop's products.

WPI at Rank-Xerox talking of future projects.

Another project was the design of a demonstration on the principles of flight at the Royal Air Force Museum. This explains the principles of flight simply and without mathematics so that anyone can gain a basic understanding. The team built many working demonstration apparatus, ran test presentations to the public and prepared a detailed instruction package for the friends of the museum, so that demonstrations can be given after the team left. This is indeed the case as, the museum runs demonstrations on 'Flight and easy' during holidays.

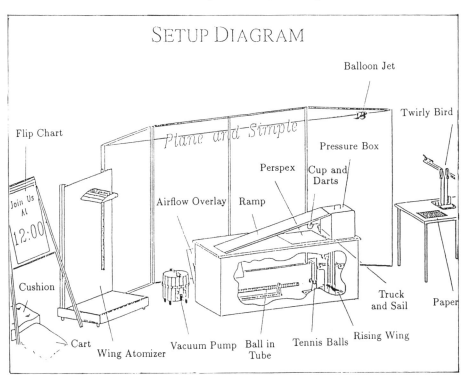

7. References

Burn B.B. and Griggs A.(1985) Study abroad: European and American perspective. urop. Inst. of Educ & Soc. Policy

Grogan W.R.,Schachterle L.E. & Lutz F.C.(1988) Liberal learning in engineering education . Jossey-Bass.

WPI Interactions No 1-12 (1980 to 1991)

WPI Innovations No 1-6 (1986 to 1991)

Schachterle L. & Watkins M. (1992) The WPI interactive qualifying project. IEE Engineering science and education journal V1 pg 49-56.

An Expansive Technique for Promoting International Industrial Exchanges

H.E. Newman

School of Systems Engineering, Portsmouth Polytechnic

Abstract
The "common sense" approach in obtaining international industrial exchanges is no longer valid especially in a time of recession. Substantial actions for change are recognised and examples highlighted to suggest a comprehensive technique which encourages placement managers to use their powers of innovation more fully.
Keywords: International Co-operation, Formation of Engineers, International Exchanges.

A Hypothesis

Throughout the eighties the world has witnessed a democratic revolution on a grand scale. Willingness to talk genially rather than argue contentiously has been welcomed by all but a few. By its very nature the popular or representative systems require all views to be heard, argued and analysed and a best resolution found satisfying as many criteria as possible. The pragmatic rather than the doctrinaire approach has been the vogue. The premis in this paper upholds this approach as being the most fruitful method in promoting and fulfilling the requirements of increasing the supply of national and international industrial exchanges. Consideration is centred on Engineering, however, the hypothesis equally applies to other disciplines.

Historical considerations

(a)   National

The early sixties saw a rise in the promotion of sandwich courses supported by the CNAA and the first steps in providing an 'Engineers

Formation' coupling academia with the 'workplace', occurred. Professional Institutions and Training Boards together with the Higher Education sector put their proposals to industry with the intention of forming a partnership for the provision of a continuing and increasing supply of well qualified and trained young people for the advancement of the United Kingdom's engineering base.

An influx of experienced and qualified recruits to Higher Education staff appointments from the industrial sector gave academia an acceptable face when in turn they were required to sell the idea of sandwich courses to their industrial colleagues. Perhaps this might be seen as the first piece of entrepreneurial activity in this field. No industrialist is willing to listen for any length of time to proposals regarding the improvement of his operation from someone who has spent his working life in an 'ivory tower'.

In these early days co-operative education had been experienced only by a small proportion of the United Kingdom's manufacturing sector and the smaller company's knowledge was even more sparse. It followed that a major 'promotions' operation was required and the 'sales representative' had to speak from experience not only from what his academic institution could supply but also from the view of the Training Boards, the professional Institutions and the CNAA.

Academia's industrial contacts were usually small in number and, in part, resulted from research associations and those firms who supported the old part-time routes to professional status via the Ordinary and Higher National Certificates.

At first these sources of support were used to promote the increasingly popular thin and thick-sandwich degree courses and to prepare the groundwork for expanding into the, as yet, untapped market of small and medium sized companies. A further problem for the promoter was to unravel the various schemes of grants and levies which accompanied training programmes so that these could be used as a type of bait to encourage the supply of training places in those companies which had previously secured their trained manpower from the open market. Within a remarkably short period, the once small list of participating companies began to grow. Many large organisations realised that not only had they to expand their training departments to satisfy their own future manpower requirements but the aspect of government introduction of levy/grant laws necessitated them in putting larger resources into this area if any profit, other than an increase in trained personnel, were to accrue.

It could clearly be seen that as well as developing a wider range of areas where training and industrial experience would be obtained the training fields could equally well provide sources of employment, profit, power and further expansion.

(b) International

The accumulation of a significant number of placement companies was achieved by this speculative investment. The groundwork had been done for the placement of unpresidented numbers of nationals - all that remained was to extend this to the international scene. France, St Etienne, was the first customer and it was fortuitous that the initial candidate was trilingual, highly intelligent and in possession of a first class personality. Personal knowledge of a local company of Anglo-French origin paid dividends and the first of the many international placements was quickly achieved. The major factor was the immense confidence one had in the students ability to undertake the job in hand and to be accepted generously. It was not training in the narrowest sense - it combined appreciation of a problem, interpretation, motivation and involvement which enabled the student to gain from the placement the training he required.

Small companies were able to accommodate the most adventurous and outward going students more readily, essentially because they were given a wider responsibility although one young man placed at multinational company who arrived in his new Porsche accompanied by his holidaying friend became the star attraction in that company for his six month stay. Very quickly word travelled between organisations through conference and visits that international exchanges could be a benefit to all involved and unsolicited calls for this service were a commonplace occurrence. Having successfully introduced the French contingent the University of Seigen presented the next challenge - a different culture with emphasis placed on the very practical engineering nature of the German system. Here again the placement technique altered and a new range of companies which linked with the disciplined approach were soon to find their names on the ever increasing 'placement register'

Reciprocation exchanges for British nationals were dealt with by French and German colleagues and attendance at World and National Conferences on Co-operative Education increased the range of personal contact made by placement staff to such an extent that as to now include Holland, Italy, Denmark, Norway, the United States, Canada and the Phillipines.

Membership of AIESTA provided a further source of placements and it was through this connection and the Central Bureau that a major exchange occurred. A chain of international hotels had advertised for final year students to apply for an eighteen month placement in Chicago where training in 'front and back of house engineering management' would be given. Building Services was the speciality and the course at Portsmouth at the time which appeared to fit the requirements was the four year thick sandwich degree in Engineering with Business Studies. The chosen applicant was successful and the course was subsequently recognised as being the most suitable for the hotel industry. Time has shown this multi-national Hotel Chain to have continually used Portsmouth for their Executive Engineers in Chicago, Washington, KL, the Camen Islands and parts of Australia.

These considerations have, in some small measure, shown the development of initially common-sense methods for securing industrial placements from which the next stage became self-evident. Build on these secure foundations a more enterprising and audacious technique.

A Developing Scheme

By its very nature, entrepreneurial activity suggests undertaking an operation often at personal risk - if one sticks ones neck out there is always the likelihood of losing ones head and this is the risk which must be taken in adopting a method of this nature, the rewards are worthwhile.

No connections however tenuous should be dismissed, no introduction turned away, no invitation brushed aside and no offer, however unlikely, repulsed until all usefulness has been strained from it.

In the synopsis the word 'belief' was employed, faith could have been used equally as it is necessary for all those employed in this area to be entirely committed to the necessity for such an activity. Any doubt is so easily perceived by the prospective customer and a valuable contact may be lost. As practising engineers, it would be conceived that the initial selling point must be the engineering speciality of the placement candidate - does this fit in with prospective placement company, if not, is there a secondary speciality such as 'computing', 'systems' or some other link? Experience has shown that most engineering courses produce a 'problem solving' mentality which can be turned to a wide variety of causes. Technology is universal by nature and it follows that this is the common factor when considering international exchanges. Emphasis must then be placed on the advantages of choosing to accept a student possessing a different culture - there are many - language, custom, routine, methodology, standards, ground rules and a host of others. By integrating exchange students with ones permanent staff there are many advantages in both directions. These features must be emphasised by the negotiator and connections found for the case in hand. As an example, a company wishing to export to a foreign country for the first time would be indebted to an employee having that country's language and not simply from the language criterion but from a knowledge of the custom and practice of the country in question.

Plates I and II have been produced to illustrate the advantages of publicity; it is not good enough simply to do a job - it must be seen to be done and seen in as many places as possible. Contacts must be made, societies joined, links formed, lectures given, brochures distributed, calls made and visiting diaries filled if a hundred per cent placement record is to be achieved and especially in the international market. The case of the Australia Link with Portsmouth Polytechnic is an ideal example of the use of these 'connections'. Itemised, the diary of main events reads:-

1. Press report "Portsmouth the Birthplace of Australia".
2. Meeting with author of the report and 'Exchange Idea' mooted.
3. Introduction to National President, Australia-Britain Society.
4. Linkage between the City, the Polytechnic and the Australia-Britain Society.
5. Introduction to Royal Australian Dockyard, Sydney who offer two exchange vacancies in Sydney.
6. Contact with immigration officials.
7. Two British exchange students located.
8. Britain-Australia Society Director General gives assistance.
9. Exchange implemented.
10. New South Wales Institute of Technology offer two Australian students for reciprocal exchange.
11. Exchange agreed with IBM and Plessey.
12. Reciprocal exchange implemented.
13. Exchange programme finalised in both directions.
14. City links secured with Bicentenary Celebrations starting from Portsmouth and the formation of Portsmouth - Sydney Sister City Inauguration.

Limited space prevents the full itemisation of events, nevertheless the nature of the operation is obvious and has been repeated for other areas.

It is to be hoped that this expansive technique should with, its obvious advantages, be followed by others seeking to develop their national and international connections and become a small but important part of the world's educational system.

314 Engineering Education

Efficiency Through Mobility: The European Networks

V. de Kosinsky

European Education and Training Programmes, University of Liege

Abstract
It is important that people are prepared for mobility when they are young, because it is not enough to dispose of opportunities and material means of mobility, there is also a human element, an attitude and disposition to mobility. But mobility cannot be improvized, it needs careful preparation, contacts, management, and this is where European networks come into light.
Keywords: Europe, Mobility, Education, Training, Networks.

1 Introduction

Why is European mobility linked to efficiency of engineering education and training? Europe is an immense reservoir of knowledge and expertise, but to have access to these assets, people have to move about in Europe, either to deliver them or to benefit from it. There is no place any more to individual, local isolationism. Efficient training for competitivity of our enterprises must be conceived within the framework of networks of centres of excellency. But mobility cannot be improvized, it needs careful preparation, contacts, management, and this is where European networks come into light.

During the past 10 years "NETWORKING" has became a much used key-word. Under the impulsion of the Commission of the European Communities, there are several European Action Programmes, where participants "must" build networks or participate in them. Everybody seems to create networks and any "self-respecting" establishment feels that it may be at disadvantage if it is not in one or several European networks.

There are European networks at different levels, looking after mobility at higher educational level to start with, other ones preparing and managing mobility during the initial professional training. There are also networks concerned with continuing education and training, with standards and qualifications, with transfer of technology, with business collaboration, etc. This means that people in the profession have at their disposal, from school age to fully qualified status, means and organizations to help them with European mobility.

It is important that people are prepared for mobility when they are young, because it is not enough to dispose of opportunities and material means of mobility, there is also a human element, an attitude and disposition to mobility. People who experience mobility from their young age, will have a cultural background and a natural inclination to move much easier across borders, which in turn will be profitable to our enterprises who seek to increase their competitivity on the European market.

The author does not pretend to know all existing networks in Europe, as there are new ones every month, but would like to take a few examples and compare their specificity. In this presentation we shall not include Associations, because on one hand they have a different role to play, and, because on the other hand, that would be too long.

2 Case Studies

We shall have a look at the following networks:
2.1 Networks at Higher Educational level:
Networks like the "*SANTANDER*" & "*COIMBRA*" Groups, comprise universities all over Europe, exchanging students and staff, developing common curricula, organizing credit transfer systems, etc.

For instance, the "*SANTANDER GROUP*", which was founded in 1988, has currently 26 members from 11 European States. It constitutes an integrated network of universities whose aim is to establish special academic, cultural and socio-economic ties and to set up specific and advanced facilities as well as privileged channels of information and exchange. The Santander Group also wishes to encourage contact between universities and their surrounding communities or regions on matters related to social and technological development.

In particular, the Group works toward the following purposes:
a) to facilitate and encourage the movement of students and members of staff within the network;
b) to promote and encourage joint research projects among members and to promote and encourage the collaboration with public and private institutions;
c) in the case of EC nationals accepted as exchange students from Santander Group universities to waive tuition fees, in the context of mutual agreement and with the aim of achieving a general balance in the number of students interchanged;
d) to recognize approved study periods spent in Santander Group universities as an integral part of students's degree programme; such periods to be preferentially of one academic year's duration, with appropriate arrangements for credit transfers;
e) to provide assistance in the appropriate foreign language, and on procedures particular to the receiving university and host country;
f) to give assistance in allocating limited residential accommodation for student interchange;
g) to promote the interchange of curriculum information and where practicable to produce interchangeable academic and distance learning materials;
h) to promote the discussion of curriculum development and the establishment of joint programmes;
i) to encourage additional social, cultural and sporting activities.

The Santander Group started as an agreement, a "Charter", signed by participating universities. Currently it is being transformed to a registered international non-profit-making association.

2.2 Networks at Professional Training level:

A good example of such a network is *COMNET* (Community Network for European Education and Training), organizing and managing professional training in enterprises, combined with European mobility, for students and young graduates.

COMNET has mainly been associated with the European COMETT Programme (Community action programme for Education and Training for Technology), whose main objective is the competitivity of European enterprises by helping to meet industry's requirements for a qualified and mobile workforce. This is done through encouraging and strengthening European cooperation between enterprises and higher education, and technology training in companies of students, engineers and technicians, as well as managers. The various "Strands" of the COMETT Programme are:

A. The development up a European network of UETPs (University-Enterprise Training Partnerships) to fur-ther transnational cooperation both on a regional and sectoral level; there are currently 158 such UETPs;
B. The reinforcement of transnational long term training exchanges between higher education and enterprises of students, young graduates and staff.
C. The promotion of joint projects for continuing training in technology.

Along with the rapid and intensive development of Europe, numerous bodies, higher education institutions, administrations and enterprises felt that it was necessary to create a network with sufficient infrastructure and with all the necessary logistics behind to organize multilateral exchanges. Created and launched in 1987, *COMNET* identified as one of its missions to federate UETPs and similar organizations, which could benefit from exchanges of experience and professionally managed international links.

COMNET was also started as an informal group, and transformed in 1988 into a registered international non-profit-making association. It is currently a federation of over 100 institutional members from 27 European countries, most of them being COMETT UETPs or similar local, regional, international or sectoral bodies. COMNET's members represent more than 2000 European enterprises and higher education institutions. In 1991 *COMNET's* members organized:
- industrial training placements for 2400 students and young graduates, representing 16800 student-months;
- staff exchanges between industry and higher education: 60 people for 270 months;
- 340 European courses in technology training.

2.3 Networks at Continuing Educational level:

These networks foster European and international cooperation at Continuing Educational level. Good examples of such networks are the *EUCEN* (European Universities Continuing Education Network) and *CEEC* (Civil Engineering European Courses).

EUCEN is being currently set up by about 30 European universities, so it is strictly European and academic. It covers all fields and sectors. EUCEN's fundamental mission is to enable its members to "serve most effectively the people of Europe, through the provision of high quality continuing education".

CEEC is a typical COMETT project in Civil Engineering. A transnational sectoral UETP was created, with the objective of promoting continuing education in civil engineering, gathering experience from various European countries for the benefit of all their civil engineers. CEEC organizes short intensive training sessions in civil engineering, structures, transport, buildings, hydraulic and maritime fields.

2.4   Networks at Institutional level:
ECCE (European Committee of Civil Engineers) was initially created in 1985 out of the common concern of the professional bodies for Civil Engineers in Spain and the United Kingdom with the implications for Civil Engineers of the proposed European Community Commission's Directive for Mutual Recognition of Qualifications. Current membership includes bodies from Spain, Portugal, United Kingdom, Ireland, Italy, Greece, Denmark, Belgium and Germany, with representatives and observers from France and The Netherlands. However, in one way or another *ECCE* has contact with all member states of the European Community. ECCE has broadened its interests to embrace all legislation emanating from the EC Commission, as it affects the Civil Engineer and the Construction Industry generally.

The *ECCE* objectives as originally drafted are:-
A.   With respect to the European Community -
(a)   to implement the general policies and objectives of the Treaty of Rome insofar as they affect the profession of civil engineering;
(b)   to promote higher standards of civil engineering in the Community.
B.   With respect to the EC Commission -
(a)   to keep the Commission informed of matters relevant to the civil engineering profession and, in particular, to ensure that the text of the Directives is in keeping with the first objectives;
(b)   to advise the Commission on the organization and methods of the industry and the framework within which civil engineers work;
(c)   to advise on research and research policy related to the industry, and on safety and the quality of civil engineering works.
C.   With respect to the EC National Governments and Institutions -
(a)   to advise and to influence individual Governments and professional institutions through national delegations;
(b)   to formulate standards to achieve a mutual compatibility or acceptability of the results of different regulations which control the profession;
(c)   to formulate standards for a European Code of Conduct of the Civil Engineering Profession and disciplinary procedures applicable throughout the EC Community.
D.   With respect to the Profession, related Organizations and Industry -
(a)   to formulate guidelines to maintain and raise standards of civil engineering education and professional competence;
(b)   to assist in achieving mutual compatibility of codes, standards and regulations in the related industry;
(c)   to encourage and improve levels of safety and quality in the industry;
(d)   to work with industrial and other related professional organizations and the industry to achieve the main objectives.

The founding of *CESAER* (Conference of European Schools for Advanced Engineering Education and Research) in 1990 was an important event in the European academic life, because it is in a good position to generate standards and proposals about the most efficient ways and procedures of European cooperation for undergraduate as well as graduate and continuing engineering education. CESAER is especially designed for exchange of information between institutions of similar high level characteristics, primarily concerned with high quality in formal and research based training.

The 40 higher education institution members, from 18 European countries, are characterized by active research and education programmes in a wide variety of engineering disciplines. Out of a large variety of engineering education institutions, a rather homogeneous group can be defined on simple criteria:
- selective admission and graduation procedures;
- responsible for engineering education of the highest level in their country;
- individual educational methods based on an important research activity well recognized internationally;
- close contacts with industry;
- a large degree of autonomy and responsibility of the students in the educational systems.

2.5   Technology Transfer Networks:
TII (European Association for the Transfer of Technology, Innovation and Industrial Information) is a European Association of professionals working in technology transfer and innovation support. *TII* operates as a service centre to help the members to establish working relations with colleagues elsewhere in Europe, to access European Community programmes, to exchange professional experience and to transfer technology across European frontiers.

TII's aims are:-
(a)   to help members promote their services, to make them more widely known and used by firms;
(b)   to facilitate European cooperation between members, to help them and their client firms benefit from the large internal Community market;
(c)   to provide training in innovation support methods and to encourage high professional standards;
(d)   to represent members' interest at European level.

Founded in 1984, *TII* today groups over 470 member organizations from all regions of Europe (20 countries) and from all areas of innovation support. *TII*'s members are innovation and management consultants, universities and technical institutes, research organizations, science parks, innovation centres and business incubators, venture capitalists and financial institutions, industrial firms, licence brokers and intellectual property advisers, chambers of commerce and industry, local and regional governments and development bodies, engineering and management consultants, etc ...

2.6   Business networks:
In 1984 a number of innovative business creating agencies from several European countries founded, with the help of the EC Commission, an international non-profit making association called *EBN* (European Business and Innovation Centre Network).

The aim of *EBN* is to provide its members, within an international network, with the means of making a more effective contribution to the development of healthy, diversified, innovative and technologically up-to-date local economies in Europe.

In practice, *EBN*:-
(a)   promotes the establishment of efficient Business and Innovation Centres (BICs) in Europe;
(b)   provides professional support and guidance to BICs that are in the process of being created;

(c) offers professional services to improve the performance of existing BICs;
(d) is a clearing house for the exchange of commercial and technical information between all EBN members and their client or associated companies;
(e) promotes public-private partnership at a European level to support the activities of BICs;
(f) provides information to its members on European Community legislation and grants;
(g) provides professional services to the EC Commission and to other parties on a contractual basis;
(h) facilitates the promotion of the enterprises created or supported by BICs on the European scene;
(i) stimulates development of links between BICs and their clients with universities, research centres and laboratories.

3 Conclusions

Having looked at different networks, their origins, aims and objectives, change in structure and activity over the years with the evolution of European integration, we have tried to show how these various networks contribute to *European Mobility*.

The *higher education networks*, like the *SANTANDER* and *COIMBRA* groups prepare future engineers for European mobility by providing and organizing such mobility already at undergraduate level. The usual duration of such mobility is between 3 and 12 months. This is a contribution of great consequence because it is important that people are prepared for mobility when they are young, as it is not enough to dispose of opportunities and material means of mobility, there is also a human element, an attitude and disposition to mobility. People who experience mobility from their young age, will have a cultural background and a natural inclination to move much easier across borders, which in turn will be profitable to our enterprises who seek to increase their competitivity on the European and World market. Networks at *continuing educational level*, *EUCEN* and *CEEC*, provide a similar contribution but mostly at postgraduate level.

Networks at *professional training level*, like *COMNET*, fully organize and manage European mobility at undergraduate, postgraduate and staff level, in collaboration with higher education establishments and industry. The usual duration of such mobility is between 2 and 12 months and contributes to the professional and cultural development of young Europeans.

Networks at *institutional level*, like *ECCE* and *CESAER*, both aiming to develop and maintain standards and qualifications through cooperation and exchanges, contribute also to the development of the European spirit, though they organize materially a more restricted number of mobilities.

European *technology transfer and business* networks play a different role in the mobility. Their place in the evolution of European integration, their function of fostering European cooperation and partnership, contribute directly and indirectly to people moving much easier across borders, which in turn will be profitable to our enterprises who seek to increase their competitivity on the European market.

Engineering Education in Developing Countries

A.H. Pe (*), A. Sevillano (**), E. Tadulan (**)

(*) School of Civil Engineering, University of Portsmouth

(**) Xavier University, Cagayan

Abstract

This paper sets out to examine the major problems confronting Engineering Education. It will relate to the courses that are especially relevant to the poorer countries in the region. The general Engineering Education will be examined and discussed together with how improvements could be made to produce well trained and tested graduate engineers. The paper is written with reference to Civil Engineering but should also be applicable in general to the overall engineering disciplines.

Keywords: Engineering, Education, Developing Countries.

1 Introduction

The advances made in Technology and Engineering in the industrial nations of the western countries have now made it necessary for changes to be brought about, especially in those regions of the world. These have already taken place in some countries but there are still some remaining which need to be brought about to meet the challenges of the 1990s.

In order to implement these changes "Engineering Education" will need to be given top priority. In addition to the support provided by both the Government of the country concerned, industrial participation in educational programmes must be extensively increased. Also it is vital that the engineering profession must respond to the changes. In this respect, effective and efficient Professional Engineering Institutions need to be formed, if not at present, to promote the status of Engineers within Society.

Last but not least, valuable help which can be provided by International Agencies for funding of educational needs and provision of advice regarding course curriculum developments by other academic institutions are needed.

Therefore it would be useful to review the present position regarding "Engineering Education" in these countries and examine how these changes could be brought about effectively to help transform them into industrially advanced nations.

2 Background

In most of these countries it will be noticed that in spite of their present inadequacies there has been some form of tradition in Engineering in the past. This is evident by the study of ancient structures that exists within a particular country. Unfortunately due to the stagnation in both the economic and political position of a country these traditions have been overlooked. Consequently whilst the industrial changes have been brought about especially in Western countries, they have remained dormant in Developing Countries. Therefore if the present educational systems are examined especially in the poorer countries then it will be seen quite clearly that the educational budget is of a very limited nature. This fact will then affect the poorly trained staff trying to provide an education with limited funds for educational courses. This is bound to have a particular effect relating to laboratory equipment and instruments regarding Engineering Training. Therefore the students, though achieving a high intellectual level, are unfortunately deprived of hands on practical experience during their study period.

Also due to political changes in these countries the status and rewards received by the graduates have not been attractive enough for the brighter students to pursue a career in engineering.

Finally there is a lack of respect and appreciation for Professional Engineering Institutions in these countries. Consequently, the profession is not able to influence the government in power regarding engineering matters and also is unable to keep a close check on the performance of its own members regarding their professional competence. Therefore it is necessary to act upon the current situation and determine the necessary actions to be taken to bring about the required changes.

3 Current Position

Normally the students who attend a Degree course tend to leave High School at 16 years old and join University courses which last six years. By the time the students graduate they are twenty-two years. The first two years are spent in providing a scientific and engineering foundation to appreciate engineering theory and practice. The next two years are then spent in learning engineering priciples and engineering applications. The final two years relate to specialisation in a particular discipline. This pattern seems to be the general trend throughout these countries although there might be exceptions in some other countries.

The theoretical concepts are taught well but engineering applications tend to be limited. Also laboratory work tends to be limited due to lack of equipment. Industrial training is available on some courses but again these tend to be rather minimal.

After graduation the young engineer is able to practice as a "Qualified Professional Engineer". There is very little postgraduate training (continuing education) requirements regarding retaining professional status. Exceptions tend to be those for departmental requirements for assessment of performance regarding promotion.

Engineering Education 323

Thus it is felt that there is room for improvement of these courses. It is necessary to identify the areas which need to be strengthened to make the course structure and training suitable for the challenges of the future.

4 Improvements to be made for current courses

The Degree courses in most of the Developing countries operate on a sound theoretical basis, but if they are to be further improved the following areas have been identified to take appropriate action.

(i) A high quality teaching staff, well versed not only in theoretical aspects but also in applied engineering.

(ii) The laboratories to be well equipped and managed by competent technical staff.

(iii) Participation by industrial representatives from all aspects of course development to implementation and training programmes should be anticipated.

(iv) Promotion of large capital funding and a close involvement of those courses to be undertaken.

(v) The training of young engineers to develop an awareness of social and environmental effects of resultant engineering projects.

(vi) There should be opportunities for graduates to undertake research and development projects.

(vii) Encouragement of career planning to enable graduates to have a clear understanding of what to expect in the future.

(viii) Strict vetting and validation of the courses such that they are of a high standard.

(ix) Encouragement of attendance and participation at Professional Engineering conferences, seminars and visits to industrial units.

These are fairly obvious but it is difficult to ensure that these really take place. Most of these countries have their priorities and problems and trying to fit them into that context is the one which needs to be examined.

5 Action Plan for the immediate future

From the foregoing it will be noticed some of these may be achieved easily whereas others could be difficult to resolve in the present situation. In this respect the major contributors to bring about these changes must be the governmental funding agencies and industrial partners.

Most governments have their own priorities but they must be made to realise the importance of engineering education for the younger people of the country. They must appreciate that the investment is well worth it and will be rewarded in later years. Engineers need to be in influential positions to effect government policies. This is when the respect by Society of the Engineer must be earned and established. Hence the need is for effective Engineering Institutions, as in western industrialised countries. This is already happening in some countries. The credibility of the engineer must be established by having good postgraduate training schemes and professional vetting. For example in some countries the issuing of a license for practice is a good example. The engineer is expected to update his knowledge by attending continuing education courses and relevant seminars, conferences etc., to renew their license at regular intervals.

Another partner who can serve a useful purpose is the participation of local, national and international industries in educational programmes. These can be at several stages in the training of an engineer.

The participation at School stage is a useful one especially as it will help the student to be conscious of the engineering profession. This will be helpful in recruiting students, especially girls, for future engineering study.

Industrial involvement at undergraduate level is also most necessary. Their contribution will be most useful regarding course development, tutoring, counselling, provision of ideas for solving open-ended real problems, and the provision of specialist lecturers. By such activities the students will be able to keep close contact with industry regarding their training and future career choice. Also, with multinational industries, students should be able to obtain their training in both their own countries of origin and overseas.

Finally, last but not least, there are several universities in the industrially advanced countries of the world who are willing and able to offer their advice and services. Such activities should be further encouraged. British Council for example have been a useful source of contribution to this programme. This type of assistance may be made by overseas Universities located nearby or in Western Europe. Hence, present problems should be overcome if the right opportunities are seized and acted upon promptly.

6 Conclusion

The problem relating to Engineering Education is a major one. The method by which this may be achieved is difficult to identify. The major problem lies with the fact that old traditions linger on and therefore engineers' status is very much undervalued.

Society must be made aware that engineering creates wealth and therefore it is a sound and worthwhile investment for the future. Once a firm basis for engineering has been set, the industrial scene will change as a result in creating more benefits for society at large. Also contribution by industry is an important one. This may not be too difficult a problem to overcome. Most industries in the regions, especially the international ones fully appreciate the likely shortages of skilled and well trained personnel in the near future. Therefore if they have a sound economic sense they will only be too willing to participate in such ventures.

Another means of obtaining financial help can be from foreign governments who may be attracted as a future investment in the training of engineers. There are several countries who will only be too happy to contribute for such purposes, if it is well administered. Last but not least it is important that the contribution may be made by several universities of the world. Such assistance must be sought. Presently there is evidence of these activities but they should be still further extended to reach most universities in the region. Thus with good planning and imnplementation "Engineering Education" in the Developing countries could be further improved in helping to promote a sound and effective training for the future needs of these countries.

7 Acknowledgements

The author wishes to thank Professor T V Duggan for enabling this paper to be written and Mrs F Crabal and Mrs J de los Santos for their assistance during the preparation of this paper.

References

FINNISTON, Sir M. (1980), Engineering Our Future, Report on the Committee of Engineering Profession, HMSO publication, 21-26.
PE, A.H. (1989) Industrial Training for Undergraduate Degree Courses in Civil Engineering, **World Conference Proceedings**, 135-141.
PE, A.H. (1991) Professional Experience Relevant to Master of Engineering Courses in Civil Engineering, **World Conference Proceedings**. 256-260.
WARD, R. Teaching Management to Engineering Students, **World Conference Proceedings**, 64-69, Hamilton, Canada.
DUGGAN, T.V. (1989), How Technology Will Change Engineering Education **Int.J. Appl. Eng. Ed.** 5. 687-690.
DUGGAN, T.V. (1989), Future International Cooperation in Engineering Education, **Int. J. Appl. Ed.** 5. 687-690.
DUGGAN, T.V. (1990), Trends and Attitude to Change in Engineering Education, **Proceeings ASEE**. Annual Conference, Toronto.
DUGGAN, T.V. (1991), An Overview of Engineering Education In Europe. **Australasian J of Engrng Ed.** 2. 155-156.

Equivalence Questions in Engineering Courses - The Effect of the Student Exchange Process

C.A. Walker

Department of Mechanical Engineering, University of Strathclyde

Abstract

The topic of the equivalence of engineering qualifications is one which is currently under review. Equivalence may be approached in a legalistic framework or as a matter of organic development. It is suggested that the student exchange schemes, which are now a common feature, will, in the long term, prove the most powerful effect in the creation of a practical system of mutual regard.

Keywords: Equivalence, Exchanges, Engineering education.

Introduction

In the creation of a single European market in goods and services, one aspect of particular interest to engineers is the question of the extent to which a qualification earned in one country will be recognised as an equivalent qualification in another country. Given the care with which titles are bestowed, it is quite possible to imagine a situation where each national entity would set up its own standard, and insist that anyone wishing to practice in a professional capacity should conform to that standard. Rather than simply protect the public, such policies are most likely to be used to restrict the entry to professional practice. The aim of this paper is to evaluate the effect of student exchange schemes on questions of equivalence, and to suggest that the unifying influence of such exchanges is such that the equivalence "problem" will vanish within a few years, without the need for a legal framework

Diversity in Engineering Education

One of the hallmarks of the British engineering scene is its adoption of overt badges of rank - memberships of institutions, chartered status, degrees with Honours classifications. Unfortunately, engineering in the UK remains a profession of fairly low public esteem; recruitment of highly-motivated school leavers is a matter of difficulty compared with law, accountancy and medicine. By way of anecdotal contrast, this author was introduced to the father of a French Engineering student in the following terms: "My father is a gynaecologist in Bordeaux. But he is very proud that I'm going to be an engineer." The situation in France, then, is quite the reverse of that which exists in the UK. The engineering profession in France considers itself absolutely to be at the summit of intellectual attainment in Europe.

Further diversity abounds across Europe - engineers derive from ancient universities, from Technical Universities, from Polytechnics (with all the bewildering connotations that the title implies), from "schools" set up only twenty or thirty years ago. Plainly, engineers come in all shapes and sizes, in curricula of extremes of length - from three years up to nine years, with academic standards which may be validated in any number of ways. This diversity poses major problems as we move towards the single market post - 1992.

In a move to impose order upon this apparently chaotic situation, the national institutions under the banner of FEANI (Federation of National Engineering Institutions in Europe) proposed that recognition should be given at the "professional engineer" level, whenever that status is deemed to have been attained - CEng in the UK, Diplome d'Ingénieur in France and so on.

The title Eur Ing has been proposed as the appropriate designation; unfortunately, as might have been anticipated, it is the monoglot, label conscious British to whom this has appealed, and so 85% of the holders of the Eur Ing title (in 1991) derived from the UK. Once again, the real diversity that is rooted in the European soul appears to be intent upon avoiding the tidy solutions decreed from the centre. Is it possible, then, that a consensus will appear as to what should constitute the components of an education process acceptable across the continent; does it matter if it does or does not?

The spirit of the FEANI proposals contains more than a grain of realism. In the years following 1992, when the Single Market legislation swings into action, it will become more and more common for companies to operate on a Europe-wide scale. Many already do: the oil business, vehicle assembly, major defence programmes and consumer durables are already accustomed to operating on a wide scale. Nothing magic will happen on January 1, 1993, but the market in goods and services will gradually move into a new plane of operation. Already, strategic alliances are being put in place; the cloying inhibition of customs regulations and local specifications will gradually fall away, and people will be able to buy and sell as freely as they could in the days of the Roman Empire. What, then, are the implications for engineering education?

At least, we must accept that the working lives of today's graduates are going to be affected in large measure by the operation of the single market. While it is unrealistic to expect small companies to carry their sales and procurement across the whole of Europe, such activities will move much further down the scale of company size. But before the benefits of large-scale operation begin to accrue, the Europeans will have to develop a finely-tuned awareness of each other's sensibilities and capabilities, in the same way that already happens within national boundaries, and, on a continental scale, in the USA. Such acute perception is not achieved simply by the adoption of a title - it demands to be firmly integrated within the engineering education process itself.

One of the biggest problems that arises is the lack of accurate information about the education in the various countries. One is not talking here about the data that appears in official course handbooks, or the admirable SEFI guide to European Engineering Education; rather, the question concerns the subtleties of the nature of the students recruited, their aspirations and the extent to which the education system fosters these aspirations, the balance between authoritarian didacticism and creative, student-centred learning. A further imponderable is the extent to which the task of education itself is viewed as primary; by contrast, are students chiefly undergoing a process of maturation, and so must be encouraged in social endeavours such as sport, drama, debating? On top of these purely educational questions, one must be aware of the ways in which graduates are set on the paths of their careers during their first few years of employment, since this will profoundly influence their development or relative stagnation.

In the last ten years the European Community has funded large scale R & D programmes (ESPRIT, RACE, BRITE-EURAM etc). As a result, consortia of academic and industrial institutions are working together on individual research programmes, and these, in their own fashion, are teaching disparate groups of researchers to work together in a fruitful manner. At a different level, though, the student exchange programmes are likely to have a more long-lasting effect in enabling the European engineers to understand each other's modi operandi.

Student Exchange Programmes

Study exchanges have been a common feature of language degree courses for many years. Individual institutions laid greater or lesser emphasis upon study exchanges for other courses, but non-linguistic activity was at a rather low level. In 1986, the EC set in motion a pilot phase of the ERASMUS programme (European Community Action Scheme for the Mobility of University Students), with the programmes moving into high gear from 1988. From the onset, Business Studies saw this as a great opportunity, and were early leaders in the field. After a slow start, (save for honourable exceptions such as the Anglo-German exchange scheme organised by Coventry Polytechnic since 1980) the benefits of participation became evident to most faculties of engineering, and engineers now share with business students the first place as most enthusiastic participants in these exchange schemes.(Erasmus Bureau,1991)

The programme has matured rapidly, with simple, bilateral links being superseded by complex networks of up to twenty institutions. The point of the networks is to establish mechanisms that will ultimately allow students to move freely about the EC. At this stage, the networks have been set up by personal contact and word-of-mouth, and this close contact has allowed the system to function fairly smoothly.

Exchanges in Practice

Students may exchange to take lecture courses, to carry out a design or research project, to concentrate upon laboratory practicals, to stay for a period of industrial training or a combination of these. The actual choice of courses is normally made with reference to the courses available at the parent institution. Some departments are keen upon a near-perfect match; others are fairly relaxed about their student's programmes of study during exchange, seeing the exchange process itself to be a valid educational experience. It is true that the success of the exchanges depends greatly upon the student's ability to integrate quickly into the educational and social scene of his host institution: this integration is greatly assisted by the arrangements made to welcome the visitors. (see Appendix).

European Credit Transfer System (ECTS)

In parallel with Erasmus, a pilot phase of ECTS is in operation, allowing students to acquire elements of courses in various countries and carry the cumulative credit with them.

Effects of Exchange Schemes

The numbers of students involved in exchanges is now very large; the overall effects of this flow will, in the long term, be to ensure that the blank incomprehension, that frequently marked previous encounters between engineers from different parts of Europe, will gradually become a thing of the past; in a word, our engineers of the future will be Italians, Greeks and Spaniards first of all, and to some degree all European.

While this mutual recognition at the personal level is very important, of equal significance is the language facility that is generated as an integral and necessary part of the exchange process. One is scarcely surprised to hear Dutch students speaking English (and French, and German!), but the fact that we now have Scottish engineering students speaking Italian and Portuguese is a truly revolutionary development.

The point of all of this with reference to engineering designations is that if the European engineers have confidence in each other's training, and can communicate freely, the need for specific "badges" is greatly reduced. With time, too, courses will be modified in the light of the exchange experience to incorporate new ideas which have been observed working elsewhere. This in no way implies that we will move towards a common curriculum; in 50 years time, the French engineering schools will still be arguing over their places in the pecking order, and we will still be marvelling at the Dane's sense of open-minded logic applied to their courses. We will, though have extended the boundaries of our awareness to know what to expect of students from Lulea and Valencia, without having to go through an extended process of enquiry. While this may sound like one of the more fanciful hopes of the founding fathers of the EC, it is rapidly becoming reality. Within ten years, several millions of students will have participated in exchanges and will be in employment at every level and in every area.

Conclusion

The exchange programmes fostered by the EC are in a healthy state of development. Engineering students now represent 27% of all exchanges; the long-term consequences are seen as an acute appreciation of the capabilities of engineers whatever their national origins, and an easier assimilation into industrial and commercial concerns. It may well be that this is a more powerful effect than the awarding of pan-European qualifications.

References

Erasmus Bureau (1991) Erasmus and Lingua Action II Directory.
Pollard, A F ,(1988) The Education and Training of Chartered Engineers for the 21st Century. Fellowship of Engineering

Appendix

Aspects of Exchange Programmes

1. **Lecture Courses**

   These are inevitably the most fruitful, but demand adequate language preparation and the necessary prequalification. Typical problems which may arise are:
   for British students, the greater emphasis placed upon mathematical rigour in France;
   courses which start at times which do not align easily with terms and/or semesters, so that visiting students either miss the beginning or hang about waiting for them to start.
   the need for visiting students to be informed ,from the outset, of the foibles of the teaching system they are about to experience.

   Most of these problems may be avoided by year-long exchanges, rather than a term or semester.

2. **Laboratory Classes**

   These can be most satisfactory, where they already play a large part in the curriculum. As an example, Strathclyde students go to DTH, Lyngby with minimal Danish (or none at all) and complete a series of practical classes.

3. **Design/Research Projects**

   Projects have the advantage of allowing an easier starting period while language is brought up to speed and social integration completed. They offer the chance of study in depth, and of being a part of a research group. Against that may be the isolation from other undergraduates, and a lesser onus on the continued language development that is forced upon students who opt for lecture courses. This is largely avoided if the visiting student can work along side a student of the home institution.

4. **Industrial Placements**

   At present these take place on quite a small scale, but the numbers are increasing. Employers of all nationalities find it quite difficult to take the plunge and give a placement to a foreign student. Just as students can exchange study places, so can placement exchanges be organised; this is a healthy development which will no doubt flourish as the current recession eases.

5. **Language Preparation**

In many engineering schools and courses, language tuition is mandatory at 6-8 hours per week. This is about the level required over a period of 2-3 years for a successful integration into a series of lecture courses. Languages remain a problem (and probably the limiting factor) in the UK, with many engineering departments making no provision in their timetabling for languages. Indeed a survey taken a few years ago gave a low priority to language tuition (Pollard,1988). If the evidence of the exchange programmes is to be believed, this situation is now changing rapidly.

6. Staff Involvement

For any exchange to be successful, the sine qua non is the dedication of large amounts of academic staff time, initially to set the exchange scheme in motion, and later to attend to the myriad unanticipated little problems that can arise. It is most important that the visiting students should have an individual member of staff to whom they can turn for assistance.

At present, for the majority of students, an exchange is for half of an academic year. This often causes as many problems as any other factor, since it involves a mid-year dislocation; a complete year on exchange is probably a preferable option, if it can be organised.

Summer Engineering Program for U.S. Students in London

J.W. Lucey, E.W. Jerger

Department of Aerospace and Mechanical Engineering, University of Notre Dame

Abstract

The University of Notre Dame offers a six week summer program for its undergraduate engineering students in London, England. Students enroll in two technical elective courses for a total of six semester credit hours. The Program includes several required trips to significant technological facilities. Students not only gain academic credit but also the invaluable experience of living for a significant period of time in a culture different than their own. That cultural exposure may well have a greater long term benefit to students' professional futures than the explicitly technical education they receive.
Keywords: Foreign Study, Summer Program, Field Trips, Engineering Education

1 Introduction

Notre Dame's College of Engineering has offered a summer program in London, England since 1988. The objective is to offer the undergraduate engineering student an enriching educational experience in a foreign study environment. Other than a program in Rome for third year architecture students, this is the only foreign study opportunity for engineers at the University.

Students live and travel in a foreign culture while enrolled in two engineering courses, obtaining academic credit applicable to their degree requirements. The program is an intensive six week session in London, England. Each course meets for the same number of contact hours as an on-campus summer session course for the same number of credit hours. In addition a number of required field trips, one overnight, are integrated into the courses. All students have been Notre Dame undergraduate engineering students in good standing. Their participation in the program allows them to make progress to their intended degree while maturing as well educated members of contemporary society.

With required field trips students visit engineering projects unique to the United Kingdom, including the Channel Tunnel excavation, the Sellafield nuclear fuel reprocessing plant operated by British Nuclear Fuels, the Thames Flood Barrier and several industrial facilities, including IBM research and manufacturing facilities in Southern England, and Brown and Root, Vickers, Wimbledon, to observe and discuss the practice of engineering within the United Kingdom and the evolving European Community. Brown and Root, Vickers have also demonstrated and given students hands-on experience with their CAD system for designing off-shore oil platforms, a system which is at the forefront of that technology. These visits allow students to witness the technical interaction of United States and foreign owned companies and the operation of United States owned companies in an international environment with local management.

Early each summer the class attends a performance at the Regent's Park Open Air Theatre. "A Midsummer Night's Dream," "Twelfth Night," and "Much Ado About Nothing" have, over the years, introduced students to available theatre in London.

Program evaluations, based on independent private student individual exit interviews with the Director of the University's academic year Arts and Letters London Program, have been most favorable. The evaluator, a different individual each year, prepares a detailed report of his interviews and conclusions.

2 Rationale

The University and its officers have long considered foreign study opportunities a valuable aspect of a student's educational program. Prior to 1988 all Colleges of the University but Engineering provided a foreign study option for their students. Notre Dame has had success in conducting foreign study programs for its other Colleges in England, France, Austria, Italy, Mexico, Japan, China, and the Middle East. The Engineering College's Rome program in architecture has been enthusiastically received by faculty and students, as well as the greater architectural community.

Typical of United States universities, Notre Dame does not require a foreign language of its engineering students, thereby restricting engineering study abroad to English speaking countries. Hence the choice of London, where the University already has an educational facility.

Rev. Edward Malloy, C.S.C., President of the University, who in his annual addresses to the faculty and elsewhere has often expressed his desire for expanded foreign study opportunities for the University's students, has visited the program and praised it as an excellent solution to the constrained problem.

The program recognizes a need to make engineering education more responsive to global technological progress and social concerns. Students in the program are exposed both to the unique technological achievements in the United Kingdom, and to the everyday life of the UK and other European countries.

Today's engineering graduates increasingly practice their profession in an international environment. A 1987 National Academy of Engineering Report developed a strong case for a new level of international cooperation on technological issues and reported a growing need for US engineers to respond to the increasing quantity and quality of engineering activity abroad. In its recommendations the report called for making engineering education more responsive to world-wide progress and concerns.

The United Kingdom, birthplace of the industrial revolution and a leading force in the European Economic Community, is a logical choice for the location of our engineering program. London, one of the world's great cities, has a long and interesting history. The development and construction of London's public projects is a significant chapter in the history of technology. Thames river development projects, the Underground transportation network, bridges and the mammoth Thames Flood Barrier (often cited as the eighth wonder of the world) are historic milestones. It is an exciting home for the program.

Economic changes in the European Community scheduled for the nineties will affect industrial economies around the world. In their visits to industrial facilities, and from the daily press and television news, students acquire a unique perspective on this transition which will have such a profound effect on their professional lives. Students based in London observe different national approaches to problems of the environment, economic competitiveness, and labor relations to mention but a few areas which will effect their professional futures.

The program is based at the existing Notre Dame Centre in Mayfair, increasing the cost effectiveness of that facility. One classroom and one faculty office are used. The existing Centre staff is utilized. During the summer the facility is shared with Notre Dame's Summer Law Program. The Notre Dame Centre is fully utilized during the academic year by a Masters of Law program, an MBA program and an undergraduate Arts and Letters program, which precludes establishing an academic year engineering program at this time.

3   Student Selection and Orientation

During the summer prior to a student's enrollment in the program a letter is sent to undergraduate engineers at their home addresses describing the program. An information session for interested students is held in early October. Applications for participation are due before the Thanksgiving holiday. All applications are reviewed for potential disciplinary problems with the Office of the Vice President for Residence Life. Students in recent years were mostly completing their sophomore (second) year of study.

On campus orientation meetings with the students and the Residence Assistants are held in the spring to answer questions regarding scheduling, appropriate clothing and conditions in London. Another orientation session is held on the students' first day in London.

4   Academic Program

Two three credit hour courses are offered. Engineering Economy has been offered each year. It was modified slightly from the course taught on campus to include examples of British engineering projects. The second course offered has changed from year to year, but has been related to professional practice and incorporates examples of British practice from projects visited on field trips. Classes meet for 2 hours each, four days a week, with one of the four days used for a required field trip. The two courses offered are acceptable technical electives in all engineering disciplines.

5   Events

We hire a coach to meet the students' flights at Heathrow Airport to transport them and their luggage to their flats. After a few hours settle-in time, a lunch is served for the class at a nearby hotel, a few doors removed from the flats. After the lunch an orientation meeting is held to outline the program and ground rules, provide information on London, and answer any immediate questions. On the next day, the first day of class, the class takes the London Transport narrated bus tour to become familiar with the city.

The first Friday the group travels by boat from Westminster Pier to the Thames Flood Barrier, a major engineering project designed to protect London from flood. The trip includes extensive commentary on the history of the river and those who lived and worked along it. The Visitor's Center at the Flood Barrier contains an operating model of the Barrier and a film show on the need for the Barrier. The class then travels up river to Greenwich to visit the Royal Observatory and Royal Naval Museum.

A trip the third week visits Ironbridge Gorge and British Nuclear Fuel's Sellafield facility. Ironbridge Gorge contains a number of sites associated with the birth of the industrial revolution, including the remains of the first coke fired iron smelter, a museum of iron and an open air museum which recreates village life during the start of the industrial revolution. From Ironbridge we travel to Lancaster University to spend the night.

The following morning we travel by coach through the edge of the famous Lake District to the Sellafield plant on the Cumbrian Coast. Sellafield has been reprocessing spent nuclear fuel since 1952 and is the location of the Calder Hall nuclear power station, the world's first commercial scale nuclear power generating station, commissioned in 1956. BNFL graciously provides us with an all day guided tour. We visit the Calder Hall reactors and the MAGNOX spent fuel storage and decladding operation, where we observe the sophisticated remote handling equipment used in the process.

We have visited the IBM facilities at Havant and Hursley, both near Portsmouth. At Havant we have been able to observe several advanced manufacturing procedures, while at Hursley we have seen the design and development process in progress. At both sites we heard from upper level managers and young engineers about the challenges and accomplishments working as British nationals in an international corporation.

A visit to the offices of Brown & Root, Vickers, a firm responsible for the design of off-shore oil drilling platforms, was arranged by a member of our Engineering Advisory Council. Located near Wimbledon, BRV gave us a series of presentations on past and present engineering projects, their corporate structure, Computer Aided Design techniques and other aspects of the design and development process. Students had an opportunity to experiment with their CAD system.

We visit the Channel Tunnel project, the largest in Europe at the present time. In the early years of the program we toured the construction site, but for the past two years our visit has been restricted to the Visitor's Centre, as the construction site is now an international frontier. All three tunnels in the project are now open from England to France. The Channel Tunnel Visitor's Center has information about previous attempts to tunnel under the English Channel, dating back to Napoleon's time, a display of work in progress and a model of the completed project, model trains and all. The return trip to London includes a short visit to Canterbury.

In previous years we have also visited, or been visited by, AT&T, Ford Motor Company, and Olympia & York.

As a supplement to the programs technical content, members of the class take advantage of the London theatre scene, attending many of the West End productions on their own. Many also manage to attend the Wimbledon Tennis Championships, athletics (track & field) competitions, football (soccer) and rugby matches.

6   Administration

The two tenured professors jointly administer the program, sharing an office at the Notre Dame Centre. The Centre's administrative/secretarial staff have been most helpful. Two Resident Assistants, chosen because of their previous experience and familiarity with London, complete the Program's staff. Their background is beneficial to the program as they are able to advise students on travel plans and everyday London life. They live in the flats with the students and are responsible for maintaining order in the flats and interfacing with custodial and maintenance personnel of the Vienna Hotels Group which manages the flats.

7 Housing

Students are housed in flats in the Bayswater (W2) area of London, adjacent to Kensington Gardens and within walking distance of the classroom building. Four or five students occupy each flat. Each flat has two bedrooms, kitchenette, living/dining area and private bath. Each is furnished with cooking utensils, incoming telephone and color television. The flats are managed and maintained by Vienna Hotels who provide a weekly cleaning and change of bed linen. Faculty flats are a few blocks removed from the student flats.

The two Resident Assistants live in the flats with the other students and are responsible for student behavior in the building.

8 Conclusion

The student fee, which covers round trip air travel, housing (but not meals) and program costs including field trips, has allowed the program to break even for the five years of its existence. Individual travel and food costs add to students' expenses. In addition students gave up the opportunity for meaningful summer employment and associated income. This does limit the number of students who are able to participate. Financial aid specifically for the program would allow economically pressed students, including minority students, to participate. It seems clear that the fee will have to be incremented annually to accommodate continuing inflation and changes in currency exchange rates.

Our judgement, based on our talks with students during the summer and on campus after our return, and on evaluations conducted in London by the Arts and Letters Program Director, is that the program is a huge success. We have had no attendance or health problems. Discipline problems have been relatively minor.

The summer engineering program in London is now a well established program of Notre Dame's College of Engineering. We expect enrollments to continue in the mid twenties for the foreseeable future and we continue to explore the possibility of an academic year program.

Transition Programs in Canadian Engineering Faculties

P.M. Wright (*), J.D. McCowan (**), J.D. Ford (***)
(*) Department of Civil Engineering, University of Toronto
(**) Faculty of Applied Science, Queen's University
(***) Faculty of Engineering, University of Waterloo

Abstract
Commencing in 1973 at Queen's University, transition programs have been established in six Canadian engineering schools, transition programs being defined as those which permit students who have performed very poorly in their first term to recover in time to proceed to Second Year with their original classmates. The paper describes these programs, outlines why they exist, and presents data to illustrate their advantages.
Keywords: Transition, Promotion Regulations, High School, University.

1 Introduction

Formal engineering education began in Canada in the 1870's usually in programs associated with existing universities. As was the custom at the time, academic staff assumed that the most effective way to motivate students who did poorly was to require them to rusticate. Those who returned usually worked harder and often completed their degree requirements but with the loss of one or more years.

However such a strategy is not the best if the student's difficulties stem from having attended high schools in which they were not sufficiently challenged. In such cases students enter universities with poor study habits, often accustomed to combining studies and part-time jobs. Studies conducted at both Queen's University and the University of Western Ontario revealed that the lack of success of new students was usually due to their inability to adjust quickly enough to the heavy workload. By the time the adjustment was made, it was too late. Rustication for these students yields a greater loss due to the forgetting of basic knowledge than is gained through increased motivation.

Another factor in Canada is that many, if not most, high schools have fewer than 100 students, and thus are usually unable to offer solid instruction in mathematics and sciences. Students from these schools may have appropriate study habits but they are often unable to compete with students from the large schools. Students who have been out of school for several years have similar disadvantages.

As a consequence of such factors, six engineering faculties in Canada have introduced programs which allow students who have done poorly in their first term to immediately repeat courses, defer others to the summer, and thereby earn the right to enter Second Year with their original classmates. Such programs are defined as "transition programs". Success rates for students who elect to enrol in them range from 50% to 70% whereas for students who do not, success rates are well below 30%.

2 A Description of the Existing Transition Programs

In 1991 the authors conducted a survey of Canadian engineering schools and identified the following transition programs.

Queen's University The oldest transition program is in the Faculty of Applied Science, Queen's University. Commencing in 1973 first year courses in chemistry, physics, and mathematics were offered in the usual one-term format and in extended versions with the latter having 50% more lecture time. Students who did poorly in diagnostic tests conducted early in the fall term were encouraged to enrol in the extended courses, as were students encountering difficulties during the term. Students who did poorly in the fall term examinations could also transfer to the extended courses. The examinations in the extended courses were held in late February and those who were successful enrolled in the second set which finished in late June.

While the above system reduced failures, it had some serious disadvantages. First, most students were unwilling to accept the evidence revealed by the diagnostic tests even though they were remarkably accurate in predicting failure. Second, the instructors in the extended courses found it difficult to cope with the constant stream of students switching into their courses. Finally, over time, most of the students in the extended courses were those who had failed the final examinations in December and they were not being well served by attending for only six weeks.

As a consequence, the diagnostic tests and the fall term extended courses were abandoned in 1981. Now students who do poorly in the December examinations can elect to immediately repeat up to three subjects (chemistry, physics and mathematics) during an intensive five week period. New examinations are written at the end of February and, as before, the extended spring term courses are offered until the end of June. As reported by McCowan (1992), the program has reduced failure rates from almost 20% of the entering class to about 5%.

It is a fact of life that most candidates for a transition program can only be persuaded to enrol after they have faced the reality of failure. Once that has occurred, they usually accept the chance to immediately repeat courses and defer others into the summer session.

University of Victoria First Year students who are in academic difficulty after the fall term are counselled to reduce their load in the spring term. Summer courses are available so that deferred courses can be cleared.

University of Manitoba Commencing ten years ago, all first year fall term courses have been offered again in the spring term and all spring term courses in the summer. Although the rules permit students who do very poorly in the fall term to continue into the spring term, most, after counselling in January, elect to immediately repeat courses deferring others to summer.

Carleton University First Year students can retake failed fall term courses in the spring term and failed or deferred spring term courses in the summer. One or two of the deferred courses are not exactly the same but that has not caused any problems.

University of Western Ontario The Faculty of Applied Science introduced in 1991 the "Extended First Year Program" for students who attain an average between 50 and 59% in their December examinations. Following counselling in January those who elect to participate sign an agreement which outlines the courses to be repeated and those to be deferred to the summer. Adjudications for progress into Second Year are made after the summer course results are available.

University of Toronto In 1990, the Faculty of Applied Science and Engineering introduced a pilot transition program, now called the T-program. Engineering students who do poorly in the fall term can elect to enrol in the T-program and by so doing can repeat up to three courses (algebra, calculus, and mechanics) in the spring term and defer a similar number to May/June. Students who enrol in the program must obtain an average of at least 60% in the repeated courses with no mark under 50 in order to proceed to the summer courses. The usual promotion regulations are applied after the summer course results are available.

In 1990/91, students with fall term averages between 50 and 59% could select the number of courses to be repeated. Students with averages between 45 and 49% had to repeat all of the available courses for which their original mark was less than 60, in most cases all three. For this latter group, the alternative was rustication.

Because the program at the University of Toronto was established on a trial basis, data has had to be collected on the relative performance of students who chose to enrol or not enrol in the program. Some of the results are included in this paper; more complete details have been presented by Wright (1992).

University of Waterloo Co-op engineering programs such as those offered by the Faculty of Engineering are unable to offer transition programs of the type noted above. However a modified transition program has been devised for students who have averages of less than 50% after Term 1A and thus would normally be required to withdraw. These students can enter a future Term 1B if they successfully complete the "Qualifying Program for Re-Admission" which involves the repetition of courses in which their original marks were less than 60. To succeed, students must obtain an average of 70% or more in the repeated courses and no mark can be less than 65. Approximately half of the eligible students enrol and, for the period 1986-1991, 54% of them succeeded in gaining admission to Term 1B.

3 Some Results from the University of Toronto

The first year class in the Faculty of Applied Science and Engineering has about 900 students and in January, 1991, 62 of the 155 first year students who had been placed on probation having obtaining an average between 50 and 59% enrolled in the new T-program. Another 33 students enrolled who had averages between 45 and 49% and would normally have been required to withdraw. As an aside, both the University of Toronto and Queen's University have found that the existence of transition programs reduces the stress on most students, and the drop-out rate in the fall term.

In Table 1, students who achieved the required 60% average in their repeated courses and thus were able to proceed to their summer courses are shown in the columns labelled "Yes". As shown, 48 of the 92 students who wrote the final examinations could take the summer courses.

Table 1   Results in Repeated Courses

Term 1F Average	1 Course Yes	1 Course No	2 Courses Yes	2 Courses No	3 Courses Yes	3 Courses No	Withdrew	Total
55 to 59	16	8	4	2	-	-	1	31
50 to 54	1	8	11	6	2	3	-	31
45 to 49	-	-	4	2	10	15	2	33
Totals	17	16	19	10	12	18	3	95

Again referring to Table 1, the overall success rates for the students in the second and third groups were similar even though the third group would normally have been required to withdraw after the fall term. Also as shown, students whose fall term averages had been between 50 and 54% and who repeated two or three courses performed very much better than those who only repeated one course. This latter observation led the Faculty to revise the regulations for 1991/92 so that such students must choose between repeating all of the T-courses for which their original mark was less that 60 or none of them.

It should be noted that the averages in the original fall term courses had been just over 40% for both those students who succeeded in being allowed to take the summer courses and for those who did not do so. The successful students had an average of 66% in the repeated courses whereas the unsuccessful ones only achieved an average of 52%. The improvement was very much a matter of renewed motivation rather than background.

By mid-July, the academic status of the T-students was

determined. Table 2 gives the success rate in terms of cleared first probation (CFP) for all students who had been placed on probation after the fall term.

Table 2 Comparison of Spring Term Performances

Term 1F Average	Students in T-Program			Students Not in T-Program		
	No.	CFP(%)	Withdrew	No.	CFP(%)	Withdrew
55-59	29	17(58%)	3	58	20(34%)	5
50-54	30	10(33%)	1	35	5(14%)	7
45-49	32	9(28%)	1	–	–	–
Totals	91	36(40%)	5	93	25(27%)	12

Students who enrolled in the T-program did very much better in their six spring term courses although admittedly under slightly different conditions.

The initial common measure for the two groups of students was their relative performance in Term 2F as shown in Table 3.

Table 3 Comparison of Term 2F Performances

Term 1F Average	Students in T-Program (CFP)					Students Not in T-Program (CFP)				
	No.	Ave.	PP	B2	Other	No.	Ave.	PP	B2	Other
55-59	16	65%	11	2	3	19*	58%	9	5	5
50-54	9*	65%	8	1	–	4	60%	3	–	1
45-49	7*	61%	4	3	–	–	–	–	–	–
Totals	32	64%	23	6	3	23	58%	12	5	6

PP = Proceed on cleared first probation
B2 = Proceed on second probation
Other = Repeat term, failed, or withdrew.

* One student transferred to part-time studies.

Students who enrolled in the T-program also did better in Term 2F than did those who chose not to do so. Five of the former T-students obtained averages of more than 70% with the highest being 76%.

Finally, many First Year students do not succeed in clearing probation in Term 1S and therefore have to seek re-admission to a subsequent Term 1F. An examination of the re-enrolments for Term 1F in 1991/92 revealed that 12% of the unsuccessful T-students had returned whereas the comparable figure was 25% for those who had not elected the T-program.

4 Other Programs to Assist in the Transition

The same factors which led to the establishment of transition programs have also led to a variety of other initiatives to assist engineering students with the transition. For example, a few faculties have reduced the number of courses in First Year from 11 or 12 to 10. Several no longer assess first year students after one term and instead consider the results for the first two terms or even the first 12 months thereby including summer courses.

As outlined by Ford (1992), the Faculty of Engineering at the University of Waterloo, has developed a diagnostic test in mathematics which identifies students who might have academic trouble unless corrective action is taken quickly. The two-hour test which is administered in the first week of the fall term is based on six aspects of the Ontario high school mathematics curriculum. Students attend review sessions during the second and third weeks of the term for those topics in which they have done poorly. Students showing continued signs of weakness are assigned to special tutorials. The results of this program have been reassuring.

5 Conclusions

To date transition programs have been established at six Canadian engineering schools and over the next few years more are likely to be implemented. The existence of a transition program reduces the stress on most new students, many of whom are apprehensive about their abilities relative to others in the class. Students who are unsuccessful in a transition program are more likely to immediately seek alternate career options. Most importantly, transition programs permit many students who have started badly to adjust to the heavy workloads in engineering without the loss of time. In summary, transition programs give engineering faculties a softer image without sacrificing quality.

References

Ford, J.D.(1992) Mathematics Assessment of First Year Engineering Students in **Proceedings of Eighth Canadian Conference on Engineering Education,** Université Laval, Québec, Canada, 7 pages.

McCowan, J.D.(1992) The Queen's University Experience with Transition Programs (1973-1991) in **Proceedings of Eighth Canadian Conference on Engineering Education,** Université Laval, Québec, Canada, 5 pages.

Wright, P.M.(1992) Survey of Transition Programs in Canadian Engineering Schools in **Proceedings of Eighth Canadian Conference on Engineering Education,** Université Laval, Québec, Canada, 10 pages.

Development and Reorganization of Engineering Studies in Italy

G. Augusti

Facoltà di Ingegneria, Università di Roma "La Sapienza"

Abstract
This paper describes, with some examples, the new pattern of Engineering studies in Italian Universities, which - starting from Academic Year 1992/93 - will include three-year "Diploma" courses besides the traditional five-year "Laurea" courses.
Keywords: Italy, University reform, Engineering diplomas.

1 Introduction

In November 1989, the Italian Minister for Universities and Research appointed a "National Engineering Committee" (NEC in the following), charged to study the structure of Italian Engineering Education and to make proposals for its development, in relation to the needs of the country and the prospects of an increased internationalization of the labour market, and in the context of the changing structure of the Italian University system.

The first proposals of NEC have been described in a paper presented at the 1990 SEFI Conference in Dublin and published in SEFI-News No.26 (September 1990). In the present paper the situation is updated with regard to the definitive proposals and their implementation.

2 General national context

The work of NEC developed while very significant modifications were introduced in the general organization of Italian Universities and their course structure. In fact, not only the Universities (which in Italy are in great majority State Universities, and all are regulated by National laws) have been joined in June 1989 with the Research system under the newly instituted "Ministry for University and Research", but a number of very important and innovative laws have since been approved by Parliament. Thus, within the limits set by the general legal structure of Italian Society (and in particular by the profession-related "legal value" of each degree, which requires general rules to be set by law on the curricula and the running of the courses), a comparatively large autonomy has been granted to the University system and the Universities.

Among the approved laws, particularly relevant for the subject of this paper is the November, 1990 Law on the "Ordinamenti Didattici", which redefined the degrees that Italian Universities can award, introducing the "Diploma Universitario" (to be obtained after post-secondary courses of two or three years) besides the traditional "Laurea" (obtained through courses of four to six years). This diversification of degrees (which should not exclude in the future other post-secondary qualifications) follows the trend prevailing in the more advanced educational systems; however, two points of the quoted law are worth underlining. Firstly, the Italian "higher-education system" will be based on the existing University network, in that the Universities (a definition which includes fully the three "Politecnici": Milano, Torino and Bari) will have the responsibility of setting up and running the Diploma courses (although they are allowed and solicited to seek external collaboration and support, organize consortia, etc.) and will grant the degrees: space does not allow to elaborate on the reasons for this choice, which is strictly related to the traditions of the Italian educational system, and in particular to the lack of high-level studies outside Universities. Secondly, while in general "University Diplomas" and "Lauree" will be in "parallel" rather than in "series", the law explicitly requires that the studies for getting a Diploma be, at least in part, recognized towards a corresponding "Laurea": this provision mitigates the "parallel" relationship and does not exclude that some Diploma and the corresponding Laurea can be "de facto" in series.

3 Comparison with other industrial countries

Indeed, when at the beginning of its work NEC had compared Italian Engineering Education with that of countries of similar industrial development, we had recognized as an Italian peculiarity the lack of alternatives to the only existing Academic Degree, the five year "Laurea in Ingegneria". Moreover, while finally in the last (1989) reorganization of the "Lauree", the Engineering disciplines have been clearly distinguished in three parallel "Sectors" (Civil, Industrial and Information Engineering), the Engineers' Professional Institutions (an "Ordine" in each Province) still list - again according to a National Law - all registrants in one single "Albo", which legally gives to every engineer the same professional competence, irrespective of his specialized qualifications.

A further quantitative comparison pointed out that Italy has by far the lowest output of engineering graduates.

While the National Council of the Engineers' "Ordini" has elaborated a Bill for the subdivision of the "Albi" of the "Laureati" Engineers into Sectors of well defined competences (a proposal which NEC has warmly welcomed), it appeared that the "Ordinamenti Didattici" Law may contribute to overcome the other two drawbacks mentioned above.

4 The proposals of the National Engineering Committee (NEC)

Therefore, even before the final Parliamentary ratification of the quoted Law, NEC proposed the institution of "University Diplomas" in Engineering . In NEC's views, these Diplomas must be aimed at the formation of qualified engineers, able to tackle immediate technical problems and to accept and utilize innovation. It is therefore necessary a high level formation in physics and mathematics (however with a stress on the application aspects), a "sectorial" engineering formation (spacing throughout the range of the "title" defining the Diploma) and a reasonably specialized professional preparation.

Thus, three-year courses were designed, more devoted to applications than the "Laurea" courses, but not neglecting the general formation that distinguishes an "Engineer" from a "Technician" and allows the continuous updating of his knowledge. In other words, the courses designed by NEC aim at forming neither "technicians" knowing only a restricted portion of technology without sound general basis, nor "general purpose" engineers without any professional training.

This view on the cultural profile of the "ingegnere diplomato", made NEC to agree fully with the provision of the 1990 Law, namely formation within an Engineering Faculty, i.e. the same establishment where "ingegneri laureati" are formed. Thus, the "ingegneri diplomati" will study in a culturally active environment, and their teachers take part in research and development programmes: indeed, we envisage a unique full-time teaching body, with possible exchanges of tasks between "Diploma" and "Laurea" courses. This organization, that will indeed be peculiar of Italy with respect to most countries, should allow a better use of human and material resources (teachers and facilities) and avoid the frustration of career Diploma teachers being cut out of research activities. Of course, the teaching body for the Diploma courses must be completed by part-time instructors coming from the productive world.

As to the actual running of the courses, we expect many different solutions: some Institutions will hold the Diploma courses on the same premises as the Laurea courses, but others will prefer specific buildings, in many cases even located in a different town. As a rule, anyway, the same Institution will award both types of degree (Diploma and Laurea), and differences in the legal status of the respective full-time professors will not exist.

On the other hand, the institution of the University Diplomas in Engineering does not exclude other forms of shorter post-secondary technical education: indeed, such courses were already hinted in the NEC's Preliminary Report, but not elaborated upon; some Industrial initiatives for "Technicians" courses (which purport to be complementary and not alternative to the University Diplomas) are currently under study.

The general design of the University Diplomas in Engineering was outlined in the already quoted 1990 paper and will not be repeated here. In the subsequent months, NEC finalized its proposals as to the list and curricula of each "Diploma" to be awarded, required - as a general rule - by Italian laws: list and curricula were then discussed and approved, with some modifications, by the National University Council and finally made official in a Decree issued in December 1991 by the Minister for University and Research.

The final list, shown in Appendix No.1 below, includes 12 Diplomas, 9 grouped into three "Sectors" (Civil, Industrial and Information Engineering respectively) plus 3 "Inter-sectorial" ones.

An example of curriculum is shown in Appendix No.2. It can be noted that - although the general structure of the course is fixed by the Decree, because of the already discussed reasons - significant degrees of freedom are left to the individual Faculties, both in the choice of the specific subjects of each module (that are defined only in very general terms) and in a number of fully "open" modules.

5 The implementation

The Diplomas in Engineering have been included, with other University Diplomas, in the first three-year plan for Universities (1991-93), approved by Parliament and Government in late 1991. Some resources in terms of teaching staff and finances (rather scarce, alas!) have been provided. Then, the Minister has issued a Decree listing the nationally approved Diplomas and the Institutions allowed to start them as from 1992/93, provided well-defined feasibility conditions are met: the Engineering Diplomas are reported in Appendix No.1.

It appears from this list that many Diplomas will actually be in locations different from that of the Mother Faculty: it is expected that most of these will be supported by industrial concerns and local Authorities. It will take a few years to evaluate the outcome of these collaborations and in general of the Diploma courses, and then possibly improve and modify the approach followed in their design.

6 The "Civil Engineering vs. Architecture" problem

The scheme formulated by NEC for the Engineering Diplomas is similar for all Sectors of Engineering, but - while in the Industrial and Information Engineering Sectors a sufficient consensus has been reached so that significant difficulties in the implementation will presumably arise only from scarcity of resources and from organizational problems - the Civil Engineering Sector is still facing major difficulties, because of the yet unclear requests of the construction industry, of the problems related to individuals' professional responsibilities, of the existence of other concerned subjects besides the Schools of Engineering. For instance, the proposed curriculum of the Diploma in Building Engineering has not been yet approved because of its possible interplay with the Diplomas in Architecture: a proposal for the latter has been finalized only very recently and follows a completely different philosophy, aiming explicitly at very specialized "technicians". Another specific Diploma in the field has also been requested by the Professional Association of "Geometri" (qualified building technicians and land-surveyors).

Therefore, it has been necessary to form one more Committee: the "Joint Engineering-Architecture Committee", which held its first working Session on 9 April, 1992. The writer hopes to have soon something to report on the work and the outcomes of this Committee.

===========================

Appendix No.1

List of University Diplomas that can be activated by Engineering Faculties during the 1991-93 three-year plan (from the Decree of the Minister for University and Research of 31.1.1992):

DIPLOMA	UNIVERSITY (in brackets: seat, if different from seat of Headquarters)
1. CIVIL ENGINEERING SECTOR	
Engrg. of Civil Infrastructures	L'Aquila; Cosenza; Reggio Calabria; Salerno; Napoli "Federico II"; Parma; Roma "La Sapienza"; Roma Tor Vergata; Pavia; Ancona; Catania; Messina; Palermo; Salerno(Avellino); Bologna (Cesena or Forlì'); Bari Politecnico(Taranto).
Building Engrg.(*)	Reggio Calabria; Roma "La Sapienza"; Bologna(Cesena or Forlì').
2. INDUSTRIAL ENGINEERING SECTOR	
Aerospace Engrg.	Perugia; Roma "La Sapienza"; Bologna(Cesena or Forlì').
Chemical Engrg.	L'Aquila; Salerno; Salerno(Avellino); Trento; Napoli "Federico II"; Ferrara; Genova; Roma "La Sapienza"; Bari Politecnico(Foggia); Torino Politecnico(Biella); Messina; Bologna(Cesena or Forlì').
Mechanical Engrg.	L'Aquila; Cosenza; Salerno; Modena; Parma; Napoli "Federico II"; Trieste; Cassino; Roma "La Sapienza"; Genova; Brescia; Padova; Ancona; Catania; Palermo; Pisa; Torino Polit.(Mondovi' and Novara); Milano Polit.(Lecco); Udine(Pordenone); Roma Tor Vergata; Bologna(Cesena or Forlì'); Cassino(Frosinone); Bari Polit.(Foggia).
Electrical Engrg.	L'Aquila; Roma "La Sapienza"; Cassino; Cassino(Frosinone); Genova; Catania; Bologna(Cesena or Forlì'); Torino Politecnico(Alessandria).
3. INFORMATION ENGINEERING SECTOR	
Electronic Engrg.	Salerno; Parma; Roma "La Sapienza"; Ancona; Catania; Messina; Bari Politecnico; Potenza; Genova; Pavia; Torino Polit.(Ivrea, Mondovi' and Vercelli); Cagliari(Nuoro); Pisa; Padova; Udine; Bologna(Cesena or Forlì'); L'Aquila.
Telecommunication Engineering	Torino Politec.(Aosta); Roma "La Sapienza"; Siena; Bologna(Cesena or Forlì').
Informatic and Automatic Engineering	Cosenza; Salerno; Napoli "Federico II"; Parma; Trieste; Roma "La Sapienza"; Roma Tor Vergata; Ancona; Palermo; Milano Polit.(Cremona); Firenze(Prato); Messina; Bologna(Cesena or Forlì').

4. INTER-SECTORIAL DIPLOMAS

Logistic and Production Engrg.	Roma "La Sapienza"; Genova(Savona); Milano Politecnico(Lecco); Bologna(Cesena or Forli').
Environment and Resources Engrg.	Potenza(Matera); Roma "La Sapienza"; Roma "La Sapienza"(Latina); Cagliari; Genova; Genova(Savona); Trento; Palermo; Pavia(Mantova); Firenze(Prato); Bologna(Cesena or Forli'); Udine.
Biomedical Engrg.	Milano Politecnico.

(*) Note: Curriculum not yet approved.

==============================

Appendix No.2
An example of National scheme of curriculum: the University Diploma in Mechanical Engineering

[Note: The scheme prescribes the minimum number of "teaching modules", i.e. one-semester single-subject courses]

Table A: <u>Common Modules for all University Diplomas in Engineering</u>
Mathematics:	4
Physics:	2
Chemistry:	1
Basic Informatics:	1
Economics and Management:	1

Table B3: <u>Common Modules for Engrg. Diplomas of the Industrial Sector</u>
Solid Mechanics:	1
Mechanics of Machine and Indistrial Design:	1
Termodynamics and heat transfer:	1
Electrotechnics and applications:	1
Energy systems:	1
Science and technologies of materials:	1

Table C.3.4: <u>Specific Modules of Diploma in Mechanical Engineering</u>
Fluid Mechanics:	1
Technical Physics:	1
Machines and Energy Systems:	1
Mechanics of Machines:	1
Mechanical design and machine construction:	1
Workshop Technologies and Systems:	1
Mechanical Industrial Plants:	1
Electrical Movements:	1
<u>To be defined locally:</u>	7

Engineering Education in the Philippines

S.P. Claridge (*), E.L. Tadulan (**)

(*) *School of Systems Engineering, Portsmouth Polytechnic*

(**) *College of Engineering, Xavier University*

Abstract

Many so called 'Developing Countries' face major obstacles in the evolution of their Engineering Education Programmes. A classic case study is the Philippines, a country with a complex culture and currently facing major economic problems. This paper gives an overview of the present state of engineering education in the Philippines and highlights the 'Macro-plan', currently being implemented to assist the Government to meet the goal of making the Philippines a Newly Industrialised Country (NIC) by the year 2000.

Keywords: Engineering, Education, Philippines

1. Introduction

Since the downfall of the Spanish at the turn of the century and until 1945, the Philippines was governed by the United States of America and hence many of the American Institutions and systems were adopted. Naturally, an American styled education system grew very quickly with the establishment of classical Elementary (7-12 year old), High School (13-16 year old) and University/College institutions.

Until the mid-1970's the Philippine education system was the envy of the region. Today the education system generally and engineering education in particular is wanton, starved of resources and plagued with the Countries overall socio- economic situation. It now stands far below the Southeast and East Asian neighbours.

Following the popular 'people power' revolution in 1986, the Philippine Government stressed the importance of Science and Technology in strategic planning by including specific reference to the subject in the 1987 Constitution of the Republic of the Philippines namely "Science and Technology are essential for national development and progress. The State shall give priority to research and development, invention, innovation and their utilization; and to science and technology education programs" (Ref 1.) Consonant with this, the Philippine Government has initiated an ambitious plan to fully harness engineering technology to enable the country to become a Newly Industrialised Country (NIC) by the year 2000. The emphasis being on the development of the engineering education system, particularly the quality, type and number of engineering graduates and the capability of the engineering education institutions.

2. Overview of Philippine Engineering Education System

In the last 20 years the number of institutions offering engineering programmes has doubled, making it extremely difficult for the Government to manage, monitor and maintain quality engineering education.

The Philippines currently has 185 engineering education institutions; 46 (25%) are State Colleges or Universities and 139 (75%) are private institutions offering in total 571 undergraduate programmes, 25 graduate programs and 2 doctoral degrees. The National Capital Region (Metro Manila and surrounding area) dominates the statistics with a greater number of institutions and offerings compared to the other 13 regions.

The undergraduate institutions operate a full Credit Accumulation and Transfer System with the curriculum and facilities being monitored and evaluated by the Technical Panel for Engineering Education (TPEE). The TPEE aims to enhance the Department of Education Culture and Sports (DECS) effectiveness in planning and supervising engineering education and consequently plays a prominent role in policy making. The TPEE ensures that minimum requirements are met by the individual institutions and subsequently recommend recognition of the engineering programs by the DECS. If an Institution is deficient of any of the requirements, and in 1988 80% failed to meet the Engineering laboratory equipment requirements, a special dispensation is required to operate the program.

The Professional Regulation Commission (PRC) also plays a significant role within the higher education structure. Twice a year a written National Professional Board exam is held in Metro Manila and if approved by the PRC, simultaneously at Centres around the country. An Engineering Graduate must achieve a pre-determined passing percentage in this exam to acquire a 'licence to practice' the profession. Consequently, the board exam result is vital to the career of any young engineer.

Between graduating from a University/College and taking the relevant Professional Board Exam an Engineering Graduate usually spends considerable time, effort and money attending one of the numerous Review Centres, the courses of which are tailor made to assist the engineer to pass the Board.

The 'quality' of an engineering institution or programme is usually measured by the success rate of the graduates in the Board exam. Especially, if one of the 'Top Notchers' (top ten percentage passes) is a recent graduate, the undergraduate institution then receives tremendous kudos and credit and will use this to full advantage when recruiting.

3. The Macro-Plan for Engineering Education

The Macro-Plan has recently been developed to strengthen Engineering Technological Education and Training through the establishment of a National Engineering Education System (NEES). The plan sets forth guidelines to identify and designate key installations to specialise and support other institutions in a higher education hierarchy (Ref. 2). The structural components are illustrated in Figure 1. below:-

NCE - National College of Engineering
RCE - Regional College of Engineering
CIE - College/Institute of Engineering

Fig 1. Conceptual Structure of the NEES (Ref. 3)

The introduction of a flagship approach as a rationalisation mechanism is a novel idea and intended to extend assistance to the deserving institutions as well as curb the proliferation of sub-standard schools and 'diploma-mills'. It will also hopefully develop an infrastructure that will produce technological/engineering manpower:-

- of appropriate type
- of appropriate quality
- in the appropriate quantity

4. Technological Manpower Supply and Demand

As previously stressed, it is vital that any engineering institution supplies manpower to the correct specification. In 1980 a Manpower Survey covering the 13 Regions of the Philippines was conducted(Ref.3). Though the predictions were limited to 1984, it shows a glaring trend of oversupply of engineering graduates and an increasing cumulative excess over the years. It is not only the quantity which is questioned but also the quality.

The distribution profile of industry and hence manpower requirements, through-out the islands is complicated by a number of factors. Firstly, the geographic situation of an island archipelago means that a small number of isolated centers of industry have emerged. Metro Manila and surrounding provinces is obviously the largest, followed by Cebu, Davao, Misamis Oriental and others. Secondly, the very nature of being an emerging industrialised country means that many of the larger companies, the principle employers of engineering graduates, are predominantly Multinational, involved in light manufacturing or material processing. It is most likely that the primary engineering functions of design, manufacture and R and D are performed at the Parent Companies location overseas.

Consequently it could be argued that the need for design and R and D specialists in this current industrial environment is limited, although within the

Macroplan a 50% increase in the number of R and D engineers/scientists is called for, over the period 1988-1992, from the present number of 8000. This new manpower resource, with the corresponding innovative, creative and analytic skills will have to be utilized within research groups in the NCE and RCE's and other private and Government research establishments. It is within this sector, that the engineers are going to be able to contribute to develop locally based and funded businesses and thus contribute to maintaining the 10% growth in GNP necessary to qualify as an emerging NIC.

5. Discussion

With the majority of engineering institutions being privately owned, the commercial considerations play an important part in policy making and implementation of Government guidelines. It is a fact of life that in this type of system Institutions look to make a profit by offering the 'popular' engineering programmes, without fully appreciating the investment required in equipment. The TPEE are often very pedantic in the enforcement of the Laboratory Equipment Requirements and take little or no account of the way in which this equipment is used/demonstrated and the learning processes involved.

Although the TPEE appear to have full control of implementing engineering education policy, the institutions have no accountability for the quality of the education programmes offered. It is the Board Exam which is all important. The Professional Regulation Commission must play a more prominent role in moderating the engineering education programs, even recognising the degree qualification as an initial licence to operate. A written arithmetic exam in itself is not a measure of the competence of an engineer. However, the finances involved in operating Review Centres and staging Professional Board Exams is probably a major cause of resistance to this idea.

If these additional resources, both in terms of teaching staff time and salary and student's time and money, could be redirected into the undergraduate programmes a marked improvement is likely.

The individual Faculty members should also be made more accountable for the courses they teach. It is not unusual for a young engineering lecturer to be given a course to teach, with only a broad outline of the syllabus. He chooses the course content, the instructional medium (more often 'talk and chalk', due to financial limitations), the assessment method and he then assesses the students work. This is a very dangerous situation. More involvement by moderators both internal and external would vastly improve the quality of instruction.

The fundamentals of education must never be overlooked, in that somehow LEARNING must take place. The Philippine's education hierarchy has proved to be excellent at producing strategies and policies to 'improve' the education system. However it may be far more beneficial if sometimes the learning processes were more carefully scrutinised, "What actually happens during that one hour in the lecture room/theatre?" A general improvement in engineering education is more likely to be forthcoming if this question is more rigorously addressed.

Finally, it must be stated that Staff of Engineering Colleges generally have a very difficult time. In a typical Provincial University an average salary of a lecturer 'teaching' up to 20 contact hours per week is in the order of US$200 per month, which is considerably less than a qualified engineering counterpart in local industry. Obviously, this does not encourage the top engineers into the academic profession and makes it extremely difficult to keep good Engineering Faculty, unless they can supplement their income in someway, often detracting from their effectiveness as teachers. A common solution is to increase the number of part-time lecturers, however this too has many disadvantages.

6. Conclusions

(I) There is already an elitist system of higher education existing in the Philippines today, with around ten 'mainstream' Colleges of Engineering, most of which are in the confines of the Manila Capital Region. These institutions dominate policy making and have most to gain from a formalised tier system. It is vital that if the Macro plan is fully implemented, the Regional Colleges and other Colleges of Engineering must not be alienated even further.

(II) It is imperative that both Private and Public funding is immediately made available for research and development work,in order to encourage the education sector to initiate appropriate research and development work. These funds should be administered through the Department of Science and Technology or similar body through research grants.

(III) The Master Plan for Engineering Education in the Philippines is a noble attempt to address some of the deficiencies within the Engineering Education System. The success will obviously depend on the availability of substantial funding and the partnership of Institution, Regulation Commission and Government.

7. References

The 1987 Constitution of the Republic of the Philippines - Article XIV, Sec. 10

The Macro-Plan for Engineering Education in the Philippines, published by TPEE, Bureau of Higher Education, DECS (March 1990)

Engineering and Technological Manpower Study, Executive Management Group Inc., Philippines (1980)

Educational Programs to Combat the Serious Lack of Professional Engineers in China's Village and Town Industry

H. Wang (*), Z. Li (**)
(*) Higher Education Institute of Fuxin Mining
(**) Shenyang Industrial Engineering University

Abstract
Having made an investigation into the serious lack of professional engineers in China's village and town industries, we put forward the following educational programs. These programs are suited to developing countries.
Keywords: Engineer, Educational Program

INTRODUCTION
With reforms and opening in China, collective and private enterprises in villages and towns are suddenly coming to the fore. There are 82,073 small towns in China; there were 1,578,700 village and town enterprises in 1987, each small town having 19. The number of people employed reached 47,024,900, and the income of these enterprises amounted to ¥ 293,412,000,000, while there were only 131,000 engineers(including temporary ones) and there was only one engineer among 12 enterprises. The serious lack of engineers in these enterprises will be increasingly acute and will seriously influence the development of China's village and town industries. According to the report in " China's Statistical Information " on February 4,1991, among China's village and town enterprises, are only 30% with good product quality and there were five ten thousandths which put overall quality management into effect. The main reason is the lack of the engineers and technicians. Thecondition similar to China's village and town industry ispresent in other developing countries.

Without engineers, these industries will remain underdeveloped. How do we solve the problem of serious lack of engineers in village and town industries? We have to resort to improving the higher education of engineers.

I. Starting Short-term Engineering Colleges

During the period of economic rehabilitation in Japan after the second world war, people with professional skills were badly needed. Short-term colleges emerged under these post-war circumstances. There were 516 short-term colleges in Japan in 1987, with 440,000 students.

Various people with professional skills are badly needed to accomodate the rapid development of China's village and town industries. We should use Japan's experience. It is necessary to start various short-term two-year college systems to meet the needs of our village and town industries. What kinds of specialities should be offered? It should be according to the actual needs of the village and town industries.

China's village and town industry can be classified as 33 types, the sequence of which is described in the following table:

Classfication of China's Village and Town Industry

Names of Enterprises	Building Material Industry	Coal Mining Industry	Engineering Industry	Food Industry	Metal Products Industry	Textile Industry
Number of Enterprises	39,322		20,085	22,814	16,973	11,138
Output Values (Million Chinese Yuan)	29,123		24,166	17,583	18,380	30,595
People Employed	3,332,800	757,500	1,328,000	623,900	849,100	1,686,400

They badly need technical personnel. Short-term colleges should be first started to meet the needs of these industries in village and town enterprises, as they are very short of technical personnel.

II. Starting Correspondence Engineering Colleges in the Country

Most administrative personnel and the skilled workers in the village and town enterprises do not have enough time to study at college because they are very busy. Then, how can we keep them on site while providing them an opportunity to increase their knowledge and skills of engineering? The correspondence course is one of the best ways.

In order to suit and to meet the needs of engineers in the rapid development of village and town industries, Rural Correspondence Engineering Colleges must be started. The Correspondence Engineering College Centre can be started in the area where the village and town enterprises are centralized. Various engineering laboratories should be built which can be used by the students to do their experiment. Correspondence Engineering Colleges can train and bring up engineering and technical personnel who do not need to be released from work. These colleges have many advantages, such as less tuition fee, short period of study and quick training. And they must be well received by the public.

III. The Institutions of Higher Learning Will Be Setting Up Extra Engineering Classes Needed in Village and Town Enterprises

According to the statistical data, there are, altogether, 503,000 scientific and technical personnel engaged in agricultural technology (including the engineers in village and town enterprises), Six out of 10,000 being scientific and technical personnel, while sixty out of 10,000 in such developed countries as America, Japan and Germany. Thus it can be seen that the number of technical personnel is not enough. On the one hand, the Agricultural Colleges will train and bring up agricultural experts, agronomist and horticul-turist of high quality, on the other hand they will set up some engineering schools needed in village and town enterprises to train and bring up the enginers and the technical personnel badly needed by these enterprises.

There are 62 Agricultural Colleges. If they committed themselves to train qualified personnel for village and town enterprises, the need for qualified personnel in management of the village and town enterprises would be satisfied. In addition, other ordinary institutions of higher learning can also train engineering and technical personnel for the village and town industries in the form of evening classes, short-term training courses, entrusment training and classes for advanced studies. Fuxin Mining Institute, for example, started a 2-year school system specializing in management of the village and town enterprises to meet their needs in 1989. And it is well received by the village and town enterprises.

IV. The Cooperation between the Village and Town Enterprises and the Institutions of Higher Learning; the Mutual Benefit

The cooperation between the productive units and the colleges, an educational form, is that the study of the basic theory at college can be combined with the working experiences from productive units. Both the productive units and the college can benefit. The new arrangement, cooperation between the productive units and the colleges, has only recently been used in China. But the philosophy that education must be combined with productive labour has had a 40-year history. In 1950, the Committee which guides the practices of colleges and universities directly under the Educational Ministry was set up, when the People's Republic of China was founded. The Chinese Government decided that " education must be combined with production labour " as the educational principle in 1958. Since the 1980's, Colleges and universities in China have made further explorations on the cooperation between the productive units and the colleges. The suitable cooperation between the institutions of higher learning and the village and town enterprises mainly are:
(1) The three-in-one combination system should be set up which consists of the productive units, the institutions of higher learning and the scientific research institutions. According to incomplete statistics, there are 570 such systems throughout the country. The village and town enterprises' participation in this combination system not only can provide the practice and experiment base for the institutions of higher learning and the scientific research institutions, but also can introduce qualified personnel to current techniques and to the advanced results of scientific research.
(2) In cooperation between the productive units and colleges, the entrusted and directed training can overcome

the lack of the engineers.

(3) The cooperation between Textile College of Shanghai Engineering and Technical University and some textile enterprises is " Sandwich " type of cooperation between productive units and the colleges. During the students' four years at college, they are required to go to the textile mill for 10 weeks. This kind of practice can replenish the technical force.

(4) The "second classroom " can be provided for students at colleges (The student's social activities during their spare time are called the second classroom). The students should practice, temper themselves and give service in the village and town enterprises.

In brief, the cooperation between the village and town enterprises and the institutions of higher learning can be mutually beneficial. The institutions of higher learning can acquire scientific research and the practice base, while the village and town enterprises can obtain the support with qualified personnel.

Besides the educational programs mentioned above, there are some other measures to overcome the serious lack of engineers needed in village and town enterprises. For example, some enterprises hire part-time engineers for high pay. There are 256 enterprises in Daqiu Village in Tian Jin City. The output value in 1990 was ¥1800,000,000. What they have done is to start entrustment colleges and universities to train qualified personnel and to hire sparetime engineers for high pay as well. Daqiu Village plans to get 1,000 experts and engineers to work there in 1992.

Though it is feasible to hire part time engineers we must rely on our institutions of higher learning to overcome the serious lack of the engineers in village and town enterprises. The institutions of higher learning must take on the heavy responsibilities to train and bring up engineering and technical personnel for the villege and town enterprises.

Major Reference

1. " The Generalization of Education Development of Seven Countries "
   The Center of Education Development and Policy

           Investigation, National Educational Committee,
           Tianjin Educational Press, 1986
2. " Japanese Education Policy for Today "
           Office of Intelligence Investigation, National
           Educational Committee,
           Peking Polytechnic University Press, 1988
3. " China Statistical Yearbook "
           The Statistical Press, China, 1989

COMNET

V. de Kosinsky
COMNET Secretary General, Liege

1 PRESENTATION

1.1 Definition

> **COMNET is an Independent International Association in the field of European Education and Training in Advanced Technology.**
>
> It also represents the only very large scale European network allowing multilateral exchanges and contacts in the field of training.

1.2 Origins

Created and launched in 1987 in Liège (B), COMNET was originally based on the COMETT Programme (European **Com**munity action programme for **E**ducation and **T**raining for **T**echnology), whose main objectives are:
- to give a European Dimension to co-operation between enterprises and higher education establishments;
- to foster the joint development of training programmes, the exchange of experience, and also the optimum use of training resources at Community level. The various "Strands" of COMETT are all founded on the development of a European Network of University-Enterprise Training Partnerships (UETPs).

COMNET identified as its mission to federate these new organizations, called to pursue very difficult objectives and which could benefit from exchanges of experience and stable and professionally managed international links.

1.3 Development

Along with the rapid and intensive development of Europe, numerous associations, higher education institutions, administrations and enterprises felt that it was necessary to create a network with sufficient infrastructure, with all the necessary logistics behind to organize multilateral exchanges. The increasing importance of the COMNET network has allowed to fill up the gap and produce the missing link for these exchanges and close collaboration in the field of European exchanges for training and technology.

1.4 Membership
COMNET has **over 105 members coming from 27 European countries**, most of them being COMETT UETPs.
Most of the members themselves are local, regional, international or sectoral associations. With its members, COMNET represents in fact **more than 2000 European enterprises and higher education institutions**.
A large number of SMEs are numbered among the enterprises involved.

1.5 A few figures
In 1991, COMNET's members organize:
- industrial placement for 2400 students, i.e. 16800 student-months;
- personnel exchanges between industry and education: 60 people for 270 months;
- 340 European courses (in 650 sessions) in technology training.

> COMNET's actions extend from student exchanges (the COMNET Student Form is used Europe-wide) to marketing of products, and from COMETT, where it all started, to many complementary programmes.

- NETWORKING AND SERVICES FOR MEMBERS -

2 COMNET'S STRUCTURE

2.1 Membership
Membership can be **either full or of associate status.** Full members only have voting rights - Associate members have the right to participate in the activities and meetings of the association, with a consultative role.

2.2 Its European management...
The General Assembly of Members is the main Governing Body of the Association.
The Board of Management comprises one member (and a deputy) from each State where the association has full members.
The 1991-1992 Steering Committee consists of the President from Lisbon (P), the Vice-Presidents from Bologna (I) and from Cork (IRL) and the Secretary General from Liège (B).

2.3 Its headquarters - the Secretariat General in Liège (B)
COMNET's Secretariat General is performed by the ALUEF UETP in Belgium. It is composed of a group of enthusiastic people able to provide assistance, co-ordination and transnational services in co-operation projects to the benefit of the members.

3 ACTIVITIES AND SERVICES

3.1 The COMNET Newsletter
Periodical bulletin addressed to all COMNET members and COMETT entities in general, with a circulation of 1000 copies.

3.2 The COMNET database
Essential for all those who communicate regularly within COMETT, the COMNET database includes all the COMNET members, the contacts for all the COMETT projects, the members of the COMETT Committee and the COMETT Information Centres.

3.3 M.I.S. (Management Information System)
Efficient, up-to-date and user-friendly electronic information exchange system based on Information Technology. It is designed to provide easy and rapid access to information on student and staff requiring work placements, job vacancies, training needs, training courses, etc. Training sessions are being organized this year by various Pilot Centers in Europe addressing all COMNET members.

3.4 TTT (Training on Technology Transfer).
Training needs analysis and courses for UETP staff covering areas of management, technology, technology transfer and complementary topics.

3.5 ACTT (Association for Cooperation in Transeuropean Training)
East-West cooperation in the field of education and vocational training.

3.6 EUROPEAN FORA
Regular major events organized in Liège and other European cities - the 2nd COMNET European Forum took place Porto (P) on 13-14 April 92 - to bring together the COMETT Community on the most relevant subjects of the moment.

3.7 In addition to this, COMNET offers technical assistance to members' projects, as well as regular contacts with the EC in Brussels.

For further information, please contact:

COMNET Secretariat General
6, quai Banning
B-4000 LIEGE
Tel.: +32 41 52 80 85 - Fax: +32 41 53 40 97

- AN INSTRUMENT IN TUNE WITH THE ORCHESTRA -

366 Engineering Education

COMNET International Association
Quai Banning 6, B-4000 Liège, Belgique
Tel: + 32 41 528085, Fax: + 32 41 534097

Membres **COMNET** *Members*

24.03.92

The Teaching of English as a Second Language for Engineers

M. Rigal, R. Suarez

Department English Language, University of Castilla-La Mancha

Abstract

The purpose is to show how the teaching of English can be oriented in an Engineering Curriculum Development. We would focus on the needs of our students: Future Mechanical and Electrical Engineers. We take into account both:
i) Their actual requirements as students integrated in their Engineering courses who not only need English to pass the English examinations but also to consult Bibliography related to other subjects which is written in this language.
ii) Their preparation for a modern society in which English is totally necessary to find and perform a job with success.
To do this, we use texts, tapes, videos, and any other material centred on technical topics. This is what many authors call English for Specific Purposes, commonly abbreviated ESP. We also pay attention to communicative approaches of the language i.e. greetings, telephone conversations, going to the bank, to the shops, and so forth.

Keywords: Authentic and Non-authentic material, ESP, Integrated Skills Approach, Syllabus Design.

1 Introduction

Due to the great number of English speakers all over the world and to the economical, political and cultural hegemony of English speaking countries, this language has become a sort of "Lingua Franca", specially for those involved in Science, Technology, Business and any kind of research.

For this reason, the study of English as a foreign language is carried out in Spain not only in Primary and Secondary Schools but also in Colleges and Universities. For instance, English is a compulsory subject in the curriculum of the following degrees: Tourism, Advertising, Journalism, Medicine, Veterinary Science, Engineering and others. That is why English is included in our careers for Engineers.

Regarding the history of our Polytechnic College (Escuela Universitaria Politécnica, Albacete) we would mention that it was founded on 18 April 1978 being Forestry, Agriculture and Industrial Engineering (Mechanics and Electricity) the branches developed here. Computer Science Studies were added in 1985. This College belongs to the recently created University of Castilla-La Mancha. In this paper we will concentrate on the teaching of English for Industrial Engineers.

2 English for Specific Purposes

We teach groups of learners with specific needs for English dealing with subjects such as Electricity, Thermodynamics, Mechanical Design, etc. Some of these may constitute their speciality, so they require some knowledge of the language related to the concepts they are studying. This is what makes this area of Language Learning Development separate from the rest of English Language Teaching. On the other hand, general concepts and structures of the Language are to be taught since they are essential for the Understanding of these Texts.

According to Strevens (1983) ESP is an "umbrella" term which can further be divided into two main areas: English for Science and Technology (EST), an area which is so wide that it is often confused with ESP, and all other courses not included in EST. The following is a reproduction of part of Strevens' classic diagram:

EST is characterized by what has come to be termed "Scientific English", which, although in essence no different from "General English", does contain certain features: Specialized vocabulary pertaining to each field, Considerable expressions of quantity, special symbols, and so forth.

OTHERS include a wide variety of courses such as Tourism or Business; each with a special vocabulary but without the heavy dosage of Scientific English. Both EST and OTHERS can be broken down into Occupational (EOP) or Academic (EAP) courses. Students of EOP require English as an instrument for professional advancement. This could include training for a waiter in a restaurant or for an architect who must attend international symposiums. Students of EAP require English for formal academic study such as Mechanical or Electrical Engineering.

These above categories of ESP are often not as clear-cut as would seem, as those studying ESP can work and study at the same time, or apply concepts learned while studying EAP directly to new-found jobs.

In short, as Kennedy and Bolitho (1984) state: "ESP has its basis in an investigation of the purposes of the learner and the set of communicative needs arising from those purposes."

Our students, who will be technical engineers in the field of Mechanics or Electricity, are EAP learners who study English as an independent course of study. Many students realize the necessity of an EAP course to help them read numerous bibliographical references required for their studies. The usefulness of a command of English in the professional world also relates our courses to EOP, and requires us to include elements not strictly related to the disciplines chosen by our students. Therefore, such things as "applying for a job" or "writing business letters" are an integral part of our students' formation.

2.1 The ESP Teacher

Because of the special characteristics of ESP courses we would like to add some personal considerations regarding the ESP teacher: The majority of English Teachers are graduates in English Philology, having had very little or no education in the sciences. This absence of familiarity comes to fore when the General English Teacher is suddenly confronted with texts on electric circuits, to mention an example. The result can be an initial lack of confidence in one's ability to deal with such subject matter.

Strevens (1983) states that ESP requires teachers of more advanced experience to cope with these negative factors. Although this is not necessarily true, ESP teachers can benefit by:
i) Acquiring a basic knowledge of some of the fundamental principles of each subject area.
ii) Cultivating a positive attitude towards the materials being taught.
iii) Maintaining fruitful professional relationships with specialist teachers.

3 Methodology

Although the history of ESP has been a short one indeed, from an early beginning, in the sixties, a tremendous amount of time and effort has gone into research to develop methodologies that correspond to the ever-expanding needs of ESP students all over the world. Five major trends in the field of ESP have been of great importance:
i) <u>Register Analysis</u> The essential concept is the use of the appropriate language for a given situation. At the sentence level, this soon gave rise to the nascent field of:
ii) <u>Discourse Analysis</u>, where Scientific English is considered as a vehicle to communicate the concepts and procedures common to universal Scientific Discourse, the emphasis is on the "use" rather than "usage".
iii) <u>Needs Analysis</u> Before preparing a syllabus, an exhaustive profile of communicative needs must be constructed.
iv) <u>Skills & Strategies</u> The stress was shifted from what should be learned to how the students could employ strategies to improve their foreign language skills in Listening, Reading, Speaking and Writing.
v) <u>Learning- Centered Approach</u> It places the importance not on language use, but on language learning.

All of them have a place in our Curriculum and none should be excluded, by using an eclectic approach, we can offer a more varied and richer ESP programme.

Summarizing, the best methodology corresponds to an integrated skills approach, i.e., extracted from all these previously alluded methods, we pay attention to the different skills according to the lesson planning and to the students' requirements.

4 Student Profile

In order to better understand what makes our students tick a survey was conducted in December 1991. This survey of the <u>background</u> and <u>opinions</u> of our students, coupled with the daily observations of their performance within a given educational framework, helped us create an English programme more in accordance with their needs.

195 students were surveyed. This number accounts for 45% of our total enrollment for English in the branches of Engineering.

4.1 Background Information
4.1.1 Age/Sex
The average age of our students is 22 years old, with a very high percentage -over 80%-

being male. Age is an important factor to take into account here as we are now dealing with bright, capable young adults able to assimilate abstract concepts and analyze scientific data. These students require well-organized lesson plans with objectives clearly stated.

4.1.2 Previous Studies

According to the information given in the survey, 10% had attended Vocational Schools before beginning their studies here. A very small percentage, 4%, attended other Universities before registering. The rest had completed Secondary School.

4.1.3 Previous Studies in English

When asked about their level of English before entering our College, the enormous differences in students' background studies of English became apparent. The number of students with no previous knowledge before coming here was 21%. The difference in level among students of the same class is one of the biggest problems we must face.

4.2 Opinions

4.2.1 Motivation

The importance of a student's motivation to learn any material cannot be underestimated. At the College, students must consider their studies relevant at different levels:

-Professionally- to help them to function in their future jobs.
-Academically- to provide access to scientific journals, reports, texts, etc.
-Personally- to widen the possibility of travel, general culture.

Theoretically speaking, our students are quite motivated on all three of the above levels. When asked if a knowledge of English was a tool useful for their studies, 85% of the Industrial Engineers answered affirmatively. They also considered English as necessary for travel and for bettering their level of general culture.

4.2.2 Skills

After having asked them what skills they considered of greatest importance for their present studies, the answers broke down as follows: 44% Reading, 21% Speaking, 12% All Four Skills, 11% Writing, 9% Listening, and 3% Don't Know.

From these responses, we can gather that students appreciate the importance of an integrated skills approach to language learning. Their opinions bear directly on the activities we prepare for use in the classroom.

5 Course Design

The studies of Mechanical and Electrical specialities are organized within three school years. To obtain their final degree students must: Pass quarterly examinations, attend laboratory activities, and present a Final Project Work. A new system of credits is being put into practice so that we follow the EEC procedure. Nowadays, English is imparted compulsorily in the second and third school years, being the subjects English I and English II respectively. Needless to say, students must pass the first level in order to attend classes at the higher one. Regretfully only two hours per week are devoted to this subject. The total of class periods imparted in each level reaches approximately 60 hours per school year, starting at the beginning of October and finishing at the end of May.

According to our College Statute each student must accomplish -within these sixty hours- the following objectives:

-Acquire a basic knowledge of grammar, and be able to identify such structures in any kind of text. (Level I).
-Learn at least 3,000 words of both general and scientific English. (Level I).
-Write on definite topics. (Level II).
-Read comprehensively scientific texts. (Level II).
-Participate in speaking activities based on every day situations. (Level I and II).

Level I students are supposed to gain the knowledge and comprehension foundations that must be performed in level II. As we have already stated, these goals were designated by the school so we consider them as general ones. Nevertheless our Department has designed concrete aims taking into account both the students' needs and their motivations:

i) <u>Functional</u>: Give and find factual information (identify, describe, etc.); Express and Ask for intellectual, moral and emotional attitudes (agreement, disagreement, likes, dislikes, etc.); Social Behaviour (greetings, introductions, etc.); Persuasion (require help, give instructions or orders, etc.); Expressions of hypothesis, generalizations, classifications, etc.

ii) <u>Morphosyntactical</u>: Recognition and Use of the grammar structures needed to achieve all the above stated functions.

iii) <u>Lexical</u>: Mastery of both the basic and specific vocabulary contained in the different units of the programme.

iv) <u>Phonological</u>: Cognition, Recognition and Reproduction of the English phonemes.

All the reported purposes are embraced within the above mentioned four skills: Listening, Reading, Writing, and Speaking.

Broadly speaking the objectives pursued are shown below:
-Being able to: Understand conversations and oral texts (Listening Skill); Express orally and Participate in talks (Speaking Skill); Write personal and formal letters (Writing Skill); Take notes (Writing Skill).
-To get to know the organization of written language: Summaries, Recensions, Compositions, Definitions, Classifications, Conclusions, Descriptions, etc. (Writing Skill).
-Attainment of reading habits by means of guided reading (Reading Skill).
-Learn how to obtain general and specific information through texts (Reading Skill).

<u>Points to bear in mind</u>:
i) Most of these goals are mainly concerned with comprehensive reading, because we try to cover the most urgent necessity of our students: Understanding, Conception, and Perception of Technical Texts (see: 4 Student Profile). Nevertheless the number one goal of learning a foreign language is the <u>communicative use</u> of that language.

ii) These goals comprise both General English and English for Specific Purposes.

5.1 Evaluation

The techniques used to test are related directly to goals, content and methodology used in the classroom as above mentioned. We test:
-Students' command of communicative skills.
-Students' command of language elements: Morphosyntaxis and Lexis.

With slight variations, the evaluation system could be summarized as follows:
-Three Examinations spaced evenly throughout the academic year;
-A Final examination for students who have not yet passed each of the three examinations.

5.1.1 Level Required

We have already dealt with the problem of the different levels of our students when entering the College (see: 4 Student Profile). Despite this diversity we expect them to fulfill the level of the Cambridge First Certificate- plus a special handling of the Reading and Writing skills- when they are tested at the end of the English II Course.

5.2 Learning Material

Although material preparation is time-consuming, it seems to be an integral part of ESP teaching. Classroom material for Engineering is often lacking or very limited, but much can be done in the way of "customizing" both authentic and non-authentic texts, videos, magazines, and tapes among others. Authentic materials are those taken directly from scientific publications and non-authentic have been previously manipulated for a specific objective. Grammar books, dictionaries and "general purpose" text books are also exploited.

6 Conclusions: The English Language in The Curriculum of our students.

The fact that we should cover all the aspects of language learning, makes of our approach an eclectic one, which is precisely what is intended. Our students, would-be engineers, will use English in many different situations either formal, such as conferences, congresses, lectures, business relations or in their jobs, or informal, for instance dealing with the people that they meet in any of these circumstances.

A general vision is proposed so that they are able to face any situation they may encounter. That is the main reason why we use a wide variety of material -all related to Scientific and non-Scientific matters. With all these and the reported explanations about the design of our courses, we try to improve the understanding and proficiency of the English Language.

References

Alcaraz, E, & Moody, B. (1983) **Didáctica del inglés: Metodología y Programación**, Alhambra, Madrid.

Allen, J.P.B. & Widdowson, H.G. (1985) Teaching the Communicative Use of English, **Episodes in English**,69-89, Pergamon, Oxford.

Chomsky, N. (1980) The Utility of Linguistic Theory to the Language Teacher, **Reading in Applied Linguistics**, 234-240 I,O.U.P., Oxford.

Dubin, F. & Olshtain, E. (1987) **Course Design**, C.U.P. New York, New York.

Harmer, J. (1986) **The Practice of English Language Teaching**, Longman, New York.

Hutchinson, T. & Waters, A. (1989) **English for Specific Purposes. A Learning-Centred Approach**, C.U.P., Cambridge.

Hutchinson, T. & Waters, A. (1985) ESP at the Crossroads, **Episodes in ESP**, Pergamon, Oxford.

Kennedy, C. & Bolitho, R. (1984) **English for Specific Purposes**, MacMillan, London.

Strevens, P. (1983) **New Orientations in the Teaching of English,** O.U.P., Oxford.

Widdowson, H. G. (1990) **Aspects of Language Teaching**, O.U.P., Oxford.

Willis, J. (1983) The Potential and Limitations of Video, **ELT Documents: 114**, Pergamon Press, Oxford.

SECTION 3: QUALITY ASPECTS

How Effective is the Teaching of Engineering?

M. Acar

Department of Mechanical Engineering, Loughborough University of Technology

Abstract

Interviews with lecturers from the engineering departments and a survey of final year students at Loughborough University of Technology showed that lectures are generally ineffective due to the large volume of information which must be taught by necessity on engineering degree courses. There is often too little emphasis placed on explanation and relevance of material, and course structure. Students would prefer a more participative role in their education with more discussion.

Lecturers would be relieved of the pressures of presenting the bulk of what is often tedious basic information, if some form of programmed learning was introduced through books, hand-outs or computer packages: more time could then be spent on seminars and tutorials. 'Modular' teaching whereby intensive teaching days are set aside for lectures, tutorials and practical sessions would allow more effective use to be made of the time available.

Although the methods used are important, the interest and enthusiasm of lecturers play a more significant role in teaching and so care must be taken in appointing academic staff.

Keywords: Engineering Courses, Teaching Methods, Effectiveness of Teaching

1 Introduction

On the one hand, at the U.K. universities, which are primarily funded by the Government, a greater emphasis is given to research, particularly as this is seen as the quickest route to promotion since the number of papers produced and money brought into the university is tangible; the other two criteria for promotion - administration and teaching performance - are considered to be difficult to gauge and assess. On the other hand, in the present climate, academic staff are under increasing pressure to rationalise teaching without losing quality.

However, the importance of teaching seems to have been recognised at last. A pilot quality assessment study is currently taking place in order to develop an assessment technique of the teaching quality, providing incentive for lecturers to take more care over their performance.

The objective of this study is to analyze the teaching methods used in engineering degree courses by lecturers from all levels, the aim being to find out how various academics teach their courses, how they differ from each other and how effective the teaching is. It should then be possible to identify the weaknesses and strengths of teaching methods used which in turn may lead to suggestions for improved and more effective methods.

2 The Method of Approach

The method of approach has been based around surveys of students and lecturers at Loughborough University of Technology which has the largest engineering and technology student population in the UK. With seven engineering departments offering over two dozen engineering courses, there is sufficient depth of engineering courses for a survey of teaching methods.

2.1 Student Survey

The most appropriate method of gauging the opinion of the maximum number of students was to conduct an opinion survey which was developed from discussion with students and reactions to an initial pilot study questionnaire. The questionnaires were sent out to all final year engineering students, who have a more mature attitude towards the teaching to which they are subjected.

To encompass the maximum number of students the questionnaire was designed to be relatively quick and easy to fill in. The main body of the questionnaire was in multi-choice answer format, with an optional section for comments at the end. A typical time for completion of the multi-choice sections was 20 minutes.

In total 146 questionnaires out of 400 were returned allowing effective analysis to be carried out. The survey was chiefly concerned with gathering opinions of students on the teaching that they receive. The open section, giving opportunity for comments by students, was filled in by half the students and provided a source of further ideas.

The student survey outcome is limited by the fact that the students are notoriously subjective and will be influenced by course difficulty and preferences but with objective questions this can be minimised. Due to their inexperience, students may not be able to discern between what they want and what they need until they have the hindsight of a few years in industry.

2.2 Lecturer Survey

Since it would be more difficult to induce the required responses from lecturers in a questionnaire, and difficult to convince them of the merits of completing it, a series of individual interviews were arranged. Of 180 requests for interview sent out, 80 were returned and of these 29 were selected for interviews of 30-75 minutes.

The purpose of the interviews was to gain a general feel for the opinions of the teaching staff and individual ideas for possible improvements of teaching methods. The approach used was that of an 'interactive questionnaire' such that although the same areas were covered with each lecturer the depth varied according to comments made.

The interview results are limited by the fact that lecturers may be tempted to appear more conscientious than they actually are in practice. It is also difficult to tell whether the methods they claim to use are effective or not since basic styles do not tend to vary a great deal: it is often the way the method is used which makes the difference.

3 Discussion of Results

Both the student survey forms and interviews with the lecturers have been analyzed and the results are discussed in this section, and improvements to teaching methods are suggested in Section 4.

3.1 Aims of an Engineering Degree Course

Lecturers see the aim of an engineering degree course as to develop an analytical frame of mind, ability of critical thinking and an approach to problem solving. Students should be educated for a general understanding of the discipline which will be longer lasting than a more specific course; the degree is a mark of how well a person can assimilate technical information and so employers feel that a good degree is the evidence of a level of intelligence or educational ability.

Since it is the foundation for a career, a degree should provide the basic building blocks of knowledge and practical experience: students are being trained to a level set by industry and professional institution (on accredited courses) and given an overview of related disciplines. In some ways an engineering degree is a passport to a career in industry.

3.2 The Approach to Lectures

Quite often it seems that the lecturing technique is not used to its full potential; the old adage "that the purpose of a lecture is to pass the information from the notes of the lecturer to the notes of the student without passing through the minds of either" is often more of a reality than a joke. Lectures are frequently a means of transferring information rather than understanding. Students often feel that there is little to be gained from going to lectures over copying up notes at a later stage and this indicates the system is not effective. A better way, given that the same amount of work must be covered, would be to use hand-outs or reading assignments given in advance which could then be explained in lectures, making better use of lecturers time. Such a method puts more pressure on lecturers to perform well as the students attention is focused on them, the burden of note taking having been relieved; speed is often a defence used by less confident lecturers who do not want questions or undue attention to be directed at themselves. A good lecturer makes time for the student to make comments, take notes and take in explanations.

If the lecture content load was reduced it would allow for shorter lectures - which are desirable for concentration - enabling more attention to be paid to structuring the material and adequate repetition. A frequent plea from students is for material to be presented in a more organised way since it is often difficult to tell what has been 'covered' by the lectures. Detailed summaries of the core syllabus with corresponding references and suggested background reading would provide much more directed learning on the part of the student.

The classical approach to engineering subjects is to convey abstract knowledge in lectures since this does not change as fast as the applications and allows a more flexible approach to problems. The aim of a degree course is accepted as being to prepare the mind to tackle new difficulties and as such there is a danger of cramping students creativity by relying on specifics to justify the theory. This approach is acceptable from the point of view of the lecturers since they are well versed in their chosen field, but for the student it is both hard to understand and hard to be motivated without specific examples and emphasis on the relevance; for subjects such as Mathematics it is often hard to see the application of the theory. From the students point of view the aim is, in the short term, to pass exams and for this they want examples to relate the theory to exam type questions.

Lecturers complain that students are not prepared to take an active role in lectures and indeed this is often the case, particularly when the numbers involved are large. The problem can partly be attributed to the assumption that lectures are a one-way communication process, an impression gained during early stages of university life. Students come to expect that they will be spoon-fed but accept that the most successful lectures are those which involve interaction between themselves and the lecturer. For effective improvements, changes must be made at the outset, although this can be difficult when covering basic information.

3.3 Personal Qualities of Lecturers
Lecturers are generally appointed on their academic rather than personal qualities, yet it is largely personality which governs the level of interaction between staff and students. Although improvements can be made in presentation skills, some people are naturally not good communicators. If student directed learning was used this would perhaps be less significant, but lectures show up any flaws which are present.

If lecturers are unable to put teaching skills into practice or are not concerned with their teaching performance, it appears that little can be done apart from taking care in appointing them initially. For those who are simply unaware of what they are doing wrong then adequate coaching and feedback would help. Enthusiasm towards research is a good thing, showing interest in the subject concerned, and there is a general feeling that researchers make the best teachers, provided they feel a duty towards the students.

Most students comment more on the personality of lecturers than on the methods they use, admitting that different techniques are equally valid. If students get on with a lecturer then they are prepared to forgive minor teaching faults which may be annoying in a less friendly lecturer. Better staff-student relations can be encouraged through departmental; lecturers and students have more personal contact through sport and various functions, paying dividends in the lecture theatre.

3.4 Laboratory and Coursework
Individual assignments are the only way of truly assessing a students work, but the benefits of team projects are overriding in developing communication skills and work organisation. Asking students to provide a break-down of project responsibilities can improve the accuracy of individual assessments as can careful attention to responses in tutorials; in practice both individual and team projects must be used.

Laboratory exercises are generally deemed to be successful although they tend to encourage cheating. Careful modification of labs from year to year is one way of discouraging students from copying as thought must be given to the application of the theory to a slightly different situation. Group reports increase the teamwork benefits and reduce the perceived workload, but mean that students do not necessarily gain full understanding of the exercise. Students would prefer more emphasis on the report marks due to the time spent on them, but often lecturers do not want such a detailed write-up; bringing deadlines closer to the laboratory exercise would reduce the content, as would specifying the actual report requirements more clearly. Often it is best to put a limit on the report size.

Research is a great help in providing interesting laboratory tests to which students can see the relevance, but are often not seen until the final year. The application of current research contracts is also particularly interesting in design projects, notably in the final years.

4 Improvements to Teaching Methods

4.1 Modular Teaching and Programmed Learning
Since lectures are largely an unsatisfactory method of conveying the volume of material required on engineering degree courses some system of programmed teaching would seem appropriate. For certain 'dry' subjects, in which lectures have little to offer in adding interest, they would be best learned away from the classroom

with lecturers' time spent on explanations and discussions. With less time allocated to putting over the subject matter greater emphasis could be given to the relationship between different engineering areas and lectures could be used as a source of inspiration rather than frustration. Whether the learning be from hand-outs, books or computer packages there is the advantage of a more flexible time approach, and if there are test exercises it can be ensured that students have understood the material.

The application of intensive modular teaching would also allow for a more practical use of time available and a greater relationship between theory and practice while it is still fresh in students' minds. The wastage of both lecturers' and students' time through individual one hour lectures cannot be underestimated and provided adequate breaks were given the benefits would outweigh the problems of maintaining concentration.

4.2 Student-based Instruction

Since the most frequently used method of problem solving is to go and ask another student it appears that more use could be made of this technique; contact between different years is generally so poor so a student 'proctor' or 'mentor' scheme would be an advantage. Such a system could be used in conjunction with tutorials but has the advantage that students are generally more willing to discuss problems with other students. The same interaction could be created in larger tutorials, encouraging students to work together on problems and combined with smaller tutorials to sort out specific difficulties.

4.3 Student Committees

Greater use could be made of staff-student committee meetings particularly in facilitating improved communications; many of the grudges and complaints which are voiced by students amongst themselves could be sorted out if they were brought out in such meetings. The poor understanding, which students generally have, of what is expected of them could be improved if the points were made clear in meetings.

The student committees would be the best way to provide constructive feedback on teaching performance. At present there has to be relatively major dissension among the students before action is taken. If students had year group meetings at regular intervals (such as mid term) they could decide as a group if there were any points which needed raising. Such an approach would encourage comments from those unlikely to see the group representative specifically, enable an exchange of ideas and allow the overall group feeling to be expressed in a tactful way. Individual students who think they are alone in their difficulties with the teaching methods often find everyone else is having the same problems if they talk to other students about them.

If staff are not prepared to generate feedback themselves then student committees should prepare a questionnaire to be made available to those who would care to use it; the results, known only by the lecturer concerned, would enable an increased awareness of what students like and dislike. This technique would be welcomed by newly appointed lecturers in particular.

4.4 Teaching Training

There appears no reason why there should not be a compulsory training course for inexperienced lecturers appointed to an engineering teaching post as part of their probationary period. A carefully devised course tailored towards the specific requirements of engineering would enable lecturers to gain experience in longer presentations as well as information gathering and the psychology of learning.

Lecturers could be encouraged to sit on each others lectures to provide feedback from a knowledgeable background and use made of feedback from videos.

'Update' courses would prevent lecturers from slipping into bad habits such as overfamiliarity with material. Lecturers who don't quite 'get round' to undertaking such courses at the present would be forced to find the time if they were compulsory. Ultimately however the level of benefit which lecturers gain is dependent upon their motivation towards teaching.

4.5 Motivation

Whatever teaching methods are used, they will only be effective if students are motivated. This motivation is greater among students who can see the relevance of what they are doing, such as those who have spent time in industry or have done more practically based courses other than A-levels; in addition those who have had to make sacrifices to take part in their chosen studies are less likely to need encouragement to work, a fact born out in the increased level of motivation among American students for whom university life is often a financial struggle.

Whatever methods are used, the motivation of lecturers is paramount since they have the power to influence students by their own attitude. Care must be taken in the appointment and training of lecturers that they feel a suitable duty towards students who are, after all, our country's future professional engineers.

5 Summary of Conclusions

Lecturing should be used as a basis of motivation and explanation rather than information. Information should be gathered through background reading, hand-outs or programmed instruction, developing students' ability to teach themselves. Student participation should be encouraged at an early stage. Student-based learning should be introduced. Emphasis should be given to individual and group projects. Groupwork should be increased with group reports for laboratory exercises.

Newly appointed lecturers with no experience should take a mandatory teacher training course specific to engineering. Staff should be carefully selected since personality and interaction are important influences in successful teaching. Staff interest and enthusiasm are more significant than the methods which they use.

More personal contact between staff and students is required to improve interaction and may be achieved through departmental societies. Improved communication may be achieved through the staff-student committees and should be the basis of feedback.

Acknowledgements

The author would like to thank Mr I G Hartley, the final year student in the 1988/89 session who undertook this study under the author's supervision. Thanks are also due to the lecturers and final year students who took part in the survey and interviews.

"As Wise as We are Smart"
Engineering - An Education in the Abstract
G.A. Hartley
Department of Civil Engineering, Carleton University

Abstract
The segregation of our university engineering programs from the older more traditional parts of the academe suggests, and may underlie, a separation of our profession from our culture. This paper proposes that we adopt a more fundamental and intuitive approach in our engineering courses helping our students toward a higher level of engineering literacy, by cutting back on numerical problem solving. It also recommends that we take a critical look at our priorities and policies in the part of the curriculum dealing with "complementary studies".
Keywords: Engineering, Education, Literacy, Humanities, Image.

1 Introduction

While the engineering undergraduate programs address well the nature and substance of the engineer's special role in society, and provide the necessary skills needed by engineers to successfully carry out these responsibilities, they are acultural. As professional engineering practioners, we work less directly with the public than do those in other professions. Rarely do students enter our programs with the committment to our academe, to the fundamental knowledge, that we would find in the students entering arts and science programs. Their committment is to a thing called "technology", and it will go unconsummated for the four years it takes to complete their programs. They enter engineering because of the promise of a good job, and because they are strong in mathematics and science. Many of them having started their engineering studies, regard their curriculum as comprising engineering problem solving courses, plus the simplest available arts electives from the non-forbidden list.

This paper is critical of this narrow view and suggests that we educators are very much responsible for it, and should foster a deeper commitment to, and fuller understanding of our corner of the academe, beyond the solving of problems. This suggestion relates to the engineering subjects themselves, apart from untutored writing assignments and the fourth year honours thesis, a major piece of writing done at the eleventh hour of the program. It is also about our lack of attention to what is actually being studied in the complementary part of engineering education dealing with the humanities, society and culture. It is suggested that the recommended changes may lead to engineering graduates who will be aware of working more harmoniously within their society rather than for it. Ultimately the paper is about literacy.

2 Our Incomplete and Unfulfilled Objectives

Our mission as engineering educators is to prepare students for a rapidly changing and increasingly multidisciplinary engineering practice. It may well happen that following graduation they will have a relatively limited range of work experiences, but we do not anticipate this; and we oppose in principle an overspecialized engineering program which we believe can result in a lack of imagination in engineering design work.

As faculty members of a university engineering department our primary educational objectives are expressed in terms of professional priorities. Safety and service to society, for example, may come first. These objectives thus relate to our responsibility to make them well-prepared as problem solvers and to be error free in their calculations.

Our educational policy is incomplete because it addresses only the needs of the engineering profession and what society needs from engineering, and overlooks the broader educational needs of the student. The courses from arts and social sciences intended to complement the technical aspects of the curriculum are not clearly specified and we apparently do not have a clear sense of the importance of this area of study. The requirement that these be "essential in the education of an engineer" CEAB(1990) has led us into our current predicament. This should be replaced by the priority of addressing the liberal education of a civilized human being.

It is futile to debate the importance, for example, of management training, the impact of technology on society, or the central issues of the social sciences. Obviously these are important. It would be more appropriate to examine how many of these "essentials" can be packed into a four year bachelor's degree program, while maintaining depth in what we claim to be a true university level education. Our sketchy policy in this regard, and the reality of how the students are now satisfying this part of the program, are consistent. The students are taking junior level survey courses in the social sciences having very broad coverage.

Our stated objectives are being only partially fulfilled because some of us place too heavy an emphasis on the solving of specialized numerical problems in the classroom, tutorial sessions and examinations. We do this in order to keep abreast of technological know-how; to provide a more specialized approach because of our awareness of the importance of job training; and perhaps because it is easier to cover the material by using numerical examples.

There is also pressure on us to do this from the students themselves. To most students, and some faculty, the objective of the course is exam preparation, and exams are numerical problems. They demand from us as many example problems as we are willing to provide, and that our old exams be published and that solutions be posted. The students see a tremendous risk in going into an exam with little more preparation than a thorough understanding of theory. Their education has become an experience involving a succession of deadlines - for assignments, lab reports, and preparation for quizzes and exams. This manifests itself in considerable pressure on them. This is one symptom of our continuing evolutionary course in engineering education toward less emphasis on fundamental understanding and more emphasis on problem solving and specialization. It is time to consider taking a backward step toward an approach to education from which we started to drift a few decades ago, when our class sizes were smaller.

In the following two sections it is suggested that the engineer and culture may be put into more intimate contact in two ways: by improving engineering literacy; and by improving the cultural literacy of engineering students. Neither of these ideas is new. Both imply basic education harking back to a more traditional approach.

3   Engineering Literacy

Engineering literacy is the ability to express the meanings of engineering principles in our native language, to define technical terms, and to write technical essays and reports. It serves educational and cultural needs, as well as the professional career. Most importantly for those of us who teach, this skill helps the engineering student to understand theoretical principles at a fundamental level. Secondly, engineering literacy leads to an improvement in the ability to communicate technology to those outside of the profession. This has wide-reaching implications, relating to more diverse role-taking by engineers in the non-engineering sectors. Finally it satisfies essential professional requirements by fostering good communications among engineers.

Problem solving is clearly important in getting the engineering design work done. The example problems engineering teachers use to demonstrate theory are also important. And numerical problems are one means for testing for competency. But this should not be the only basis for testing because it is only indirectly related to the knowledge, and because it can lead to a very restrictive style of learning by the student. Students should also be encouraged to express what they are learning in a more intellectual way using the language of their culture. This implies that we carefully limit the number of solved numerical problems in the classroom, and their availability, in order to force the students to face this greater educational challenge. It is recommended that more of these problem assignments relate directly to the first principles; and that more writing assignments be included with these other exercises. The brighter students will probably improve their ability to solve a wider range of problems, because they will apply first principles rather than procedures.

In any attempt to reinforce this literary side of engineering in our courses, we should do this while appreciating the difficulties that our students will face especially as fledgling undergraduates, where engineering writing is a new mode of communication for them involving both new language and conventions, and one that seems to be more difficult than those that have been previously practised and reinforced in their educational experience, Hartley (1990).

In engineering writing generally, mathematical and scientific concepts are used to describe the physical behaviour of systems. Some of our mathematical models are given a physical sense, such that they

can be discussed in a non-symbolic way. In the Engineering laboratory
report, the student is writing from the point of view of the gatherer
and presenter of data, in other words the person controlling the
experiment. This is different from earlier high school exercises in
scientific essays in which the student was the collector of the
writings and findings of others, and in which the task was to put it
all together into kind of a collage, summarizing the important
elements of this recently acquired knowledge. The engineering data is
often inconclusive, and the inconclusiveness itself is important
information to be communicated. Writing around the scatter inherent
in data, or highlighting experimental error within a discussion of
results is yet another challenge for the student, and one which we
often take for granted.

Students' reports indicate problems handling the vagueness about
tense, and use of the passive voice. The data was collected in the
past and the writing is being done in the present, apparently by a
narrator who was not directly involved, about results which are true
for all time, but will be read by someone else in the future. Writing
of this kind for the novice must be reinforced through practise,
preferably with a tutor, who is preferably the engineering professor.

All of this suggests that we become more involved in grading these
reports ourselves; do more in-class and tutorial instruction; include
more writing as part of the testing; and try to motivate students by
stressing the importance of engineering communications. We should also
be mindful of the risks the student feels with engineering communications, and avoid overpenalizing.

4 Humanities

Studies in the liberal arts relate to the pursuit of a higher life
through education, and in the present context, to engineering itself
as a more integral part of our culture, Florman(1968), Cross (1952)
and Schaub and Dickison (1987). We refer to the studies, external to
engineering-mathematics-science, as "complementary" to engineering and
there is indeed a connection, going both ways.

What we call our intelligence is largely rooted in our cultural
heritage. Intelligence is clearly a requisite of engineering training, but this training does not give rise to intelligence in the
traditional meaning of the word, for example, in the sense of our
being able to communicate with other educated individuals, or in the
sense of creating things which inspire because they make a cultural
statement, or in the sense of pursuing a more intellectually enriched
lifestyle. Too heavy a concentration on engineering subjects and non-humanities electives leaves little room for cultural diversity in our
engineering work. The professional arrogance that any specialized
education may instil can cause otherwise intelligent people to ignore
the cultural aspects of their work. In addition to this, where professional training addresses too narrow a role, individual professional responsibility for work that goes wrong is forfeit, and
collective responsibility is no responsibility at all.

What we do in engineering is culturally determined to some extent.
For example it relates to our lifestyle and life quality, and exists
very prominently in our surroundings. Many classical engineering works

stand out not because of technical complexity, but often because of a cultural connection. This implies that, in other cultures, engineering will be somewhat different. In several ways the education of engineers, and much of engineering practice, appears to have become detached from its cultural origins.

University education has the capability to increase the base of shared information in the educated populace, and open up lines of communication. This is called "cultural literacy" by Hirsch (1987). Literacy goes beyond the meaning of words and the mechanics of grammar. Liberal education also provides shared symbols which allow educated people to communicate at a higher level. What we read and write goes beyond the encoding and decoding of words; there is also a shared knowledge which does not have to be defined in our communications with those who share our culture. This includes knowledge about our past, political systems, philosophical points of view, our legal system, ancient and modern civilizations, the literary classics, mythology, and so on. This suggests that the courses needed are those that are the core of most arts and science programs. The direct benefits of cultural literacy for engineering students also relate to a further broadening of their engineering communication skills.

Should an education in engineering not suggest a broader range of roles for its graduates? Why is engineering not a more logical stepping stone to statesmanship than the study of law, for example? We have established a very strong academe in all areas of engineering, and society has an interest in what we do. Why are we not a stronger presence in teaching at lower levels of education, journalism, the literary arts? Has this something to do with the fact that our skills relate to the solving of problems at more detailed specialized levels using a language that has become too abstract for anyone but an engineer to understand?

What arts courses are most beneficial in a liberal education? In the author's view we have already made mistakes in engineering by applying criteria that these have some professional utility. A preference is given here for courses in English literature, and courses dealing with one's culture and history. Courses dealing with other cultures can be valuable as reference points for one's own. The course reading materials will have to be selected very carefully, and should probably be modern, if they are to be taken seriously by the engineering students.

5 Recommendations and Conclusions

Do we really believe that our students want to choose their own non-engineering electives? Most students see this as an exercise in finding the simplest courses to ease a heavy work load. Students will not oppose the removal of the elective category from "complementary studies". They will see this as educational guidance that we should be providing. Our faculties need help from our colleagues in arts, both to help us to better understand the purpose of these courses and to select study materials which are appropriate for this special category of non-majors.

The author has reservations about some typical courses now filling these elective slots. Technology Science and the Environment is one of the most recent accreditation requirements. TSE relates to our moral sensitivity but TSE awareness should be extracurricular and involve all of society. Moral sensitivity should not be taught, it develops as a bi-product of a strong educational experience. In referring to the broader aspects of our engineering curriculum, we should cease to rank the social sciences. These are as valuable for first year social sciences students attempting to find their way in a strange new place, as our professional practice courses are for our students. These courses have very limited value for engineering students and belong in the forbidden category.

Students ought to be given more opportunity to provide non-numerical responses to questions, not just in engineering project courses but all engineering courses. An atmosphere must be created where students realize the importance of this aspect of the course, and the likelihood that it will form part of the testing for competency. This will be sufficient motivation to alter study habits.

Is 12.5 % of the program devoted to the complementary studies enough? If engineering undergraduate programs are increased in length, the first priority for this additional breathing space ought to be in this category. One year of general arts in five (20%) is not excessive in a university baccalaureate program.

It is unprovable at this point that such changes in our educational approach, will have any effect on the engineering profession, for example, a change in image for the engineer, or more rounded engineer. Even stronger changes along the same lines might suggest more interesting consequences involving engineers with increased sensitivity toward culture, enjoying a higher quality of life and taking on non-traditional engineering roles, perhaps in the arts. This is indeed what is under speculation in this paper, and for now we can only guess at possible side effects.

Acknowledgement

The words "As Wise As We Are Smart" in the title are from page 11 of Samuel Florman's "Engineering and the Liberal Arts" cited in the references. The full sentence is "When we are as wise as we are smart, we will be qualified to talk about new goals as well as methods, ultimate ends as well as means".

References

Canadian Engineering Accreditation Board. (1990) "The Canadian Engineering Accreditation Board 1989/1990 Annual Report".
Hartley G.A. (1990) "Writing Skills of Engineering Undergraduates" in **Seventh Canadian Conference on Engineering Education**, University of Toronto, Canada.
Hirsch, E.D. (1987) Cultural Literacy: **What Every American Needs to Know**, Houghton Mifflin Co., Boston.
Florman, S.C. (1968) **Engineering and the Liberal Arts**, McGraw-Hill, New York.
Cross H. (1952) **Engineers and Ivory Towers**, McGraw-Hill, New York.
Schaub, J.H. and Dickison, S.K. (1987) **Engineering and the Humanities**, Krieger Publishing Co., Florida.

A Non-Vocational Approach to Development of Engineering First Degrees

H. Cawte

School of Systems Engineering, Portsmouth Polytechnic

Abstract

The general assumption that engineering degrees are vocational courses leads to overcrowded syllabuses and overworked students who nevertheless fail to acquire many of the important skills of the professional engineer. Such courses also exclude students who could benefit from an education in engineering but who do not intend to make it their career. This contributes to the isolation of engineering from the general public. This paper argues that a three year course designed on a non-vocational basis could offer a more rounded and rewarding educational experience, with an intellectual demand more in balance with that prevailing in other disciplines. Such a course could form a basis for formation of engineers, while at the same time providing a general tertiary level education for some who have other ambitions. This would be a useful step in increasing the understanding of engineering in the general population. A structure and content for the course is proposed.

Keywords: Formation, Non-vocational, First Degrees, Engineers, Profession, Creativity.

1   The Vocational Degree

Engineering is a purposeful discipline. It is endemic in the culture of educators in engineering that their courses must be vocational, and that a first degree should provide a relatively complete and balanced education for the professional engineer. As presently formulated, engineering degree courses, which in England and Wales are still predominantly three years long, are intended only for those planning a career in engineering.

However, while vocational training is essential to the practice of engineering, as a model for first degrees it is unsatisfactory because it compels us to fit into our courses all the factual knowledge which we think an engineer should know. This now includes not only science, knowledge and skills in engineering, but also much that is not engineering and which should be taught in other places at other times. A further pressure has been added by increased emphasis on the 'engineering dimension', a valuable concept widely misunderstood in academia.

The result is that engineering courses now include elements such as business studies, computer programming, CAE and languages. British industry has drawn back from its commitment to training, reducing sandwich opportunities and forcing degree courses to include incongruous craft training modules. Design elements have been expanded in the often mistaken belief that they are more 'applied' than the traditional sciences, whereas in reality syllabuses are heavy on philosophy while many of the problems set lack rigour and fail to challenge the student to apply knowledge and principles learned in other parts of the curriculum. Coursework and project commitments have grown in importance. At the same time, engineering science teachers have compressed and edited their material, so that the applied science base has been eroded by time constraints and the haphazard dropping of topics, without regard to the educational value of the remnants.

Consequently, there has been an increasing work-load on students coupled with a reduction in the intellectual achievement expected of them. A degree is now very much more difficult to get in engineering than in most other subjects, yet it still yields a graduate ill-equipped to tackle the many-faceted problems which face the professional engineer, and unversed in the means to advance his own education. To quote one of our recent first-class honours graduates, the three year BEng is "a test of endurance rather than of ability".

2 The Educational Degree

Now there is no axiom which dictates that degrees in engineering must be very difficult to obtain, or that engineering may only be studied as vocational training and not, as is the case with many other subjects, for its general educational value. However, the status quo dictates that no student will elect to study engineering unless he seriously intends to make it his career.

The isolation of engineering from the general population which this engenders is a serious cultural handicap, and is in some large measure responsible for the long-term decline in prosperity of the UK relative to its neighbours. The UK is unique in offering 40% of all graduate job vacancies to applicants with a degree in any subject, and far too many of our industries are led by accountants and generalist managers with scant knowledge of technology.

It is clear now that a three year degree course can be only a component of an engineer's formation, because engineering is a unique blend of science, knowledge, organisation and creative art which cannot be taught or learned in only three years. The whole process should begin in school with a good general education and continue beyond first degree into a master's programme and continuing education. The recent trend towards longer courses for engineers is a welcome sign that this truth is beginning to be accepted. However, there remains a place for the three year course as a component of an engineer's education or as a course for any numerate person wanting to fill his university years with challenging study. That study should not be a dry cramming of factual knowledge, nor an endless drag through coursework, but should seek to combine a sufficient science and knowledge base with practice in the arts of the engineer.

3 The Non-vocational Degree

The proposal is, then, that we abandon any idea that in three years we can 'turn out' an engineer. Instead, we look upon the three year first degree as an education in technological creativity, comprising three elements:
i)   An understanding of mathematics and engineering science sufficient to support the course aims.
ii) A limited knowledge base of existing, past and possible materials, technologies and production processes.
iii) Structured and progressive practice in creative problem analysis, synthesis and diagnosis, and in design and manufacture.

Deciding how much mathematics and science is appropriate is likely to be difficult and contentious. The aim should be to give the student the ability to obtain approximate solutions to an appropriate range of problems. However, there is a need to nurture the ability to develop conceptual and mathemetical models which have an appropriate degree of complexity for solving the problem in hand. For this the student needs to understand the implications of any simplifying assumptions, which requires a greater understanding of science than is needed to pass a typical examination, but does not necessarily call upon more complex models. For example, it is enough initially to model gases as having constant specific heats, because the logical extension of the model to specific heats varying with temperature comes easily a year or two later. Likewise, asking a mathematician to teach integration, without telling him the range of integrands to be considered, is likely to end in his burying his students in irrelevant complexity.

The knowledge base of materials and technologies need not be so carefully chosen, but it should be relevant to current practice in industry. It is not possible for us (the academics) to know what knowledge (facts) each graduate will need in his career. Nor is it possible, because of the vast range and rate of change of technology, for us to acquire the knowledge or disseminate it. We do, however, know some things about it:
i)   it will be greater in quantity and complexity than can be included in a first degree course;
ii) it will be continually changing as technology and the engineer's career advance;
iii) it will be largely outside the experience of the educators.

Only a few percent of the facts (as opposed to scientific principles) taught on an engineering degree course are eventually relevant to any particular job or career. Teaching facts is therefore largely a waste of time, except in so far as the knowledge base supports the teaching and practice of creative engineering. It is vastly more important to teach how to find out facts, and how to use them in accordance with sound science to obtain results.

It follows that in a non-vocational course, the factual content would be reduced and the creative content enhanced. Creativity would be a theme throughout the course and syllabuses would be designed to support the creative elements. This needs cooperation between teachers in various years and subjects so that skills of synthesis and analysis, and talents of inspiration and creativity, are nurtured. Some material would be included primarily for its interest and entertainment value, rather than for its direct relevance to the practice of engineering. For example, occasional lectures on major engineering and environmental disasters would both entertain and act as cautionary tales for the new generation. There would be increased emphasis on the role of the engineer as an environmental and social force.

The creative element of the course would be a more rigorous discipline than Design currently is, and many of the problems would be smaller and more detailed than has tended to be the case in the past. Appropriate analysis would be encouraged, designing by drawing discouraged. Examples would be drawn from the consumer goods industries as well as commercial and defence applications. All lecturers and all subjects would take part in this, and it is even possible that Design as a separate subject would disappear.

The intention is to greatly diminish the time spent teaching hard sums and details of processes. At the same time, the ability to manipulate the easier mathematical expressions and interrelated physical concepts would be greatly enhanced. The students' contact hours and total workload would decrease somewhat, and the work they were given would be rather different in type. A large measure of self-management would be needed on the part of the students, and this type of course would, one hopes, attract students from a wider range of backgrounds and, on average, of higher ability.

The result of such a course should be a thinking technologically literate graduate who would be ready for a range of careers. Those wishing to pursue engineering as a profession would require at least a year's industrial experience and a further two years of carefully directed study on one of a range of MSc or equivalent courses.

4 Designing the Timetable

Naturally, subject divisions would not disappear, primarily because learning is best achieved when there is a clear structure to the knowledge and because the subject names are a useful shorthand for the sets of assumptions which are appropriate to the problem. For example, Thermodynamics generally deals with problems in which internal energy changes are important but potential energy may be neglected. In Fluid Mechanics the contrary assumptions apply. In combined Thermofluids classes students often ask "is this a thermo or fluids problem?". The best, if unhelpful, answer would be the further question "What set of assumptions is appropriate?", and one test of success on this course would be that graduates would be able to answer such questions.

The normal vertical striping of timetables, in which all subjects are taught side by side throughout the course, would be replaced by subject sequencing so that concepts and facts are presented in an order which allows cross-fertilisation between subjects. Horizontal striping (in which the timetable would be filled with sequential short blocks of different subjects), would be inappropriate, but there is merit in *diagonal striping*, in which the first term or semester would be heavily loaded with Mathematics (45%), Elementary Statics and Dynamics (25%) and Electrical Principles (25%). Introductory creative work (5%) can be used to establish method and maintain enthusiasm and motivation. The second semester could reduce the Mathematics element, add some Fluid Mechanics and Electronics and perhaps an introduction to Manufacture or Materials. In the second year, Maths would be reduced further, Thermodynamics and Control would appear, Manufacture and Materials would acquire greater prominence and the new rigorous creative element would be emphasised. The final year would be at least 50% creative project-style work, with relatively little new factual material being taught. Computing is endemic to the course, so it is treated simply as a tool and not as a separate subject. Principles of digital devices would form part of the mathematics and electronics base.

5 The Creative Element

The following modest example will give a flavour of the kind of creative work which is intended. It was set for final year students studying energy systems on our Systems Engineering degrees.

The subject is a wall-mounted convector room heater comprising a copper tube fitted with steel sheet fins. Hot water passes through the tube and the heat output is known to be sensitive to water flow rate as well as the water and air temperatures. To minimise his costs, the client (this originated as a real consultancy) had a test carried out over a range of water temperatures at one fixed flow rate, the result of which was given in the form of the top equation in Fig. 1. The task was to estimate the additional effect of varying flow rate.

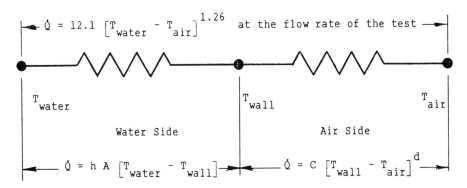

Fig. 1. A Complex Situation Simplified

The heat transfer processes in the heater are a complex of forced and natural convection and conduction, and detailed modelling would be prohibitively difficult. Clearly, too, there is no information about

the effect of flow rate in the test results. However, by making
intelligent approximations, the situation can be simplified into just
two thermal resistances in series (Fig. 1), air side and water side,
connected at the tube wall, the resistance of which may be neglected.
   On the air side, the resistance depends on natural convection over
the fins, for which published empirical equations are of the form shown
in Fig. 1. Within the pipe on the water side forced convection occurs,
for which standard textbook equations can be used. As the rate flow
varies, so does the water side thermal resistance and hence the wall
temperature. Students had to determine heat output as a function of
wall temperature from the test results, and then apply that information
to operation at other flow rates.
   The task was specified using the original client's fax request, and
while the specification was incomplete in some respects it also
contained some irrelevant material. Students were encouraged to elicit
extra information by questioning the lecturer, as a substitute for the
client. They were not initially shown how to solve the problem.
   Most of the students' time on this was spent in working out how to
simplify the problem to make it yield to analysis. For this to be
possible, they needed to know, and be able to link up and manipulate, a
number of elementary principles and empirical equations, all of which
were in their course textbook. They also needed to know how to solve
an equation by iteration. They did not need (and this is most
important) a solution for the temperature profile in an idealised fin,
or any other mathematically complex methods. Neither did they need a
detailed knowledge of the production process, though if they had been
interested this could have come up in discussion. In fact only one
student (from Germany) out of thirty was able to formulate a method of
solving the problem, and the others had to be given the technique at a
late stage when they had proved themselves unable to make progress.

6 Conclusion

The course described here should open up a new field for higher education in the gap between vocational engineering and general tertiary
education. It is intended to make engineering accessible to a wider
range of students, and to provide its graduates with skills and
attributes which will fit them for a wide range of futures. Above all,
it should capture the excitement of exercising creative skills in a
technological environment, and be fun!
   These graduates may not be engineers, but they will be interesting,
valuable people with the ability to become engineers if they wish.
They will be technically literate but able to tackle the commercial
world. They will be artists in a more rigorous world than the arts.
Such people should excel.

Quality in Engineering Education: Changing the Culture - Time for a New Paradigm

S.R. Cheshier
Southern College of Technology, Marietta

Abstract
Higher education has begun to look to the many industrial models[1,2,3] for strategies for implementing total quality concepts into the educational environment. This paper addresses the issue by identifying those characteristics of our present educational culture that make it difficult for us to break through the quality barrier, and suggests needed changes.
<u>Keywords</u>: Educational Culture, Engineering Education, Paradigm Shift, Quality, Total Quality Management (TQM).

1 Introduction

There is really no way to talk about implementing real quality in an academic environment without understanding that a significant cutural change has to occur. It is surprising how easy it is for engineering faculty, who spend much of their time teaching total quality management (TQM) and various other topics related to customer satisfaction and service, to totally ignore their own teachings when it comes to dealing with their students as customers.

Certainly, there is no environment in which a commitment to quality ought to be more obvious, or the buy-in more complete, than in a community of scholars. Yet, we educated technical people, who constantly emphasize to our students the need for adaptation and change, are unfortunately often the last to see the need for a paradigm shift in our own thinking and methodology.

2 The Quality Culture

I suspect that few service organizations are as insensitive to their customer's needs as are today's colleges and universities. In colleges, thinking about customers is especially challenging since there are so many constituencies. Also the collegial environment is both a strength and weakness in changing the culture. On the one hand, we have a higher percentage of professionals who should be easily persuaded, yet, on the other hand, autonomous professionals often think their way is best and are resistent to change.

Any change is usually tramatic. Maciavelli said it as well as anyone when he wrote back in 1513:

> " There is nothing more difficult to take in hand, more perilous to conduct, or more uncertain in its success than to take the lead in introducing a new order of things. . . an innovator has as enemies all of the people who were doing well under the old order, and only half-hearted defenders in those who hope to profit from the new."

Faculty may perceive "quality control" as forcing them to uniformity in their teaching and evaluation processes, while as faculty members they believe in their rights to diversity. I'm sure that some of this attitude is evidenced as colleges talk about articulation and incoming transfers. There is often the perception that, if it wasn't done here, by our faculty or our way, it can't be real quality. Yes, the need for cultural change is real.

3 The Campus Political Environment

Of course there is also the campus political environment that must always be considered. This is not insignificant. Woodrow Wilson described it well when he said, "I learned my politics on the campus of Princeton University then later I practiced in Washington among amateurs". Of course the politics on most campuses is such that the administration is looked upon with suspicion by many of the faculty. Faculty often think that their administrations practice the "Mushroom Theory of Management" which says 'keep your faculty in the dark and feed them manure'.

I remember years ago, as a young faculty member, saying to a colleague, that I couldn't do something because I wasn't an administrator. His response was that "administrators are to colleges what pigeons are to buildings". That was probably a little harsh, but it emphasizes the fact that there is no such thing as an administrative solution to an instructional problem.

4 The Faculty (Academic) Role

Although we certainly need to break down barriers between faculty and administration, the academic development of the student will occur primarily in the classroom under the influence of his teachers. Often they alone really make the difference in the quality of the process.

We make sure that our faculties are sufficiently prepared technically to teach their subject, but often we require no preparation in pedagogy, classroom management, dealing with student diversity, instructional technology, counseling/advising, etc. These skills may ultimately be the most important to both the student and teacher's success - especially for teachers coming from an industrial background. We must find ways to either require these skills at the beginning (albeit unlikely), or provide sufficient in-service faculty development in support of teaching.

5 Total Quality Management (TQM) in an Academic Setting

Most readers are already very familiar with the theory of TQM. We know how important it is to identify and listen to the customer, and to find every way possible to satisfy him concerning the product. We know about the need to establish a vision, mission statements, and goals. We know about the need to access the outcomes of our educational processes to close the quality loop. I believe that one challenge for us is to implement measurable quality standards into our academic programs - and then carefully assess the results of our processes. The academic area is where we can make the most important changes, since all of the other things we do are supportive to academics. We should, of course, make every effort to provide a supportive, hassle free, customer service environment in these service areas, but, if the student has a great academic experience, he will likely feel good about his entire college experience.

We can generally use successful business and industrial quality models in many of our support services. Certainly, areas like our business offices, bookstores, food services, physical plants, computer centers, admissions offices, and libraries function enough like most typical service industries to benefit from similar improvement strategies.

Not surprisingly, many of the TQM programs on campuses (including Southern Tech) have focused first (or even only) in these areas. Clearly the academic areas are more challenging.

Certainly, as academic institutions we are in a service industry, but we also produce a product, the graduate. Students come to us from a variety of "suppliers", and they are very diverse in quality. It's much easier for industry to ensure that the quality of their input meets an acceptable standard. There are many good initiatives in place around the World to ensure that high school graduates have a reasonable level of mathematical, scientific, and even technological literacy, but there is still not a lot of consistency in our intake to engineering programs.

At the college level itself, though, what are some of the issues relating to quality in academics? Certainly our engineering faculty are very knowledgeable about the concepts and principles involved in TQM - and this is an advantage over those in many other disciplines. Yet, knowing about quality does not necessarily mean that we practice its principles in our own daily classroom activities. Historically, we have just not been used to thinking of our college students as our customers - as those who ought to feel privileged to learn at our feet, yes; as our customers that require satisfaction as buyers of our services, no.

6 Engineering Faculty Issues

Frequently, engineering faculty are not professional educators, often coming to the classroom from industry. As a consequence of this background, they have been working in a somewhat uniform environment where their colleagues looked a lot like them - that is, they share the same general educational background, motivation, work ethic, etc. As a result, they may have little patience in dealing with the diversity that makes up today's university student body.

Many of our engineering faculty were educated in the old days when engineering professors tried to find ways to fail students ("look to the right, look to the left, only one of you will graduate"). Today if you said that, probably all three students would get up and say "then why waste my time, I'm going to the business school". Of course, we can't afford to waste these valuable human resources. We pride ourselves in how few make it through our programs, yet we hand pick the most capable students, most of whom should certainly be successful. In engineering colleges, the old weeding out philosophy is all too slowly being replaced with a nurturing and cultivating philosophy. This is the hardest cultural change for many of our faculty, who themselves were neither nurtured nor cultivated as students.

7 The Instructional Process

Of course, at most of our colleges, our instructional process itself is far from perfect. In education we still make the mistake industry made for years - trying to inspect quality into the product. In today's more progressive industries, there is no longer a quality control division - quality is incorporated into every step of the process, after the culture was changed. In fact along this line, since letter grades can become an end unto themselves for some students, I would support doing away with letter grades in favor of "pass/fail" grades, perhaps with descriptors about characteristics like initiative, teamwork, motivation, knowledge, creativity, perseverence, etc. appearing course by course on the transcript.

When students don't succeed, we rarely see it as at least a partial failure in our process, which often remains unchanged. We must get more attuned to examining the process and constantly assessing it and our educational outcomes so that continuous correction and improvement result.

This assessment and correction has to come from the faculty. The current reward system often recognizes research and scholarship more than undergraduate teaching, and institutions must better balance this so that faculty have incentives to spend the hours necessary to improve classroom instruction. Under the current system, it is too easy for us to get complacent about the fact that over half of our students are not going to make it, rather than to be deeply concerned that this group of capable students is being lost to the profession, such that every imaginative solution is brought to bear to save them. Students are so different that we must find ways to help each one to be successful at their own most comfortable pace.

8 Changes From The Past

One of the great strengths of engineering education over the years has been its focus on application and the development of state-of-the-art technological knowledge as re-inforced through a variety of appropriate laboratory experiences.

In the past, the student who was interested in engineering already had a pretty solid knowledge of things mechanical, and even electrical. That is, the student perhaps was already an electronic kit builder, an amateur radio operator, an automobile tinkerer, etc. Today, students come to us having no idea of the inner workings of things technological. This presents some new challenges in the way we explain things in the classroom. Students in general don't have the facility with lab experiences that we had growing up as technically interested persons. This is certainly true for all students today, but especially our growing populations of female students. It's easy for faculty to forget this in planning their courses.

Sometimes even we administrators and faculty view the college as existing to serve us and to meet our needs as the end goal. For example, many faculty senates spend much more time dealing with issues that will enhance their position, quality of life, or standing, than they do issues that will improve the quality of education for their students.

The faculty work ethic seems to be changing, with fewer faculty viewing their role as full-time teacher as their only (or even their predominant) professional priority. They want class schedules that leave ample time free for consulting and other outside activities, while viewing office hours, student interaction, and committee assignments as an imposition. Some of today's faculty need to be encouraged to spend more time on campus, so that they can interact more with students outside the classroom. Students badly need to get to know good role models one-on-one, in a non-threatening environment, and to benefit from their advice, guidance and experiences. Advising, counseling, and career planning activities must cease being annoyances to some engineering faculty, being viewed more as a key part of the profession of college teaching.

9 Conclusion

In the college environment, we need to spend more time listening to our customers and our employees. That means communicating with our students, their employers, the faculty and staff. I'm sure that many of us believed the oft repeated adage that "90% of my problems as a manager are caused by 10% of the people". I certainly have had many days when I felt that way. Yet, I think the more we get into the examination of our institutional problems, with an eye toward improving upon them, we learn that really 90% of the problems are caused by the process, not the people. As someone wiser than I has said, "It's so easy over the years to add on to the current processes, spending our time thinking we are improving them year by year - yet never stepping back to take in the big picture to see if that process needs to be done at all - or at the very least, if it needs to be significantly changed".

At Southern Tech, we are just several years into this process and we have a long way to go to get where we want to be. We are committed to the principles of TQM and to providing the highest quality of education possible to a customer base that is very satisfied with every aspect of their educational experience. We are not there yet, but with the kind of people we have at the college, and the efforts underway, we certainly will get there.

10 References

1. Crosby, Philip (1979) A classic treatise on the importance of quality in business and industry, Quality is Free, McGraw-Hill, Inc.

2. Deming, W. Edwards (1986) The "Father of Quality" discusses its importance, Out of Crisis, MIT Center for Advanced Engineering Study.

3. Juran, Joseph & Gryna, Frank (1980) A text on implementation of the quality process, Quality Planning and Analysis, McGraw-Hill, Inc.

Quality Measures in Engineering Education in Australia and the U.K.

R.K. Duggins

Department of Mechanical Engineering, University College (UNSW), Australian Defence Force Academy

Abstract

The paper focuses on engineering education at university level and describes how engineering compares with other disciplines when subject to performance measures in a wide range of activity. The information has been distilled by the author from a recently published government report and is important not least because of the increasing extent to which funding levels are tied to quantifiable institutional performance. Continuing to concentrate on quality-related matters, the paper goes on to make comparisons between Australia and the UK regarding recent and on-going reforms in the higher education sector; they are shown to be strikingly similar in the two countries.

Keywords: Engineering Education, Quality Measures, Performance Indicators.

1 Introduction

Recent developments in higher education in Australia and the UK have been similar but there have been some interesting differences in both content and chronology. The abolition of the binary system and the merging of the separate sector funding councils have already occurred in Australia but are being undertaken only now in the UK. Conversely, whereas quality assurance has only recently been intensively focused upon in Australia, there is already a strong and increasing linkage between an institution's performance and its funding allocation in the UK. In spite of the chronological differences however, there is in both countries an increasing importance being attached to the measurement of quality, and institutions need to be fully aware of the measures already being used and others being considered for the future

The paper focuses on engineering education in particular and assesses how the profession might be expected to fare with respect to the developments taking place in the higher education system as a whole. The assessment has been done by distilling and analysing information contained in a report published in August 1991 by the Higher Education Performance Indicators Research Group of the Australian Government's Department of Employment, Education and Training. The report had described the outcome of trials of proposed new indicators of institutional performance, covering teaching quality, student progress, research achievement and staff professional service, that had been carried out at fourteen higher education institutions. The distillation of the data contained in the report has enabled the present comparisons to be made between performance in engineering with that in other disciplines.

2 Recent Developments in the UK

2.1 The UK White Paper on Higher Education

The most conspicuous recent development in the UK was the publication in May 1991 of the *White Paper on Higher Education* [Incidentally a *White Paper* bearing the same title was published in Australia in 1988.] Its reforms, all of which have parallels in Australia, included
(i) the abolition of the binary system enabling polytechnics to become universities,
(ii) the merging of separate sector funding councils, and
(iii) increasing the emphasis on cost efficiency through greater competition and an improved use of resources.

In some of these reforms, the arrangements in our two countries are similar, for example with respect to the probable replacement of the binary system by a *de facto* two-tier system of research and teaching universities (as predicted by the UK Secretary of State for Education and Science). In some of the other reforms however there are differences, for example in the provision of measures for quality assurance and the account to be taken of assessments in funding decisions.

Quality matters comprise a major part of the *White Paper* and the new framework which the latter proposes includes *inter alia* the following three features:-
(i) external scrutiny of the quality control arrangements of UK higher education institutions by a nationwide quality audit unit developed essentially by the institutions themselves,
(ii) quality assessment units within each of the three funding councils (for England, Scotland and Wales) to advise on relative quality across the institutions, and
(iii) co-operation among the councils to maintain a common approach to quality assessment.
The Government stated that it intended to introduce legislation to implement the above changes as soon as parliamentary time permitted.

That the new arrangements for quality assurance should be deemed necessary by the Government clearly is related to its stated aims of improving cost efficiency and taking quality into account on a common basis in the funding of all higher education institutions. For teaching and general research, funding will continue to flow through the same channel allowing quality in the two components to be assessed on an equal footing (See 2.2). Numerous concerns have been expressed about the new arrangements but perhaps some solace can be gained from the Government's reassurance that grants for individual institutions should normally be a matter for bodies other than itself to determine.

The casual observer may find it difficult to distinguish much difference of substance between the new arrangements in the UK and their counterparts in Australia. The Minister for Higher Education in Australia sees the situation differently however and has recently stated that his Government has specifically rejected the UK type of approach, contending that it would be counter to Australia's traditions of institutional autonomy in the extent of central intervention it involves. Perhaps he had in mind (i) the British Government's stated insistence that a more flexible salary system be linked to performance and (ii) a recommendation from the Australian Industrial Relations Commission that salary levels should not be affected by any staff appraisal that is undertaken.

2.2 Other Developments

Shortly before the *White Paper* was published, Sir Bruce Williams completed a report on related matters for the British Universities Funding Council. Grants for teaching in the UK are based on calculations similar to those used in Australia (the so-called relative funding model) but, evident from the Williams report, the criteria used for calculating grants for research are somewhat different in the two countries. The UK research grant criteria have in fact been changed and the report describes the process that is in train which is linking, to a progressively increasing extent, the provision of funding and assessed performance. Whereas in 1986-7, the floor provision component SR (related to the numbers of staff and graduate students) was slightly greater than the component JR based on assessed research performance, by 1994-5 the situation will be dramatically reversed. It is expected that by that time JR will be twice as large as SR.

Quality assurance was another issue that was covered extensively in the *White Paper* and it is interesting to see that two colleges in the UK have responded by adapting and adopting the BS 5750/ISO 9000 system for the quality control of their education and training activities. Imported from manufacturing industry, the system requires practices and procedures to be implemented aimed at yielding quality outcomes and enabling any failings to be identified and traced back to the persons responsible. Because of the intervention and bureaucracy that are involved however it seems very unlikely that such an approach will gain widespread acceptance.

3 Proposed Performance Indicators in Australia

Although the Australian Government has said that it supports the concept of quality assessments being taken into account in funding decisions it has added that it has no intention of <u>prescribing</u> performance indicators to be used by higher education institutions. Nevertheless it established a research group with the task of developing and testing such indicators of institutional performance including indicators relating to teaching and learning, and to academic staff achievement in research and professional services. Fourteen institutions were involved in the trials, including many of their engineering schools, and the results appear in the research group's final report published in August 1991.

A secondary outcome of the trials of the performance indicators was that the results enabled interesting comparisons to be made between the performance in engineering with that in other disciplines. The author has distilled such information from the Government's report and a summary is as follows:-

3.1 Perceived Teaching Quality

A new questionnaire was developed to quantify students' evaluations of the average teaching performance in their courses and, as indicated in the table, five aspects of performance were assessed by this means. The results show that engineering fared badly in each category and was below the average for the nine disciplines that were trialled. Student workload in engineering was judged to be the least appropriate and, in the other aspects of performance, engineering was spared the wooden spoon only by health sciences.

Scale	Engineering	All Disciplines
Good teaching	2.79	3.21
Clear goals	3.18	3.35
Approp. workload	2.22	2.94
Approp. assessment	3.09	3.33
Emphasis on independence	2.21	2.62

Similar statements of student opinion had been made earlier to the Williams Review, for example that engineering was seen as an "unexciting hard slog". The overall picture is that both a majority of engineering students and the Williams Review itself believe the quality of teaching in engineering needs to be improved, particularly in the early years of courses where student wastage is greatest.

Digressing for a moment, the Australian Government has recently announced that from 1994 it will provide $70 m per year over and above normal operating grant funds [i.e. about 2% extra] to support and to reward measures adopted by higher education institutions to enhance quality of teaching. [In providing such incentive funding, it claims to be taking an international lead; it will be interesting to see if the British Government follows suit.] Although peer judgments and an independent authority are to play key roles in the allocation, potential applicants might consider that it will be in their interests to support their case by using the newly developed questionnaire and indicators of teaching performance.

3.2 Student Progress Rate
Student wastage was rightly seen as an indication of educational inefficiency and this, or rather its converse, was quantified by means of a student progress rate parameter. It involved measuring the percentage of subjects passed at different stages of the course, electrical engineering and ten non-engineering disciplines having been trialled for this purpose.

The table reveals that electrical engineering's performance improved as the course progressed; it was well below average in the first year but scores of almost 100% were attained by final year.

	Elec Eng	All Disciplines
First year	76.5	82.2
Mid year(s)	85.0	89.1
Final year	98.2	97.1

The outcome is broadly in line with the results of the earlier Williams Review which concluded that, of students who commence engineering degrees, little more than half eventually graduate. It also found a moderately strong link in engineering between standard of entry and program completion rate. A minimum entry score centile rank of 30 resulted in a completion rate of 48% whereas a rank of 85 was associated with a completion rate of 70%.

3.3 Number and Value of Research Grants
Success in securing external research grants in open competition was correctly seen as another important part of an institution's performance and accordingly steps were taken to measure it.

[Needing to be borne in mind here is the relevant Government minister's stated intention (25 Oct.1990) to allocate future research funds by performance-based measures.] Tabulated values are per staff member and indicate that mechanical engineering not only received considerably more grants than the other disciplines, but also they were of much greater value.

	Number of grants	Value ($000)
Mechanical Engineering	0.63	44.6
Five non-engineering disciplines.	0.27	8.4

3.4 Publication Rate

	Books and monographs	Refereed journal articles	Published conf. papers
Civil and Structural Engineering	0.03	0.62	1.47
Seven non-engineering disciplines	0.11	0.70	0.23

The table lists the three types of publication that were considered, the tabulated values being the number of equivalent sole-author publications per staff member. The results show that civil and structural engineering published few books and monographs, an average number of refereed journal articles, but surprisingly as many conference papers as all the non-engineering disciplines together.

3.5 Consultancies

Only consultancies valued at a minimum of $1000 were considered, the presented data being the number of such consultancies per staff member. For the discipline of engineering the surprisingly high figure of 1.01 was obtained, compared to an average of only 0.26 for five non-engineering disciplines.

3.6 Professional Service

This indicator quantifies the percentage of staff involved in four activities - (i) executive officer for professional organisations, (ii) executive responsibility for editing journals, (iii) appointment to expert bodies and (iv) external examination of research higher degree theses.

	(i)	(ii)	(iii)	(iv)
Civil and Structural Engineering	108	23	43	58
Six non-engineering disciplines	12	14	14	16

It is gratifying that civil and structural engineering achieved by far the highest score in each of the four activities. Disappointing however was the non-inclusion in this list of professional service activities of continuing education offerings; engineering would almost certainly have scored very well here too if the latter had been considered.

Summing up, the findings of the Government's research group have revealed that engineering fared well with respect to external research grants, conference publications, consultancies and other professional service activities but it fared very badly concerning the retention rate of first year students and perceived teaching quality.

4 The Present Quality Debate in Australia

Mention was made earlier of the response of the Australian Government to its research group's report on performance indicators. It said that it has no intention of imposing performance indicators on higher education institutions but intimated that the institutions might choose to use them in the course of their self-assessment. The 1988 *White Paper* had foreshadowed that such indicators would be incorporated in general funding arrangements for the universities and it comes as a welcome relief to many members of the academic community that the Government now appears to be addressing the matter differently. It seems that, at least in the short term, an institution's performance and output will continue to play only a minor part in the Government's determination of funding levels.

Nevertheless the Government continues to be concerned with the quality of the system and the need for measures to be taken to ensure that it is recognised, encouraged and rewarded. To this end, the Government has instigated a further national enquiry and has asked its Higher Education Council (HEC) to examine

(i) the characteristics of quality and its diversity in higher education;
(ii) the strategies that may be developed by Government and the higher education system to encourage, maintain and improve the quality of higher education;
(iii) the relative importance of factors affecting quality, including student mix, teaching, and research, in furthering the quality of higher education;
(iv) the nature of the relationship between resources and quality; and
(v) the means by which changes in quality over time may be monitored and evaluated.

The approach which the HEC is adopting, in making its examination, is to look at

(i) the particularly difficult task of assessing competency, i.e. what is expected of graduates in terms of skills, knowledge and other attributes and whether graduates are being adequately prepared for the workplace,
(ii) the characteristics of students entering the higher education system, together with other input parameters, and
(iii) those aspects of the education process that lead to quality outcomes.

The Government has to some extent given direction to the HEC by stating its preference for the establishment of a national quality assurance structure, independent of Government, responsible for reporting and commenting on the adequacy of quality management arrangements at the institutional level. The Australian Vice-Chancellors Committee, in its response, has stated that such a move would be counter productive and that the system would be better off retaining the status quo with responsibility for quality assessment continuing to reside with academic staff and their peers. It fears excessive interference from outside the system, notably the move towards competency assessment which it believes could lead *inter alia* to uniform monolithic reporting arrangements, narrow curricula and overly rigid practices.

Notwithstanding the Vice-Chancellors' objections, the HEC appears to accept the Government's direction and is expected to recommend the creation of an external quality assurance agency. To support the position it is taking, it alleges that the experience of most OECD countries is that a self-evaluation strategy does not satisfy all the stakeholders in the system and does not permit the monitoring of changes of quality over time.

Doubts persist whether the proposed external agency will improve the situation and, more generally, whether the HEC enquiry will lead to sufficiently specific and acceptable measures to both improve quality in the higher education system and assist in the allocation of government funds. Since the HEC is due to publish its report in June 1992, we should have some indication of the answer by the time of the conference.

5 References

Williams, Sir Bruce (Chairman) (May 1988) **Review of the Discipline of Engineering**. Australian Government Publishing Service (AGPS), Canberra.

Australian *White Paper* (July 1988) **Higher Education**, AGPS,Canberra.

UK *White Paper* (May 1991) **Higher Education: A New framework**, Cm1541,London.

Linke, R.D. (Chairman) (August 1991) **Performance Indicators in Higher Education,** Report of a Trial Evaluation Study commissioned by the Australian Department of Employment, Education and Training. AGPS, Canberra.

The Function of Examinations
R.H. Dadd
School of Systems Engineering, Portsmouth Polytechnic

Abstract

The paper surveys the requirement for examinations and describes a novel structure and examination system adopted for the course leading to the award of a BEng in Electrical and Electronic Engineering at Portsmouth Polytechnic. The system was designed to overcome the shortcomings of more conventional systems and was based upon experience of designing and operating honours engineering courses going back to the late fifties. The results to date are presented.

Keywords: Engineering Education, Degrees, Assessment, Learning Environment.

1 Introduction

The main objective of all degree courses is to develop the students' intellectual capability. In engineering degree courses this involves the students acquiring knowledge of and skills in one or other branch of engineering which then provides the vehicle for the intellectual development. The knowledge falls into two categories viz either fundamental and enduring principles or ephemeral technology. The skills likewise may be of enduring wide application or specific to the technology of the day.

In the design of engineering degree courses academics strive to establish a learning environment which enables students to develop their abilities, knowledge and skills and examinations are part of the overall system.

Examinations are traditionally used to provide a measure of student's abilities, however they may also be used both as an aid to establishing a learning environment and as a measure of the quality of the overall design and delivery of a course.

In designing the BEng course in Electrical and Electronic Engineering much time has been spent over the years by the course team on consideration of students' learning and studying problems. This has resulted in the realisation no single method of assessment is applicable to all subjects and the design of the course structure must take into account optimum methods of assessment. This has produced a course whose structure and assessment methods is markedly different from its predecessors of even 20 years ago. Each faltering step in its development has apparently resulted in an improvement and the paper reports its present state.

2 Course Description

2.1 Assessment Philosophy

The broad assessment philosophy is based upon the perception of an engineer's function in the world. An engineer is expected to be ingenious in the analysis and solution of practical problems on the basis of scientifically established principles, competence in certain skills and a broad knowledge of the technology available. Now no-one expects an engineer to know the designation and characteristics of every transistor or bearing in the catalogues nor to be familiar instantaneously with every instrument that might be needed. However a knowledge and understanding of the enduring fundamental principles is expected.

As a result the components involving knowledge of enduring basic principles and skills of wide fundamental applicability (like mathematical skills) are assessed by an appropriate formal written examination. Course components involving a substantial knowledge of ephemeral technology or transient skills are assessed continuously by an appropriate method.

Another issue involves recognition of the reaction of students to what is universally recognised as a gruelling course of study. In the United Kingdom students are prepared for entry in schools and colleges where class sizes are relatively small and it is something of a culture shock to find themselves in classes of a hundred or more. It is therefore important not to make too many changes in the other aspects of the course from those with which they are familiar in the first year of the course. Students enter the first year "bright-eyed and bushy-tailed" excited that they are on the first stage of becoming a professional engineer. The second year is often regarded as more of the same and just another year on the same treadmill. It is essential therefore that somehow the assessment procedures should apply a gentle pressure to study, not to present such high hurdles that they become regarded as insurmountable with the result that potentially valuable engineers are lost to society. It is nevertheless necessary that a proper foundation of knowledge is gained to provide the foundation for the final stage work. In the final year (presently the third or fourth depending upon whether a year of industrial training has been undertaken) the student sees light at the end of the tunnel and this provides the impetus to study effectively. Assessment procedures at this point in the course may therefore be rigorous.

Finally the assessment procedures adopted must be economical in academic staff time to allow staff to undertake the scholarly activity essential to maintain course quality and furthermore the studying and assessment system should form a coherent whole which makes efficient use of the students time on the course.

2.2 Structure and Assessment Procedures

Figure 1 shows the structure of the full time bachelors course and indicates the assessment methods used.

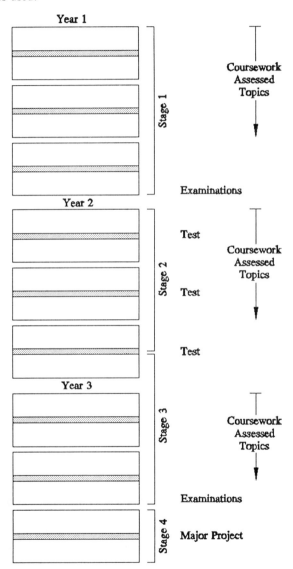

Figure 1. Structure of Full-time Bachelor's Course

A feature of the course is a week free from all regularly time-tabled classes in the middle of each term. This allows students to catch up on assignment work, working through tutorial example sheets and if necessary repeating experiments that somehow went wrong when first carried out.

Stage 1 occupies the whole of the first year and is used to level up the knowledge of all entrants while at the same time introducing new skills, knowledge and a more rigorous professional engineering approach to the knowledge of fundamental matters covered in the qualifications under which students are accepted onto the course.

In the mid seventies a change was made from conducting the final degree project on a one-day-per-week basis to devoting the whole of the final term to this component. Final examinations are held just prior to the Easter break in the final year. This ensures that students do not misjudge the fraction of their time that should be allocated to studying for the examinations. It leaves them free to devote the whole of their time and effort in the final term to the task set and more closely reflects the industrial situation than heretofore. In order for this to operate all continuously assessed practical work is undertaken during the first two terms thus freeing the laboratories and equipment for project students. Each student is allocated a laboratory position and collects the necessary test equipment as the work progresses.

Initially there were problems in reconciling the marks awarded for quite disparate projects. This has been resolved by establishing a Project Assessment Group of experienced academic staff who each monitor about twenty projects as the work is conducted. The marks awarded by the individual project supervisors are compared with those of the assessors and if there is a serious discrepancy a discussion between the Assessment Group and the Supervisor takes place to ensure that students are not unfairly treated.

This arrangement leaves five terms in which most of the professional engineering material is covered. This is divided into two stages (2 and 3) each consisting of two and a half terms. This arrangement was adopted to match the aspirations of staff teaching in the third stage in which the only optional subjects were chosen from subjects investigated by the two major research groups in Control and Robotics and Microwaves and Telecommunications.

The second stage is regarded as a foundation stage for the third stage and all students follow the same diet of subjects. New topics such as control and telecommunications are introduced and the emphasis is upon students gaining a basic level of knowledge in these and the other subjects studied. In order to assess this without presenting students with too high a hurdle and at the same time applying gentle pressure to maintain a steady level of study a novel method was adopted. Instead of assessing by an end of stage examination a test is taken in each "mids-week". Initially these tests covered each of the five subjects assessed in one three-hour paper in which all questions had to be answered. This has now been replaced by two shorter tests, one covering three subjects and the other two; students are still required to answer all questions. As the first test covers only five weeks of study the marks for each subject in the tests are weighted as being out of 20, 40 and 40 in the three tests. After each test the students are advised of the aggregate mark achieved in each subject and to pass into the third stage are required to gain at least 40 marks in each subject tested. Thus students can identify any area of weakness and adjust their study procedures accordingly. Students failing to gain 40 in any particular subject are reassessed and the methods of carrying out this reassessment is still a matter of experiment.

It must be emphasised that these tests are not phase tests, there is no question of students being required to undertake a single repeat test in matters in which the required mark is below par.

The difficulties of this method of assessment for the student are that a knowledge of a range of subjects at one time are required. Of course this more nearly reflects the real life situation, few engineering problems require knowledge of just a single subject studied. Setting the tests presented problems for staff inured to a system of five-out-of-eight examinations in a single subject and initially failure rates and reassessments were high. This problem has been addressed by establishing an internal group of staff with across the board expertise tasked with moderating the questions set bearing in mind that students are being required to switch from one subject to another under examination conditions. This has reduced the problem to manageable proportions.

Another feature of the course is that in addition to being allocated to a member of staff as a personal tutor for welfare purposes, students are time-tabled to meet their tutors on a regular basis. At these meetings academic problems are discussed and students guided to manage their study procedures. In the second stage the tutorial meetings allow discussion of "mids-tests" results and individual student's problems to be given attention.

3 Results

There is much anecdotal information from students on the efficacy of the structure and assessment procedures. Basically the students find the "mids-tests" very arduous albeit the questions set are not especially difficult. However when students reach the third stage the feedback is that, though these tests are demanding, they find they are well prepared with a sound foundation of knowledge and analytical methods. Prior to adoption of this scheme it was common to find substantial gaps in what students had learned after having crammed for an end of stage test. Additionally by and large students feel that the course designers have given some thought to their likely problems, it was flattering to hear a student on another course in the School asking why the course he was following was not similarly arranged.

However some more tangible information is available by examining the number of compensations required by students in the final examinations. At the time of preparing this paper only two cohorts of students have been through the complete course and these results together with those of the two years on the previous scheme are tabulated below. The figures are given for those students awarded a lower second class honours degree and above, those being awarded a third or unclassified degree and an overall cohort figure. Even these figures do not reflect the full situation. In 1988 the average marks in the optional subjects were very low and after much agonizing at the examination board were adjusted in accordance with a formula proposed by the external examiners; the figures in parentheses are those following adjustment. Those for 1989 were similarly adjusted.

Class of Award	Percentage of Students Requiring Compensations by Years			
	1988	1989	1990	1991
2.2 and above	26%(8%)	15%	3%	8%
3rd and Pass	66%(62%)	76%	73%	64%
All Students	40%(26%)	26%	12%	17%

So far on the new scheme it is gratifying to report this has not been necessary. The marks on which the data for the years 1990 and 1991 is based, are the raw marks gained by the students. It is clear that the students' performance now more closely matches the aspirations that the staff have for them. As expected the effect on weak students is less marked than that on the stronger students although since the numbers in the third class honours and below group are small, it may not be wise to draw too firm a conclusion.

4. The Future

The course having been in operation for five years is now being reviewed.

The teaching team are under pressure from the management of the School to avoid course stages which overlap calender year boundaries. Therefore a block of material from the final year options that depends more on ephemeral current technology than fundamental principles has been extracted from those syllabuses. This appears as an isolated module, assessed by coursework, taken by the students after the last "mids-test" and is accounted as part of the second stage which now embraces the whole of the second year.

Other new developments which will alter the structure involve CAT rating, modularisation and a Polytechnic-wide requirement for all courses to be based on 15 week semesters rather than terms. Additionally the course will offer a four year MEng option.

Both the IEE accreditation committee and the external examiners have encouraged the teaching team to continue the experiment into novel assessment methods.

Encouraging Ingenuity in Civil Engineering Students

S.G.D. Johnston, C. Williams

Department of Civil and Structural Engineering, Polytechnic South West

Abstract

Higher Education in the UK is rapidly undergoing major change. It is generally accepted that fundamental shifts in attitudes held/techniques adopted are necessary to accommodate the change.

This paper presents the view that urgent action is required to:-
(i) provide enhanced staff development programmes and (ii) rationalise the external pressures on course development .... if civil engineering courses are to take maximum advantage from the opportunities generated.

The paper then briefly outlines and reviews trials of non-traditional course delivery, student ownership of courses and student assessment that have been attempted in topic areas "read" during the B.Eng(Hons) course in civil engineering at Polytechnic South West.

1   Introduction

Educationalists and industrialists speaking at a conference on "Civil Engineers for the 1990s" (Kingston, 1985)[1] consistently underlined the need for technologically competent but adaptable graduates, capable of meeting the varying demands of a changing industry. A plethora of later reports has called for innovative and creative graduates and diplomates with well developed interpersonal skills. Sparkes[2] provides a reflective commentary on how to develop means of assessment more appropriate to these objectives.

The changes already underway will lead to Higher Education Courses with different emphases presented in different and changing educational environments. There is therefore heavy and immediate pressure to develop different course presentation and assessment procedures.

2   The Need For Enhanced Staff Development

The context in which lecturers present courses is rapidly changing. A developing commercial ethos, rapidly increasing student numbers, an ever widening profile of course entrants, decreasing contact time per student and changes in the structure of courses each reinforce the need for <u>lecturers</u> to be adaptable. They also lead naturally to greater emphasis being placed on student led learning activity. Implicitly the lecturers' role is moving away from teacher/provider towards facilitator/tutor. To operate confidently, creatively and effectively in this changing role many lecturers must develop skills additional to those for which they were appointed. Although academic staff are commonly stated to be the major resource available to Higher Education institutions, little provision has been made to regularly update their lecturing expertise or to broaden their awareness of learning techniques. There is a need for improved immediate and continuing support in the form of staff development programmes which (i) recognise the consequences and requirements following from the nature and timing of the changes and (ii) are appropriate to the particular background of the individual lecturer.

3   The Need To Rationalise External Pressures

Most lecturers are now alerted to the dangers of over-assessment. In addition to inducing student fatigue, even disinterest, peak achievement levels are liable to be subdued to a lower average performance which is then monitored and recorded. The peak levels become masked and may go unrecognised.

Regrettably it appears that the lesson learned by individual lecturers has still to be appreciated at Course and Department level. Departments come under immense pressure when attempting to achieve a balanced and academically viable approach to course development which complies with the views/concerns of many external, interested and influential bodies. As these views may conflict the net effect can be to inhibit change.

Development of a Higher Education civil engineering course may have to take account of the views of:-

(i)   academic colleagues of other disciplines, at the course validation stage;
(ii)   H.M.I.;
(iii)   The Joint Board of Moderators;
(iv)   B.T.E.C.;
(v)   External Moderators;
(vi)   External Examiners;
(vii)   Practising Professional Engineers; acting as advisors to Course, Department or Faculty Committees.

Professor Cusens in his Presidential Address to the Institution of Structural Engineers[3], suggested a similarity between the development and operation of Higher Education Courses in Civil and Structural Engineering and the negotiation of a maze inhabited by a number of influential acronyms. The implied hazards of progressing through the maze neatly illustrate the need to rationalise the many external pressures on course development. As the pressures may conflict their net effect may be to mitigate against change.

4   Background To Change In Engineering Courses

Current changes in Higher Education bring not only the necessity to rethink the character of engineering courses but also the opportunity to redesign/reconstruct them.

Morice[4] notes that engineering courses have traditionally sought to update their content by progressively adding on coverage of advances in technology as they occur. He warns that one result has been that students perceive such courses as overloaded and often dull and boring.

However if the chance is taken to identify and set clear, revised, realistic and achievable course objectives and aims and matching initiatives on resource provision are taken early enough the nature of the resultant courses might stimulate Evans' "creative engineering mind"[5].

5   Innovations Introduced on B.Eng(Hons) Course in Civil Engineering

5.1 Resources Utilised

Many H.E. lecturers are drawn directly from industry and have not attended comprehensive courses in educational philosophy/psychology; however they do have extensive experience of project based work in industrial practice. The authors suggest that student-led, project-based learning can encourage the students' development of interpersonal skills and creativity whilst drawing on two under-utilised areas of resource; (i) the lecturers' industrially acquired technological and interpersonal skills and (iii) the only H.E. resource that is growing, the enrolled students themselves.

5.2 Common Themes

A number of innovations have been introduced with the intention of broadening and enhancing the students' education by providing encouragement and opportunity for the development of interpersonal skills, adaptability and creativity whilst technological knowledge and understanding and technical skill are simultaneously developed. A common theme to the innovations has been the emphasis placed on student led, project based, group activity undertaken by students in response to an industrial style brief.

It is believed that this approach has led to greater academic maturity in students and has introduced them to a style of work that they will encounter later, during their professional training. Both student: student interaction and the breadth of awareness and skill in student Year groups are deliberately utilised as educational resources.

5.3 The Context

The innovations have been introduced progressively within a climate of overall course development which has gradually placed greater emphasis on recognising and meeting the needs of "the customer" whilst insisting that undertakings once made are perceived as contractual and are respected. For example students are required to submit coursework to a pre issued schedule of deadlines. A sliding scale penalty is paid for late submission, the reduced mark being determined by the delay occurred.

Student ownership of their course has been carefully reinforced using routine, informal corridor, semi formal and formal contact and a variety of feedback opportunities. Using Year One as an example, Student elected Year representative(s) attend Course Committee meetings; irregular informal meetings between the Year Group and Year Tutor are reinforced by semi-formal termly meetings; a Year Council meets regularly with the Year and Course Tutors and the Student Year representatives from each year of both of the Department's courses are encouraged to meet as a group. In addition a Student-led Proctoring Scheme has been initiated and a further feedback system based on confidentially completed questionnaires has been developed which seeks the student's perception both of the relevance of topics covered and of the performance of individual lecturers.

5.4 Outline Approach

A common outline approach has been adopted when introducing the innovations. The year group is divided into lecturer determined — usually arbitrarily selected — work teams of 4 - 6. Students are briefed to act as junior members of a consultancy; as a team they elect a 'communicator' who acts as a link between the team and the outside world — through a "Partner". The team presents written reports to the Partner at the end of the Autumn and Spring Terms.

Each report contains a number of Appendices which when appropriate include laboratory/tutorial/research work and a Project Diary. The Diary indicates details of:-

meetings held/those attending
topics discussed/decisions made
actions taken/contacts made
time taken individually and collectively.

Students are required to initial each sheet of the report to which they contributed. The Diary and the initialled sheets provide information on which to base mark allocation within a work team.

Each work team appraises and assesses other teams' work — using standard check sheets — and then reviews their own submission.

5.5 Detailed Procedures

The innovations have been introduced in

    A.   Year 1   Geotechnics   and

    B.   Year 2   Construction Management.

Some of the detail involved is indicated in the following paragraphs.

A <u>Year 1</u>   <u>Topic</u> Geotechnics
   <u>Students</u>   60 No 1990/91; 92 No.1991/92

Geotechnics is now a compulsory free standing, $\frac{1}{2}$ credit topic, taken in each of the three academic years of a modular format course. During the first year students are introduced to a wide range of basic concepts; their relevance to construction activity is stressed. Emphasis is placed on integrating lectures, tutorial/worked example sessions and laboratory work.

Objectives include (i) developing the students' awareness of geotechnics as a framework for understanding and predicting the response of the earth's crust to construction activity and (ii) extracting the maximum value for each individual student from the timetabled laboratory and tutorial based work.

In their work teams students investigate and report to a Partner on factors influencing earthworks construction. This theme is supported by a programme of lectures, videos, worked example classes and laboratory work. The large year group means that it is not possible to programme each student to undertake all the laboratory activity. Therefore emphasis is placed on work team members comparing/explaining/discussing each individual's laboratory test results/worked solutions.

As members of work teams, students contribute to 2 No termly reports; in doing so they identify, collect, collate and appraise relevant theoretical concepts/

practical applications. Each report includes as appendices laboratory reports, worked tutorial examples, data collected from press/published articles and a Project Diary.

A multiple choice/selected answer/guided solution type assessment is taken by individual students after each term's lecture programme. Peer marking of these assessments followed by rapid self reappraisal gives students immediate feed back on their progress. Students also have to face an end of year examination.

After the programme of end of year examinations students have to contribute to a further report on a week long period of group based fieldwork. Students undertake work designed to demonstrate the value of visual appraisal in predicting ground conditions.

The above assessments are weighted continual assessment 25, end of year examination 75 – to give a students overall percentage mark in this topic.

B   Year 2   Topic   Construction Management
    Students 48 No. 1990/91;   66 No. 1991/92

Construction Management is one of the topics contributing to the assessment of 2nd Year B.Eng(Hons) students. An outline feasibility study serves as the total coursework required for the topic. The study is group-based and student led; it prepares students for a project orientated final academic year. Formal contact time is limited to two, hour long, briefing sessions, a two hour seminar and a final three hour workshop.

The objectives set for the study include developing (i) the students' awareness of the social, environmental and technological contexts within which construction projects are designed/constructed/operated/maintained and (ii) their creativity and interpersonal skills.

Reporting to a "Partner" individually and as a team students undertake a feasibility study based on a prestigious local project – real, proposed or imaginary. Using essentially non-numerical methods each team researches, develops and defends their optimum solution.

A two stage (termly) written report is submitted by each team – Individual students submit a further report researching and proposing a solution to a construction problem that they themselves identify as relevant to their team's proposals.

Teams critically appraise and rank other team's reports and then assess their own.

At the end of the study period outline drawings are prepared and a studio presentation is made by each team. Other teams offer their views and a panel of Assessors offers adjudication – the panel has included consulting engineers, local authority engineers, a planner, an architect and a contractor as well as members of the lecturing staff.

In addition to lecturers' formal assessments each team receives copies of the relevant check sheets completed by other teams.

Students also face an end of year examination in this topic. The assessment is weighted continual assessment 50, end of year examination 150 to give the student's overall mark in this topic.

5.6 Conclusions/Implications
(i) The use of group based project work provides a good basis for open-ended, student led learning.

(ii) Maximum learning benefit occurs when a student is able to conduct an early review of his/her own work — ideally after assessing others' work.
(iii) Peer assessment and marking can ensure rapid feedback to individual students on level of achieved performance whilst providing some relief for lecturers from routine work.
(iv) Adoption of work team self direction has begun to utilise the student body as a resource to the benefit of both students and staff. Minor misunderstandings are identified by and discussed between team members; major problems are highlighted and referred to the Partner. In this context the role of academic staff emphasises tutoring c.f. lecturing skills.
(v) When student perception is that assessment is realistic, fair, demanding, relevant and interesting it may cease to be only an end in itself. It may become useful at least as a reinforcement of earlier learning, possibly enjoyable and even a means of enhancing motivation to study further.
(vi) As student intake groups increase in size and broaden in background the potential benefits to be derived from group based work, student led learning and peer assessment will also increase.
(vii) A single industrially set, group based project could provide the basis for all continual assessment work required by a Year Group. Individual topic requirements would form subordinate parts of an overall portfolio presentation.
(viii) In meeting the requirements of the learning and assessment strategies set out by the innovations, civil engineering students develop and utilise a wide range of communication skills. They collect, collate and evaluate technological and non-technological information and debate, assess and present individual and group views. In doing so students have designed and conducted polls of public opinion and presented their views in written and spoken word, in sketching and outline drawing, in model making and in video format.

References

1. Civil Engineers for the 1990s. Proceedings of 7th Conference on Education and Training, Kingston (1985); Thomas Telford (1985).
2. Sparkes, J. J. (1989) Quality in Engineering Education. Occasional paper, Engineering Professors' Conference, No. 1, July 1989.
3. Cusens, A.R. (1991) Concrete Steps to Construction's Future, Presidential Address: The Structural Engineer, 69, No. 21, pp 365 – 368.
4. Morice, P.B. (1991) Education and Europe. The Structural Engineer, 69, No. 23, pp 400.
5. Evans, F.T. (1991) The Creative Engineer; Innovative Teaching in Engineering, Editor R.A. Smith, Ellis Horwood, pp497-502.

What about the Teacher?

R.G.S. Matthew (*), D.G. Hughes (*), R.D. Gregory (**), L. Thorley (***)

(*) Department of Civil Engineering, University of Bradford
(**) Division of Mechanical and Aeronautical Engineering, University of Hertfordshire
(***) Enterprise Unit, University of Hertfordshire

Abstract

Increasing demands are being placed on teachers in Higher Education. In engineering schools the move toward student-centred and group based learning methods places new demands on teachers. In particular, group based activities present challenging new ways for teachers to interact with students.

Bradford University and the University of Hertfordshire (formerly Hatfield Polytechnic) have both responded to this challenge by devising a series of staff development courses and workshops. This paper describes the issues surrounding a 5 day group skills and curriculum development workshop which was initially run jointly by the two institutions. Subsequently, it has been run at the two institutions separately.

The reactions of both staff and tutors to the workshop will be described along with the benefits and problems of such developments.

Keywords: Teachers, Learning Managers, Group Skills, Staff Development.

Introduction

Increasing demands placed upon teachers in Higher Education is leading, in some institutions, to increased use of student centred and group based learning (RSA, 1991). Since the responsibility for learning is firmly placed upon the student the *teacher* needs to develop new skills so that they can adequately respond to the difficult demands placed upon them.

For those who regard themselves as *teachers* letting go of control can be very difficult - but if they regard themselves as *learning managers* it is much easier to achieve. Emphasis should be on the learning process rather than on the locus of control. However, the shift in control is often the thing most obviously experienced by those involved.

In this paper we describe the activities of two Higher Education establishments, Bradford University and the University of Hertfordshire in the field of staff development and in the resulting enhancement of a professional attitude to engineering education in the two establishments.

So What are these New Learning Activities ?

The Department of Civil Engineering at Bradford and the Engineering School at Hatfield have responded to the demands by introducing various new initiatives.

An example of what is happening at Bradford may be found in the undergraduate civil engineering degree course. Two of the core curriculum subjects are presented to the students through Problem Based Learning (Matthew & Hughes, 1991; Hughes & Matthew, 1992). The students work continuously in small groups of 6-8 for 5 weeks learning the academic subjects through working on a number of graded problems.

Problem based learning is founded on the premise that the problem comes first. A problem can be defined as a situation in which an immediate solution is not apparent. Rather than applying a readily available algorithm, a strategy must be developed to acquire and use the appropriate knowledge in order to synthesize a solution to the problem.

In order to do this successfully the students must use the higher level cognitive abilities as described by Bloom (1956), e.g. analysis, synthesis and evaluation. This type of approach requires the students to take a deep approach to the learning and understanding necessary to solve the problem.

At the University of Hertfordshire, group work has become well established in some areas, for example design work. Three hundred and sixty second year engineering students from all disciplines undertake a week-long design project in multidisciplinary groups of 6-8. The aims of the project are to learn about the process of design in engineering and the process of working in groups (Hamilton & Gregory, 1991). This makes considerable demands on group supervisors, who themselves come from all areas of engineering and need to develop a common approach.

New demands are placed on both the students and staff. Whilst the students have to accept responsibility for their own learning, the staff have to *manage* this process. Group management and facilitation are not functions common to all engineering educators. In fact they are skills which take considerable time and effort to develop and are essential to the successful implementation of a group, problem based learning course. Hence, it is essential to develop new facilitation skills to complement those required for conventional lecturing.

How do we Help the Teacher ?

The staff development associated with such innovations is carried out in a multidisciplinary environment (across faculties). In order to help the *teachers* at both institutions develop appropriate learning management skills a series of staff development courses and workshops are run.

At both institutions the major course/workshop is a 5 day residential course on group skills and curriculum development. Initially these courses ran independently at both institutions with input from CRAC (Careers Research and Advisory Council). However, as a result of personal contact between the authors it was decided to experiment and run a joint course for colleagues at both institutions. Why did we do this? Part of the reasoning behind the experiment was to give the staff participants a greater cross-section of experience to draw upon. Tutor teams included representatives from both institutions and from CRAC. This enhanced the experience for all.

The course was presented in two distinct parts. In part one the participants were exposed to a series of group development exercises which gave them first hand experience of what it is like to actually participate in group learning activities. This was the first time that many of the participants had been exposed to this learning methodology. The second part encouraged course members to develop, run and debrief exercises for their peers. These exercises can be related to the academic discipline as learning tools or as group skills development aids. The intention was that by the end of the course the participants would feel able and relatively comfortable with running learning groups in their subject areas.

The course was designed such that the role of the tutor could change throughout the week. Initially, the tutoring was fairly directive which quickly decreased and was replaced by a supportive role. The intention was to show, by example, the changing roles of the academic within the classroom.

Course Team Development

The joint course only happened once although each university has continued to offer its own versions of this course. The reasons for this are related to institution differences in the implementation of staff development; the geographical situation of the two institutions and the problems this caused the tutor team in the planning of the event; the development of the tutor team. We are a group and as such need coherence and meeting once or twice a year makes this difficult.

The relationship between the tutors themselves is crucial to a successful outcome of the event. The tutor group is itself a team and needs to *team-build* in just the same way as does the participants team. When running a joint course it is essential to build in time for this to happen.

This implies that meeting the evening before the course starts does not give the course team adequate preparation time. Indeed, groups go through a well documented process of development (Mulligan, 1988). This process takes time and it is only in the later stages of the development that the group performs as a cohesive unit. Our problems with this only go to show how difficult inter-personal skills are to develop.

This paper itself is evidence of the necessary and long term team building required by training teams.

Participants Reactions

Staff development in Higher Education departments is not an easy issue. Some staff have developed lecture/tutorial/laboratory materials over a number of years and feel that they have nothing left to learn about teaching, or that the materials are fine as they are. Indeed, many staff have been using group based activities for a number of years without any clear idea of the process they are putting their students through or how to use the skills developed by the students to maximum effect, including issues of group dynamics and experiential learning. Both of these require careful reflection by the participants before the meaningful learning occurs.

Use of the complete experiential learning cycle allows for powerful reflection to occur (Jaques, 1991; Kolb, 1984). An awareness of the emotive issues facing students and the need for structured skill development is raised. In reviewing their reactions to the course many staff found that after the first part of the course their initial doubts about the value of groups for learning high level cognitive and inter-personal skills had evaporated as they realised the REAL difficulties which THEY had encountered in working effectively in a group. For example, listening is a skill that really needs to worked at - how many times have you seen engineers have meetings where NOBODY is actively listening to what is being said.

It is probably true to say that only ready converts are prepared to institute group based learning by the end of the course. This highlights the need for further support and follow-up when the tutors and participants are back in the University and on an equal footing. It should not be forgotten that tutors and participants are work colleagues which adds another dimension to traditional tutor/participant relationships.

One area where difficulties can arise is in the giving and receiving of feedback. In developing group skills, or for that matter any other skill, feedback on what and how you are doing gives valuable insights and aids reflection. The giving and receiving of feedback is a difficult skill to develop. This becomes particularly apparent once staff begin to realise the role that emotions play in group dynamics. This, in our experience, is true in all groups, including student learning groups. How the group learns to handle emotions, and how we, as learning managers, handle or facilitate this in student groups is one of the major learning points from the course. Student groups are often much better than staff groups at handling and solving their emotional problems. However, learning managers still need to be aware that these problems exist for students. Academics have to learn to cope with feelings as they affect groups. This can be a major factor in decisions to continue with the safe practice of lecturing where the lecturer is in full (?) control.

What About the Tutors ?

The tutors' backgrounds are varied: civil/mechanical/aeronautical engineers and an enterprise tutor. Initially unhappy about learning outcomes in Higher Education, particularly engineering, we experimented with our own teaching, sought help in the form of staff development and finally saw staff development course tutoring as a means of passing on the benefits of our expertise to colleagues.

To those colleagues who may be considering running similar courses for interested staff, we recommend that the tutors certainly try multi - institutional courses. A very productive part of the course that we ran was the plenary review day which allowed the tutors to reflect on what happened and plan how to go forward. For example, at the plenary day we decided on the idea of a *taster day* for future courses. This gives possible participants a taste of what is involved in the course, what aspects are non-negotiable, and what is negotiable. For subsequent courses at both institutions this has been a very successful introduction. At the same time we largely restructured the course and developed a proper rationale and philosophy to underpin it. This process was facilitated by the cross - institutional input and by the team building efforts we had previously made.

The experience was an education for us and the lessons will not be forgotten.

Conclusions

From our experience on this and other staff development courses we feel that :
1. Support for staff implementing major cultural changes in education is paramount.
2. Group based learning is difficult, but it is a very powerful vehicle once running properly.
3. Learning to handle the emotional aspects of group work is a key skill in the successful implementation of group based learning in the curriculum.
4. Feedback, debriefing and self reflection are major skill areas to be developed by staff in working with groups.
5. The multidisciplinary nature of the staff development course has lead to a great deal of creative cross-fertilisation, leading to higher staff motivation.

References

Bloom, B.S. et al (1956) **Taxonomy of Educational Objectives,** volume 1, Longman, London.

Hamilton, P.H. & Gregory, R.D. (1991) Interdisciplinary design project, in **Innovative Teaching in Engineering**, (ed, R.A. Smith), Ellis Horwood, Ch 67, 409-415.

Hughes, D.C. & Matthew, R.G.S. (1992) Skill and cognitive development - An impossible pairing? Paper presented at this conference.

Jaques, D. (1991) **Learning in Groups**, Kogan Page, London 2nd ed.
Kolb, D. (1984) **Experiential Learning**, Prentice Hall, New York.
Matthew, R.G.S. & Hughes, D.C. (1991) Problem based learning - A case study in civil engineering, in **Innovative Teaching in Engineering**, (ed, R.A. Smith), Ellis Horwood, Ch 54, 330-336.
Mulligan, J. (1988) **The Personal Management Handbook**, Sphere Books, London.
RSA (1991) **Further education for capability** Summary Report, RSA, London 1991.

The Future Pattern of First Degree Courses in Engineering in the U.K.

J.J. Sparkes
Open University

Abstract

This paper explains and extends Occasional Paper No.3 of the Engineering Professors' Conference, with the above title. The main proposals are that:
(i) 3-year, 1st degree courses should concentrate on teaching fundamental principles and transferable skills, plus some degree of specialisation;
(ii) some of the normal content of mathematics, basic science, factual information and specialist skills should be transferred to fourth year courses, which can be taken elsewhere or as part of a 4-year degree,
(iii) teaching and assessment methods should be adapted to match these different educational aims, and
(iv) students should be enabled to become independent lifelong learners.
Keywords: Engineering degrees, Educational methods, Assessment methods.

1 Introduction

The Engineering Professors' Conference (EPC) has, for some years, been discussing the problems of quality in engineering education (Sparkes 1989) and has decided that there is a need for change This is partly because most present courses are severely overloaded, and partly because employers are showing some dissatisfaction with present graduates. The request nowadays is mainly for flexible engineers who can 'think' and have a firm grounding in basic principles and transferable skills. The EPC discussions have resulted in Occasional Paper No.3 (EPC3) with the same title as this paper. The aim of this paper is to outline, explain and extend the proposals of EPC3.

Although it is clear that nowadays a professional engineer needs more than 3 years' education and training, EPC3 accepts that a 1st degree course of only 3 years duration can provide an excellent grounding, provided effective educational methods which match the stated educational aims are used. A fourth year can then be taken either immediately or after gaining some experience, and can be in industry, in a different country, at another institution, or simply as a continuation of the first degree. In this way each individual's interests can be better served.

In order for this strategy to be effective it is necessary, even within a given subject area like electronics or civil engineering, to make distinctions between the different kinds of learning we expect of students. The distinctions used in EPC3 are, in the 'cognitive domain', between 'knowledge', 'skills' and 'understanding'. These are chosen because they not only refer to very different kinds of learning, but they can also be matched directly to different methods of teaching and assessment. The 'fourth dimension' - the 'affective domain', especially students' motivation to learn - is also of great importance, and needs attention throughout a course since it affects all aspects of teaching.

Unfortunately the terms 'knowledge', 'skills' and 'understanding' are often used in normal discourse too imprecisely to be of much value in course design. So, for the purposes of EPC3, they are given greater precision than is customary. This a familiar enough strategy in science, where many everyday words, such as 'work', 'energy', 'force', etc, have also been given precise scientific meanings without causing any serious confusion. The same approach is adopted in EPC3 because it is just as important to be clear in education as it is in science.

2 Knowledge, skills and understanding

Knowledge is regarded as information that has been memorised and can be recalled. If students are interested and understand, their learning of knowledge can be almost instantaneous. To teach knowledge it is necessary to present information that students can comprehend in an optimum way (e.g. verbally or visually or practically etc). It is also sensible to help students to remember it, by making it interesting and relevant, by frequent testing, by teaching study skills, etc.

Skills are what people can do without thinking too much about it, like speaking, walking, doing sums, solving familiar equations, playing tennis, touch-typing, etc. Although some skills are called 'manual' and some are called 'intellectual', all are 'mental' in the sense that the learning occurs in the brain. They are all taught by giving instruction and/or demonstrations, and can only be acquired by practice. Unlike knowledge, skills cannot be learned instantaneously however interested the learners might be. (Note that skills do not necessarily involve much understanding: people don't understand how they speak or walk and yet they can perform very well.)

Understanding is difficult to define, but can be described as the ability to use concepts creatively in explanations, in new designs, in correcting errors, in fault diagnosis, in asking searching questions, in argument and discussion, and so on. Understanding is the key to 'thinking' and to the ability to tackle new and unfamiliar problems. Acquiring understanding is not so straightforward an activity as learning knowledge and skills; and, like skills, it cannot be achieved instantaneously. The difficult part is grasping the abstract concepts (such as force, energy, magnetism, quality, productivity, etc) upon which understanding depends. Concepts need to be defined, put in context, analysed, justified, talked about, read about, written about, applied in modelling and problem solving, and so on. Once acquired, understanding - like skills - tends not to be forgotten.

Knowledge and skills are also referred to as 'surface learning', whilst learning which emphasises understanding and conceptual development is often called 'deep learning'.

Experience indicates that for most people the distinctions between these terms are meaningful enough, though, as in any model of a complex process, there are borderline ambiguities. Since the teaching and assessment strategies are different for each kind of learning, this analysis is useful as a basis for course design.

However, it has become clear since EPC3 was written, that it is important to distinguish between two meanings of 'understanding'. One meaning refers to understanding which is acquired experientially; the other meaning refers to understanding based on a firm grasp of the key underlying concepts. Thus:

(i) the experience of working alongside an expert, or working on a job for some time, seems to develop a kind of 'understanding'. People learn to make sense of their experiences, and find that they can tackle new, but not very different, problems sensibly and successfully without having grasped the underlying concepts which explain them. For example, people could obviously throw stones before Newton established the concept of 'inertia' which explained how it could be done. Similarly people can build simple bridges successfully without understanding the underlying physical principles upon which bridges depend. Nevertheless, in some sense, they 'understand' what they are doing. They acquire what might be called 'experience-based understanding' or 'know-how'. However, it is important to note that progress based on this kind of understanding is generally by trial-and-error (or, preferably, trial-and-success!), which may not be good enough in engineering. Testing alone is no guarantee of the safety or reliability of a product or process, or that the next design will be successful. It is a truism in engineering that quality cannot be 'tested into' a product or system - it must be designed in. So experience-based understanding has severe limitations. But in the absence of reliable underlying principles it may well be the only route to success.

(ii) The other kind of understanding is the ability to grasp and successfully apply basic principles and concepts to new problems, even when practical experience of dealing with such problems has been limited. In engineering, these principles come mainly from science and technology; and since it took geniuses like Newton, Maxwell, Faraday to discover them in the first place, we cannot expect students to re-discover them for themselves. That these principles are important is evident from what happens when an engineering disaster occurs. If it is found that any of these well-established design principles have not been used in the design, the engineers responsible are rightly criticised. So, if the aim is to produce well-grounded graduates, it is this kind 'concept-based understanding' which degree courses must teach, though as much experience as possible is also important.

There are, however, some employers who prefer to hire graduates with specialist skills, and there are also students who prefer to specialise. So it is not a foregone conclusion that concept-based understanding is an essential part of every engineering degree; much excellent engineering can be done, based only on knowledge, skills and experience. It is necessary therefore, before designing a degree course, to be clear about one's educational aims.

3 Assessment methods.

It has been well said that "assessment draws learning through a course", That is, students generally want to get high marks and good degrees and therefore tend to learn what brings them good marks. So it is sensible to ensure that the ways in which students are assessed encourage the kinds of learning implied by the course aims. Typical present-day exams do this rather badly.

The testing of knowledge and skills is a relatively straightforward matter. Students' knowledge is tested simply by their ability to recall information. Similarly, tests of skills simply involve observing students' performance at tasks which exercise the required skills. In principle this is simple enough, though performance at some skills, such as interpersonal skills or communication skills is not too easy to judge reliably. All the same, it is clear what needs to be done. But understanding cannot be tested directly through student performance, it can only be inferred from what they know and do. Only inputs and outputs can be observed directly, the mental processing which occurs in between has to be inferred.

In typical present-day exams in the UK, knowledge, skills and understanding are tested together by means of 3-hour exam papers. It is therefore impossible to identify the kind of learning that students have achieved. Indeed research shows that even when the intention is to test students' understanding, students manage to perform better by revising knowledge and rehearsing the skills which examinations demand, than by relying on understanding. So, in order to assess students' levels of understanding it may be better to separate it from tests of knowledge and skills. This has the further advantage that the tests of knowledge and skills can be made more rigorous, with a pass mark of 90% or more for really essential knowledge and skills - as in such subjects as navigation or pharmacy. Then, in another form of assessment, understanding can be graded in accordance with the levels achieved. In other words, if several exam papers are to be set within one subject area, it may be better to differentiate between different kinds of learning than between different topics.

However if understanding is to be assessed reliably, the problem remains of devising ways of assessing it which eliminate the possibility of success through the exercise of well-revised knowledge and well-practised skills. EPC3 suggests the following possibilities:

- the widely used final-year design projects or problem-solving projects (not simply projects requiring information gathering and comment). But there is a need for mini-projects in earlier years too, so that students can learn problem-solving skills before they are assessed on them.
- formal examinations in which students are asked: (a) how they would tackle a problem, rather than to actually tackle it; or (b) to correct errors in explanations and analyses; or (c) to compare and comment on given solutions to design or other problems; or (d) to compose questions themselves, and explain why they would be good tests of understanding.
- Essay-type examinations in which key technological issues are discussed, such as modelling, safety, reliability, quality, effects on the environment and society, etc.
- Oral examinations (as widely used in some continental countries).

4 Teaching strategies

Teaching methods should, like assessment methods, be matched to the knowledge, skills and understanding expected of students. They should also cater for the students' different learning styles (e.g. Pask 1976). The basic approaches to the teaching of knowledge and skills were described in Section 2. These strategies are quite straightforward. The teaching of understanding, however requires a richer learning environment and more challenging activities for students. That is the lectures, tutorials, laboratories, projects, computer-based methods, etc, need to be designed specifically to include concept development in their aims. It is helpful to think of the concepts taking shape in students' minds being illuminated from a variety of directions by the various teaching methods used.

Lectures, for example, are not a very effective teaching activity where understanding and skill development are the aim, though they can convey information very effectively. Students don't leave a lecture with any significant internalisation of the concepts which have been explained to them or with any development of the skills that may have been demonstrated. Most of the conceptual learning and practice of skills takes place later. Nevertheless lectures can (but often don't) fulfil various useful functions, such as 'starting students off', introducing them to new concepts, pacing them, creating a sense of belonging, spelling out the syllabus, enthusing students, etc. But these functions require many fewer lectures than are normally given.

Similarly, practical activities in a laboratory can be used for a variety of purposes: to exercise practical skills; to enable students to confirm and reinforce the theories presented in lectures; to give students the opportunity to 'discover' facts or laws of behaviour; to help them design experiments; to provide facilities for projects; to help students grasp difficult concepts, etc. The actual purposes chosen depend on the educational aims of the course. If practical work is to be used to help understanding, the kinds of experiments which, like many scientific ones, only require students to follow instructions, obtain data, plot curves and draw conclusions in well specified ways, are not challenging enough. Laboratory activities concerned with conceptual development need, for example, to include open-ended design tasks or mini-projects, which demand the use of relevant concepts for their successful completion. On the other hand, if the purpose of the lab. work is to develop skills, the activities should be designed differently.

Again, the meetings between a tutor and a small group of students in 'tutorials' can take various forms. They may be remedial in nature, concentrating on correcting students errors or misconceptions; or they may be strongly tutor-directed, as in a school classroom, with the tutor setting tasks and helping students who get stuck; or they may become a forum in which students can express their own understandings of the problems presented to them. (Abercrombie 1979). The third type is the most effective in helping students with their conceptual development. For example, it is usually better for students to talk about what they believe they <u>do</u> understand than for them to listen to the tutor explaining again the latest thing that they <u>don't</u> understand. Whilst it may be better when developing knowledge or skills for tutors to correct students' mistakes, it is usually better for tutors to help students sort out their own misunderstandings than to sort them out for them.

Other teaching methods, especially computer-based ones, can similarly be adapted to the different possible educational aims.

5 Conclusion

Under the pressure of overloaded courses it is likely that many students will only try to grasp the concepts presented to them if they see that their level of understanding (in addition to their knowledge and skills) is specifically to be tested. So long as assessment methods reward 'surface learning' as effectively as do typical engineering exams, teaching strategies aimed at conceptual development are likely to be ineffective, and may even be resented by the students. In other words, improved teaching methods are only worthwhile so long as the assessment methods are matched to the same educational aims. The proposals of EPC3 are focussed, therefore, not only on changing the educational aims of engineering degrees in the direction of emphasising flexibility and transferable skills, but also on how both the teaching and assessment methods can be matched to these revised aims.

References

Abercrombie, M. L. J. (1979), Aims and Techniques of Group Teaching (4th Edn.) Society for Research into Higher Education, University of Surrey, Guildford.

Pask, G. (1976) Styles and strategies of learning, Brit. J. of Educ. Psychology, 46, pp.128-148.

Sparkes, J. J. (July 1989) "Quality in Engineering Education". International Journal of Continuing Engineering Education (1990) 1, No. 1, pp 18-33.

Improving Student Learning: Some Contextual Dimensions

B.L. Button, R.M. Metcalfe, I.P. Solomonides
Nottingham Polytechnic

Abstract
There is an implicit assumption in some Higher Education courses that the process of learning is exclusive to the student and hence there are degrees of better and poorer learners. We will explore how students have personalised styles of learning, and how these relate to the learning material presented. This relationship emphasises a need to examine the learning experiences we offer students and the effect these have on the approach students may take to their learning. Learning is described as having a qualitative component seen to be affected greatly by multiple factors, many of which are not attended to, are outside the learners control and therefore the responsibility of tutors, academics and institutions. To improve quality the contexts of learning should be recognised and attended to. The contexts in this case are described as all aspects of the learner's personality that enable learning, the effects thereon of the learning environment, and the realities of engineering as presented to students.

Keywords: Student Centred Learning, Learning to Learn, Learning Styles, Curriculum Change.

Introduction

Teaching and learning are not the same thing. An obvious statement perhaps, but one that needs making to combat the more intransigent who say, "if we are teaching then they must be learning". The reality is that some students arrive into higher education with no real idea of the difference between teaching and learning. More alarmingly, some lecturers have insufficient understanding of the individual differences in student learning. We need to re-evaluate this relationship; to publicise our understanding of good engineering learning outcomes and how to obtain them. The aim of this paper is to propose a description of some learning contexts and the way in which engineering education tends to relate to them.

What is engineering learning?

The objectives and aims of engineering education have been extensively discussed elsewhere by the likes of Sparkes (1989) and Carter (1985).

They both make effective distinctions between the types of information and attitudes emerging engineers should acquire. But how is this information learnt? Consider the diagram below (Fig. 1.) The three boxes represent aspects of the learner he may use to relate to, and to process information from the learning environment provided. They are interrelated. Everybody has them. They are a system, and as such individuals will attend to information in quite unique ways.

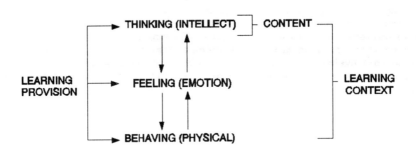

Fig. 1. Aspects of the personality interacting with the learning environment.

At a content level, a discrete engineering topic might appear to only demand use of the 'thinking system'. The learner is said to use his 'cognitive structure' and to display 'cognition'. This might be when attending to information that demands logical thought and linear procedures such as strength of beams. Hence the analysis related to this area might be delivered through lectures in a very precise, sequential manner.

But where is the proof positive that this is the way to present the information? Much of this proof is based in assumption, culture and history. In other words we have a 'tradition' of teaching and learning. Although there may well be a sequence of information to be learnt, the way in which this information is presented can have a profound influence on the quality of learning.

At one level the quality is affected by thinking or cognitive style and the capability to match or mismatch the learning environment to the learner. There are many descriptions of these styles, for example some learners prefer to 'see the big picture' and use a global approach to problem solving before fitting in the details, while others work step-by-step through topics and cannot draw together the argument until forced to do so. Clearly there are opportunities to present information sympathetic to one style or another, or for the learner to fail to be able to be adaptable to the differing presentations.

At a deeper level the quality of learning is affected by the other influences playing on the thinking system. In our model these are described as the behaving and feeling systems. There is an interplay between thinking or cognition, feeling, and behaving. This explains how learning is dependent on other factors besides 'ability' such as motivation and self-concept. Any education system that intentionally or unintentionally attends only to one of the three items in the model is likely to be creating a mismatch between what is taught and what is learnt. Engineering education is very good at attending to the thinking system, but to be even more effective it should now turn its attention to include the areas of feeling and behaviour. To teach only to the thinking system implies that content is all, and that the context of learning is unimportant.

We believe there is a strong link between the content and context of learning which demands that all Higher Education and not just Engineering be seen as a system (Entwistle 1992) where all the factors relate towards the learning outcome. This inter-relationship of factors is represented in the diagram below.

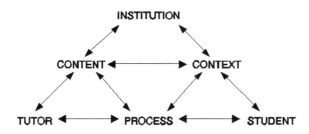

Fig 2. The elements of learning in higher education.

Carter (1989) and Sparkes (1985) have both implicitly emphasised this relationship. Carter describes objectives for learning that range from facts to integrity. Sparkes (1985) describes how,

> "... attitudes values and personal qualities... (are) of great importance in engineering and might even be regarded as a further aspect of engineering capability."

If we deem these qualities important then, we can begin to see how the context of learning influences their development. These are not qualities that can be mechanistically learnt as items of content in the same way as a computer might 'learn'. They are learnt as we apply our content knowledge in respect to social feedback, our goals, motivations, and perceived expectations of significant others. These influences can be seen as profoundly affecting attitudes, values and personal qualities and hence are very much part of the learning context.

So, learning is contextually dependent for the development of the whole individual, but at a much more practical level we believe that context is important in relation to content. For example: assume that there is a tendency to teach theory before its application. This would imply that students will apply the theory to practice, when in fact theory that grows out of practice may be much more personally relevant and therefore retained. In other words, it is important to learn concepts in context with examples of modern and exciting technology. To learn in this way is to build toward understanding rather than memorisation. We tend to be easily convinced by a student's display of knowledge and skill that he can be said to have understanding. The teaching and memorising of procedural facts can easily cover a conceptual vacuum, particularly when the methods of evaluation are not designed to elicit true understanding and personal qualities. Sparkes (1989) makes the point,

"It has been found that helping students to appreciate the differences between memorising, doing and understanding, and between creative thought and analysis, has very positive results. They can come to realise for example, that trying to grasp a concept requires a quite different approach from that needed to learn a fact."

This could suggest that some attention to the process of learning now has to become the legitimate concern of engineering education. The process of learning is something students can take with them to cope with the changes of content they will inevitably encounter.

Conclusions

We now know (see for example Entwistle 1990 and Eley 1992) that the awareness a student has of, and his involvement in, active and effective learning is much more context specific than we have previously given credit for. It is the learning environment, as well as the beliefs a student has about engineering education that will become the primary determinants of his level of understanding. This relationship the student has with the context of learning is called his approach to learning and it predisposes his involvement, described as deep or surface. The distinction is simplistic, but is useful as a means of describing the learner's changeable approach to each learning experience.

This change is significantly affected by the student's perceptions the course evokes. Therefore to facilitate the preferable deep approach there are actions that institutions, departments and tutors can take. There are direct interventions that can be made in relation to students. These might be:

1. Providing means of making students more aware of their learning and the approach they take to engineering education.

2. Enable students to see the purpose of engineering education and the specific concepts to be learned, by when, and to what standard.

Ultimately the aim is to coax students into taking more responsibility for their learning by weaning them off the tradition of providing packaged sets of knowledge and skills. That tradition can induce a passive, surface approach to learning. Perhaps it is now more appropriate, especially with the increasing student numbers, that students see tutors as facilitators and managers of learning rather than sole providers of knowledge and skills. It is important therefore that tutors consider their role in modelling appropriate methods and processes of working and problem solving to their students. The personal enthusiasms and insights of tutors are easily picked up by students, as are their attitudes towards methods of working. This powerful line of communication, often called the covert curriculum, could substantially affect the students perception of working, both as an undergraduate and emerging engineer.

We feel that there are steps that might be taken at a course and departmental level to facilitate a greater emphasis on context and process than is the present norm. The two points highlighted above might be considered aims. The objectives that go with them are:

1. To avoid overloading students with factual information and attempting to force them to learn too much too quickly.

2. Reduce the content of courses as recommended for example, by the Institution of Electrical Engineers (1991)

3. Present content in a variety of ways and consider carefully how efficient and valid lectures are in presenting content only.

4. Use the released time to concentrate on the process, context, and implications of learning.

5. Assess to evaluate understanding rather than memorisation by making sure deep level responses are elicited.

6. Prepare learning materials that attend to more than just the thinking level by including an 'experience' whenever possible.

7. Introduce new knowledge by linking or building on the student's previous knowledge, skills and personal qualities.

8. Select new entrants with appropriate motivations, expectations and beliefs about engineering education.

9. Adopt best practice from any source, and be prepared to act on valid and reliable educational research.

We have to aim to empower students with the ability to learn how to learn. This is qualitatively different to allowing staff to teach to a passive audience, and implies that students should become more pro-active in assimilating knowledge, skills and understanding. Indeed, true understanding can only be achieved if students are prepared and equipped for suitably active learning and depth of involvement. There are several contextual influences playing on this quality of learning and we now have a responsibility to positively change those within our influence.

The strategies to do this have only been mentioned here, and there are other documents that are more comprehensive, but, if we bear in mind that we tend to operate a now outmoded tradition of teaching and learning, then we are beginning to accept the changes that must come.

Ultimately the success of any interventions stand or fall with the department and institution concerned. A call for an improvement in student learning will only be accepted by students if they themselves perceive the ethos displayed by teaching and assessment to be coherent with this call. Incongruities are quickly picked up; if we want to be seen as facilitators of learning rather than as instructors we must bring this ethos about.

References
Carter R. (1985) 'A Taxonomy of Objectives for Professional Education', Studies in Higher Education Vol 10/2
Eley M.G. (1992) 'Differential Adoption of Study Approaches Within Individual Students', Higher Education Vol 23.
Entwistle N., Thompson S., Tait H. (1992) Guidelines for Promoting Effective Learning in Higher Education, Centre for Research on Learning and Instruction, University of Edinburgh.
Entwistle N. (1990) 'How Students Learn, and Why They Fail', Proceedings of the Conference of Professors of Engineering, London, Jan 1990.
Institution of Electrical Engineers (1991) Accreditation: Some Policy Statements. Course Content of Electrical/Electronic, Computer RelatedEngineering and Other Relevant Courses.
Sparkes J.J. (1989) 'Quality in Engineering Education', Engineering Professors Conference Occasional Papers No 1.

Enhancing Engineering Education using Experiences in Medical College

Z. Xiao (*), B. Chen (**)
(*) Dalian Institute of Light Industry
(**) Shenyang Aeronautical Institute

Abstract

The medical education system in China has been established and stabilized over a hundred of years. It has accommodated itself to the real situation of social demands. In contrast with medical education, the engineering training seems to be some what lagging behind.

In this paper, the comparison between the Chinese engineering and medical education is detailed. It reveals the urgent need of reforming for the Chinese engineering schools. Evidently, to enhance the engineering education in the light of the experiences in medical colleges may be an effective solution. In addition, this reforming assures the prospect of eliciting the enormous potential of engineering schools.

Currently, we have taken some exploration steps along this direction. The preliminary results are also related briefly in this paper.

Keywords: Engineering Education, Reforming, Practical Training

1 Introduction

In recent years, there has been significant improvement in the Chinese engineering industry, since it has been brought into the circumstances of changing and competition. On the other hand, the body of knowledge on which the professional training is based grows rapidly. It requires the engineering schools to afford a more active and knowledgeable younger generation. It is a sudden urge for the Chinese engineering education to keep pace with this trend.

The basic problem, perplexing the engineering teachers, is how to integrate the Theory and Practice in engineering education. The key point is that the necessary ingredients of theory or practice should be in conformity with the development of science and technology. In addition to this old poser, the economic problem manifests itself to be of no less importance. Traditionally, almost all the Chinese engineering schools are funded officially. At present, this limited funds would not be sufficient for necessary expenses and sustained prosperity for the school. Seeking for diversified financial resources becomes an indispensable responsibility. It is also an impetus to propel the Chinese engineering education.

The medical colleges seem to have solved the above mentioned problems more effectively. We have deliberated their experiences and tried to put it into practice. The initial results are encouraging.

2 Comparison of Chinese Medical Education and Engineering Training

The Chinese medical education system has been established and stabilized over one hundred years. China is the most populous country in the world. The public health condition is improved remarkably, since the late 40's. It owes much to the contribution of the Chinese medical education. There are about one hundred and thirty medical colleges and universities in China. The status quo for each medical school may be diversified, but they keep the same tradition in the principal respects. Its salient feature lies in the fact: "Sound Foundation and Rigorous Practical Training".

2.1 Sound foundation and rigorous practical training

Just as the same in the other countries, medical courses are rather tough. Students have to pass through a series of theoretical curriculums from freshman to senior years. Simultaneously, practical training is considered in top priority. Usually, students are gathered in the classroom in the morning and distributed to the laboratories in the afternoon or evening. They are busily occupied six days a week. In addition to laboratory works, clinic practice is of prime importance. In the senior year, students have to serve as an intern in the hospital. They have to learn every medical process in an environment of reality. In the activities of outpatient service, bedside lecture, ward inspection, physical examination, diagnose as well as surgical operations, they are taught to be familiar with the hospital care to a minute detail. In so training, after graduation, the students are well prepared to be assigned as residents.

The situation in most of the Chinese engineering schools may be different. The theoretical courses are emphasized in the first two years. Only one-fourth to one-third of the credits are arranged for experiment or practical operations. There has hardly had systematical and standard practical training. Most of the engineering students, after graduation, are not familiar with the practical work in machine shop or engine room and they have to learn their professional life from the very beginning.

2.2 Teachers in medical school are also qualified physicians

Without exception, the well-known doctors are famous professors. The dual-capacity of the medical teachers guarantees the close integration between theory and practice in medical education. But in our engineering schools, a considerable number of the faculty of professors are lacking in practical experience. Some of them are restricted in a narrow field. They are primarily interested in the courses they teach. How could they train their students to be skilled engineers?

2.3 Affiliated hospitals are officious base for medical training

In a medical school, about half of the proficient personnel, half of the financial resources and most of the precious equipment are installed in the affiliated hospitals. They are generally the most distinguished hospitals in the city. The hospitals supply the school with experienced staff, powerful material base, financial support and various educational facilities. A great part and most important period of school life for a medical student is expended here.

Although, there are also affiliated factories in an engineering school, but usually they are not profitable or even rely on the official fundings. The well equipped machine tools and apparatus are not fully utilized.

3 Urgent Need of Reforming for the Engineering Education

It is evident, there is much to be learned from the medical education for the purpose of improving the engineering training. The urgent need is also obvious. In addition to the causes mentioned above, special attention must be paid to the real situation in the Chinese industry. In the developed countries, most of the enterprises have established a complete system of permanent continuing education. Students after graduation, have to serve as an assistant in the machine shop or to practice as a salesman. The young engineers are nurtured systematically according to the tradition of the enterprise. This continuing education system has not yet been well founded in most of the factories in China. Especially, the majority of the enterprises are small. There is hardly any strict and normal assistantship. The new-comers begin with their job in a timid emotion, since they are unfamiliar with the daily life in a machine shop. Some of them are even reluctant with the practical work. Consequently, people are complaining of the retrogression of the quality of the engineering education.

The serious problem that the engineering education comes apart with industrial demand, has been recognized by teachers and administrators. Measures to deal with this problem have been proposed. One of the effective means is to enhance the engineering education in accordance with the experiences in the medical schools.

4 Enhancing Engineering Education Using the Experiences in Medical Schools

4.1 Develop the affiliated factories into a productive unit
In recent years, we have tried to change the management system of our affiliated factories. Reliable and profitable products are arranged. Just as the teaching hospital, it serves the society as an ordinary factory. But it supports the teaching process with technical staff. Engineers in the machine shop may be teachers in the classroom as well. This factory also plays an important role as a financial prop for the school.

4.2 Expand the laboratory into high-tech center
We have a small CAD/CAM laboratory equipped with some NC machine tool, workstation for numerical computation and necessary apparatus for measurement

and control. Utilizing these facilities, we have developed some kind of new products, such as indexing-cam, jet vacuum pump and manipulators for automatic assembly. These products have ready market and the profit amounts to hundreds and thousands of dollars a year. It improves our research work and strengthens our economic situation. This high-tech center is also an appreciable base for teaching activities. It becomes accessible and most enjoyable for the sophomore and graduate students. This center links the teaching, production and research work in a unified entity.

It is worth while mentioning that most of the small enterprises in China do not set up their own develop-research department. This center reveals to be very much helpful to these enterprises.

4.3 Reform the teaching-research staff and curriculums
The reforming of the teaching process would inevitably require the corresponding promotion in teaching-research staff and curriculums. Although it may be the most difficult problem, however it must be carried out step by step, simply because it is the decisive engagement for the reformation of the engineering educational system.

5 Concluding Remarks

It is obvious that engineering production is an integral part of the social economic activity. The engineering schools are responsible for preparing engineers to shoulder the mission of promoting the social welfare. Of course, training the young generation in the surroundings of real production and research work is a natural process. The essence of the experience of the medical school converges to this very point. Previously, we may have stressed too much scholastically. Considering the real situation in China, to amend this defect is too impatient to wait. It will not only improve the engineering education system but also will release the various potentials of the engineering schools.

We are just on the threshold of transition. There is still a long way to go. The school management system should adjust itself to realize the model of teaching-research-production entity. The faculty and staff of the school have to modify their inherited ideas. It is also not a viable course of action.

The Chemical Engineer in Society

S. Ruhemann

*Department of Chemical Engineering,
University of Bradford*

Abstract
The paper describes a course which has been developed in the author's department for the implementation of the concept of "integrated complementary studies". The course seeks to inculcate communication and co-operation skills, to teach subjects relating to conservation, health and safety and to convey awareness of professional ethics in order to help students on graduation to function as responsible chemical engineers in harmony with their colleagues and with the public. It was awarded the Esso Prize for 1990 under the Partnership Awards scheme.
<u>Keywords</u>: Chemical Engineering Education, Complementary Studies, Social Acceptability, Process Safety.

1 Introduction

When this author was a student in the early 1950's, the educators of engineers were just becoming sensitive to allegations that they were bringing up their charges in an excessively narrow technological culture. Modestly overlooking the total absence of any technological material from the great majority of arts courses, they engaged over the following two decades in strenuous efforts to remedy this deficiency - see e.g. Venables (1959), Christopherson (1967).

The earliest attempts consisted in the provision of "liberal (or general) studies", typically taking the form of a weekly lunch-time lecture on some subject of interest from the arts, or perhaps current affairs. If the institution lacked arts and humanities faculties able to provide speakers, these were engaged from outside. In some cases, these provisions developed into extensive programmes offering students a wide choice of subjects.

It soon became apparent, however, that students were not responding well to this attempt to broaden their outlook. Some of those whose seed fell upon stony ground attributed this to philistine indifference on the part of the students, but it was soon realised that the problem lay in the fact that, the engineering curriculum being very exacting, students simply could not afford to devote time and effort to activities which, however rewarding in the abstract, did not contribute tangibly either to their understanding of their subject or to their degree class.

This insight led quickly to the establishment of a new kind of policy based on two principles: the vehicle for "broadening" the students' outlook had to be in some sense vocationally relevant; and the students' participation in the associated work had to be formally assessed and to contribute to their degree classification. At the national level this concept was enshrined in the CEI's compulsory paper "The Engineer in Society" (CEI (1970)). At Bradford University it was expressed in the adoption of a policy requiring departments to include in their curricula schemes of "integrated complementary studies".

The course to be described represents an attempt to apply these principles to a specific discipline. It also attempts to treat in an academically coherent way the formerly neglected but increasingly important subject recently characterised by Marshall (1990) as *Social Acceptability*, and in so doing to satisfy the academic requirements for corporate membership of the Institution of Chemical Engineers as outlined in IChemE (1989).

2 Objectives

The Department operates several overlapping undergraduate programmes: the *Chemical Engineer in Society* course is common to all of them, but is principally orientated on and supported by the BEng and BEng-MEng sandwich courses, which provide respectively two and three industrial placements. Many of the students undertake one or more of these placements abroad (mainly in continental Europe), and foreign language studies are provided to support this.

The general objective of the course is to diversify students' studies in a purposeful manner that contributes materially to their professional formation, and on that basis to engage their attention and commitment.

In pursuit of this general objective, the course seeks to help them: to develop a mature understanding of the relationships between the central techno-economic function of chemical engineers and the society within which they operate; to acquire skills in communication and co-operation which will enable them to function effectively within both their specialist groupings and the wider community; to maximise the benefits they derive from their industrial placements and to build on these to optimise their pursuit of employment on graduation; to acquire specific knowledge of the risks which are posed by the process industries to the health and safety both of their employees and of the public, and a sense of the professional responsibility of chemical engineers for minimising these.

3 Delivery

For simplicity, we will refer the description of the course to the four academic periods of the BEng programme, which comprises one full academic year of a nominal 30 weeks' duration, two semesters nominally of 15 weeks each alternating with industrial placements, and a final full academic year. The course occupies one hour per week of class time throughout this period, saving only the final summer term. Including a supervised syndicate exercise in the final term, the total contact time is about 75 hours.

3.1 The First Year
In the first period, three themes are addressed: communications, including instruction in the use of the library, practice in understanding and summarising a technical article, practice in preparing and delivering a short talk; instruction in the preparation of technical reports; preparation for industrial placement, including completion of a "real-life" application form and briefing for interview by the company; a course of lectures on Occupational Hygiene and Laboratory Safety, in part also directed towards the first placement. A satisfactory standard in the coursework is among the conditions for progression to the following year.

3.2 The Second Year
The second period begins with a four-week syndicate exercise on Industrial Employment. The class is divided into teams of five or six students; each team is required to study a specific industrial sector (such as oil, fertilisers etc.) in order to find out how it employs chemical engineers, and then to report its findings in a promotional presentation on behalf of its industry on the career opportunities it provides for chemical engineers.

The remainder of this period is devoted to a course of lectures. The first of these addresses the professional responsibilites and resources of the individual chemical engineer; the remainder discuss the rôle of chemical engineering in relation to the conservation of natural resources and the protection of the environment (the programme varies, but typically includes issues such as energy, mineral and food resources, the water cycle, air pollution, acid rain and global warming).

In order to avoid the danger of formalism which might arise if this series were to be given synoptically by a single individual, the course organisers have deliberately arranged, while maintaining a coherent programme, to have various topics presented by different lecturers with a specific interest in them.

Performance in the second year is assessed partly on the basis of the "industrial" presentations (including an element of mutual assessment of individuals within each team) and partly on two essays on a choice of topics from the lecture course. The marks contribute up to two per cent towards the aggregate for the BEng programme.

3.3 The Third Year

In the third year the employment theme is continued with a business game supplied by the Unilever Company. This is a rôle-play exercise lasting three weeks, in which teams of six students each represent one of a number of competing companies in the same industry (manufacture of soap). Each student plays a particular part such as sales manager, production manager etc. and is individually briefed on the rôle. The teams are set a series of problems (e.g. increasing production capacity, reducing staffing levels, improving industrial relations) in each of which the various rôles are involved. They are required to propound strategies which will enhance their company's competitive position, and the team which best fulfils this objective is declared the winner.

In this exercise, emphasis is placed on a "real-life" approach involving identifying with a rôle, careful study of facts, persuasive advocacy of one's views, recognition of those of others, reconciling conflicting objectives and optimising achievement within resource constraints.

The remainder of the third period is devoted to the first part of a course (which continues into the fourth period) on Safety in the Process Industries. While the first-year course on occupational hygiene focusses mainly on personal safety, this course attempts a coherent presentation of the philosophy and practice of safe operation of chemical processes from a more strategic and large-scale perspective.

In the first part ("Fundamentals") students are first shown schematically the great variety of ways in which hazards may be realised and may develop in chemical operations. There follows a study of the hazardous features of systems, such as the chemical properties of individual materials, the reactivity of mixtures (flammability is an important example), and the thermal characteristics of systems which may give rise to runaway reactions. Consideration is then given to the various ways in which losses of containment of hazardous materials can occur, from the blowing of a relief valve to the total failure of a large vessel, and to the subsequent dispersion of the released material. Short essays and numerical examples are set to ensure that the students engage with the ideas involved.

3.4 The Fourth Year

The first four weeks of the fourth year are devoted to further discussion in the sphere of employment, focussed now on the search for suitable appointment on graduation. Students are encouraged to formulate their aspirations and study the labour market, and to direct their search so as to match these as well as possible. They then receive advice and practice on how best to present themselves through the media of application forms and interviews.

The fourth-year lecture course is divided into two parts. In the first part, entitled "Phenomenology", the actual realisations of the three principal major hazards - fire, explosion and toxic release - are classified and described, both in general terms and by reference to case studies of a number of major incidents.

The lessons of these case studies significantly underpin the second part of the course, which treats the "Management of Process Safety". Here there is presented a broad strategic approach to the identification and control of hazards in the framework of the elaboration of a process project from conceptualisation through initial and detailed design to commissioning and operation. Among the topics discussed are the quantification of hazards, the identification and assessment of risks, the consideration of alternative processes, inherent safety, the safe lay-out of plants and the principles of state regulation.

One half of a fourth-year examination paper consists of three questions on topics from the third- and fourth-year lecture courses on Process Safety, of which students are required to answer two, contributing two-and a-half per cent. to the total available mark for classification.

The course finds immediate application in a major process design project which is undertaken by all students. For this purpose they are divided into teams of six, each of which designs a different plant. At all stages of the project, including an oral presentation, they must demonstrate that they have considered safety and environmental issues. In the final stage (in the summer term), each team carries out a *Hazard and Operability Study* ("HAZOPS") on a small section of their plant selected by their project supervisor. The technique for this is introduced in the lecture course described above, and the project teams undertake a "dry run" on an academic problem beforehand. The applications to the Design Project of various parts of the Chemical Engineer in Society course contribute about ten per cent of the assessment for the Project, and thus about 1.8 per cent. of the overall classification mark.

4 Concluding Remarks

The course commenced its sixth year in October 1991. Much of its content and methodology are new and experimental, and it is still developing. It is perhaps open to the criticism that it tries to spin too many different strands into a single yarn. The author might well have taken this view in the past, but the pressures under which higher education is labouring at present make it essential to serve as many objectives as possible with each project.

The degree programmes incorporating the course have been accredited by the Engineering Council through the Institution of Chemical Engineers. Student response has as yet been estimated only in approximate ways, but appears to be generally favorable - a more systematic evaluation will be conducted in the near future. In the meantime, it is pleasant to be able to record that the venture has received the accolade of the 1990 Esso Prize under the Partnership Awards scheme of the Council for Industry and Higher Education.

Acknowledgements

The author is a member of a course team which also includes P.J. Bailes, M.J. Rhodes, D. Geldart and B.J. Carney, Head of the University's Careers Advisory Service.

Acknowledgement is also due for contributions from other colleagues in both his and other departments.

References

Christopherson, D.G. (1967) **The Engineer in the University**, The English Universities Press Ltd., London.

CEI (1970) Statement No. 7. **Rules for the Council's Examinations; Exemptions; Syllabuses.** Council of Engineering Institutions, London.

IChemE (1989) **The Content of First Degree Courses to Satisfy the Academic Requirements for Corporate Membership of the Institution of Chemical Engineers**, The Institution of Chemical Engineers, Rugby.

Marshall, V.C. (1990) The Social Acceptability of the Chemical and Process Industries, **Transactions IChemE (Part B)**, 68, 83-9.

Venables, P.F.R. (1959) **Sandwich Courses**, Max Parish, London.

Learning by Doing: WPI Engineer-in-Society Projects in London

L. Schachterle (*), R.D. Langman (**), W.R. Grogan (*), M.L. Watkins (***), G.L. Watson (****)

(*) Undergraduate Studies, WPI Worcester, Massachusetts
(**) R.D. Langman & Partner
(***) Interdisciplinary Studies, WPI London
(****) Urban Affairs/Higher Education, USMG Xerox Corporation

Abstract
Since 1974, Worcester Polytechnic Institute (WPI--the third oldest private technological university in the US) has required all students to complete a major interdisciplinary project (equivalent to nine credit hours) concerning the interaction between technology and societal needs and values. The intent of this requirement is to insure that all engineering students at WPI, as an important part of their formation, recognize their future responsibilities as professionals and as citizens in managing technology as creatively and effectively as they make it.

Keywords: engineers in society; project-based education; experiential education; interdisciplinary studies; engineering education reform; corporate support for education.

1 The WPI Project Program: Learning by Doing

Educating engineering students about the crucial societal contexts of their work--managerial, ethical, economic, cross-cultural, and esthetic--has been and remains an important challenge to all educators concerned with the preparation of future scientists and engineers for an increasingly interdependent global environment. Like many American technological universities, Worcester Polytechnic Institute (WPI) confronted this challenge head-on in the late 1960's when engineering suffered a severe "image problem" in America as the result of dramatic layoffs after the Sputnik boom era, combined with the widespread linking of technology to the unpopular Vietnam War and the beginnings of the pro-environment Green political movements worldwide. However, unlike most of their US colleagues, the senior engineering faculty at WPI elected to respond to these criticisms of their discipline not by piecemeal tinkering, but rather by totally recasting the century-old WPI curriculum, based on conventional classroom pedagogy and on sequences of required courses whose necessity was rarely reviewed by the faculty or often comprehended by the students.

After several years of intense study and deliberation, the WPI faculty in 1970 accepted an entirely new educational program: a new set of degree requirements, a new calendar and grading system, and--most important--new expectations for both students and faculty. At the core of the new program was the belief that students, especially in applied disciplines like engineering, learned best by doing, that the abstract principles of mathematics and science could most effectively be translated into engineering practice by making project work the focus of study. Reasoning that students were best motivated to learn fundamental principles when they were eager to apply them to solving a problem--a "just-in-time" approach to education, if you like--the faculty designed culminating or "capstone design" experiences in all three areas of central importance in the new program: the disciplinary major, humanities and arts, and "engineer-in-society" or "interactive technology/society" studies.

In keeping with the stress on "learning by doing" rather than on passive classroom pedagogy, WPI determined to promote the understanding of the societal consequences of engineering through a required project (equivalent to nine credit hours) called the Interactive Qualifying Project (or IQP). The IQP demands that students define, investigate, and report on some topic demonstrating their understanding of some model of interaction between technology and society. Students are encouraged to work in teams (both to draw upon different disciplinary strengths and to learn teamwork) and to select projects from sponsoring agencies external to the university (to provide invaluable "real-world" learning in which the documented results make a difference to the sponsoring organization, usually a government or educational agency, or a corporation or business.)

Since this interactive project became a requirement for graduation at WPI almost twenty years ago, more than half the living alumni of the university have learned about the societal contexts of their profession by completing a project assignment. A survey of graduates completed in 1986 indicated that students regarded the IQP as very important for introducing them into the world of team project assignments for which determinant numerical solutions provide only part of the answer to the kinds of real-world problems engineers face in balancing technical constraints against other limitations (such as social and financial). Interactive project topics cover a wide range of areas including environmental, urban planning, health care, risk analysis, humanistic and ethical studies, managerial and economic applications, and legal and social services. WPI is especially pleased that throughout the program, its students and faculty have consistently selected, as the most active project area, topics aimed at improving mathematics and science education at the pre-college level, which has led to the formation of the "WPI School-College Collaboration in Mathematics and Science Education." For more background on the history of the program, see Reference 1 and the annual WPI publication _Interactions_, which lists abstracts of all completed projects.

2 Global Perspectives in Engineering Education

An unanticipated but most welcome result of the IQP (which requires a quarter year of full-time effort within a four-year curriculum) is the unprecedented opportunity the interactive project offers to engineering students to obtain a "global perspective" on their education and future professions. Because WPI's calendar was re-formatted into five seven week terms (two each replacing the traditional semesters and one corresponding to the optional summer school), students may elect to leave campus to complete their IQP in a single term. Approximately one-third of students currently complete the IQP off-campus, for which advanced preparation (usually including submission to the sponsoring agency of a formal project proposal) is required.

To encourage engineering students to begin learning as undergraduates about the global scope of their professions, WPI strongly encourages international study through interactive project participation as well as through longer exchange programs. WPI's "Strategic Plan 1990-2000" commits the university to sending fully half its students off campus by 1995. To assist in this substantial logistical challenge, the university currently operates both residential student and faculty project centers, and more than a dozen student exchange programs. Consequently, WPI, which graduates well under one percent of the science and engineering students in the US, accounts for about ten percent of US engineering students who experience significant international pre-professional study.

Beginning in the early 1970's, WPI began to create "project centers" in which faculty join students to work on interactive projects submitted by local professional agencies. At present, the university operates such centers in Washington, DC; San Francisco; London; San Juan, Puerto Rico; Guayaquil, Ecuador, and Bangkok, Thailand. The last three are especially noteworthy for providing experiences in Latin and Asian cultures, which we believe will be increasingly important for American technological industries. WPI has also negotiated an extensive network of university exchange partners, at project center locations and elsewhere. We believe WPI at present provides more opportunities for engineers to study abroad than any other US technological university. Current partners include major universities in England, Ireland, Scotland, Canada, Sweden, Germany, Switzerland, Italy, Belgium, France, Russia, Thailand, Hong Kong, Taipei (Republic of China), Ecuador, and Puerto Rico.

3 The London Project Center (LPC)

The earliest experiments in interactive learning by doing off-campus projects occurred in the mid-1970's in the first project center in Washington, DC, and in the first exchanges with The City University (London) and the Eidgenossische Technische Hochschule (ETH-Zurich). Successes with both types of programs led to the first international WPI project center being established in London in 1986-87, directed by Professor Maria Watkins, electrical engineering senior lecturer emerita of The City University.

The London Project Center (LPC), which is reviewed at length in Reference 2, succeeded handsomely in that an initial student intake of twelve students annually soon rose to nearly fifty. Interactive projects have been conducted with dozens of different agencies, many of whom--like the Institution of Electrical Engineers, the Health and Safety Executive, and the Patent Office--supply one or more topics per year. The Center is directed largely from London, with Professor Watkins as resident co-director, assisted by several other professional staff with affiliations at Imperial College and elsewhere. In addition, the LPC has been recognized by the Charity Commission as an educational institution with legal standing in the UK, and the three Project Center Charity Board members (currently Professor Charles Turner of King's, Mr. Henry Strage, formerly of McKinsey Co., and Mr. Douglas Read, Lucas Ltd.) provide very important advice in all aspects of supporting the program.

The LPC is presently planning to form project teams consisting of students from WPI, the UK, and from the European Community. Our connections with technological universities in Germany, France, Belgium, and Sweden will provide groups of students able to interact productively across both disciplinary and national boundaries on a common project goal. We also are examining possibilities of UK "sandwich students" joining our projects, with the support of their sponsoring firms; and of extending the project program into all three types of projects that WPI offers, so that disciplinary "capstone design" projects in students' majors as well as projects in the humanities and arts will soon become common in London. New experiments in scheduling may also open up opportunities for American engineering students to participate as well.

4 Recent Projects in London

Because students are required by the faculty accompanying them to their project sites to complete a formal written and oral presentation of their work before they leave, sponsoring agencies receive excellent returns for their investments of time and resources in providing topics and in guiding them to fruition. Consequently, our experience has been that, once started, successful project activity quickly breeds new contacts and sponsorship. Happily, London has proved no exception.

One growing series of interactive projects has been guided by Eur Ing Professor R. D. Langman, who began serving as a consultant to the LPC in 1987. With his assistance, WPI students and faculty conducted a project (under the sponsorship of the IEE M2 committee which he then headed) on a question literally vital to the UK's survival as an industrial nation--was the country training enough power engineers to provide the generating capacity needed for near- and mid-distant future needs? The student results (pointing to a potential serious lack of power engineers) stimulated considerable discussion, including several presentations organized by M2. Their work was even referenced by a Parliamentary Select Committee!

The value of the first manpower project led to a succession of related topics: (1989) "Career concerns of mature engineers in the UK"; (1990) "An examination of the status of engineers in the UK";

(1991) "Issues in the energy engineering field in the UK" (with the Watts Committee); (1991) "Engineering awareness at London teacher training colleges"; (1991) "An overview of the British foundry industry in terms of the educational background and projected needs" (with the Institute of British Foundrymen), as well as several other topics on registration and status issues.

Youthful enthusiasm brought a fresh point of view to all these projects, and provided some comparisons and insights that observers within UK frameworks might be less likely to offer. "Engineers are non-entities....and if attitudes are not improved, Britain will become dependent on other industrial countries for its technological advancement"; "the engineering institutions are too closed-minded, old fashioned and bureaucratic....their secretaries hold their positions for too long, inhibiting new ideas"; "bankers embrace short termism, tax incentives on new investments are meagre, and the educational system steers children away from engineering" were some of the comments from these youthful analysts. With equally youthful enthusiasm, the trans-Atlantic engineering students contrived remedies such as privatizing schools and universities, creating better career structures (especially through continuing education), improving lobbying and marketing of engineering as a career, and creating an "environmental engineer" to present the profession as solving problems rather than making them.

5 Interactive learning by doing as viewed by a multi-national corporation

The WPI engineer in society program is quite different from the usual ways in which industry interacts with universities to promote the professional development of its future technological employees. Unlike cooperative education programs, the project program is clearly focused on academic objectives, and supervised (and graded) by faculty. And unlike "shadowing" or internships, the emphasis is on solving a specific problem that has been presented in advance, rather than just getting to see what the flow of work is like, day to day, in an office. In fact, the WPI students work more like young consultants than like students. Their faculty advisors indicate that establishing high expectations leads to very useful results for the sponsor, as well as to a unique learning opportunity for pre-professional degree candidates.

To any corporation, one of the most important objectives in bringing students into an office, a manufacturing site, or a research facility is to introduce them to the culture of the corporation, and hopefully to help recruit them as future professional employees. The very close interaction that comes from trying to solve a problem of interest to the sponsor provides better insight both for the student into the realities of professional life, and for the company into the potential the student demonstrates. We are especially pleased that WPI and other students have the opportunity while working at Xerox to learn about our managerial culture, which emphasizes "Leadership through Quality," and about how a global company plans, manages, and operates.

Major multi-national corporations like Xerox rarely if ever hire entry-level professionals for immediate assignment abroad: we simply have so many qualified and experienced employees from around the world that foreign assignments usually go to natives of the region. But in deciding about an entry-level hire, if all other factors are equal, the determining factor could well be the initiative and willingness to accept risk that these WPI students demonstrate by pursuing a degree requirement in a foreign country. And, if the new recruit progresses well in the company, after two or three years he or she may get preference for an overseas assignment. To Xerox, quality professional performance abroad is one of the three or four major factors we assess in selecting employees for "high-flier" career development within our increasingly competitive global environment.

6 Conclusion

The WPI London Project Center, as one of the most important components of the university's "Global Perspective Program," provides opportunities for American engineering students to learn though projects about the societal contexts of their future professional careers and about the roles they will play as citizens of an increasingly interdependent global environment. For many WPI students, being responsible for these projects in London is their first extended experience abroad, and thus prepares them for careers where they will increasingly be called upon to work in cultures new to them, and with people who hold viewpoints and values different from their own.

References

1. Grogan, W. R.; Schachterle, L.; and Lutz, F.C. (1988) Liberal learning in engineering education: the WPI experience, in New Directions for Teaching and Learning No. 35, Jossey-Bass, San Francisco.
2. Schachterle, L. and Watkins, M.L. (1992) The WPI interactive qualifying project--a model for British engineering education?, Engineering Science and Education Journal, 1, 49-56.

Engineers, Societies and Sustainable Activity

J.R. Duffell

Division of Civil Engineering, Hatfield Polytechnic

Abstract

This paper is concerned with engineering and sustainable activity. It presents a synoptic review on current environmental issues and the response of the engineering profession. It highlights the impetus for greening undergraduate curriculae and the response by industry to green issues. The importance of engineers adopting a high profile in public affairs to convey concern and committment to sustainable development is stressed. It calls for opportunities in engineering education for synthesis of rational and intuitive thinking in solving problems.

Keywords: Environment, Ethics, Sustainable Development.

1 Introduction

The term environment is high on the political agenda even given the present world recession. McCormick (1991) reports a Mori/Times poll in July '89 which found 35% of people interviewed regarded green issues as more vital than the National Health Service (29%) and unemployment (24%); he reported a 1988 Department of the Environment survey indicating 55% of the population favoured environmental protection over economic growth. Doubtless a repeat survey today would produce different figures but the question of economic growth and environmental protection appears never to be put in such surveys.

Concern on global warming and the ozone layer has increased since the ozone conference in London in February 1989. The repercussions of the Exxon Valdez oil tanker spillage in April '89 and the fires in the Gulf Oil fields in 1991, have done much to focus attention on environment. The pace of change over the last three years has heightened graphically the effects of pollution in Eastern Europe. The words "sustainable development" are now well understood and should guide individuals and societies to ethical human activity. Meanwhile, the greening of industry continues as firms seek 'market share' against a background of an articulate and discerning public. The passing of the 1990 Environmental Protection Act and efforts to value the environment in economic terms, are also matters of concern for engineers in their initial formation and continuing education which this paper addresses.

2 The Engineering Council, the Fellowship of Engineering and the Environment

The Engineering Council (hereafter EC) (1983) advised those devising enhanced or extended degrees thus:- "It is becoming more widely accepted that the solution to engineering problems cannot be separated from other economic and environmental consequences and account must be taken of the non technical disciplines which contribute to the success of an enterprise".

This postdates similar sentiments by the Institution of Civil Engineers and the Joint Board of Moderators (ICE, I Struct E, I Mun E). The FEANI Code of Conduct (1988) includes within its terms "strive for a high level of technical achievement which will also contribute to and promote a healthy and agreeable environment ... be conscious of nature, environment, safety and health and work to the benefit of mankind".

The EC published a booklet (1990) Engineers and the Environment and asked individual engineers to study and take account of issues such as global warming, energy efficiency and waste production. The EC challenged engineers in looking to the 21st Century to update themselves on environmental issues, work to develop a 'green' code of practice, apply that code as a professional discipline, set environmental objectives, design for the environment, specify energy efficient equipment, contribute to the environmental debate and promote sustainable development in order to safeguard the future.

Partly to meet the foregoing, the EC jointly with the Fellowship of Engineering sponsored a high profile conference (1991) 'Growth and Prosperity in a Green and Pleasant Land'. The prime objective was to demonstrate that economic growth and environmental protection can complement each other. The author draws on the paper by Peter Chester of National Power "Industrial and Engineering Perspectives" in which he suggests engineers should help "fashion a technologically sensible future not simply respond to current pressures and expedients". He also asked engineers to get on top of the science, to think and act globally and to seek best practical environmental options rather than the best technological ones.

Currently an EC Working Party and Task Group is formulating a Code of Practice on Environment, the starting point for which will be the International Code of Environmental Ethics for Engineers to be presented to the UN Earth Summit in Rio de Janeiro in June this year under the signatories of FEANI and WFEO (World Federation of Engineering Organisations). In the light of the foregoing, attention is now turned to valuing environment.

3 Valuing the Environment

Those involved in transport infrastructure planning, have long grappled with the elusive problem of putting a value on the environment. More recently Pearce et al (1989) have attempted to develop an environmental economics framework drawing on a number of case studies involving water pollution, and traffic noise effects on property values. They conceived a pay off matrix based on first optimistic and pessimistic policies towards problems in the environment and anticipatory or reactive action to confront such problems. They proceeded to show that anticipatory actions would entail less financial cost in the long run against six criteria: time preference (discounting), cost escalation, uncertainty, information by delay, irreversibility and sustainable development. The value of the environment can be seen in either the costs of anticipatory measures or the costs associated with cleaning up pollution for want of a better term. Broad brush calculations suggest environmental protection measures costing between 1% to 3% GDP.

The acronyms BATNEEC (best available technology not entailing excessive cost) and BPEO (best practicable environmental option) are assuming a common currency, involving not only economic and engineering judgement, but also environmental sensitivity. Clift (1992), chairman of the SERC Clean Technology Unit's management group, has said: "A clean technology is a technology which provides a service or product in a way which reduces costs (as opposed to "Clean Up technology" which may reduce environmental damage but increases costs). The ICI 'LCA' Ammonia Process is a well known example of a Cleaner Technology; Flue Gas Desulphurisation is an obvious example of a Clean Up Technology. Profligate Environmentalism is the opposite of Cleaner Technology; it describes any practice which increases costs and also increases environmental damage or resource consumption ...usually arises where a pressure group addresses a single perceived environmental problem but fails to think through the implications of its proposed "solution"".

Sir John Fairclough Chairman of the EC and Scientific Advisor to the Cabinet Office (1987-1990) in his paper 'Technology Opportunities' to the 1991 conference closed with these words 'Economists talk of pollution as an external diseconomy. For too long all of us have seen environmental matters as of concern only to others. There is a simple definition of marketing which states that no decision should be taken without an assessment of the likely impact of the decision on the customer. For customer read environment. We have to internalise the environment so that everyone becomes aware of the impact of their own actions and of the contribution they can make. Internal charging systems for pollution loads are a way of bringing the message home.... My concern is that developing a system for valuing the environment, which is sufficiently robust and simple to apply, will take some time".

4 Greening the Engineering Curriculum

This section is about "greening" of the undergraduate curriculae where there is a need to include environment in all disciplines across the arts and science divide. The PCFC (Polytechnics and Colleges Funding Council) has asked all institutions to embrace environmental quality in their strategic plans. H M Government set up in 1991, an Environment Committee to develop policies for "greening the curriculum" and a working document with that title was produced by Dr Ali Khan (1991) at Hatfield Polytechnic under the auspices of the CDP (Committee of Directors in Polytechnics). Translating its sentiments into the engineering curriculum is to say the least challenging, principally because of the pressures already on the curriculum, and attitudes amongst engineers themselves.

There is much already of an implicit nature on environment: engineering science modules and applied subjects in civil, mechanical and building services engineering, contain environmental matters, such as energy, efficiency, water pollution control, highway and public health engineering and so forth. Within the author's own discipline and until the current session, all students studied Planning and Environmental Technology in the second year embracing: land use planning, social cost benefit analyses, environmental economies, law and pollution of air, water and land. It is now a paired option with foreign language study because of pressures on the curriculum. Currently the School of Engineering hopes to embrace environmental thinking through its final year group design and individual honours projects. At second year level environment will be addressed in the product design specifications for the Year 2 inter-disciplinary design project activity. The Planning and Environmental Technology course will be made available for all Year 2 engineering students as an option. Finally, there does appear to be a need for instruction in physics, chemistry and biology to underpin in an informed way scientific understanding of environmental issues.

EC's Code of Practice on Environment will in due course, give guidance on such approaches and include case studies drawn from engineering practice; this will be apposite in terms of continuing education up to and beyond the CEng and IEng qualifying stages. The author can visualise the coming together of engineers in weekend workshops/ master classes on environmental topics along the lines of his fifteen successful residential schools for civil engineers to date (1984). Held under the joint auspices of the ICE (and former IMunE) and the Field Studies Council, over 200 graduate and chartered engineers have engaged in lectures, workshops and field work (over 3/4 days) on topics including: landscaping, land reclamation, river pollution, urban regeneration and so forth supported by around 50 organisations: local authorities, public bodies and consulting practices. The schools could easily be adapted to cater for the needs of other engineering disciplines.

5 The Greening of Industry, Public Relations & High Profile Activity

By being seen to have an environmental concern, industry in selling its products and processes, increases market share, aided by growth in financial investments with an environmental emphasis. Elkington and Burke (1987)chart the growth of green capitalism and conclude by suggesting ten steps to environmental excellence. Most of these embrace a more ethical stance on the part of employees and employers and place environmental affairs under a member of the Board. It is fair to say that most major companies now have an environmental policy statement and public bodies such as Nuclear Electric plc have gone to considerable lengths (under statute) to consult with bodies such as the Countryside Commission and English Nature (formerly the Nature Consultancy Council) on any proposals or actions likely to affect the environment.

Environment now permeates most engineering thinking and the Construction Industry Environment Forum has been set up principally under the arm of CIRIA supported by the Building Services Research and Information Association (BISRIA) and the Building Research Establishment (BRE). It is operating under six technical areas: energy use, global warming and climate change; resources, waste and recycling; pollution and hazardous substances; internal environment; planning land use and conservation and, lastly, legislative and policy issues.

Further sources of information are the various awards containing successful case studies incorporating environmental protection, enhancement and conservation through engineering activity. The EC's own Environment Award Scheme was instituted in 1991 with sponsorship from British Gas "to encourage good practice in the application of engineering for the protection of the environment". The EC's November 1991, Newsletter gave brief details of the winners: first was the development of a commercial fuel system cutting toxic emissions by 50%, second the development of an innovative water treatment project for use in developing countries and third the design and manufacture of a heat recovery unit and economiser.

The Royal Society of Arts runs annually its Better Environment Awards for Industry which embraces four categories for Pollution Abatement Technology, Green Products, Environmental Management and Appropriate Technology. The "Civils" Infrastructre Planning Group produced its third report, the subject being Pollution and its Containment which drew on 40 case studies of environmental conservation.

The annual British Construction Industry Awards were instituted in 1988 and are sponsored by 100 organisations, The Daily Telegraph, New Civil Engineer and New Builder and attract well over 100 entries for the coverted: Civil Engineering, Building, Small Projects awards. Besides the winners, many submissions receive commendations having satisfied in whole or in part the design criteria "fitness for purpose, performance, appearance and environmental harmony, economy of labour and materials, benefit to the community".

6   An Educated and Ethical Way Ahead

Population growth, people's material expectations and their realisation through technology and engineering, provide an overwhelming challenge to engineers to devise ways of promoting sustainable development. Much of the foregoing has been about ways and means of achieving this , but the most important aspect is in the formation of attitudes, where education has a vital role to play. In this synoptic review of developments the author has been influenced by Capra (1983) who suggested "What we need, therefore, is not a synthesis but a dynamic interplay between mystical intuition and scientific analysis . . . To achieve such a state of dynamic balance, a radically different social and economic structure will be needed; a cultural revolution in the true sense of the word. The survival of our whole civilisation . . . will depend ultimately on our ability to adopt some of the yin (female) attitudes of Eastern mysticism; to experience the wholeness of nature and the art of living with it in harmony".

Much the same thinking pervaded Frye's paper (1991) on a 21st Century perception of ethics. He suggested that "learning is, or at least should be, something that happens throughout life ... It raises those issues some five to ten years ahead with the express purpose of stimulating, creating and facilitating change". He foresaw two levels of meaning and perception - lower (direct experience) and higher (intuition).   Somehow engineers and their educators/ facilitators have to be involved in these developments if they are to play their part in this new age of enlightment. Hopefully, this paper will contribute to that process and demonstrate that ethical thinking and action is already at work in the profession.

References

McCormick J (1991) **British Politics and the Environment** p151 Earthscan Pubs London

Engineering Council(1983)**The enhanced and extended undergraduate engineering degree courses** EC London

FEANI (1988) **Code of Conduct** tract 04.04.88 Paris ref DB/cc No 242

Engineering Council (1990) **"Engineers and the Environment - some key issues"** EC London

Fellowship of Engineering/EC (1991) **'Growth and Prosperity in a Green and Pleasant Land"** EC London

Pearce D, A Markandya and E B Barbier (1989) **"Blue print for a Green Economy"** p 11 & 19 Earthcan Pubs London

Clift R (1991) **"Profligate environmentalism"** The Chemical Engineer p 3 February

Committee of Directors of Polytechnics (1991) **"Greening the Curriculum - Working Document"** May CDP London

Duffell JR (1987)**"A balanced environment - civil engineering in its social and political context"** Municipal Engineer Vol 4 Feb p 50

Elkington J & Burke T (1987) **"The Green Capitalists"** pp 228-237 Victor Gollancz London

Capra F (1983) **"The Tao of Physics"** pp 339/340 Famingo/Fontana Books London

Frye M (1991) **"A 20th Century perception of ethics for the individual, the community and the RSA"** p 17 and p 21 Royal Society of Arts Journal December 1991 London

Teaching Engineers to Break the Rules is Rational and Desirable

M. Hancock

Construction Study Unit, School of Architecture & Building Engineering, University of Bath

ABSTRACT
Contractual claims and disputes in engineering projects reduce the efficiency of the industry and the cost of this is usually borne by the client. This polemic argues that the problem is rooted in the education of engineering professionals which is currently of a technical and unquestioning form and which encourages blind adherence to sets of rules and procedures that are both inflexible and often irrelevant. It is argued that engineers must face up to fundamental moral responsibilities in carrying out their work. It is proposed that this is essential even if it requires them to break the rules and procedures (both explicit and implicit) in forms of contract and as issued by professional institutions
The paper concludes that the education of engineers must be changed to one of a more Socratic and dialectical nature if clients for engineering work and society at large are to receive maximum benefit and value from the industry.
Keywords: Rationality, Rules, Roles, Decisions.

INTRODUCTION

Disputes in the course of engineering and construction projects are an all too common occurence. Regardless of who is appointed to resolve the conflict, their modus operandii will have a number of things in common, one of which will be an assessment of each party's deviance from the "rules and roles" as defined by the law of contract and established professional practice.
 In making this assessment the individual will rely on a combination of his/her education, training and experience in the industry to investigate, evaluate and pass judgement on the dispute.
 The underlying theme of this paper is a feeling of discontent with our current education system and a belief that little or no attention is given to the influences

that condition the perceptions and behaviour of those involved in the construction process.

The critical issues identified above, are those of rule and role. Consideration of these concepts forms the main body of this paper which concludes that without a true moral content, the education of engineers can only be of an unquestioning and technical nature, which prevents or disallows the imagination, creativity and understanding necessary for the resolution, or more importantly the avoidance, of dispute and its resulting claims.

Rules, Roles, Rationality and Decisions

Frequently disputes in construction revolve around matters financial. Perhaps the cost of a delay, or a disagreement over the price of items of work and/or materials. Differences of opinion as to the method of and need to calculate amounts of money abound. We might ask why or how such disagreements arise. Weber (1947, p.185) wrote that "...it is necessary to take account of the fact that economic activity is oriented to ultimate ends of some kind: whether they be ethical, political, utilitarian, hedonistic, the attainment of social distinction, of social equality, or of anything else. Substantial rationality cannot be measured in terms of formal calculation alone, but also involves a relation to the absolute values or to the content of the particular ends to which it is oriented."

Engineers exercise what Thompson (1983) refers to as "responsible autonomy". Workers in this category are trusted to work without direct supervision and are allowed or required to exercise judgement in the course of that work. Mintzberg (1983) refers to organisations that operate through the employment of workers with this responsible autonomy as being Professional Bureaucracies and accepts that in understanding their operational structure we must reject the notion of an integrated pattern of decisions, common to the entire organisation. Furthermore he believes that it is an actual requirement of the professional that s/he should make decisions based on judgements formed on the basis of previous experience, a long period of training and the appraisal of a given situation in order to effect the required result. This constitutes a subjective rationalism.

Earlier work (Hancock, 1991) has suggested that Weber's definition implies that rationality separates the individual from the community and subordinates the decision maker to legal, political and economic regulation at all stages of life: that all organisations are rational-legal and that rules are directly linked to rationality.

Both Weber (explicitly) and Mintzberg (implicitly) seem to accept the need for rationalism in decision making and therefore as a justification of adherence to the rules that embody a particular working environment.

Rules, by their nature impose moral obligations on people by making sure that they do what is "right" and in order to identify the points at which individuals do "wrong". The question here would seem to be..are these moral obligations the result of individual choice and concensus, or is there something about them that goes beyond individual choice? The fundamental argument seems to be this: do we insist that individuals can personally decide upon agreeing to have a commitment of some kind, whether personal or social, to a set of rules?...or is it that individuals develop within a social, political and moral context which forms their personalities and over which they have no control. Most people believe that they know what is right and wrong, what is moral or immoral (or even amoral). Whether they actually do know is questionable. Theologians might assert that morality arises directly from divine or natural law, but there is more than one version of what this law is and as the philosopher Kant (1786, p.29) wrote....

"...EVEN IF THE HOLY ONE WERE TO STAND BEFORE ME, I WOULD STILL HAVE TO DECIDE WITHIN MYSELF, ON THE BASIS OF MY OWN IDEA OF MORAL PERFECTION, THAT THIS WAS INDEED THE HOLY ONE OF GOD."

The point here is that before accepting the truth of any moral judgement and therefore, before accepting any of the established rules and procedures laid down by the governing institutions within the construction industry, each participant must exercise his/her own judgement of the situation and act accordingly. The part played by human judgement is, I would suggest an example of a genuine morality and rule interpretation. Submission to imposed rules constitutes an evasion of a fundamental human responsibility for the evaluation of matters moral. Morality and adherence to rules are then the result of an autonomous decision making process and not, as we are often led to believe, pronounced by some higher authority.

How then, is an ajudicator to determine whether one or other of the parties in dispute has breached the rules, that is the boundaries of acceptable behaviour? Has one of the protagonists acted outside the scope or contrary to what we might reasonably expect of someone in that position and carrying its associated responsibilities? Here we must consider the concept of role.

Much of the behaviour which we suppose to have come from within the individual and based on his/her personal character is (according to many industrial psychologists) actually a function of the individual within a group.

Brown (1980, pp.65-66) cites professional examples and concludes that they "...behave in certain respects according to their social role - the concept society has given them of how men in their position ought to behave..."

For our purposes there are two significant areas of concern in the above..
a) The place of individuals within groups and
b) The concept of role play

Salaman (1980 p.128) argues that organisations can be considered without regard for the individuals within it because the regularity of organisation "..is taken as a consequence of, either individuals' conformity with shared norms, or their commitment to an emergent series of negotiated definitions and meanings."

The conformity mentioned is to rules of some kind, but why and how the individuals within an organisation behave in this manner is because of the concept of role.

So, individuals behave in a certain manner due to the role they are playing at a given time, but..why? Social theorists appear divided on this issue. Katz and Kahn (1966, p.175) believe that the answer lies in the complexity of the relationships within which individuals are involved and on which they depend for assistance and approval. "All members of a person's set role depend upon his performance in some fashion; they are rewarded by it, judged in terms of it, or require it in order to perform their own tasks. because they have a stake in his performance they develop beliefs and attitudes about what he should do and should not do as part of his role.....the crucial point (for our theoretical view) is that the activities that define a role are maintained through the expectations of members of the role set, and that these expectations are communicated or sent to the focal person."

Elsewhere, Kahn et al (1964, p.13) say,"Our first requirement in linking individual and organisation is to locate the individual in the total set of ongoing relationships and behaviours comprised by the organisation. The key concept for doing this is office....Associated with each office is a set of activities, which are defined as potential behaviours. These activities constitute the role to be performed, at least approximately by any person who occupies that office."

Giddens (1979) however, argues that there are major objections to the use of the notion of role in social analysis.

The very idea of a fixed role for each position in society is ill-founded in that it assumes stability and order to be natural and definite, and the concept of change to be non-existant, or at best, slow. Several role theorists (notably Merton, Dahrenhof and Goode) in considering the consensual issue have been concerned about the notions of conflict and strain which they have observed between the individual occupying a role and the expectations of the society which imposes the rules pertaining to that role. This strain, (according to Giddens) is derived from disjunctions between an individual's psychological traits and role.

Giddens (1979, p.117) disputes the idea that social systems can be usefully understood as consisting of roles or their conjunction; "and the associated thesis that role (quoting Parsons)....is the primary point of direct articulation between the personality of the individual and the structure of the social system..........it is fundamental to affirm that social systems are not constituted of roles, but of (reproduced) practices; and it is practices, not roles which....have to be regarded as the points of articulation between actors and structures."

OBSERVATIONS & CONCLUSIONS:

This paper has highlighted some of the difficulties faced by any claims assessor in getting to the root of such disputes which (it is believed) have become increasingly common in the industry.

Standard forms of building contract concentrate on the distribution of risk throughout a project, rather than on the definition of the roles to be adopted by those responsible for its administration. However, these contracts do imply that it is possible to set down detailed rules etc..that will cover virtually every eventuality and which take little or no heed of circumstance. The (quite reasonable) underlying principle is the apparent need for some kind of order, agreed by the parties concerned, which functions as a clarificatory code and minimises confusion and abortive or contradictory effort. The rules and roles which are contained both implicity and explicitly in forms of contract stem from the theoretical considerations with which this paper has dealt. It is maintained however, that the theories of rationality, rules and roles cannot exist in a vacuum and are bound to circumstance and context.

Weber's rigid form of management control assumes permanent organisation, whereas all forms of engineering contract accept, ipso facto, that building works are executed in temporary multi organisations. Whilst we may assume that the contracting parties share a common objective in striving for the successful completion of the contract works we must also realise that the parties will not share certain other objectives within that framework. This is precisely a rational response to the

"lack of fit" of the models exemplified in standard forms of contract.
I would contend that (at least) part of the answer lies in the education of those responsible for controlling engineering projects and the need for changes therein.

If we are to "educate" engineers, rather than simply training them, we must understand the differences in the terms. It is not enough to simply have a mere knowledge or technical ability. The professional engineer must have conceptual understanding of the "reason why" of things. S/he must be able to see the connections between engineering and the wider aspects of life. The way to achieve this kind of education is through a more dialectical approach. This will enable the engineer to make decisions and to behave in a manner that will not have the profession stagnating and hidebound by sets of irrelevant and inflexible rules. If society is to gain the maximum benefit from the engineering profession, then engineers must learn how to break the rules and infuse their role in the building process with imagination.

References:

Brown J A C. (1980) *The Social Psychology of Industry*, Penguin, Harmondsworth
Carr E H. (1962) *What is History?* Macmillan, London
Giddens A. (1979) *Central Problems in Social Theory*, Macmillan, London
Hancock M R. (1991) *Improving Construction Industry Performance by Simplifying Standard Forms of Building Contract - A Theory Based Case,* in the Proceedings of the 4th Yugoslav Symposium on Organisation & Management in Construction, Dubrovnik, (Ed. Marko Zaja), 581-588
Kahn R L, Wolfe D M, Quinn R P & Snoef J D. (1964) *Organisational Stress: Role Conflict and Ambiguity*, Wiley, New York
Kant Immanuel. (1786) *Groundwork of the Metaphysics of Morals,* 2nd German Edition, Riga
Katz D & Kahn R L. (1966) *The Social Psychology of Organisations*, Wiley, New York
Mintzberg Henry. (1983) *Structure in Fives: Designing Effective Organisations,* Prentice-Hall International, New Jersey
Salaman G & Thompson K. (1980) *Control and Ideology in Organisations,* Open University Press, Milton Keynes
Thompson Paul. (1983) *The Nature of Work,* Macmillan, London
Weber Max. (1947) *The Theory of Social and Economic Organisation*, Free Press, New York

The Role of the Study of History in the Formation of Engineers

W. Addis

Department of Construction Management & Engineering, University of Reading

Abstract

Although many engineers, especially senior ones, extol the value of knowing the history of their own branch of engineering, very few engineering degrees include purposeful and challenging courses in engineering history. To resolve this contradiction, the author reviews a number of arguments which have been used to justify a study of engineering history and puts forward some of his own. He concludes that, if the aim of engineering courses is to produce high quality designers, then history is a necessary part of an engineer's formation.

Keywords: History, Engineering, Education, Formation.

1 Introduction

In addition to learning new facts, concepts and ideas, one of the principal aims of education is to make us aware of what we already know and to formalise and deepen this knowledge. Thus, for instance, mankind has always known that wood is more flexible and less dense than stone but it was only in the last century that such qualitative knowledge could be formalised and structured in such a way as to be used consciously and quantitatively in engineering design.

In a very real sense we are all historians and make use of history all the time – our own experience and memory are essential to our everyday lives as well as our engineering activity. An awareness of more experience and of more memory – that of our contemporaries and, especially, of our predecessors – must surely, therefore, be of great benefit to us, both as persons and as engineers, if we can only use it. This alone, however, seems not to be an argument of sufficient strength to ensure a place for the study of history in engineering courses. This paper presents a number of additional arguments which may assist the cause at hand.

Before continuing, it should be noted that in many other arts – for engineering *is* an art not a science – great importance is placed on a knowledge of their history. What student of painting, architecture, music, English or sculpture would be satisfied with the engineering student's typical awareness of the history in their chosen field of study: many would argue that to study an art *is* to study its history.

2 The benefits of knowing (some) history

In order to achieve excellence engineers need to muster every means at their disposal. A fundamental goal of an engineer must surely be to act intelligently. It is important to understand our past in order to act intelligently today, and also to be able to defend these actions and propose future actions.

In such a short paper it is only possible to outline the different ways in which history does contribute to an engineer's life, and ways in which a deeper knowledge of history could contribute much more.

2.1 Knowing

Perhaps the most obvious benefits of being aware of the past are to be found in simply knowing what has happened, in order to copy or to avoid what has been seen to work or not to work:

- **Awareness of precedent**
  Most engineering is tackling problems which are not new. The past contains a wealth of design and manufacturing solutions which, if used carefully, can be a fruitful source of ideas.
- **Knowledge of success & failure**
  Reports about major disasters can have far reaching effects on design practice in a whole industry (e.g. Comet air disasters).
- **Standing on the shoulders of giants**
  Newton was one of several great men who was well-aware that his contribution to learning was wholly dependent on the achievements of many before him. Engineers too are in this position and can only evaluate their own contribution and importance in the light of past achievements. It is also important to understand which direction 'forward' has been in the past, and is in the present.
- **Repair, maintenance and refurbishment**
  Some industries have a legacy of artefacts which are inherited from previous generations. As these die out, so younger people must learn and understand the old techniques of design and manufacture in order to be able to repair and maintain them [Pugsley et al. 1974, 1975].
- **Avoiding rumour, suggestion and untruth**
  At the most fundamental historical level, an accurate record of historical data is needed. A knowledge of these can avoid such commonly-held erroneous beliefs as the idea that Eiffel's Tower is built of steel and that Whittle invented the jet engine.

2.2 The nature of progress in engineering

It is sometimes claimed that engineers do not need to know about history because they must look towards the future in order to provide the progress which society demands. This is a very short-sighted attitude. Without a long-term view of what progress is and the patterns it follows, it is easy to lose an overall sense of the context in which engineering takes place, both relative to other activities and between different branches of engineering. Three widely-held beliefs about progress are, at best, misleading, at worst, simply wrong:

- The first is the naive idea that history progresses like a river growing as more and more streams add to the main flow. The consequence of this view is the mistaken belief that every

development in the past was somehow leading towards our present
position, a view which imposes the idea that our present state is,
by definition, the best and that all previous positions were, to a
greater or lesser extent inferior.
- The second is the belief that engineering progresses by means of a
series of inventions. While some steps of progress have indeed been
associated with inventions, the inventions have usually required
enormous development by equally ingenious people in order to make
them viable. (On the other hand, inventors can be heroes!)
- Finally, there is the view that developments in 'theory' precede
and, indeed, bring about developments in 'practice'. This is, in
turn, based on the idea that engineering design is a matter of
'putting theory into practice' - an idea which, under examination,
can be shown to be meaningless [Addis 1990]. Only be studying the
separate development of each of the three strands to engineering
history - engineering design, technology and science - can the
relationship between the three be understood.

2.2.1 Evolution and revolution in engineering progress
Developments in engineering fall into two broad types - those which
result in a gradual evolution of a branch of engineering, and larger
changes (revolutions) which result in a wholly new way of looking at a
problem, even a redefinition of the problem itself. It is important to
learn to recognise the signs which characterise the end of a line of
engineering development and lead to the state of crisis which precedes
a revolution in engineering design [Addis 1990].

2.3 Nature of design
Out of the previous discussion of engineering progress there needs to
be abstracted the way in which an understanding of engineering history
can illuminate the nature of the design process itself. Far too many
people nowadays confuse the ability to calculate, using the equations
and formulae of engineering science, with the skill of design. This
understanding can be particularly developed by a study of the history
of engineering design methods and procedures and their relationship to
parallel developments in engineering science and technology.

2.4   Learning from the past
Hegel reminds us that 'The one thing that one learns from history is
that nobody ever learns anything from history'. In order to avoid the
nihilism to which this observation could lead, we need to ask why it
might be the case. Two answers are relevant in the present context:
- we are, often, ignorant of the past;
- we are not taught how to learn from past.

The engineers who are aware of their history are also often the ones
who are able to draw lessons from it. The lessons which may be drawn
from successes and failures have already been mentioned. This can be
quite straightforward when a report into a failure exists. It is a more
difficult skill when faced with an engineering artefact of any kind, to
unlock the engineering knowledge which is contained in that artefact.
This is especially so when it involves imagining the nature of the
engineering problem and the intellectual and other tools which were

available to the engineer. These are skills which must be learnt if they are to be useful.

2.5 Character

Most of the points mentioned previously relate to technical matters. There are also several issues which touch on what can only be called the character of engineers, and its development. During the Victorian age and the early part of the present century this rôle for the study of history in general was considered most important [Titley 1921, Pendred 1923/24]. Several different facets can be distinguished:

- **Heroes and rôle models**
  It is still considered that young people can be motivated by heroes and this has long been the reason behind so much biographical study of famous people. Smiles' biographies of Victorian engineers were, first and foremost, lessons in the virtues of 'self help'. Nevertheless, in both Westminster Abbey and the National Portrait Gallery, engineers are outnumbered many times over by politicians and poets.
- **Humility**
  many people believe that our present achievements are the best there have ever been, based on the 'river' model of history. This is arrogant as well as inaccurate and also leads to the tendency to devalue past achievements and the engineers who brought them about. Even a little history can teach some much-needed humility.
- **Cultural context and status**
  In the last century it was unquestioned that engineers played an essential and major part in shaping our world and these achievements were, in general, widely publicised in all sectors of society. The same can hardly be said of the late-20th century - it is widely agreed that the engineer has lost status, both financial and relative to other professions. This change has arisen both from society's perception of the engineer and the engineers' perception of themselves. It is in this last area that history has a rôle. To some extent, our self esteem depends on a knowledge of our past, both at the level of culture and society, and in our own work. Over the last 100 years or so an engineer's education and training has changed from being largely by apprenticeship (and thereby empirical and historically rich) to largely based in academia (and thereby theoretical and historically poor). Many engineers now see themselves as little more than technicians and calculators - this is demeaning and in need of enrichment: an awareness of engineering history can provide such enrichment [Hamilton 1945/46].

2.6 Rhetoric and growth

An important difference between education in the humanities and engineering is that the former, in true classical tradition, still cultivates the skill of rhetoric. This enables students of humanities the better to put forward arguments, defend opinions, criticise and evaluate alternative proposals and persuade others to change their minds. By comparison, engineers, who do need such skills in their jobs, are given little opportunity to develop such abilities as students. Their ability to persuade hence often relies only on the results of calculations. Small wonder they often have difficulty in influencing

politicians. Historical study could provide useful skills in two areas:
- **Criticism**
  By studying existing engineering artefacts students can develop their critical abilities by learning to explain why certain works were good exemplars and others not, and why certain engineers were important in terms of the contribution they made to their field. In this way the essential idea of quality can be introduced and engineers become aware of what good quality engineering design and production in their own field were, and are. An idea of the qualities of excellent design can be developed. This type of understanding is essential in developing both self esteem and confidence.
- **The art of persuasion**
  By being encouraged to have to explain where we are (in engineering history), and how we got here, as well as why we did not end up somewhere else, engineers would develop their ability to argue using 'soft' data rather than only numerical data. With this skill they would also be better able to argue, with confidence, about where they might go in the future and to persuade others of their views.

3 Concluding remarks - What to do?

If it be assumed that the aim of engineering education is to create articulate and confident graduates who will wish to continue to improve the quality of engineering design, production and progress in their sector of industry, as well as to seek to raise the status of their profession, then, based on the arguments given above, the author would suggest that it is **essential** that engineering students study some history. The questions remain 'what history?' and 'what to teach?'.

3.1 What history?

The distinction must be drawn between the methodology of history and the raw material. There would be little point, let alone time, to furnishing students with an encyclopaedic knowledge of the history of design, technology and science in many branches of engineering.

Rather, students should be introduced to the different sources of historical data, such as artefacts, the printed (academic) word, periodicals, biographies, official and public records etc. The different problems posed by artefacts which survive (e.g. bridges) and those which don't (e.g. machinery) can be discussed. The importance of preservation, libraries and museums must be appreciated [Morice 1978].

In particular it should be learnt that history is an approach to handling information, a way of thought and a methodology for creating knowledge out of raw factual data; records and artefacts are primarily valuable for the knowledge they contain and which can be worked upon to yield this knowledge using a variety of historical techniques.

3.2 What to teach?

As engineering courses are already full, it would be unrealistic to suggest more than a small historical input, at least at the beginning. Sometimes, such as in the course introducing the behaviour of engineering materials and structures or the principles of mechanisms, historical examples can be used extensively - they are often easier to 'read' than modern examples.

A dedicated historical course of perhaps only 10 hours of lectures can introduce a number of case studies from the history of engineering and illustrate the variety of subject matter and intellectual approaches to the subject. Biography is not recommended since it generally has a low density of intellectually challenging material.

Most importantly, students should be asked to undertake some investigation (from secondary sources) in order to develop the skill of selecting and interpreting factual information: even non-engineering topics could be tackled - Why were skyscrapers developed? How do cities feed themselves?

In the engineering field the true complexity of technical developments such as the steam, internal combustion or jet engine, could be studied, rather than treating them as single inventions. This could show the inter-relation of technology, design methods and engineering science.

Finally, the skill of engineering criticism (analogous to the skills of literary and music criticism) can be introduced at all stages of an undergraduate course. It should be the principal process by which an appreciation of skill in engineering design and manufacture is developed and both self-awareness and self-confidence are built up. Practice can be given on a range of artefacts, some of undisputed excellence ('classics' such as the Forth Bridge or the centrifugal governor), others of lesser quality, which may even have failed in service. These skills can then be applied by the students, with rigour and confidence, to their own work.

Thus might the author's observation, that engineers who are interested in history tend also to be more interesting as a whole, come to apply to more of our younger engineers than at present.

References

Addis, W. (1990) **Structural Engineering: the Nature of Theory and Design**, Ellis Horwood, Chichester.

Hamilton, S.B. (1945/46) Why Engineers should Study History, **Transactions of the Newcomen Society, XXV**, 1-10.

Mainstone, R. J. (1977) The Uses of History, **Architectural Science Review, 20 No.2 (June)**, 30-34.

Morice, P. (1978) The Role of History in a Civil Engineering Course, **History of Technology**, 3, 29-34.

Pendred, L. St. L. (1923/24) The Value of Technological History, **Transactions of the Newcomen Society, IV**, 1-11.

Pugsley, A., Mainstone R.J. & Sutherland, R.J.M. (1974) The Relevance of History, **The Structural Engineer**, 52, No.12 (Dec), 441-445; **53**, No.9 (Sept), 387-398.

Titley, A. (1921) Presidential Address, **Transactions of the Newcomen Society, I**, 65-75.

Psychology Teaching in the Undergraduate Electrical Engineering Curriculum : Prospects and Practice

J. MacDonald, D. Van Laar
Department of Psychology, University of Portsmouth

Abstract

A concern of the professional bodies who validate and accredit the undergraduate electrical engineering curriculum has been the aim of broadening education and training. In the United Kingdom, the Institute of Electrical Engineers (IEE) guidelines include providing the potential Chartered Engineer with an awareness of the range of constraints imposed on engineering activities. Some of these issues are explicitly psychological and behavioral, the human side of engineering practice.

This paper presents the results of a survey of UK higher education institutions to gauge the extent of Psychology teaching in undergraduate Electrical Engineering courses and finds rather piecemeal provision. We then present an outline curricula for courses, constructed on a flexible modular basis. Finally we describe an account of the implementation of a Psychology module in the first year of the BEng degree at Portsmouth Polytechnic and draw some conclusions for the future.

<u>Keywords</u>: Engineering, Education, Curriculum Design, Human Factors

1 Introduction

In 1989 an initiative by the British Psychological Society (BPS) aimed to determine the degree of support for the teaching of psychology within other professional bodies and the extent to which it was translated into practice. We report here on the findings concerning Electrical Engineering. In order to assess the possible and actual contribution of Psychology, information was gathered from a number of sources. These were the IEE guidelines on accreditation and content of courses; a subject review of Electrical and Electronic engineering courses (Polytechnics and Colleges) conducted by the CNAA (January 1991); and a brief questionnaire sent to a sample of course leaders asking about Psychology on their degree programmes.

Undergraduate education in Electrical and Electronic engineering in the UK is strongly influenced by the Institution of Electrical Engineers (IEE) guidelines. Courses are accredited by the IEE and graduates meet one of the requirements for Chartered status. The guidelines make reference to desirable features in accredited courses. In addition to the expected aims and objectives of providing a thorough grounding of scientific and fundamental engineering principles and practice there are others cited which appear to have specific relevance to the discipline of Psychology. For example, the IEE refer to

- a need to broaden the engineering dimension of the potential Chartered Engineer with a knowledge of subjects such as Health and Safety, Management, Finance and Industrial Relations
- to generate an awareness of the constraints imposed on engineering activities by physical, human and financial resources and by economic, environmental and safety considerations

The CNAA review states that courses should enable students to develop

- an awareness of engineering in the total business environment
- an ability to communicate with members of an organisation at the same, higher and lower levels of expertise or responsibility.

The IEE also require that every accredited course have a detailed statement on its Engineering Applications (EA) content, which is designed to introduce the skills and attitudes needed to complement academic study in the process of becoming a Chartered Engineer. Two of the four aspects identified are

EA3 - introduction to industry under supervision and involving a range practical assignments (including ... the development of personal skills of working with other people at all levels in an organisation)

EA4 - preparation for and a period in, a responsible post under decreasingly close periods of supervision.

From these objectives and guidelines two themes emerge with psychological import. The first is that the education of prospective engineers should equip them to take a broad view of the design and production of engineering artifacts. This is so that they should not only aim to meet a functional specification, but should also consider the usage of such artifacts. Appropriate topics come under the heading of **'ergonomics'** or **'human factors'**. The second is introducing students to the non-engineering aspects of the job functions and roles they will be required to perform; the areas of **'organisational psychology'** and **'personal and social skills'**. It would appear therefore that Psychology can and perhaps ought to be making a relevant contribution to electrical engineering education and training.

2 Current Practice

To determine the current position and extent of psychology teaching a brief questionnaire was sent to a sample of institutions selected from the 1990 list of IEE accredited courses (31 Universities and 29 CNAA Institutions). Courses in this area come under a wide variety of titles from BEng (Electrical Engineering) through to more specialist courses eg BEng (Engineering Science and Industrial Management) and many institutions offer MEng degrees and have a number of related courses. It was clear from the IEE documentation that related degrees tended to share courses in the 1st and 2nd years. It was judged that if there were courses containing psychological topics they would be likely to occur early on in the degree program and be shared across related degrees. Therefore even when an institution offered more than a single course, only one questionnaire was sent. The return rate was modest, (Universities - 17 from 31 (55%); Polytechnics - 15 from 29 (52%)).

2.1 The Teaching of Psychology

Question 1 asked whether any courses on the degree had a Psychology component. Only two answered with a definite YES, one University and one Polytechnic. (The courses were described as Industrial Psychology and Human Factors respectively). Four others (3 Univ, 1 Poly) responded with a sort of YES. All mentioned courses that they thought might have had psychologically related topics, eg three mentioned management and organisational behaviour modules and one a Communication Skills module. The remainder (26) responded NO. It appears that there is comparatively little psychology being taught or there is a lack of knowledge about what constitutes psychologically related topics.

Responses to the remainder of the questionnaire can be briefly summarised. If psychological topics are covered then they take up less than 10 hours of the timetable; are taught by someone other than a psychologist; are taught by lectures; assessed by essays and assignments; do not contribute to the overall classification of the degree; and the courses are designed for engineers and not shared with other students.

Respondents were also asked if there were plans to incorporate any psychology teaching on the degree courses. No one had any plans to increase the amount of teaching in this area. Three respondents said that it was something they might consider, but that time constraints on the timetable would make it difficult. A further question asked if any staff were engaged in interdisciplinary research projects with psychologists. Five said YES and surprisingly all of these were from institutions where there was **no** psychology teaching on the undergraduate degree.

3 Potential Curriculum

From the areas highlighted above a number of key topics can be identified to form the basis of short modules. Each could be presented as a 10 hour course comprising a mix of lectures, examples and practical exercises. There is a wealth of material to support such developments and an indicative reference is provided as a starting point for each.
- human factors: perceptual processing, memory, skill and performance, displays and controls, accident and safety etc.; eg Burgess (1990)
- organisational psychology: group processes including leadership, decision-making; intergroup processes, conflict and power; work-related stress; industrial relations etc.; eg Robbins (1991)
- Personal and social skills: interpersonal perception, self-awareness and self-presentation; empathy, emotions; social behaviour; interviewing etc.; eg Hayes (1991)

4 The Portmouth Polytechnic experience

The School of Systems Engineering at Portsmouth Polytechnic has run a course entitled 'Human factors in product design' for the past five years. This is a taught by a member of the Psychology department to HND and BEng Electrical engineers for 10 hours in the second term of year one. Approximately 60 students attend this as an option as part of a Communications course in the first year. It competes with a 10 hour German language course. The module is assessed by coursework at the end of the term. The course has been in a process of evolution, becoming ever more practical and case-based. The current version is divided into 10 hourly sessions in the following order.

Introduction
1. Why study human factors in a electrical engineering course?

The biological machine
2. Basic psychological and biological background to perception and information input.
3. Characteristics and limitations of the user.
4. The way in which people respond, information processing and memory.

5. Internal mental (cognitive) workings of the person, problem solving and thinking.
Safety, and environmental issues
6. Safety and designing for errors.
7. Environmental effects - noise, stress, health
Designing with human factors
8. Control and display design. Ergonomic guidelines.
9. Task analysis and case Studies
10. Designing for user expectations, natural mapping.

A portfolio of case studies has been assembled to illustrate points made in the lectures. Engineering students often ask for immediate 'proof' of the value of an idea, rather than relying on textbooks or theoretical points made in class. Case studies based on video recorders, remote controls, televisions and walkmans, etc are all used and found to be relevant. The more cases from modern, even fashionable, electrical goods that can be generated the better. The initial part of the course is based on the psychological characteristics of the product user. This is important because it emphasises the need to take into account topics that non-psychologists take for granted (eg that perception is not a passive process). Because of the abstract nature of this area, it can be difficult to justify directly to the students, but practical demonstrations and examples aid the learning process.

The second section explains how outside influences, affecting the user must be taken into account in the design of any good, usable product. Again, as many concrete examples as possible are used. Part three concentrates on ergonomic skills and practices. In order to emphasise the practical application of ergonomic techniques, task analysis and methods of human factors product evaluation are taught using appropriate examples. The course assessment requires each student to write a 1000 word human factors evaluation report on a group of products such as walkmans or remote control devices. Overall the School is pleased with the quality of the reports produced and feels that this also provides a good vehicle for the practice of general communication skills.

Students evaluate the teaching program by completing a brief questionnaire which are then used as a basis for course development. Apart from the usual 'witty' comments student responses are generally positive. The practical demonstrations are very popular and students always request more. One of the challenges of teaching this course is its positioning in the second term of year one. At this time the students have limited knowledge of electrical engineering and some express frustration at learning psychology when they should be doing more 'real' engineering. However, as the course progresses, many come to appreciate it's relevance and usefulness. Perhaps the real value of the course should be assessed by asking working graduates whether they felt that the knowledge and skills they acquired had been beneficial in the working environment.

Indicative reading for those wishing to set up similar courses are as follows:

Burgess, J.H. (1990) **Human Factors in Industrial Design**, TAB Inc, New Ridge PA

Cushman, W.H. and Rosenberg, D.J. (1991) **Human Factors in Product Design**, Elsevier, Amsterdam.

Gopher D. and Kimchi, R. (1989) Engineering Psychology. **Annual Review of Psychology**.

Norman, D.A. (1988) **The Psychology of Everyday Things**, Basic Books, New York.

Oborne, D. (1987) **Ergonomics at Work**, Wiley, Chichester.

5 Conclusions

It is clear that there is a role for aspects of Psychology to be included in Electrical Engineering education, and that there is support for this within IEE thinking. It is also clear that this is rarely translated into practice. In order to promote course development in this area a number of actions are needed. First heightening awareness among engineers as to how Psychology could be relevantly incorporated into degree courses. Second provision of information about how to implement such changes. This paper is an initial attempt to do both of these. The BPS is likely to encourage the development of information packs for other professions where key topics, details of support materials, references, videos etc and examples of current schemes will be offered and circulated.

We feel that the two most important points are first the applicability of Psychology's contribution and second that an effective introduction can be given within a relatively short course.

References

CNAA Publications, (1991) Electrical and Electronic Engineering: Review of Subject and Course Development. Gray's Inn Road, London.

Hayes, J. (1991) **Interpersonal Skills**, Routledge, London.

IEE (1989) Accredited Degree Courses 1990: Degree Courses in Electrical and Electronic Engineering. Guidelines on Accreditation: Booklet M10. Guidelines on Engineering Applications EA1 and EA2: Booklet M12. Institution of Electrical Engineers, Savoy Place, London.

Robbins, S. (1991) Essentials of Organizational Behavior, Prentice Hall, Englewood, N.J.

Formation of Engineers for the 21st Century
W.J. Plumbridge
Materials Department, The Open University

Abstract
The attributes of current engineering graduates are not matching the requirements of Industry. Their morale and ambition are low. A two-pronged strategy, involving a shorter Degree in General Engineering as a foundation for later diversification, together with substantial changes in course content, style and philosophy is proposed. The role of the Engineering Professoriate is pivotal and significant attitudinal changes in Academic Staff are required. It is argued that the present instability in Higher Education provides a rare opportunity for introducing an improved system for the initial formation of engineers.
Keywords: Graduate Qualities, General Degrees, Staff Attitudes, Sharing Facilities.

1 The Current Situation - A Cause for Concern.

In terms of social status and remuneration, a career in Engineering has never been highly regarded within the U.K. The principal purchaser of engineering graduates, Industry, is frequently unhappy with their non-technical attributes such as initiative, communication skills and commercial awareness. Even the morale of engineering undergraduates is depressed since they perceive themselves to be inferior to their management, financial and administrative counterparts in these important areas (Industry Ventures, 1991). With such an outlook, together with a lack of ambition, it is not surprising that an air of resignation pervades much of the Profession. Unfortunately, the expectation is usually that "**Someone**, or **They**, ought to do something". It is symptomatic that it is far less commonly heard that "**We** should do something about it". There is no evidence that the transformations to undergraduate degree courses in Engineering, brought about the by Finniston Report, have improved this situation. More than a decade later, at a time when the first engineers of the next century are beginning their careers, new dimensions have been added to the scene, and this paper argues that we should take advantage of the present instability to introduce changes which might alleviate the shortcomings cited above.

There can be little doubt that Higher Education is in a state of unprecedented change. With a Government objective that almost one in three young people will enter that system by the year 2000, (HMG White Paper, 1991), the disappearance of the 'binary line' and the establishment of a completely new family of universities, the trend towards a broader, non-traditional, access and the ever increasing demands to 'do more for less', the entire system is volatile. (Fig. 1). When one considers, in addition, the surge towards Quality Assurance that has now begun, it seems entirely appropriate to institute further radical changes before equilibrium (or inertia) is re-established.

Figure 1 The turmoil of Engineering Education

2 The Objectives

As far as engineering education is concerned, it is strongly contended that there should be a single, prime, objective which is a **significant enhancement in the quality of the graduate.** Current initiatives to develop quality assurance systems for the whole of Higher Education are laudable but the kinetics of such a strategy appear too slow to produce the desired changes in a realistic timescale.

It is expected that there will be little debate about the interpretation of quality enhancement in the engineering graduate since the desired characteristics have been cited on numerous previous occasions. However, to maintain focus, the principal attributes of the '*compleat*' engineering graduate in this writer's opinion include:

1. *An understanding of the scientific principles of Engineering*
   Science and Mathematics are the cornerstones of engineering. Without a sound grasp of them and their appropriate application, progress in engineering will be iterative, empirical and fortuitous.

2. *An appreciation of the specialisms in Engineering*
   Today's graduates are mostly produced as specialists in a subdivision of engineering such as Mechanical, Electrical, Civil, Materials etc. Many would argue that this is an artificial and unnecessary separation. What we call our graduates is unimportant as long as they appreciate the contribution of, and are able to communicate with, colleagues from across the spectrum of engineering activity.

3. *An awareness of the potential consequences of their actions*
   Engineering neither exists nor operates in a vacuum, so it is becoming increasingly important that graduates become aware of the social, economic, legal and environmental consequences of their actions. That such considerations may often over-ride the purely technical factors of an engineering decision is a realisation held by too few of our new graduates.

4. *Good personal skills*
   Success in the majority of engineering activities relies upon good communication between all parties - from the supplier of raw materials via the designer and manufacturer through to the after-sales service. Not only should engineering graduates have the ability to do this in all forms and at the appropriate level, by displaying integrity and realism, they should also be capable of inspiring confidence in those around them. Engineering is about achieving targets in terms of time, cost and performance, and this requires personal initiative on the part of the practitioner. Educators should strive to inculcate self-confidence, self-

motivation and the ability to self-educate in their students. These qualities are a far more important than blocks of specific knowledge *per se*.
5. *Competitiveness*
   Engineers have to harness, conserve and utilise the resources of nature. The survival of a company depends upon producing goods which are better/cheaper than those of other manufacturers. In both activities, competition is an essential ingredient. The days of the 'gifted-amateur' approach are long gone, if they ever existed in engineering.

3 Attaining the Goals

To achieve the necessary qualities, and to produce engineering graduates more appropriately skilled for the realities of the 21st century, two broad strategies are proposed in terms of (i) the **system** of educating engineers and (ii) the **people** involved in their education. Policies requiring substantial levels of investment of public funding have been rejected as unrealistic in the present climate.

3.1 The System Solution
To meet the requirements of breadth and depth in engineering graduates, a radically different educational system is proposed. Quantitatively, it is based upon the current 'norm' of a three degree course with an entry of a good 'A' level grades in Physics and Maths. Should these conditions not prevail, as is increasingly likely to be the case, timescales could be adjusted accordingly. The principal components of the system are:
1. Degree courses in Engineering would be in *General Engineering* and encompass all subdisciplines such as Civil, Electrical, Mechanical, Materials etc. They would be of two years duration (Figure 2).

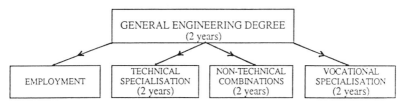

Figure 2 The System Solution

2. Subsequent education for those wishing to receive it would be in one of three general directions. For example, *Technical Specialisation* into an engineering subdiscipline, such as electronics, structures, materials etc; *Non-Technical Combinations* with such subjects as law, accounting, management etc and *Vocational Specialisation* such as bridge building, gas turbines, or computers. The duration of this second phase, which need not necessarily immediately follow the general engineering study, would again be two years.

The global benefits of such a system are that it acknowledges the significant proportion (around 40%) of graduates who require a minimal level of engineering expertise, who enter quasi-engineering posts or leave the profession altogether. It facilitates final career choices based upon a better appreciation of engineering. It provides greater flexibility with regard to market demands for employment and via the combined studies cohort it enlarges the domain of engineering awareness and influence. With a reduced number of students studying at an advanced level, it lends itself to a two tier system with concentration of expensive resources in fewer institutions. This, in turn, produces an incentive for greater mobility. Consequently, it is probable that the number of sub-discipline departments actively involved in advanced level teaching and research will diminish. The essential correlation claimed between research and quality in teaching will be challenged as it has been already by the UK professoriate (NCUP, Policy Document, 1991).

A more detailed model of the proposed degree course in General Engineering has been described previously by the author (Plumbridge, 1991a). Briefly, it would comprise of five technical strands such as (i) Chemical/Processing, (ii) Civil/Structural, (iii) Electrical/Electronic, (iv) Mechanical/Manufacturing and (v) Materials. These would constitute about half the Course with the remainder devoted to the basic sciences and the broadening subjects such as Professional Studies, Communication, Economics and Management. Formal Teaching would be reduced significantly to provide opportunity for self-learning and group activities. Whilst it is intended not to be over-prescriptive at this stage, closer inspection shows that the demands of the purists for advanced level study in a particular speciality are met by the appropriate 2 + 2 combination, with the added benefit of a base of broad engineering awareness (Plumbridge, 1991a).

3.2 A People Solution

Prior to embarking on a solution to the present malaise via the people concerned, it is helpful to outline those aspects of the problem which are people-based. To assist in this process the caricatures in figure 3 will be employed. No doubt some, if not all, are recognisable! However, the serious aspect of the problem is that engineering educators have been producing inappropriately educated graduates for many years and that they, themselves, are products of the same, incomplete, system. As far as the attributes of the ideal graduate are concerned, we have a *circle of deprivation* within the system. How can we expect the 'average' educator of engineers to produce what, for the most part, he does not possess himself? (the masculine gender is retained as a matter of general accuracy!) Neither is the obvious alternative of utilising industrialists substantially more in engineering education likely to guarantee success. If so, why isn't UK industry more successful? It is axiomatic that many of the deficiencies cited for new graduates persist through much of industry itself.

35 years in Engineering Departments still pursuing Ph.D Topic.

15 years in Department after 15 years in Industry - a member of the *when we* fraternity.

10 years in Department (student, postgrad and lecturer) - an intensive, person.

'New developments are either reinventions of the Sixties or beyond my comprehension'.

I've given good service to the Tea Club and the Library SubCommittee and will support only those changes which take place after I retire.

Sex is OK but computer simulation of the hydrodynamic drag on screw dislocations in a reciprocating electro-magnetic field really turns me on.

Figure 3   Some 'typical' engineering educators

Despite their inadequate training, the onus for producing graduates appropriately skilled for the 21st century lies with the current educators, and for many, this requires a marked changed in attitude and philosophy. While the umbrella of quality assurance will eventually pervade Higher Education, sceptics expect a prolonged gestation period as an administrative monolith before achieving quality enhancement. Truly 'top-down' dissemination cannot avoid being too general and too slow for the urgent problem in engineering education. Rather, we should adopt a more specific approach (a middle-down?) and concentrate solely upon engineering matters. Here, the Engineering Professoriate are key players, for it is they that should possess the broader perspective and it is they that have the influential powers of persuasion. While the UK professoriate, as a whole, accept their pre eminent role in the introduction of quality into Higher Education (NCUP, Policy Document, 1990) the reality of the situation is less optimistic. Estimates of the proportion of Professors of Engineering actively involved in the Quality Initiative (PIQUEE) of the Engineering Professors Conference suggest a figure of about 1 in 15 (EPC, Private Communication, 1992). If this is even an approximate indicator of the commitment to quality enhancement on the part of the expected leaders of the movement, then a monumental task is ahead.

Even when a significant core of converts is built up, dissemination and inculcation of the engineering ethos will not be easy to achieve. The majority of engineering educators are basically Engineering Scientists rather than Engineers, and are quite content to stay that way. The typical characteristics shown in figure 3 have been described more fully elsewhere (Plumbridge, 1990b). Although the educational system in which they operate may be highly unstable, the majority of our engineering educators have, in many respects, attained a near equilibrium condition. Although they may justifiably complain of being under increasing pressure, it is likely that most of them have been teaching the same subjects, working in the same research areas and even being responsible for the same administrative duties for a number of years. In such circumstances, small iterations in course content are entirely inappropriate considering the radical alterations that are required. A twin-engine for change, internally driven by the professoriate and externally reinforced by the Quality initiative, is necessary. Within two years, completely new courses could be on offer having a more appropriate balance between the technical facets of engineering and with the non technical components identified earlier. All staff might expect to be teaching quite different topics than at present, and also to change again within three or four years. This will produce a higher level of freshness of approach and enhance integration between individual subject areas. Rather than adopt a similar loading profile (lectures, laboratories, research and administration) for all staff, people's individual strengths should be recognised and more fully exploited. This would facilitate more valuable contributions from the archetypal staff of figure 3 and release more time for all to operate in their optimum mode. Above all, the transition should be led with professionalism, encouragement and ambition - all qualities of good management. Until quite recently, the training received by the Engineering Professoriate in management skills was non-existent. Since such attributes usually rank low in the selection of a professor, perhaps this aspect also should be a feature of the new system?

A second component of the People Solution lies in the interaction between teacher and student. While there exists no universal panacea for the transference of knowledge, skills, understanding and personal attributes, there is scope for more variety in how it is done. There is general agreement that too much formal lecturing still persists, and in this passive role the student is rarely stimulated or stretched. Alternative learning strategies such as group activities, projects, self tuition exercises, personal tutorials, peer assessment and senior-junior student teaching all have their supporters (Smith, Ed, 1991). No doubt, some methods are more successful than others for particular staff/student/subject combinations. The important point, however, is that these alternative approaches should be explored but it must be remembered that the critical ingredient in any initiative is the enthusiasm and attitude of the staff involved.

Intimately connected with the People aspects is the time aspect, and the question must be addressed whether teaching is carried out efficiently in this respect. As a rough estimate, it is likely that numerous basic engineering phenomena are described in a hundred or so different locations (Universities, Polytechnics and Colleges) each year. This process is repeated each year since the fundamentals of engineering are constant. A situation such as this screams for collaboration and sharing of teaching strategies and aids between the lecturers involved, yet little exists. Why should so much time be wasted in doing the equivalent of re-inventing the wheel? For example, sophisticated modelling of common engineering phenomena and interactive computer aided learning systems are basic technology but, of course, the restraint on their use is the enormous investment of time required for even a small package. Surely, the solution lies is the exchange and sharing - a library of engineering teaching facilities? That the goal here is not a series of long (about 1 hour) programmes since these would tend to be over - prescriptive. Instead, teaching aids of 10-15 minutes duration are recommended which simply concentrate on the basics of a particular phenomenon so permitting the lecturer the flexibility of his/her own interpretation. The existence of such a system on a national scale would result in a significant time saving and an improvement in teaching quality, both of which would subsequently contribute to the desired outcome of a higher calibre graduate engineer.

4 Concluding Statement

If the engineers practicing at the end of the 20th Century are to leave a worthwhile inheritance to their successors, then dramatic and revolutionary changes must be made in the very near future. The current instability in the Higher Education system is conducive to such change and it behoves us as educators to accept the challenge, even at the expense of established systems and strongly held attitudes. Engineering is about doing, achieving and flexibility. It will be many years before similar opportunities present themselves and it may be disastrous if we fail to take advantage of them.

References

Industry Ventures, 1991, Survey of Final Year Undergraduate Engineers.
HMG White Paper, Higher Education: A New Framework, Cmd 1541, May 1991.
Nat. Conf. of University Professors, Policy Document No. 3, Nov. 1991, Higher Education: A New Framework.
W J Plumbridge 'Engineering education: Time for a revolution', Inst. of Eng. Australia, National Conf. Pub. no 90/1 (Canberra) 1990, 50-53.
Nat. Conf. of University Professors, Policy Document No.1, Oct. 1990, Quality Control of Teaching Standards.
Engineering Professors Conference - Office, Private Communication.
W J Plumbridge, 'Engineering - The People problem', Inst. of Engineers, Australia, National Conf. Pub. No. 90/1, (Canberra), 1990, 123-126.
R A Smith, Ed., Innovative Teaching in Engineering Ellis Horwood, (Chichester) England, 1991.

Formation of the New Engineer

D.G. Elms

*Department of Civil Engineering,
University of Canterbury*

Abstract
Many of the newer demands on engineers require a broader range of skills than before. Breadth and flexibility are required in addition to more traditional technical skills. The engineer with both sets of skills has been called "the new engineer". The nature of the new engineers is discussed. It is argued that the fundamental discipline underlying their work is a set of basic modelling skills. A categorisation of models is followed by a list of rules and desiderata for their formation, together with a set of auxiliary skills. A course attempting to teach such matters is described and examined critically.
Keywords: Engineering, Education, Philosophy, Curriculum, Fundamentals, Flexibility.

1 Introduction

Engineering education is coming under increasing pressure from the profession to reconsider at a fundamental level the skills and knowledge given to engineering graduates. Traditionally, these skills have tended to be narrowly technical, with a high degree of emphasis placed upon mathematics, quantitative analytical techniques and design skills. However, in many areas such skills no longer provide an adequate preparation for the requirements of a career in professional engineering practice.

There are a number of reasons for this: technology is changing fast and the half-life of knowledge has become very short; newer engineering problems which are often more diffuse and intractable than those of earlier times; sociopolitical demands are having a greater interaction with technical problems; and engineering tasks are becoming more complex, often needing an interdisciplinary approach.

Clearly, further skills are needed in addition to the traditional requirements of technical competence. They are sufficiently different at a fundamental level that the professional engineer possessing them is quite distinct from the older style of engineer. The difference is sufficient to justify the name "new engineer" for a person possessing such additional skills.

This paper considers the skills needed by the new engineer, and ways by which they may be taught.

2 Characteristics of the New Engineer

The new engineer has a number of characteristics in addition to the traditional requirement of technical competence. I will mention six.

Firstly, the new engineer must have flexibility. This is necessary not only to deal with diverse, complex and often novel engineering situations, but also to be able to cope with the increasingly short half-life of knowledge.

The second requirement is for ubiquity, for generalised capability. The new engineer must have skills which are sufficiently basic that they can be applied to any problem. The focus of training must be on generality rather than on technical specifics, with a strict requirement for underlying discipline.

Thirdly, there is a requirement for creativity, a skill which is present in all engineers, but which is particularly necessary where the engineer is moving away from well-trodden and well-understood pathways.

The fourth requirement is for ethicalness and an understanding of ethics. Professional ethics set priorities and give procedural guidelines in novel situations. They give a foundation on which to build. Ethics must be understood, too, because an engineer may well have to deal with people whose ethical standpoints are very different.

Fifthly, the new engineer must have good communication skills. This is not only a matter of being able to write well, listen carefully and speak articulately, but also a need to communicate internally with oneself; in other words, to think.

Finally, there is a need for breadth. Breadth of experience and knowledge is essential when dealing with complex and novel problems. Quite apart from being a necessary source of general knowledge, a wide experience helps creativity by providing a source of analogy, and analysis by providing the metaphors which are often so helpful in synthesis.

3 The Fundamental Discipline

3.1 Models and constructs

The fundamental discipline of the new engineer is the synthesis of models and constructs. In most situations in life we do not consider in any specific way the models that we use. Rather, we use them without thought, particularly in familiar situations. However, the types of problem tackled by the new engineer are systems problems. Such problems have the three characteristics of novelty, complexity and significance, which, together, demand that close attention be paid to the appropriateness and quality of the models used at every level. A fundamental skill of the new engineer is thus the ability to use and develop models. This in turn demands an understanding of the nature of models, and of the laws, rules and desiderata that govern their formation and use.

It is useful to think of three fundamentally different levels of model: operative, system and world models. They form a hierarchy. Operative models are immediate models, the quantitative models from which, if appropriate, action can be taken. Working drawings are operative models, and so are structural analysis models, operating on load and structural data to produce load-effects and displacements. Simulation models, too, are operative models, producing specific and quantified results from which action can be taken. They are models for doing; hence the word "operative".

System models, on the other hand, focus on the totality of whatever is being dealt with, on the components and the connections between them. However, the emphasis is on careful definition of both the parts and the whole, and on the logical relationships involved. System models give a context for operative models: the former set the structure of a problem, while the latter are concerned with details and results.

World models provide a context and an environment for system models. They are the sets of assumptions within which the engineer works and produces the system models. The assumptions are not always obvious, and may often be unconscious, or unquestioned cultural assumptions. Not only does the world model give a context for the system model, but it also provides the purpose which must control its definition and the way it is generated.

The new engineer must be able to deal with all three model types. However, the focus must be on system models, for operative models are usually context-specific, and world models are general and not applicable only to engineering. The engineer's skill must be primarily in developing and using good system models.

3.2 System models - nature

Four fundamental statements about system models are:

1. A system is defined
2. The definition is driven by a purpose, outside the system
3. The system contains subsystems, and is itself a subsystem of a supersystem. It is thus part of a hierarchy
4. A system has an environment, which is that part of the world outside the system which affects, and is affected by, the system.

3.3 System models - desiderata
System models must obey certain rules and laws, discussed in the next section. Beyond the rules, there are also six criteria (Elms,1989) which must be met to produce good, as opposed to mediocre, system models. They are:

1. Completeness - all relevant elements must be included
2. Consistency - all elements must in some sense be of the same kind. This can most readily be forced by requiring all elements to be defined within the same grammatical framework
3. Balance - all elements should be of roughly the same magnitude
4. Discrimination - the boundaries between elements should be clear, and there should be a minimum of overlap
5. Coherence - the structure of the whole should be clear, for a system model is more than the sum of its parts, just as a book is more than the sum of its disconnected words
6. Parsimony - unnecessary and irrelevant matters should be omitted

3.4 Rules and laws
System models must obey certain rules and laws. These are both external, defining the bounds of the system, and internal, relating to its structure and behaviour.

The primary external laws are conservation laws. In formal physical terms, conservation of mass, momentum and energy must hold. Other and less physically strict conservation requirements can be applied to flows of non-physical quantities such as people or money. The second law of thermodynamics, too, applies to the system as a whole, both in terms of energy, and in terms of information.

Internally within the model there are other requirements in addition to the conservation laws. The two main sets are Kirchhoff's laws and the rules of logic. Kirchhoff's laws apply to many dual variable pairs, such as voltage/current in electricity, displacement/force in solid mechanics, and pressure/flow in fluid mechanics. The rules of logic govern the structure of the system model, and are particularly important for more qualitative and descriptive system models.

The laws and rules governing system models do not themselves form the models. Rather, they are bounds and limits which may not be transgressed.

4 Supporting skills and knowledge

The skill of the new engineer in developing and using system models is not sufficient by itself. It needs to be supplemented by other general skills and knowledge. Skills need to be developed in learning ability, communication, creativity, ethics and the handling of uncertainty. The reasons for this are as follows.

4.1 Learning ability
One of the first things an engineer does when acting in an unfamiliar area is learn about the situation. The ability to learn about systems is itself a learnable skill. Contributions to the theory of learning have been made in a number of disciplines, notably cognitive psychology, artificial intelligence, education, sociology and management science (Elms, 1992). It comes down to a matter of the engineer constructing an appropriate system model of whatever situation is being learned about. Therefore strategies are required for discerning the underlying structure of the situation, and also for knowing what data to look for.

4.2 Communication
The learning process often requires good communication skills. Communication skills are also crucial to other aspects of the work of the new engineer. There are interactions with the client, the public and other professionals throughout a project. All are important but the latter especially so as the sorts of broad-based problems often handled mean a close working relationship is required with professionals from often widely differing disciplines. A frequent difficulty is that each discipline will have a usage of language particular to itself. A further and most important reason for good communication skills is the need to be able to communicate well internally, with oneself. It is not possible to think without the use of language. The better one's skill with language, the better one can think.

4.3 Creativity
Developing good system models needs good creative abilities. There is a wealth of literature on the subject of creativity, indicating that it is a learnable skill. However, creativity is not only improved by learning a number of specific techniques. Almost more importantly, it is necessary to have a breadth of experience to act as a source of metaphor and analogy, and to practice thinking laterally as much as possible.

4.4 Ethics

Ethics is included here because, when the engineer is working in an unusual area where there is no established precedent, ethics becomes important both as one of the foundations on which to build, and also one of the constraints specifying bounds of possibility. This refers primarily to professional ethics. There is, however, another reason why the new engineer should have a good understanding of ethics. When dealing with others, engineers need to be aware that people operate on very different ethical axes. It is not a question of bad ethics rather than good, but rather one of underlying world views. An engineer, for instance, with a utilitarian ethic, could find it difficult to deal with a member of a pressure group with a teleological, end-oriented ethic.

4.5 Uncertainty

Skills in handling uncertainty are added here because they are often needed, and because the engineer, always planning for future action, is inevitably dealing with uncertainty, at least covertly if not overtly. It is better by far to be overt: in the writer's experience, risk and uncertainty produce by far the most difficult conceptual problems when dealing with system tasks.

5 Teaching Experience

A course based on the ideas outlined above has been taught for some years to final year civil engineering students at the University of Canterbury.

The course has now been taught for sufficient length of time that there are now clear indicators of the success of the approach. These are threefold. Firstly, the students themselves are enthusiastic and rate the course highly. Secondly, past students in professional practice have said they have found the course to be one of the most valuable they have taken, in retrospect. Finally, there has been good feedback from the profession in general, and strong indications of support for the initiative. In any case, from a teacher's point of view it is a delight to work with students who are interested, enthusiastic, and quite evidently thoroughly enjoying themselves.

7 References

Elms, D.G. (1983) From a structure to a tree, **Civil Engineering Systems**, 1, 95-106.

Elms, D.G. (1989) Wisdom engineering - the methodology of versatility, **Int. Jour. Applied Engineering Education**, 5, 711-717.

Elms, D.G. (1992) Learning about complex systems, **Civil Engineering Systems**, (in press).

Future Engineers, Are We Trying to Attract the Right People?

K. Travers

School of Mechanical and Manufacturing Engineering, Queensland University of Technology

Abstract
The general public is unaware of the necessity of engineering to their wellbeing, and the engineering profession has failed in the past to promote and encourage engineering as a desirable career. There have been changes in recent times and the engineering profession has made increasing efforts to publicise the efforts of engineers and to encourage school leavers, particularly women, to undertake studies in engineering courses.
The result has been an increase in many colleges and universities of students enrolling for engineering courses. The question arises, however, are these students that are being attracted to engineering necessarily the right ones?
Keywords: Engineering Education, Women in Engineering, Secondary Education.

1 Introduction

Engineering is necessary for almost everything that is in daily use. There are very few aspects of civilised society that do not rely on engineering to some extent. Despite this, it seems to be generally accepted that there are insufficient engineers either in practice, or being trained. The academic institutions of the world are failing to produce sufficient numbers of engineers to maintain the existing standards of life at present enjoyed by the developed countries, and to raise the standards demanded by the developing countries. Ellis (1991), for example, shows that bachelor awards in engineering in America have been declining since 1986. This is not necessarily all the fault of the universities, political decisions and financial restraints also play their part, It is also a result of students unsuited for a career in engineering enrolling in engineering courses and, more importantly, appropriate students being unaware that such courses exist. Engineering is a cornerstone of everyday life for most of the population and should be actively promoted as such. Engineering needs to be defined. The benefits of such an approach is that engineering will not only attract those students that want to become engineers, but that the community at large will also understand and appreciate the role of engineers in society. In this manner, engineering will become accepted as a desirable career, the engineer will be more readily identified in society, and potential engineers will be more aware of the necessary requirements.

2 Engineering Students

The demand for tertiary eduction is such that not all of those students eligible to enter a course are able to secure a place. Prospective students entering courses in engineering form two distinct groups;
 1. Students who have some understanding of engineering, and have the necessary entry requirements for the course chosen.
 2. Students who have the necessary entry requirements but have selected engineering as a profession by default, occasionally in error, but more often because other preferred courses, usually medicine, law, or a business subject, were unattainable. These students have no history of an interest in engineering and often have a very limited perception of the role of engineering in society.

There is a third group of school leavers that would be suitable candidates for engineering courses, mentioned by Travers, (1990). These include school leavers with eligible to enrol for an engineering course but unable to secure a place, and also those that have received very little information of engineering, have little or no idea of the skills needed, and therefore lack the entry qualifications for an engineering course. This latter group contains many of the potential women engineers who are lost to the profession.

It is important that prospective students to engineering must have been made aware of the entry requirements sometime prior to entering the university, and have some understanding of what the profession of engineering is. They must be able to evaluate their employment and development prospects in that career. They need to be aware of the skills and experience that are required for engineering.

Why are students are enrolling in engineering courses? Is it only because they have the correct academic entry qualifications, are they enrolling because they failed to gain acceptance to other disciplines, or are they enrolling because they want to be engineers and deal with engineering problems? Many students now enter engineering without this basic understanding of engineering but because of a high rating in science and mathematics and due to limited entry quotas or other restraints, engineering became a second choice.

3 Defining Engineering

To establish the correct requirements for these future engineers, and to attract the right people to engineering, it becomes necessary to define engineering. A definition is needed to establish the attributes required for an engineer to be competent in their chosen profession and provide a basis for the primary, secondary and tertiary

education procedures to be applied. A second definition is needed for the potential engineer and this definition should be short, simple, and self explanatory.
So what is engineering?

A typical definition will state that engineering combines an intuitive knowledge based on practical skills, and a knowledge of science, such as the physical and chemical laws, applied to solving problems. This gives a flat impression of what engineering is really about, but it will establish the requirements for enrolling on an engineering course. These are basic sciences and mathematics. Communication skills are also important. The intuitive knowledge obtained from practical skills is more difficult to quantify. In many respects it is a pity that students embarking on engineering courses no longer have a background of playing with train sets, with flying model aircraft, of sailing model yachts, as this type of activity provided an interest where basic, instinctive principles of engineering were learned.

The second definition should be to convince people about the importance of engineering and why it is so necessary to them. The average parent has little or no idea of the importance of engineering and does not recognise engineering as a desirable career for their children. It needs to be emphasised that engineering is concerned with making things happen, that virtually every appliance, every mode of transport, every type of food eaten and anything connected with entertainment relies on some engineering input. This wider explanation of engineering will inform people of engineering in general.

Complementary to this second definition, the general public should be continually made aware of the necessity of engineering to their everyday standard of living. Only by this continual reference to the engineering content of the objects they use, the pleasures they enjoy and even to the food they eat will the general public accept that engineering is a necessary profession, and a desirable profession.

Having defined engineering to inform and attract the correct students, it is desirable to define who should be engineers and how should they be trained and educated.

4 Primary and Secondary Education

The problems are not the result of later education errors, they begin at the primary school stage. The Professional Engineering magazine (1992), referred to national tests carried out in Britain to assess the literacy and numeracy of six and seven year old children. The tests revealed that 20% of the children could not read without help, almost as many could not count to 100, and 30% failed the test completely. These children are part of the basic pool from which future engineers will be drawn.

At secondary schools, the average teacher and the average student adviser has little or no knowledge of engineering. It is unfortunate that secondary education in mechanics, physics and chemistry contains virtually no engineering or practical components. Problems are directed to solving equations with a unique answer, without redundant or missing information and with a limited time to find a solution.

Kenaway (1986), concluded that a PC, correctly programmed to accept formal and tutorial questions of typical university examinations should achieve at least a lower second degree. An analysis of such examinations had shown that 71% simply required a numerical answer, only 2.5% required some explanation of assumptions made, critical analyses of the solutions or engineering conclusions.

The students undertaking this educational system are the same students electing to enrol in engineering studies where they discover engineering is not a series of problems with all the required information readily available and a concise, unique solution. Engineers are dealing with redundant information, multiple solutions and probabilities and need to be able to make subjective as well as objective decisions.

5 The image of engineering

During the nineteenth century, engineering was a widely appreciated profession, and to some extent still is in Germany and Japan, but this is no longer the case in most other countries. Engineering is not seen as a glamorous profession and remuneration is not as competitive as many other professions. Robert Jackson, the then UK minister for employment, was quoted (1991), as saying that the supply of engineers at all levels was vital and that for there to be success against international competition there will have to be changes in attitudes towards engineering, and moves to make engineering higher education and careers more accessible to those with the appropriate abilities. Richards (1992), refers to a survey conducted by Industry Ventures of the attitudes and intentions of 2700 final year undergraduates. The majority of these students believed that most people have no idea of what engineers do. Of 20 engineering manufacturers assessed, showed only four had relatively good images, that only 24% felt that mechanical engineers were well paid, that only 17% felt that production engineers were well paid, and that graduates in both areas had to put in a lot of time before they were appreciated.

Many students, having enrolled for a course of study in engineering, have a very limited view of the role of an engineer and of how an engineer operates.

6 Women in engineering

The move to attract women into engineering has been a policy of governments, institutions and many companies for some years now. The usual reason given is that as women are approximately 50% of the population and less than 5% of engineers, there is a large untapped market of future engineers available. This equates with a generally recognised deficit of engineers available to maintain the technological progress enjoyed so far.

Science has been suffering from the same problems as engineering, with a dull image, lack of inducements to school leavers and has embarked on many similar ways to attract students into science. This has had some effect as recent studies in

Australia have shown that females retained in high schools has increased from 26% to 70% in 20 years. This contrasts with consistent male retention rates of approximately 60%. Female students now account for 46% of enrolments in mathematics, chemistry, (43%), biology, (65%), and physics, (29%).

Travers, (1991), quotes that participation rates in Australian undergraduate courses in engineering has increases to approximately 8% in 1990 and a target of 20% has been set by 1997. Predictions have been made that 25% of practicing engineers in 2000 will be women.

7 The solution

Much progress has been made in recent times to correct some of these problems. Neighbourhood schemes, Women into Science and Engineering, Women in Engineering, National Engineering Week, Young Engineer's awards and public statements on engineering concerns such as nuclear power and the environment, transport, toxic emissions, etc., these have all helped to raise the profile of engineering and inform young students and the general population that engineering exists and is concerned and involved with decisions that affect them.

In Britain, new general national vocational qualifications are to be offered to 16 year old students in 1992. Included will be production, maintenance and technology. Victoria, Australia, has recently revised its curriculum for secondary students and now includes Technology Studies in Victorian Certificate of Education (VCE). This includes materials, engineering, energy, and systems and should introduce the students to professional engineering practices. Unfortunately, this type of development is dependent on there being sufficient numbers of qualified teachers to lecture in the courses, and there may be some delay in fully implementing these programmes.

Most of these actions and ideals are excellent. Unfortunately many of them are aiming at students already interested in becoming engineers.

Engineering needs to be introduced to children in primary school, by visits to the schools by engineers, by visits to engineering concerns by the children, possibly by introducing simple engineering concepts into the lessons and stories. This process should be extended to secondary schools before subject preferences are made. There is already a limited amount of contact with schools but the programme needs to be increased and given a much higher profile. Visits by representatives of industry and academic institutions would help by informing the children, and more importantly, the guidance officers, of the requirements needed to enter the engineering profession.

It is unfortunate that the profile of engineering is low. It needs to become more visible and project a more suitable public profile, not that of a backroom worker, or a noisy factory worker, the producer of odd, boring metal objects. The study of engineering is perceived to be long, hard and ultimately unrewarding, both in society's expectations and in financial terms. Most of the major engineering achievements are perceived as only of indirect concern to the general public. When a large, important project is of some concern, the social and political aspects are identified as the most important. Technical achievements are received as scientific breakthroughs. The engineering profession communicates to a large extent through magazines, catalogues and conferences to other engineers but fails to enlighten the layman. By associating engineering and engineering projects with everyday experiences at primary school, as occurs naturally with teaching and medicine, for example, young children will not regard engineering as an unknown career or something outside of their experience. Moves have been made to form associations between secondary schools and engineers. This needs to be encouraged. Engineers have to be seen as normal people. Visits to engineering establishments, even young engineers competitions, as is done with mathematics and science, help to form an image of engineering. This image is important. Most secondary school children have some concept of a doctor's work, of teaching, even a scientist. These images may not be factual but they have an impact on the social standing and the desirability of following that profession.

Engineers are perceived by the community as honest and ethical, they are rated as better than lawyers, slightly less than teachers, and below pharmacists, doctors and dentists so they are regarded with some standing in the community. They also tend to be quiet and introverted, and reluctant to voice opinions.

Surely, with these credentials already established, engineers can raise the level of awareness in the community about their profession and encourage those who would enhance the profession to take up engineering as a career?

References

Ellis R.A. (1991) Engineering and engineering technology degrees, 1990, Engineering Education, January/February 1991, 34–44.

Jackson R. (1991) Tunnels and Tunnelling, November, 5.

Kenaway A. (1986) Chartered Mechanical Engineer, 40–41.

Travers K. (1990) Secondary education in engineering and science, in Proceedings of the 2nd Annual Convention and Conference, Australasian Association for Engineering Education, 2, 413–417.

Travers K. (1991) Are we aiming at the right market for our future engineers? in Proceedings of the 3rd Annual Convention and Conference, Australasian Association for Engineering Education, 1, 245–248.

Professional Engineering, (1992) January, 3, Mechanical Engineering Publications.

Richards J. (1992) Poor image prevails, Professional Engineering, January, 9, Mechanical Engineering Publications.

Optimisation of Fundamental, Applied and Humanitarian Principles and their Supporting Structures as the Basis of Future Engineers' Training

A.A. Minayev, E.S. Traube
Donetsk Polytechnical Institute

Annotation
The quick change in modern know-hows and a possibility of their detrimental influence on the environment have their effect on the strategy of higher engineering education; they make it necessary to strengthen the humanitarian approach for protection of man and the environment, to ensure a sufficient level of the fundamental training of engineers and a possibility of their adaptation to the changing practical conditions through a periodic retraining in the course of their professional career. The report (1) was devoted to the analysis of these problems and the methods for their solution, using the example of the Donetsk Polytechnical Institute, a higher educational establishment known by its traditions, the Institute that had trained N.S.Khrushchyov and other prominent figures. The present report is developing these theses.
Keywords: Engineers' Training, Fundamental Training, Humanitarian Training.

1 Introduction

The Donetsk Polytechnical Institute (DPI) is one of the biggest technical higher educational institutions of the Ukraine, known by its academic and social traditions. During the time of its existence since 1924 over 85 thousands engineers graduated from the Institute. Many of them made a great success in their career. Suffice it to say that the dynamic leader of the Soviet Union N.S.Khrushchyov who started transformations in this country was among the graduates, as well as such prominent industry organizers as the deputy chairman of the Council of Ministers of the USSR A.F.Zasyadko, the Minister of Coal Mining B.F.Bratchenko, the Minister of Power Engineering Yu.K.Semyonov and many others.

However, after the Ukraine had changed its status, its higher school faced new tasks. The transfer to the market economy, the quick change in know-hows and complicated ecological problems make it necessary to renew the purposes, structure and contents of higher education, including that at the DPI.

2 Purposes and methodology of engineering education

Forming a modern specialist who is to live, work and create at the borderline of the second and the third milleniums is a very complicated task. It is not reduced to the processes of mastering professional information only, it is a complex problem of bringing up a many-sided personality with a wide range of spiritual needs. Under the action of social surrounding (state structure, family, school, higher school, reference groups) a person masters the social experience that had been accumulated before him: knowledge, forms of behaviour, ways of life, skills, traditions.

It is far from being a secret that a considerable part of applicants try to enter higher schools in order to just get a diploma, not because of their vocation.

According to results of sociological research, only every tenth student chooses his future speciality under a decisive influence of an educational institution's activities directed at professional orientation. There is a distinct tendency towards a lower competition among those entering technical higher schools. To a considerable extent the profession of engineer lost its prestige because of low wages and lack of effective methods of stimulating creative activities of specialists.

One should believe that development of the technical sciences methodology will make it possible to change the distribution of creative forces even in the near future.

The methodology of technical sciences so far has been developed to a much lesser extent than that of fundamental sciences. This not only restrains growth of possibilities for creative work in technical sciences as such, but creates difficulties for transformation of new knowledge obtained by fundamental sciences into more perfect know-hows, equipment, production organization and management.

Methodology of any field of knowledge is characterized by a number of certain attributes including the subject-matter of studying, principles, methods, laws, theories, etc.

The subject-matter of technical sciences is transformation and improvement of production.

A principle is usually thought of as a certain starting point of a doctrine, a theory. Such a determination usually does not cause any objections but at the same time it does not answer the question: "Where do these principles arise from? The most accurate answer to this question was found by the academician S.I.Vavilov: "Principles are generalized experimental data". In the light of this definition a principle is not any longer seen as something frozen and given once and forever. New factors arise, which is inevitable with the present variety and development of life, and this may lead to a necessity to formulate a new principle that had not been known before or to change the form of the established principle.

In compliance with the objective natural laws, the transfer to the modern stages of technological transformations determines a reduction of the detrimental influence on the environment. The ecological nature of new designs is associated with technological processes that mostly take place in closed mediums, with lesser losses where different technologies intersect, with reduction of the need for initial raw materials and with obtaining several products in one process.

The above attributes of technical sciences - methods of creative engineering and laws of know-how and equipment development - serve as corner stones for forming the theory and practice of engineering education. They formed a basis for developing a model of reforming engineering education in the DPI.

3 New approaches to engineering education in the DPI

On the basis of the theses described in Clause 2 in respect of methodology, a number of principles to be followed and proposals on reforming engineering education in the DPI were worked out, viz:
 - interaction between higher educational institutions and industry must be determined by the concept of demand and supply;
 - different types of engineering activities require different levels of training which must be lesser for production engineers who follow the established rules and larger for creating engineers (designers, researchers);
 - an opportunity must be ensured of continuous education going over from lower stages to higher stages, provided the student has a desire and abilities for that;
 - a sufficiently high quality of the primary basic training must be ensured;
 - selection must be ensured when going over to each subsequent level, the byproduct of which will consist in an increased motivation for learning at the previous level;
 - a certain degree of social protection is necessary allowing the engineer to find application after each stage;
 - a certain correlation of the system adopted with that existing in the West is necessary, which will make it possible to raise a question of convertibility of the DPI diplomas;
 - the name of each level of engineer's training must correspond to the generally adopted meaning. This is especially important in the conditions where competition may arise between higher educational institutions.

On the basis of these principles and the Ukrainian law on education a concept of multi-stage training of engineers has been developed. The following levels are provided for:

The first - bachelor's degree. The student is taught for 4 years and after that has a sufficient fundamental training in the field chosen by him, and some professional training. The graduate may start working at the industrial enterprises for which his theoretical training suits and which are ready to continue his specialization in his working place. A possibility of getting such a diploma reduces expenditures for engineering education. The graduate may continue his education, too, and depending on his success (and desire) he may be allowed to go over to the second or third level of learning (the principle of selection).

The second - engineer's degree. After being awarded the bachelor's degree, the student studies for another year and is trained to the extent allowing him to find a job of engineer in production. This degree suits most industrial enterprises, because they prefer to employ an engineer who has undergone a specialization without the enterprise spending its time and effort, and it is ready to pay for the training.

The third - master's degree. After being awarded the bachelor's degree, the student learns for 2 years and is trained to the extent allowing him to be employed as a creating engineer.

The fourth - degree of candidate of technical sciences (corresponds to the degree of Doctor of Philosophy in the West) for those who choose a scientific career. Training of candidates of sciences requires studying more courses on the speciality, performing and defending a qualification research work. This kind of teaching is provided in the DPI in the following forms:

full-time post-graduate course (up to 3 years);

correspondence post-graduate course (up to 4 years);

presenting one's thesis for a degree (a form of studies when the work itself is done outside the DPI).

The fifth (highest) - degree of doctor of technical sciences. This degree is awarded after the candidate's degree for significant research work, making a noticeable contribution to a certain field of knowledge. The honorary degree of doctor of sciences of the DPI is also awarded to prominent foreign scientists who make a considerable contribution to the development of science and creative links with the DPI.

4 Optimization of the contents of teaching

For the multi-level training of engineers it is required to optimize the selection of disciplines and their distribution between the above levels. It is also taken into account that separation of fundamental disciplines and humanities and socials studies from the professional contents of education reduces the students' interest. Inter-subject interaction of departments and chairs serves this purpose: examples from professional courses are introduced into fundamental courses, terms and standards adopted in the professional courses being widely used

Great attention is paid to studying programming and computer operation from the first year of studies and to continuous use of computers when doing calculations and processing results of experiments.

Preforming laboratory work with real industrial equipment or with physical models that are sufficiently similar to it is also important. Great attention is paid to design (one or two works per term).

A teacher of any of the engineering disciplines must be the main champion of ideas of engineering education humanitarization: he must show the orientation of his science at meeting man's needs, protecting the environment against detrimental technogenic phenomena, set an example of culture of interests and behaviour.

Along with this, special humanities and social studies are provided for. The number of foreign language (English or German) classes has been doubled. After the Ukraine was proclaimed an independent state and the Ukrainian language its state language, training in this language should be intensified.

Though managerial training has started recently, a considerable number of engineers in industry deal with problems of production organization to a lesser or larger extent, which makes it necessary to train students in the field of law and theory of economics, and particular emphasis must be placed on practical problems of management, accounting of movement of material and financial values, record keeping, technical means and equipment for accounting, planning and analysis of operations, engineering ethics. Though a new speciality has been introduced in the Institute - Safety engineering and Environment protection, this course is also provided for students of all specialities.

5 Retraining of engineers

The quick change in know-hows requires a periodic retraining of engineers working in industry. Earlier such retraining took place in branch institutes for qualification improvement. At the moment a question is raised on creating such structures for retraining of engineers in the DPI, which corresponds to the tendency now observed throughout the world (2). Retraining of engineers differs from training of students in that its contents is more concrete, therefore the teachers must be well familiarized with the needs of production, its current state and new tendencies of development. To ensure this, a good interaction between the teachers of educational institutes and industry is required as well as their periodic work at industrial enterprises, employment of industry workers as teachers (first of all from those who create new know-hows), assistance of industrial enterprises in equipping laboratories of higher educational institutions. Such joint work with industry will make it possible to raise the level of students' training, equipment of the laboratories and teachers' knowledge.

6 Structures for discussing problems and decision-taking

To discuss problems that may arise and to take decisions, the DPI has appropriate permanent or temporary structures. The leading role belongs to the Rector's Office and the Academic Council of the Institute. The immediate role of creators of new curricula is allotted to the respective colleges (of electrical engineering, metallurgy, etc.) and to graduation departments and chairs of individual specialities, at which commissions on teaching methods have been created uniting teachers of all disciplines taught to the students of a certain speciality. Questions of teaching methods for the whole Institute are dealt with by the Institute's Council on teaching methods. To coordinate the humanitarian training, a special Council for Humanitarization of Education has been formed, consisting of representatives of technical, general engineering and humanitarian departments and chairs.

A special centre has been organized at the Institute to deal with retraining of industrial engineers.

7 Conclusion

The transfer of the Ukrainian economy to the market system and its desire to join the European Community requires an appropriate reaction on the side of the engineering personnel training system. The change of this system is directed at a transfer to the multi-stage training of engineers and the timely retraining of the working engineers for adaptation to the quickly changing conditions of production. This makes it necessary to optimize the curricula, to change the academic structures and to intensify interaction between the educational institutions and industry, as has been considered in the report. Besides, it is necessary to broaden cooperation of Ukrainian higher educational establishments, including the DPI, with foreign higher schools and associations for a joint work on improving the level of engineering education.

References

1. Minayev, A. and Traube, E. (1991) The problem of higher Engineering Education Improvement at the Donetsk Polytechnical Institute, in Proceedings of an East-West Congress on Engineering Education, (ed. Z.Pudlovsky), School of Electrical Engineering, The University of Sydney, NSW 2006, Australia, 194-196.
2. Pudlovsky, Z. and Messerl, H. (1990) Making the infrastructure for development of Engineering Education, Australian Journal of Engineering Education, 1.1, 81-89.

Changes in the Management of Engineering Education within a New and Complex U.K. Environment

D.A. Sanders, G.E. Tewkesbury, D.C. Robinson, D.G. Sherman

School of Systems Engineering, Portsmouth Polytechnic

Abstract

Universities and Polytechnics in the UK were once mainly concerned with students but now the business aspects of education must be taken more seriously. They are in a new and dynamic environment. Management and structures will have to respond quickly and positively in this new environment. This paper will consider the new external environment and the changes taking place to the culture and structures within institutions. Changes will be analyzed using the methods of Alexander(1985), Pugh & Hickson(1989) and Child(1984).
Keywords: Change, Management, Engineering, Education, Structure, Culture.

1. Introduction: The Environment

Where UK Universities and Polytechnics were concerned about services and products, we must now be concerned with our markets and customers, Sanders(1991.a). The situation needs to be viewed from an external position (Foy, 1981) and this paper attempts to achieve this.

The UK government is attempting to meet the needs of its potential customers by monitoring the changes taking place and publicising league tables of education institutions. Products and services are being adapted and expanded, and we must hope to achieve this without diluting the content and worth of the qualifications offered, Sanders(1991.b).

At one time the Polytechnics relied on applications from individuals. This situation has changed and applications are passing through a system where preferences are identified and students are allocated to places. This situation will change again as Polytechnics join the Universities and a new quality assurance (QA) structure is put in place. At the same time, some individual applications are still permitted. Publicity departments have been created and they are often seen as the main vehicle for marketing. The actual situation is more complicated. At a lower level each Faculty, School or Department markets themselves through varying place, prices and promotion strategies, as well as the overall course content. The different departments may tend to concentrate on different geographical areas outside the UK. The only similar part of the marketing mix is the product. This remains as:- higher and undergraduate degrees, diplomas, research, consultancy and short courses. PCAS and UCCA accept applications from prospective students. These applications include a list of preferences and it must be a major aim to <u>persuade students to name the institution on their application forms.</u> Courses offered to industry, foreign students and mature students are being expanded. Minimum prices are often set by committees but actual prices and courses are often negotiated within departments.

Schools and departments are concerned with attracting customers from around the world. Their first objective is to inform people that they exist and to persuade them to find out more. In the case of students this extends to enroling. The design of the range of courses must be key to the ability to supply the range within a budget.

The external customers may be regarded as the Government, potential students, their advisors, industry, consultancy contractors, the professional institutions and the Local Education Authorities. All of these need to be informed of our existence and our abilities. The internal customers are the departments, the faculties, non teaching staff and students. The Government are looking for a good return for their overall investment. The local education authorities are more concerned with a local return and may be especially important if they are to be provided with limited budgets. The advisors of potential students, (such as parents and school teachers), industry and the professional institutions, are all looking for quality and a good track record. The final decision made by students may be based on a number of factors other than the quality of the institution and the courses. These might include geographical location (by the sea, on a campus etc), the research reputation or the sports facilities.

The staff tend to require a pleasant place to work within a modern institution with resources and facilities for research and their other personal interests. They also require adequate training and education. The different benefits are summarised in Sanders (1992.a). The qualifications and training demanded from institutions are constantly changing. Examples of these are the recent trends towards IT, management and business qualifications, Robinson *et al*(1991). Promotion by means of the media appears to have an influence over the institutions favoured by students. Generally it is the research, project and consultancy work which creates this type of publicity and this work must be encouraged.

The most important recent legislation concerns funding arrangements and the coming change of Polytechnics to University status. Institutions will be funded in relation to the number of students attending the institution. The cultural tradition of the United Kingdom has placed Universities on a higher level than Polytechnics. This tradition may continue as ex-Polytechnics may be regarded as second class Universities. This will be discussed further in section 3.

2. Structures and cultures: How do we work together?

Having considered the environment, it is necessary to investigate the internal cultures (Marshall & McLean, 1988) and structures. From these, it is possible for institutions to extract their strengths and weaknesses and to develop the organisation (Harrison, 1972). Dramatic changes have taken place and many images and logos are being created, but there may be a discrepancy between the new public images and the reality.

The goals of institutions are changing in this new environment and new objectives are being set to achieve these goals. Of the standard business goals identified by Handy(1985); survival, profit and market share are becoming meaningful, where reputation, prestige and a good place to work were important before. Achieving these new objectives may require changing the core beliefs and assumptions of staff and the role of management will be critical during the later process of change. Changes may be necessarily made over a relatively short period of time and at all levels. This will inevitably create difficulties. It should be noted that changes may adversely affect other departments. Differences are inevitable and necessary, but changes must be carefully monitored. If the changes can be seen as being initiated by staff, at least in part, then the changes will be more successful. Unfortunately the change strategy may have to be partially directive as reaction time may be short. The best that may sometimes be hoped for is to move with the general support of the staff.

In the past, cultures and sub cultures of institutions have reflected the history and origins of UK education. This has been modified by the increasing rate of change and complexity of technology. Attempts are being made to change the cultures to reflect mixes of Role Culture and Task Culture. To achieve this, a Power Culture may be temporarily imposed. Individual teaching loads are often decided by individual lecturers in consultation with a time-tabler, and not by a line manager. This can create conflicts within a new and developing structure.

For research, fluid groups tend to exist which are adaptable and change depending on the task. The largest research groups must be nurtured if institutions are to prosper and many Governments are declaring an intention to fund national and European centres of excellence. During scholarly activity, lecturers tend to effectively manage themselves and they are the focal point for the culture. Structures used to be seen as serving the individuals within it and management was often by mutual consent. At one time managers were elected by other members of staff for a fixed period. These aspects reflect the Person Culture described by Handy(1985), but all aspects of the work are now project orientated and the emphasis is on getting the job done. This suggests the cultures will at least become a mix between a dominant Person Culture and a secondary Task Culture.

Management are recognising that the old structures may be too large to continue to function well with a dominant Person Culture. Power Cultures may be introduced as a transition to introducing a Role or Task Culture; control may be centralised and key individuals selected for promotion. Power moves away from committees and towards these individuals. This may only be a transition however and once key individuals are in place, a Role Culture may be developed as the new structures become fully developed around them.

Historical Person Cultures have spawned inappropriate structural arrangements for the new climate of competition. These structures can make work frustrating and reduce effectiveness. The most suitable structure and associated culture may be a Matrix structure with a Task Culture. In this case the emphasis is on "getting the job done" and the structure seeks to bring together the appropriate resources and the right people. In imposing the vogue structural changes, management may not be giving enough consideration to the people within the cultures and their ability to restrain the forces for change. These may be the most important influencing factor and the high calibre and well qualified staff may have joined and have stayed with the School because it was dominated by a Person Culture. As people are given new responsibilities within a Role Culture, conflict will occur. Within a Power Culture, people are in competition for promotion and scarce resources. During this change, one individual can only win at the expense of others. While the structure is being imposed, the reality may be different; boundaries are often unclear, lecturers and researchers strive for autonomy and the new structures can often be seen as a violation of their territory.

The process of changing structure will inevitably create some conflict. At present people do not fully understand their roles and responsibilities and the Role Culture must be frozen into place quickly to provide an atmosphere of collaboration. Some conflict may exist between members of different departments which are split or amalgamated. This may include some limited covert plotting and exaggerated inertia. It is vital to introduce an atmosphere of open discussion to remove this opposition to change. This may be helped by asking the staff to help to formulate and agree a strategic plan during the unfreezing stage.

3. Discussion and Conclusions: What are our strengths and weaknesses?

Having considered the external environment and internal cultures and structures, some strengths and weaknesses can be inferred. These will be considered in relation to product, price, place and promotion.

Product; Although the competitive price for services does not affect the main core of the business, students do attend an institution to obtain benefits. Their needs must be monitored to ensure we are still supplying their wants and we must constantly review our courses. The total number of 18 year old students is declining and macro forecasting and aggregate product/market estimates suggest the whole UK market may be in decline. In the short term, the A, B and C1 classes of UK 18 year olds is constant, and this may protect our main market for a while - but in this environment the institutions should build on existing A GAP analysis shows a gap in market potential as compared with Germany or Japan and therefore a usage gap in the national market. This suggests we could improve our position and close the product line gap and the competitive gap with other institutions. Market penetration may be increased by diversification and product development to modify existing quality, style and variety. This may be achieved as new courses are introduced. It may also be necessary to find new markets in other countries and other age groups. The Engineering degree has moved from the mature stage to the saturated stage on the product life cycle and with the Ansoff Matrix, this also suggests a strategy of diversification, but any new products should complement the existing product range.

Analysis with a Boston Matrix suggests the courses must be managed effectively and products must be distributed through all of the available channels. Any trading surplus can be used to finance new courses and balance the new courses which are at a risky stage with mature products. Integrated degree schemes may help to spread costs and rationalise resources and other methods of teaching such as distance learning and partnership degrees must be considered.

Price; Although price generally affects the quantity sold, the price for the UK student market has traditionally been set by the Government. This may not be the case in the future as UK students may lose their grants and other Government policy takes effect. Prices may be undercut by other institutions. This has always been possible for foreign students and for industrial courses. In all these areas we must decide if UK institutions are to be quality institutions and if they are, then they must charge quality prices. Four factors influence our pricing strategy; the student needs, the market environment, the institutions and the markets. Low prices are unlikely to yield an increased market share and competition should be avoided on price alone, especially as the market may be inelastic to price. Attendance on courses may be more dependent on customer benefits and their perception of us. We should concentrate on these!

Place; Is the product available to students when and where they want it? Could we take the courses to them? This may be partly achieved with the new Partnership degree schemes, the introduction of study packs and the CAT rating system.

The geographical position and the image of the institutions are important. The location may not be the best for the UK student market, but it may provide other benefits such as access to Europe. Strong links are being developed with European institutions and research groups are bidding for European projects. This may be too insular and we may have to look further afield!

The place of effective sale may become less satisfactory to UK students as Europe becomes more accessible in 1992 and a foreign language becomes more desirable. Conversely we may become more attractive to the rest of Europe.

The flow of information through institutions, including the processing of prospective students, demand forecasting and loading must be efficient. There are benefits in using an intermediary with local knowledge, especially in foreign markets such as Hong Kong and China. These "sub contractors" should be considered in the marketing mix and we should ensure a good image is being passed on by the schools ambassadors and liaison officers. Unfortunately we must recognise that this reduces the level of our control.

Promotion; Promotion by means of the media appears to have an influence over the institutions favoured by students. Generally it is the research, project and consultancy work which creates this type of publicity and this must be supported. The aim must be to create a favourable image and to reinforce positive attitudes. The corporate image is communicated via qualifications, courses, promotion and the place of sale, including buildings and equipment.

Promotional methods include personal selling through school and industrial visits, advertising in national and local media, and articles in magazines and research journals. (It is difficult to consider sales promotions in the usual sense). Advertising should inform, persuade and reinforce messages. Establishments must be visible in the market place and foster good relations with the press and professional bodies. Attendance at research conferences and research activity can be especially useful for attracting publicity and reinforcing our message and these must be encouraged. Secondary sources must give favourable recommendations. If this is to be achieved, then customer servicing and marketing must be correct. Courses will be recommended by word of mouth.

Objectives must be closely related to an overall plan, including the mix of courses, the total number of students and the market mix. These lead to the formulation of an advertising plan. Success in these areas can be measured by considering the reaction and behaviour caused by exposure.

Institutions tend not to employ salesmen as such, but all staff and students are acting as salesmen all of the time. They must understand this!

References

Alexander L.D. (1985). Successfully implementing strategic decisions, **Long Range Planning**, Vol 18, No 3.

Child J. (1984). **Organisation: A guide to problems and practice,** Harper and Row.

Foy N. (1981). **To strengthen the mixture, first understand the chemistry,** Guardian, 2nd Sept.

Handy C. (1985). **Understanding Organisations,** 3rd edn, Penguin Books.

Harrison R. (1972). How to develop your organisation, September issue of the **Harvard Business Review**

Marshal J and McLean A. (1988). Reflection in action: exploring organisational culture, **Human Enquirey in action**: edited by P Reason, Sage, London. pp 129-220.

Pugh D.S & Hickson D.J. (1989). **Writers on Organisations,** 4th edn, Penguin, UK.

Sanders DA. (1992.a). A new engineering school in a new and competitive environment. **Feb edition of Network,** published by Ed Devel' Unit, Portsmouth, UK, No 7, Feb 92, pp 19-23.

Sanders DA (1991.b) The Use Of Simulation in Teaching Automation & Robotics. **Proc of the East/West Congress on Engineering Education,** Kracow, Poland. Sept 91, pp 379 - 383.

Commercial Applications of Co-ordinate Measurement Equipment for the Manufacture of Three Dimensional Sculptured Surfaces

W.W. McKnight (*), D. Crossen (**)
(*) Department of Mechanical and Industrial Engineering, University of Ulster
(**) Crossen Engineering, Belfast

Abstract

Crossen Engineering is a sub-contract tool making firm located five miles from Belfast City centre. It employs thirty tool makers specialising in conventional, progressive and fine blanking press tools, injection and blow mould tooling and general tool making.

This paper describes how Engineering Education has benefited from an Industrial / Academic project aimed at increasing the manufacturing potential of this company with specific emphasis on Computer Numerically Controlled (CNC) milling operations.

<u>Keywords</u> : CNC, 3-D Machining, Surface Generation, Probing, Industrial Collaboration.

1 Introduction

The specification of the project was to develop a manufacturing procedure for the manufacture of non-geometrically defined three dimensional components.

The initial requirement was to produce a press tool for the manufacture of three dimensional dimensional sheet metal component identical to a hand produced sample (Fig.1).

No geometrical drawing specification was available for the component.

2 Manufacturing Procedure

The basic manufacturing requirement was to establish geometric information from the sample component, construct a three dimensional CAD model of the component, produce CNC programs to cut the surfaces and manufacture the press tooling.

To establish geometrical information relating to the component it was fixtured onto a Wadkin V4-6 CNC machining centre equipped with Renishaw touch trigger spindle probing system (Fig. 2).

Geometrical information had to be established in a form suitable for the construction of a three dimensional CAD wire frame model of the component.

A wire frame model developed utilising lofted sectional data was the most appropriate for the current task and probing procedures to obtain this information were developed.

CNC probing programs developed utilised the report feature of the GE2000 Controller enabling sectional co-ordinate data to be transmitted via the RS232 communication port to an IBM micro-computer system - thereby storing the data in an ASCII format.

The ability to write CNC probing programs utilising parameter data to control incremental probing distances, start positions etc. proved very successful enabling the machine operator flexibility to capture data in the most economic way. Additionally the program capability of the GE2000 controller to control the RS232 communication port essentially transformed the machining centre into a programmable co-ordinate measuring machine.

A computer program was developed to manipulate the format of the ASCII data file into a form suitable as input into Auto-Cad via the script file procedure.

Therefore the co-ordinate data obtained from the Wadkin Machining Centre was reproduced as 3-D Polylines within the Auto-Cad database without any requirement for manual data input - thereby ensuring data integrity.

The Auto-Cad file enabled the sectional data of the component to be manipulated via a CAD / CAM package. The DXF (Data Transfer File Format) of the resulting CAD file was the linkage between CAD and Smart-CAM a computer assisted CNC Programming package.

Both Auto-CAD and Smart-CAM offered a range of 2-D and 3-D curve fitting routines including :-

> bi-arc
> quadratic b-spline
> c-spline
> Bezier

This allowed the geometric data to be reconstructed into a form suitable for the generation of three dimensional crossectional profiles.

Both Auto-CAD and Smart-CAM offered the following wire frame surface modelling features :-

1. Ruled Surface
2. Lofted Surface
3. Translated Surface
4. Form Patch

However within Auto-CAD these surface construction procedures were limited in the number of sections that could be used and in the geometrical construction of the sectional data.

Smart-CAM was much less restrictive in the composition of sectional geometric data and the number of sections required to define the surface and was used for the construction of the basic wire framed model of the component.

When the basic surface information was established Smart-CAM offered surface trim and extend features which enable a true wire frame model of the component to be constructed.

The completed model was transformed into various work planes for the production of CNC programs to suit a pre-define ball nose milling cutter.

3 Conclusion

This technique clearly demonstrated the versatility of the Tough Trigger Probing System and how such data may be geometrically manipulated via standard features incorporated into CAD / CAM systems to reproduce sophisticated wire frame models of complex three-dimensional components.

Additionally the versatility of the GE2000 controller was demonstrated enabling direct data communication of geometric data to an IBM computer via the RS232 communication port thereby eliminating any potential data transfer errors.

The combination of Auto-CAD / Smart-CAM successfully re-produced the required three dimension surface profiles and the technique has recently been adapted for the reproduction of a complex three dimension core component of an injection mould tool, tooling for the automotive industry and bottle manufacturing companies.

The three dimensional surface construction techniques were easily manipulated to generate models of the component and allowed the user flexibility to modify and revise data while insulating them from the complex mathematical task.

Both Crossen Engineering and the University gained a substantial amount of manufacturing experience and a greater appreciation of programmable co-ordinate measuring, CAD/CAM linkage and computer assisted CNC code generation. Effectively the transformation of a conventional CNC Machining Centre into a three dimensional copy facility.

This bi-directional flow of information has increased the range of sub-contracting services that Crossen Engineering can provide without the company having to invest in expensive turn-key systems. Additionally this project kept the University in close contact with the current reality of engineering needs which is further transmitted to the students.

Engineering Education can only benefit from joint Industrial / Academic collaboration and the experiences gained by this University have certainly improved the content and quality of manufacturing lectures.

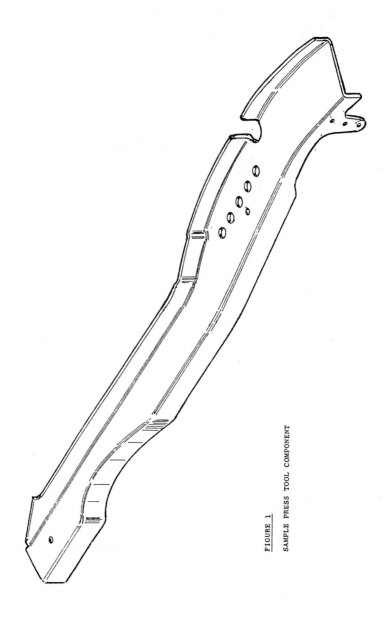

FIGURE 1
SAMPLE PRESS TOOL COMPONENT

Engineering Education 509

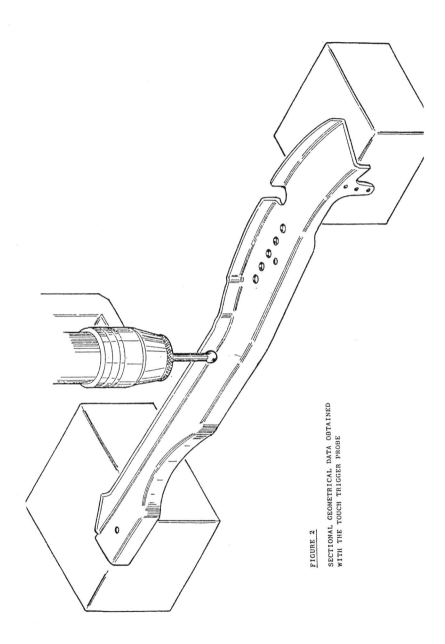

FIGURE 2
SECTIONAL GEOMETRICAL DATA OBTAINED WITH THE TOUCH TRIGGER PROBE

Testing and Advisory Centre for Industrial Trucks

J. Mather

Department of Mechanical Engineering, Design and Manufacture, Manchester Polytechnic

Abstract

This paper describes the setting up of a Test and Advisory Centre for industrial trucks such as counterbalanced fork lift trucks. The need for the Centre is explained, followed by a description of the major test facilities. NAMAS Accreditation and the running of the Centre are covered, including reference to commercial charges. The considerable educational benefits are highlighted and the paper concludes with a review of the whole project.

Keywords: Testing, Industrial Trucks, NAMAS Accreditation, Educational Benefits.

1 Need for the Test Centre

In the UK the Department of Trade and Industry (DTI) is the body responsible for ensuring that industrial trucks such as counterbalanced fork lift trucks covered by the EEC Directive 86/663 EEC comply with the Directive which came into force in 1989. The DTI does not itself have test facilities or expertise in this field; it relies upon suitable organisations seeking DTI recognition as an Approved Body. In 1988 the DTI was faced with imminent new legislation but did not have an Approved Body able to undertake the necessary tests to determine whether or not a truck complied with the Directive. The opportunity therefore existed to set up a national Test Centre on a commercial basis. A number of organisations in the UK had declared an interest in setting up an independent Test Centrte but none had taken the idea further. It should be noted that the relevant EEC legislation permits self certification of vehicles. The major manufacturers, have their own facilities but low volume manufacturers, dealers and importers can neither afford, nor justify the heavy capital expenditure. An independent Test Centre is obviously favoured by these groups and is also appropriate for handling technical disputes.

2 Designing and Building Test Facilities

Consideration of the scope and requirements of the EEC Directive led to the identification of a need for 5 major test rigs:

tilting platform for instability testing,
static vertical loading operator overhead guard,
dynamic falling object operator overhead guard,
draw bar drag brake performance,
driver visibility evaluation.

No additional space was available in the Department; the Thermo-Fluids laboratory had to be reduced in size and rearranged, including the relocation of a wind tunnel, in order to create a working space of approximately 150 m^2.
The roof had to be elevated from 5 m to provide 10 m headroom above the 5 m square tilting platform so as to accommodate trucks with triple lift masts. Major ground excavations were carried out for both the tilting platform rig and the strong floor for the static vertical loading of operator overhead guards. A 2000 kg overhead crane services the whole area of the Test Centre. The Design concept adopted was aimed at providing as large an unobstructed floor area as possible and multi-test use of the limited floor area.

Tilting Platform
Many alternative designs were considered. The design chosen had the following main features:

5 m square tilting table of rectangular hollow section fabricated with 50 mm thick steel upper slab and perforated plate surface,

3 worm screw lifting jacks with idler pulleys and chains driven by a single electric motor,

triple cylinder pivot support system 1 m from one end,

maximum angle of elevation 40 degrees,

maximum gross vehicle load capacity 25000 kg.

The platform permits both longitudinal and lateral instability tests to be carried out.

Overhead Guard Static Vertical Load

The design concept chosen was based on the use of a strong floor, permitting "point" vertical upward reaction forces. Load application is achieved by use of a single hydraulic cylinder anchored to the floor and attached through a pivot pin to one end of a loading beam. An adjustable tie bar at the other end enables the loading beam to be set horizontal above the guard and vertical loading of the overhead guard achieved. A load distributor in the form of a 50 mm thick steel plate sits on top of the guard and a load cell measures the hydraulically applied vertical force.

Static permanent deformations are measured by reference to a datum plane established beneath the roof of the guard, measurements being taken before and after loading.

Operator Overhead Guard Dynamic Falling Object Rig

A 45kg standard "missile" is dropped 10 times, using an electro-magnet, from a height of 1500 mm onto the overhead guard. Permanent deformations are determined in the same way as for the static test.

Draw Bar Drag Brake Performance

Truck draw bar drag brake performance is established by means of an electric winch, attached to a bollard, hauling the braked vehicle. A tension load cell between the winch and the truck measures the draw bar drag brake performance.

The load cell signal is fed into a computer for processing and display.

Driver Visibility Evaluation

A bank of nine halogen lamps is located in the truck, with the aid of a mannikin, to correspond with a standard person's eye position.

Analysis of the shadows cast by the truck mast, lifting cylinders and chains on a screen 4 m from the truck establishes the quality of driver visibility.

3 National Accreditation

DTI Approved Body status is conditional upon achieving NAMAS accreditation. The NAMAS accreditation procedure requires the setting and maintaining of strict quality control policies and procedures including auditing systems. A feature of the NAMAS Accreditation system is the need to have highly detailed Test Procedure documents for each test.

4 Financial Aspects

The Test Centre cost some £200 000 to set up. Tests are carried out on a strictly commercial basis. Test fee income during the first year of operation exceeded £70 000. Payback should be achieved with approximately 5 years. The charge made for the 5 major tests depends on truck size. Discounts are given for batch testing. Typically a charge of £3000 is made for the set of 5 major tests.

5 Staffing and Running of the Centre

The bulk of the planning and design work was carried out by the author supported by a Research Assistant. In the installation phase extensive use was made of technicians in the Department. Major fabrication and ground excavation work was subcontracted.

The Centre is currently headed by the author as Technical Manager with a full time Deputy Technical Manager (a temporarily seconded Departmental technician). Dutch students have spent periods in the Centre as part of their degree course industrial training element. Auditing is carried out by a member of academic staff.

6 Carrying out Tests

Tests are conducted strictly in accordance with the NAMAS Accreditation making use of the highly detailed Test Procedure Documents which define precisely the step by step test procedure to be followed. One of the challenges faced was that of writing these detailed Test Procedure documents prior to having all the test rigs fully operational.

7 Educational Benefit

Without significant educational benefit it would not be possible to justify the initial expenditure and even the relatively small accommodation which the Centre occupies.

The educational benefits have been and continue to be very substantial. Over and above the air of engineering reality brought to the Department by the arrival for testing of a wide range of trucks there are more specific benefits. As a direct result of setting up the Centre we have:

final year student projects related to stability, visibility and overhead guard strength, first year free body diagram major assignment (and examination questions) involving a force analysis of the tilting platform,

foundation year mini project in which the students use NAMAS standard procedures to determine driver visibility,

research into truck stability and overhead guard structural design,

Dutch degree students spending periods of 5 months in the Centre doing their industrial training.

The setting up and running of the Centre has provided excellent case study material for lectures in project engineering and project management, with the highly motivating advantage that the students can easily relate to the project.

Technical auditing is an essential requirement of the NAMAS quality control system. Auditing is carried out by an academic member of staff in the Department who is then able to make use of the audit procedures and outcomes in Quality Assurance lectures to the students.

8 Review of Operation and Conclusions

The whole project has been extremely successful. It has enabled academic staff in the Department to practise their engineering profession.

Relevant aspects have been brought into lectures, tutorials, assignments, practical classes and examination questions, further enhancing the engineering reality of our courses. Student project work and research has developed also. All visitors, from the industrial truck industry itself and others, have been highly impressed by the project, the quality of the rig design and the engineering professionalism of the project. Every customer for whom we have done work has been highly satisfied. Therefore overall the project has helped to enhance the high quality reputation which the Department has earned.

9 Acknowledgement

The author wishes to acknowledge the contribution made by Mr H Davis (Deputy Technical Manager TACIT) and the assistance of colleagues in the Department of Mechanical Engineering, Design and Manufacture, Manchester Polytechnic.

CAE/CAD/CAM Education for Industry
B.T. Cheok
CAE/CAD/CAM Centre, National University of Singapore

Abstract
In Singapore, the government has provided large sums of monies for the polytechnic and universities to purchase computer equipment for engineering education. Very comprehensive under-graduate, post-graduate, and continuing education curricula have been implemented by the educational institutions to support the industry. However, it is felt that the conventional courses offered by themselves are inadequate. What is needed is customised in-house training in CAE/CAD/CAM that will help individual companies improve their productivity, and develop a work culture that will enable the companies to cope with the rapid changes in technology. The author has experimented with such training projects and found them to be very effective. Case studies of projects for CAE and CADCAM training are presented in the paper.
Keywords: CADCAM, Computer-aided Engineering, Continuing Education.

1 Introduction

In countries such as the U.S.A., most colleges and universities are struggling to catch up with developments in CADCAM [1]. The main reason is budget constraint: Inadequate funds to acquire equipment and to offer industry competitive salaries to attract trained staff. Fortunately for Singapore, the Government has placed emphasis in the use of Information Technology. In the early 1980s, a National CADCAM Training Committee was set up to formulate broad policies and guide-lines for the various tertiary institutions to follow. A large sum of money was set aside for the tertiary institutions to acquire CAE/CAD/CAM equipment for educational purposes. At the same time, more funds are made available from time to time to upgrade this equipment so that the Polytechnics and Universities can always provide the latest equipment for their students.

The support provided by the Government has led to considerable success in the implementation of CAE/CAD/CAM curricula. It can be claimed that every engineering graduate from a Singapore polytechnic or university is proficient in the use of at least one CADCAM system for drafting. Most of them will also have experience using CAE/CAD/CAM related application packages pertaining to their fields of engineering. Some of these graduates may be involved in developing specialised CAE/CAD/CAM programs as part of their research projects. The approach adopted by the National University of Singapore in its CAE/CAD/CAM curricula implementation is reported in reference [2].

2 Industry Level Problems

The success of CAE/CAD/CAM implementation in the educational institutions is not matched by the industry. In 1988, a survey of local CAD/CAM users indicated that while many companies have started to computerise their design office and shop-floor, most of the equipment usage is restricted to straight forward drafting and machining automation [3]. As a young country, Singapore does not yet have the strong engineering tradition of the developed countries. Hence, the wide use of CAE/CAD/CAM and robotic systems at the industry level by itself must be recognised as a success.

However, the full benefits of computerisation can only be reaped when the expensive equipment is organised in such a manner so that it will contribute more to engineering data integration and management, and to perform more of the higher value design analysis process. In most cases, the returns from using the CADCAM system simply to automate the manual drafting and machining process alone can hardly justify the investment. Hence one of the main thrusts of training for the industry must be to help the local companies realise the full potential of their CAE/CAD/CAM investments.

This problem highlights one of the weaknesses of the continuing education programmes provided by the institutions. Currently, there is a wide variety of CAE/CAD/CAM related courses offered by the various Polytechnic and Universities. Many such courses provide an excellent introduction to computer applications in Engineering, especially for the older workers. However, they tend to concentrate on teaching the use of the commands of a particular CADCAM system, or concentrate on broad fundamental approaches to a particular field of computer assisted engineering. These courses together does not equip the participant with sufficient know-how to do specialised computer assisted engineering analysis, or to program and manage his CADCAM system effectively.

In the long term, the industry in Singapore must acquire a strong desire to continuously exploit the latest design and engineering methods in order to stay competitive in the world market. Improvements to productivity gains via computerisation can be achieved through education and training, however not just in the traditional sense. The educational institutions must take a leading role and make use of its resources to support CAE/CAD/CAM education beyond the conventional under-graduate, post-graduate and continuing educational system.

3 Customised In-house Training for Industry

Many of the companies that have invested heavily in CAE/CAD/CAM equipment will be challenged by a common scenario at one time or another:

"The Engineers are now competent users of the system, how to be more organised in order to cope with the information explosion caused by computerisation, how to be more productive, how to ensure that design data generated by the system is accurate, and how to improve the moral of the staff?"

Many recognise that the company may not have the expertise to answer these questions and will wisely employ the assistance of external consultants. In most cases, the consultant may be able to identify and solve some of the outstanding problems. However, some time after the departure of the consultant, it is most likely that the same company will again be faced with the same scenario.

What is lacking from such an approach is the education element. To ensure that a company is able to adopt to the challenges caused by the rapid changes in CAE/CAD/CAM technology, the management and staff must be educated and trained to overcome the challenges themselves.

Since every company's CAE/CAD/CAM implementation is unique, it is difficult for the educational institutions to design a general course that can help these companies directly. What is needed is a system to provide turnkey in-house training customised to meet the exact requirements of individual company seeking assistance. To ensure that a company can be self-reliance in the long run, such in-house training should include the following components:

a. An in-depth study and analysis of the company's current computer usage and its long term goals. Identify problems that will overcome existing problems and activities that can help to achieve the long term goals.

b. Design courses that will equip participants with the skills and knowledge to overcome the problems and carry out the activities identified. For courses that include hands-on, the exercises should be designed such that the solutions can be used directly by the companies wherever possible.

c. Initiate in-house projects that uses tools learnt from the training courses to solve engineering problems identified.

d. Devise a monitoring and guidance system for a period of time to ensure that the enthusiasm generated during the training is sustained by the company. In this way, it will cultivate a working procedure that will enable the company to overcome future challenges.

e. Help develop training materials and documentation to the enable company to train new staff independently.

The author has experimented with such an approach in CAE and CADCAM for a few companies in Singapore and find them to be very effective. They are also very well received by the participating companies. The case studies of two of such training projects are given below to illustrate the degree of involvements and the results.

4 Case Study One: Training in CAE

4.1 Scenario

Participating company is a subsidiary of an American multi-national company specialised in the manufacture of components for electrical and electronics products. Its parent company has identified a new PC based finite element (F.E.) package for the Singapore design department to use in product design. The company does not has any F.E. expertise.

4.2 Factors taken into account

After a preliminary study, the factors that will affect the conduct of the training program were identified:

    a. The long term objective of the company is to have all design engineers proficient with the use of the F.E. package. The design engineers will be supported by a CAE specialist who will provide guidance in model construction, lement selection, checks and interpretation of results.

    b. The design department consists of engineers with a wide range of educational background and work experience.

    c. Majority of the engineering problems can be solved using linear/non-linear static analysis of models comprising of shells, thick shells, or solids.

4.3 Courses

Three short courses were designed and conducted for the company:

    a. A short course in "Strength of Material" targeted to engineers with lower academic qualifications and the older engineers.

    b. A short course in "Finite Element Technique for Linear and Non-linear Static Analysis".

This course was attended by all design engineers. The course syllabus include: behaviour of the various types of finite elements, modelling of engineering problems, checks in pre- and post-processors, evaluation of results, non-linear analysis. In this course, mathematics is kept to a minimum and engineering problems commonly encountered by the engineers were used as illustrating examples.

    c. Hands-on training on the F.E. package. This course is aimed at instilling confidence with the F.E. package and exploring various important theoretical aspects taught in the courses. The exercises selected were standard benchmark tests to ascertain the accuracy of the elements to be used by the company, and to illustrate the effects of certain F.E. techniques, e.g. convergence, sub-modelling concept and etc. Non-linear analysis exercises were included to help the engineers establish the software's sensitivity to the stress-strain curve of the material, as well as its sensitivity to load step size and mesh size.

4.4 Projects

After the courses, two projects were designed for the company to enhance the engineers' confidence level when handling large analysis problems. The projects were based on two existing components manufactured by the company. The engineers conducted the study under the supervision of the consultants. Data obtained from the study were compared with data obtained from similar analysis conducted using an established F.E. package, and with actual experimental results provided by the quality control laboratory of the company.

4.5 Guidance

After the training, the F.E. package was used by the design department for production work. A hot-line service was provided for the company's CAE specialist to provide guidance as and when needed.

4.6 Training Material and Documentation

The training materials are left behind for the CAE specialist to train new design engineers. A documentation procedure was also recommended to the design department. In the long run, these documents will form part of the engineering knowledge database of the company and will help to shorten the time taken to do an analysis. Finally, the company was introduced to F.E. resource materials that they may need to make reference to in the future.

5 Case Study Two: Training in CADCAM

5.1 Scenario

The participating company is a medium size engineering contracting company. It has implemented a PC based CADCAM system consisting of a network of 35 work-stations. The engineering staff received their initial CADCAM training by attending the continuing program provided by the polytechnic. The computerisation process is reaching a bottle-neck. At the same time, the company has recruited 10 new design draughtsmen and engineers, and decided to add another 15 work-stations to its computer network.

5.2 Factors taken into account

After a preliminary study, the factors that will affect the conduct of the training program were identified:

a. The long term objective of the company is to achieve a high degree of automation in the use of its CADCAM system, and a high degree of integration for its application packages. It has employed 3 systems programmers to help achieve this objective.

b. The company lacked common standards and procedures for its CADCAM activities.

5.3 Courses

The following short courses and seminars were designed for the company:

a. A one day seminar for all engineering staff illustrating how the programming and management components of the CADCAM system can be used to solve some of their problems and improve drawing productivity. At the end of the seminar, a series of meetings with the section managers were held to establish common standards within the individual sections, and a common standard for the drawing office.

b. A short course on "PC and Network Evaluation and Selection" was conducted for the managers and systems programmers. At the end of the course, the tender specifications and equipment evaluation documents were designed and used by the company for the acquisition of the new computer hardware.

c. Hands-on training on the CADCAM system was provided for the new staff.

This course does not only concentrates on the use of the commands of the CADCAM system. It also emphasised the standards to be adopted by the drawing office. The hands-on exercises for the course were based on drawings provided by the managers.

d. An orientation course for existing staff on the new CADCAM standard and practises to be adopted by the drawing office was conducted. Efficient CADCAM practises were recommended while some bad habits were highlighted and corrected.

e. A short course on "CADCAM Programming and Management" was conducted for the managers, senior engineering staff and systems programmers. This course teaches the participants how to use the various utilities provided by their CADCAM system to improve it performance. It includes macro programming, menu and icon customisation, manipulation of neutral files, and linking CADCAM and Database systems together. The course exercises were designed such that the participants created symbol libraries, special menus, macro programs that can be put to use immediately by the drawing office.

5.4 Projects

After the courses were completed, several projects were carried out concurrently:

a. Implementation of the drawing office standards. The various section managers were assigned the additional task of monitoring the work of their staff to ensure that all drawings are produced according to the standards promulgated.

b. Set up a system to control and monitor the drawing files generated by the drawing office. The system administrator implemented a file management system for the section managers to manage the drawing and design data files generated for the various projects concurrently undertaken by the company.

c. Develop in-house programs. Three different projects were designed for the system programmers to develop bigger programs that will enhance the drawing office productivity.

5.5 Guidance

All the projects mentioned above were supervised by the consultant. The engineers are encouraged to propose to their managers areas of their work which can be improved by programming. The consultant helped to assess the feasibility of these suggestions and design the specifications for the feasible projects.

5.6 Training Material and Documentation

The training materials used in the courses can now be used by the company to train new staff. The drawing office standards and practises were documented and updated regularly. User's guides for each and every program developed by the company are written to ensure that all users know how to use them properly. Programmers guides are also produced so that it will be easier to make changes to these programs at a later stage.

6. Conclusion

The approach used to provide specialised CAE/CAD/CAM training for individual companies described in this paper is actually rather expensive and time consuming. However the long term benefits will more than compensate for the initial costs and efforts. While the initial impact on the industry is negligible, it is an excellent starting point to cultivate companies and individuals as effective users and managers of computer technology in engineering.

It is unlikely that profits motivated private consulting companies will be keen to offer such a service to the industry. This is because the entire training process involves a considerable amount of planning and preparation work. Therefore, it is up to the educational institutions to assume the responsibility and offer such a service to the industry.

References

1. Richards, L.G. (1985) Engineering education: a status report on the CADCAM revolution, **IEEE Computer Graphics Applications**, Vol 5 No 2, 19-25.
2. Nee, A.Y.C. and Hang, C.C. (1989) CAE/CAD/CAM curricula implementation - experience at the National University of Singapore, **Computer Aided-Design**, Vol 21 No 10, 649-653.
3. Cheok, B.T. (1989) A Survey of industrial CADCAM users in Singapore, **Singapore Polytechnic Technical Report Series**, No SP/2/89, 1-21.

The Marketing of Higher Education Engineering Courses

R.A. Otter

School of Civil Engineering, University of Portsmouth

Abstract
This paper considers aspects of the marketing of engineering courses in higher education. The measures which were taken to launch the innovative HITECC and Manufacturing Systems Engineering courses are considered and the results contrasted. Based upon this experience, it is argued that enhanced marketing of existing engineering provisions is unlikely, of itself, to result in dramatic improvements in recruitment to Higher Education engineering courses.
Keywords: Marketing, HITECC, Manufacturing Systems Engineering, Recruitment.

1 Introduction

A typical engineering faculty or department in higher education undertakes a diverse range of functions and activities in addition to managing its courses. Consequently, it employs varied methods by which it markets its range of services and establishes or maintains its reputation. This paper does not attempt to consider the full range of marketing methods. It concentrates upon aspects of the means used to publicise and market a range of undergraduate courses to potential candidates in the United Kingdom.

2 General Marketing of Higher Education Courses

There exists within the United Kingdom a very well established marketing arrangement between higher education providers and their potential students. The overwhelming majority of these are found in schools and colleges attempting to obtain their entrance qualifications. In a school or college, it is normal to find at least one person with specific responsibility for careers advice, often this is a member of the school's staff but in larger colleges, County Careers staff may be based within the institution. Normally, there is to be found a careers room or area in which is made available the plethora of careers related information sent by individual firms, coordinating industrial and trades organisations, professional bodies and institutions offering further and higher education courses. Much of this information, quite properly, is descriptive of the very wide range of available career opportunities and only a proportion relates to courses offered by Higher Education.

For a University or Polytechnic, the most important document lodged with the school or college is its prospectus. This is generally packed with both general institutional information, together with that related to individual courses. Probably of equal importance in conveying the range of courses available are the UCCA and PCAS handbooks which comprehensively list all courses within their respective sectors. For many areas of study, compilations are available of higher education provision satisfying particular industrial needs. For example, the Institution of Civil Engineers compiles and distributes details of all University and Polytechnic degree courses which have been 'accredited' as satisfying aspects of the profession's education requirements. In addition to the collation and availability of such information, there is also a well established 'circuit' of 'careers events'. Some, like "Directions" at Olympia are held annually and on a national basis. They provide the opportunity for an Institution to present itself to its potential students by means of a 'stall' or small display area. Others, like for example, the annual Higher Education Conference at Truro, organised by the Cornish County Careers service, enables representatives from Higher Education to give short presentations on specific areas of study rather than directly marketing their own Institution. There are many schools, colleges and other organisations which arrange similar events, albeit on a smaller scale. Portsmouth Polytechnic, in common, with many other Higher Education Institutions has a central unit responsible for this kind of schools publicity and liaison activity. It currently organises representation at some 350 events per year and commonly draws upon the assistance of faculties and departments for staff to attend the range of activities. It also relies upon similar sources for information for the prospectus. Often individual departments produce their own course literature which is used by the central unit. Over recent years, an annual Polytechnic Information Day has been organised for students on the first year of their A-level or BTEC Ordinary level studies. This has proved increasingly popular with over 2000 staff and students attending the 1991 event. Visitors normally have the opportunity to visit two departments in addition to attendance at more general sessions including, for example, one on the admission system's procedures. There is also a regular programme of visits to the Polytechnic by careers staff totalling around 250 in a given year.

When viewed from an institutional perspective, these arrangements are largely satisfactory. By carefully monitoring the activities of others, it can be reasonably assumed that marketing activity is at an appropriate level. By successfully utilising these arrangements and without dramatically increasing institutional marketing effort, a significant increase in Higher Education enrolments has been achieved over recent years. Those departments and courses which are adequately subscribed and in some cases, totally oversubscribed, will inevitably regard the marketing arrangements as wholly satisfactory. However, not all sectors are in such a healthy situation. Engineering courses, in particular, have experienced significant difficulty in recruiting sufficient, acceptably qualified students to fill available places and consequently have perceived the necessity for greater marketing effort.

Genuine difficulties are faced by a department or faculty which wishes to significantly supplement the established arrangements. The heart of the problem is that there is little direct contact between the potential student and the Institution until the candidate applies for admission. Nearly all effective marketing contact is dependant upon the cooperation and activity of the school's careers staff. They are responsible for disseminating information within the school or college and they generally organise their pupils' attendance at careers events. Almost all would view their job as providing impartial advice across the widest range of careers opportunities and certainly would not regard themselves as agents for a particular sector. So, even if for example, an engineering department produces and widely distributes an elegant eyecatching poster for its courses, it would be very optimistic to think that it will have a substantial impact upon the potential students it is attempting to influence. There is a strong possibility that it wil be 'lost' amongst the clamour of other material eligible for precious little available wall or noticeboard space. If an enthusiastic department successfully negotiates a special opportunity to talk to a school or college, it runs the risk of finding the event in competition with the hockey practice, choir practice etc., and end up preaching to a small number of already converted. Even if such talks did prove successful, financial restraints would tend to confine them to a department's local area and it remains the common practice for most students to attend higher education courses away from their own area!

Such a perspective has been arrived at after a number of years of actively working in the field of marketing engineering courses, it is believed that it is illustrative of the difficulties likely to be encountered in achieving significant marketing impact. Brief consideration will be given to two marketing initiatives related to new engineering courses which substantially enhanced routine market arrangements.

3 Marketing the HITECC Initiative

More fully described by Otter R.A. and May A.J. (1992), Portsmouth Polytechnic commenced its one year H(igher) I(ntroductory) T(echnology) and E(ngineering) C(onversion) C(ourse) in October 1988. This was part of a nationwide initiative coordinated by the N(ational) A(dvisory) B(ody) for Higher Education which established a small central coordinating unit. Amongst other activities it arranged a national publicity campaign which included the production of a video, poster and leaflet which were widely distributed, with particular attention paid to careers staff in schools and county authorities. They also funded a series of national newspaper advertisements and arranged attendance at a range of publicity events. Each Institution offering a HITECC course was expected to devote some proportion of its course funds for publicity. Over the initiative's early stages their task was made easier because most Institutions were only permitted to recruit from a reasonably well defined catchment area. This was because the Manpower Services Commission, responsible for student training grants, required attendance at a candidate's nearest institution.

One of Portsmouth's first tasks was the production of appropriate publicity material. Essentially the same course information was reproduced in three forms. A four sided, A4 glossy colour brochure, a coloured leaflet folded from an A4 sheet printed on each side and a single A4 sheet printed on one side with text only. These, along with course posters, were distributed to a wide variety of individuals and organisations which could represent sources of potential interest. Included were Engineering Faculty industrial contacts, adult education services, including WEAs, county library services and most importantly schools, colleges and county careers staff. The county careers services were believed to be of key importance in publicising the course and numerous visits were arranged to give presentations and answer questions. Arrangements were made with the Polytechnic's schools central liaison unit to give presentations and/or to provide information to all the Polytechnic's careers visitors. Schools were invited to send for additional literature and/or agree to a presentation. The response was limited, few requested extra copies of course publicity and only a few visits were negotiated to talk about the course. Such were welcome opportunities but the generally low attendances illustrated the difficulties described above. Arrangements were also made for a mailshot to careers advisors to arrive around the time of publication of A-level results when a number of potential candidates inevitably would be reassessing their plans for higher education following receipt of their results.

In parallel with this activity, permission was sought from relevant colleagues to send HITECC information to candidates whose application to other courses had been rejected. Because of the gross over-subscription to some courses, permission was readily forthcoming and a total of about 1500 potential candidates were informed directly of HITECC's existence. Opportunity was taken for comparing the effectiveness of the three types of publicity material. Positive response was received from a little under 10% of those circulated and perhaps surprisingly, there was no measurable benefit to be gained from despatching the expensive glossy colour brochures, rather than the very modestly priced single monochrome printed sheets. Based upon this experience, neither the glossy brochure, nor the colour leaflet have been reprinted whereas the cheaply produced single sheet has undergone many revisions and reprints and has been extensively used in subsequent marketing.

Around half the initial HITECC intake of 30, resulted from the latter initiative which meant that about 1% of information letters had ultimately resulted in an enrolment. The remaining admissions resulted from a combination of contacts made initially through NAB and direct responses to local publicity measures. Similar marketing measures were undertaken for the second year of course operation, but subsequently, the NAB central marketing ceased and the special local initiatives were reduced and ultimately replaced by the normal methods of conveying course information described above. The initial marketing and recruitment effort had been concentrated within a period of only about 6 months and the recruitment of 30 students, whilst exceeding the target number, seemed scant reward for the time, effort and cost involved. However, a sound foundation had unquestionaly been laid, Portsmouth's recruitment has increased to a current figure of 150 and HITECC or its subsequent 'Foundation Year' has become a well known and popular feature of higher education engineering provision.

4   The Manufacturing Systems Engineering Initiative

Concurrently with HITECC, NAB coordinated the launch of BEng courses in M(anufacturing) S(ystems) E(ngineering). This was an attempt to respond to the dramatic changes that were being experienced in production processes in large sections of the manufacturing industry. Like HITECC, there was a comparable national publicity campaign with the production and distribution of a videos, leaflets and posters. The twenty or so institutions offering the course supplemented the national activity with their own local publicity initiatives. Regretably the MSE initiative has not been rewarded with the level of success enjoyed by HITECC. Despite significant marketing effort, recruitment has been disappointing and after three years, the special NAB/PCFC funding ceased.

5   The Initiatives Compared

HITECC has made a dramatic impact on recruitment to higher education engineering courses, by comparison, MSE's contribution has been modest. The level of extra marketing activity associated with the two initiatives has been comparable. By every measure, they both significantly enhanced the normal higher education marketing arrangements which suggests that the different outcomes has more to do with the products than the marketing methods. It is believed that the key to HITECC's success lies in the fact that it offers an opportunity to every single person deemed to have the potential for higher education studies.   Otter R.A (1992) discusses HITECC entry criteria more fully. The MSE initiative unquestionably broadened the traditional entry requirements to higher education engineering degree courses. Whilst this broadening was opposed by some members of staff who viewed it as a radical departure from the traditional entry standards, from an outside perspective the changes may not have appeared to be substantial and certainly could not be compared with the openness of HITECC entry criteria.

It may be argued, therefore, that HITECC was seen by careers staff and through them potential applicants, as a genuine innovation in the engineering education field. It was seen as opening doors which had formerly been tightly shut. There was a potential market and when effective additional publicity was given, it was perceived as very relevant by careers staff, received with interest and enthusiasm and consequently disseminated, which ultimately led to full HITECC courses. By contrast, MSE was seen as making little progress in broadening access to BEng courses. Whilst unquestionably innovative in an engineering context, those outside the world of engineering viewed it as little more than a re-vamp of traditional provision. Despite the substantial marketing effort, it was not received enthusiastically by careers staff and hence did not ultimately have a major impact on engineering recruitment.

The foregoing views are substantially based upon feedback obtained from careers staff after a number of presentations and discussions involving both HITECC and MSE at which the different receptions described above were readily apparent.

6 Conclusions

It is quite common amongst engineering staff in higher education to believe that more and better marketing is the solution to recruitment deficiencies. The experience gained from the HITECC and MSE marketing campaigns strongly suggests that such confidence could be misplaced. Effective marketing, over and above the normal provision, is of major importance but only if an attractive relevant product can be publicised to a much broader range of potential candidates than the approximate 15% of candidates undertaking combinations of A-levels or BTEC units likely to satisfy traditional engineering entry requirements. This has been recognised by many within the world of engineering and for example, has given rise to the Engineering Council's strong support for the replacement of the narrow traditional three A-level course by a much broader programme of say 5 or 6 AS level equivalent subjects. Such a move would inevitably lead to a significant increase in candidates qualified for BEng or HND engineering studies and it is believed, enhanced marketing efforts would almost certainly result in significant improvement in recruitment to engineering programmes.

References
Otter R.A. and May A.J. (1992) Widening Access to Engineering Courses in Higher Education - the HITECC Initiative at Portsmouth Polytechnic. Proc. **World Conference on Engineering Education, University of Portsmouth.**
Otter R.A. (1992) The Establishment and Application of Selection Criteria for Applicants to Higher Education Engineering Foundation Years. Proc. **World Conference on Engineering Education, University of Portsmouth.**

Commerce in the HE Sector
R. Fletcher
Department of Mechanical and Manufacturing Engineering, Brighton Polytechnic

Abstract
The paper outlines the progress of a commercial unit operated within the Mechanical and Manufacturing Department from its tentative beginings to its current high profile status. The paper seeks to discuss the contribution such a commercial unit can make to the operation of the department in providing an alternative platform for operations and explores the relationship with the learning environment.
Keywords: Industrial funding, consultancy, integration, industrial experience, commercialisation, HE funding, DTI, contracts.

1. History and Development

For a number of years the capital provision within Polytechnics has been totally inadequate to fund a rolling programme of equipment replacement let alone the capital that is required to invest in new technologies to keep abreast of the changes in manufacture. It was quite obvious in 1988 that we would progressively be unable to offer the necessary facilities for curriculum development. A bold plan was formulated to set up a cooperative industrial plant using modern equipment..

Promotional activities within small to medium companies resulted in the setting up of an industrial board who advised on the equipment purchases. It was at this point that the first basic lesson had to be learned.

Lesson 1. Companies in general do not understand the concept of cooperation.

In general companies buy facilities and services at predetermined prices and so if we wanted industrial involvement it had to be on their commercial terms. As a result the Polytechnic was forced to seek grants and eventually obtained a 50/50 grant under the Manufacturing Systems Engineering initiative to purchase an FMT100 four pallet machining centre. Operator training was sponsored by FMT Ltd. with initial manufacturing contracts coming from the industrial board companies.

Lesson 2. Companies do not view higher education places as being capable of delivering the goods.

Breaking through this credibility barrier is very difficult. The perception is that educational establishments are incapable of keeping the price, quality and delivery. It is a very big step for a company to become dependent on a HE institution for supplies of essential parts.

Whilst the Polytechnic was willing to allow the commercial unit to proceed it viewed the departure from the well trodden academic path with some alarm. Any entrepreneurial spirit evaporated when faced with the responsibilities and the Mechanical and Manufacturing Department was required to underwrite the venture from its own resources.

Lesson 3. Do not expect any help from the institution when the going gets tough.

Lesson 4. The venture must have positive support from the managers of the institution.

Determination has now resulted in the commercial activities as being given the same status as a research unit and the demands of the commercial activities on the central financial system have brought into being a new set of Polytechnic accounting procedures for 'trading'.

The initial limited view of the venture was soon seen as being capable of extension. The structure for commercial interaction was in place and within a year other activities came within its organisation as a matter of convenience for the staff and the department. Calibration services for BS5750, materials testing, product testing and industrial consultancy. These activities contributed to the revenue since October 1991 when the manufacturing recession took its toll.

In such a commercial venture there is no funding organisation to offer grants to help with overheads, no departmental or faculty funds would be made available to tide it over the 'bad times' as may be the case with research units. A commercial unit lives or dies on its profitability.

Lesson 5. No favours! Every undertaking must be entered into on a business basis.

This is foreign to many in the lecturing profession but it has been found that commerce respects and expects this outlook. A business-like approach will achieve more and create more opportunities than working for love or interest in the topic. Any services, supplied to the department by the commercial unit, for the benefit of students labwork, assignment studies, projects etc. are also costed at the same rates as commercial services. Although money may not change hands the degree of support that the department derives from the commercial activity should be seen in cash terms.

2. Excellence or Bust

As soon as a higher education institution steps out into the business environment in research or commercial trading, excellence is the objective. The institution enters a world where quality and competence are essential if success is to be achieved. There can be no hiding behind repetitive student exercises, limited facilities or excessive time resulting from incompetence. Consultancy work taken on with commercial objectives is particularly demanding. It is no good saying the software almost works or the lecturer had to go and present a paper.

Lesson 6. You are ' The real world'.

The 'real world' requires viable solutions not blue skies dreamings, the practical not the ethereal. This disciplined approach concentrates the mind and has been a breath of fresh, but cool, air to those who are involved. Reality bursts in with its demands and our competence rises to meet the challenge.

Lesson 7. What actually do you have to sell which is of commercial excellence?

The question reveals our weaknesses more than our strengths. The objective should be no less than excellence in two areas and they are interdependent:-
- The industrial platform
- The educational provision

3. The Industrial platform

The unit must have all the features that would be expected by an enquiring customer.
- All the normal paperwork systems, delivery notes, invoicing, record keeping, order status files etc. In general the infrastructure of a small company.
- Permanent full time staff.
- A recognisable centre of operations to which to apply within normal working hours.
- An identifiable paperwork style for reports, certificates of conformance, delivery notes, quotations, letters etc.
- A specific point for collection and delivery of goods.
- Intelligence networks to keep in touch with business opportunities and price variations.
- Costing structure to enable the evaluation of the return on business and identify costs.

It can be argued that most departments have two platforms of operation already in place. These are firstly research and consultancy and secondly teaching. Each platform has its own organisation, its own funding and in many ways they live their own lives and they fit where they touch. I am suggesting that yet another platform may be put in place and operate to the benefit of the department. This is commercial activity. It would generate its own income, have its own organisational structure and seek to provide facilities and liquid cash for educational enhancement. It would be able to handle all activities except those funded by the SERC or other grant awarding bodies. What is a consultancy if it is not a business deal?

Lesson 8. Academic staff will go for the easy option.

It has become apparent that the rules imposed by the Polytechnic on doing consultancy work cause significant overheads to the staff involved. For example, prior approval, cost estimates, overhead calculations, equipment and computing charges etc. The process may be simplified. We agree a price, the staff get paid on a signature and the overheads are handled globally within the accounts to a fixed ratio. Simplicity of operation is the key. The full time staff of the commercial unit coordinate the activities.

A typical case is where a company approaches the department for paid assistance in some area of work. The enquiry is automatically routed to the commercial unit and it is dealt with in a business-like manner. Appropriate personnel both inside and outside the Polytechnic are approached about the work. Its feasibility and required delivery are discussed. Consultancy fees are agreed, other charges are added and a formal quotation is sent to the client. If an order is received the organisation is handled by the commercial unit. No order-no work! Currently we are able to offer expertise in :-
- Horizontal machining centre work
- Calibration to line, form and angle for BS5750 requirements
- Materials testing
- Microscopy including electron beam
- Welding certification
- Prototype machining
- Product testing and certification
- Specialist design projects.

This may be seen as a fairly ordinary set of facilities to offer in a quite ordinary way but it has made a significant contribution to the handling of enquiries. The fact that some twenty five companies now depend on us for their calibration needs, and that two companies accepted us as sole sources for some of their key manufactured products indicates the success of the approach.

Lesson 9: Credibility opens doors.

It has been found that success in one activity demonstrates the confidence of one commercial customer, and leads to the transfer of that confidence into other activities. This opens doors which have some commercial implications. This does not just mean more manufacturing and related work; it means more teaching companies, more industrial student placements, more companies prepared to offer software, more companies prepared to offer equipment on loan, more consultancies, and more industrial contact in general. All this because you have been able to demonstrate an ability to deliver.

Lesson 10: Expectations are high.

It takes time to establish an adequate system with trained and competent staff, and to put in place procedures within the finance office of the institution and the department. It is estimated that some two years is necessary. The expectation is that the system will start to deliver profits in the first year and all the associated spin-offs. It must be remembered the department is building a business.

4. The Educational Base

The mainspring of the venture at the start was to provide equipment and resources for the educational need of the department. Unless the industrial centre fulfils this objective it may be considered to have failed. Unlike research, which seems to be mainly motivated by the ambition of staff to build a CV. Commercial work seems not to have this clout. Even consultancies dealing with complex issues requiring answers, not papers, fall into a second class category.
The educational objective may be defined as:-
a) provide the best equipment,
b) place it in its correct environment,
c) expose as many students as possible to the experience,
d) show it warts and all.

Lesson 11: You cannot duplicate industrial experience on a desk top.

There is a continuing discussion about purchasing policy. Some companies are in the business of providing educational equipment to demonstrate industrial products, desktop industry. If there is no intention to use the equipment for real, or the utilisation is so low that the investment can not be justified, this is a route which may be taken.
The policy decision to purchase industrial or desktop manufacturing equipment may be resolved in a department with a commercial section. The purchased equipment needs to be industrially capable and the planned utilisation demolishes the desktop argument. The projected industrial use ensures a return on the capital and a capability to demonstrate the full range of its facilities on real work. With the inadequacy of funding equipment purchases have now to be justified on two counts, eduction and earning potential. The amount of equipment in laboratories throughout the land waiting for next year's group to run the same tired demonstration is incredible. For the best learning situation the equipment must be seen in its natural surroundings doing the things for which it was intended.

Lesson 12: Reality brings understanding and destroys the magic world.

It is important that student engineers should see things as they are. The academic exercises carefully crafted to avoid embarrassing difficulties lead students to have a rose coloured view of life. Such exercises give rise to the view that skill is a low priority and that people who are good with their hands make things. That components always fit properly, that temperature is not important or that leakage is insignificant, the world is perfect. One contribution that a commercial centre can make is an injection of this reality. Students may be brought face to face with genuine situations.
There is a magic world in the minds of students which assumes that life is easy and that all you have to do is press the correct button, normally on a computer. A dose of reality is of vital importance. Let's give credit to those who can produce results in the face of reality.

Lesson 13: The direction of resource flow can change

Are departments honest in their evaluation of research? I remain convinced that resources flow from the student body into the research effort. This is especially true in cases where the institution has not received any money over and above the research grant to help support the system. Technician help, time out for staff, materials "borrowed", machinery and equipment permanently on loan. On the other hand it is reasonable to expect that the commercial activity should be a net contributor to the student resource and be expected to provide liquid cash. During the last financial year two major items of equipment were purchased for the department, a Computer Measuring Machine (CMM) and a prototype injection mould making facility. As the equipment was purchased from earned money we were also able to recover the outgoing VAT.

Other direct contacts with students include:-
 a) Handling student consultancies, paid reports etc.
 b) Excellent material for quality studies and statistical process control projects.
 c) Paid employment on a part time basis.
 d) Costing data for assignments.
 e) Management experience.
 f) Good source of industrially relevant project titles.
 g) Commercialisation of student projects especially at MSC. level.
 h) Business intelligence on jobs and opportunities.
 i) Instruction by industrially capable operators and technicians.

5. The Future

Opportunities seem to continually present themselves. We already have MSc. students working on a feasibility study on automating the assembly of a teaching company product. Who knows where that may lead. Especially as another teaching company is likely to be using the same technology on the development of its products.

The Polytechnic has announced a scheme to offer interest-free loans for departmental use. This, coupled with some innovative product design, could lead to genuine business experience for students. The department is currently actively involved in the development of two new products which it intends to manufacture and sell. The involvement in a genuine product cycle with marketing and control has great potential for the students.

There seems to be one major growth area at present. This is certification. Companies want to have third party certificates of quality and performance. We predict that this aspect of the work will continue to expand as almost weekly we are receiving enquiries. The selection of the correct staff is therefore of vital importance. They should be energetic, flexible and capable of operating at a professional level and willing to work on practical assignments.

The commercial platform may eventually find applications as yet not visualised but with the continuing stress on revenue earning it is one which must not be ignored. It is likely that the purchase and servicing of future equipment will have to be justified on two grounds, commercial application and student use. Student use on its own may not be sufficient.

SECTION 4: ENVIRONMENTAL ENGINEERING

Environmental Education in Engineering Courses

R. Van Der Vorst, F. Schmid

Department of Manufacturing and Engineering Systems, Brunel University

Abstract
In this paper the authors give a critical review of existing courses in environmental engineering, in Britain and one in Germany. Furthermore the development of a new course at Brunel University is described.
Keywords: Teaching Methods, Course, Engineering, Environment.

1   Introduction

The importance of the interaction between mankind and environment was recognised by Papanek (1974), and the economist Schumacher (1973). These authors and others pointed out that infinite expansion of economy and industry, based on the use of finite resources, was a physical impossibility.

Over the last five years environmental consciousness has developed throughout society. Today words like "green" and "alternative" are popular, and there is a general understanding that environment is more than just the surroundings of the town in which one is living. Environmental protection is understood as both a task for everybody in her or his day to day life, and the product of an environment oriented industrial management. With the common European market, and the introduction of environmental legislation industry will be coerced into acknowledging the necessity of change. Every company has to rethink its attitude towards the environment. The evaluation of the environmental friendliness of their products and production processes should become a corporate ambition. Industry will be led by economic forces to abandon what is potentially harmful to the environment.

To cope with these new requirements, and to respond to the increasing relevance of environmental problems, engineers with a broader education and a special knowledge of environmental aspects will be needed. The "new" engineer needs a basic knowledge of existing technologies and the skills to evaluate the advantages and disadvantages of technical solutions to both the natural and man-made environments. Engineers for tomorrow should also be aware of themselves and the influence of their possible role on the environment. Their understanding and their evaluation skills will take the possible desired and undesired effects into account.

Universities and other Higher Education Institutions have reacted to the already increasing demand for environmental engineers and have begun to design new courses. This paper contains a general overview and a brief evaluation of different existing courses in environmental engineering. Furthermore it leads through the process of developing an environmental engineering course established by the Department of Manufacturing and Engineering Systems at Brunel University.

2   Demands on Environmental Engineering Education

> *"The future engineer must study science and their application, resources and their conversion, and man and his needs."*   Encyclopaedia Britannica

Environmental problems demand more than just scientific solutions. If a scientific solution is to be implemented in society, then the solution must reflect the daily activities of people in their environment both for the present and in the longer term. Pure "Science cannot produce ideas by which we could live. Even the greatest ideas of science are nothing more than working hypotheses, useful for purposes of special research but completely inapplicable to the conduct of our lives to the interpretation of the world." (Schumacher, 1973, p.78). Therefore environmental education, the education for effective problem solving should be interdisciplinary. Subjects like ethics, law, psychology, management, and management of change will be suitable subjects to broaden the education and to teach working with values, ideas and theories.

The further important demand on environmental education is the education of 'whole men' and 'whole women'. To use Schumacher's phrase: "Education can help us only if it produces 'whole men'. The truly educated man is not a man who knows a bit of everything, not even the man who knows all the details of all subjects (if such a thing were possible): the 'whole man', in fact, may have little detailed knowledge of facts and theories, ... *but he will be truly in touch with his centre.* He will not be in doubt about his basic convictions, about his view on the meaning and purpose of his life. He may not be able to explain these matters in words, but the conduct of his life will show a certain sureness of touch which stems from his inner clarity." (Schumacher, 1973, p.85-86) Furthermore Schumacher writes "The true problems of living ... are always problems of overcoming or reconciling opposites. They are divergent problems and have no solution in the ordinary sense of the world. They demand of a man not merely the employment of his reasoning powers but the commitment of his whole personality." (ditto, p.89) A 'whole person', a person with their own personality will be able to evaluate the different effects on the environment, results of her or his work, and will come to decisions which are unique. These, for instance, can be decisions to reach cleaner, more social and environmental friendlier ways of working and producing.

To satisfy these demands for environmental education, courses of study offered in Higher Education should have an interdisciplinary aspect and a structure which allows personality education. Development of personality can be supported through alternative teaching methods. These include attitude teaching, but also student centred approaches like student activity learning in projects and discussion groups.[1] Attitude teaching, and teaching to evaluate should help the students to find their 'Leitbild', as the Germans say, their guiding image, in accordance with which young people can try to form and educate themselves. (Schumacher, 1973, p.90)

In the following I will look at different environmental engineering courses in Britain. For comparison purposes a German course will also be described. This survey was done in the process of designing the SEE-course at Brunel University which will be described later.

3   Environmental Engineering Education in Existing Courses

The first environmental course in the United Kingdom was started in 1965 by the Department of Mechanical Engineering at the University of Strathclyde. It is called Environmental Engineering. Since that time other Higher Education Institutions have followed Strathclyde's lead. At present, about some 20 "environmental" courses are in existence, they differ in intention, orientation, student intake and entry requirements. Four different categories of courses presently exist:
- environmental engineering,
- environmental technology,
- environmental science, and
- environmental studies.

The orientation of the courses seems to reflect the department offering the course. Engineering courses are presently based in departments of chemical engineering, civil engineering or mechanical engineering. Other departments involved in such courses are science, as biology, geology, chemistry and physics.

Five different types of undergraduate environmental courses have been examined, offered at Universities and Polytechnics in the UK. Furthermore the author looked at a German further education course offered at the RWTH-Aachen designed for students of different faculties and employees of different branches of industry.

The British courses cover a wide range of subjects. They differ in their orientation on environmental matters and entry requirements. All courses ask for A-level in mathematics and two other subjects relevant to engineering (they can be made up with AS-levels). The grade requirements are between pass and three Bs. However, the course structure is more or less the same for all courses. Lectures in the appropriate subjects provide the knowledge base. In seminars and through projects, the knowledge gained is applied. Subjects like Mathematics and Computing are generally part of a compulsory block. There is a large number of courses for the Department specialisation. The subject "Environmental Legislation" is considered to be an important area for all environmental engineers. However, this subject is the only one of the courses belonging to a discipline other than engineering. The courses are well structured, but don't seem to offer free space for students' self-development.

---

[1] Based on Gagné (1977) teaching models for special purposes can be chosen out of a range described by Joyce and Weil (see also Yorke (1981)). Attitude teaching refers to student activity learning and group work (Jaques (1984) and Hill (1969)).

	orientation	A or As-level subjects	grades	course
A	chemistry	maths chemistry (AS)	passes	4 Year or 3 Year
B	energy control	entry to 2nd year: maths or physics	BBC	4 Year
C	env. studies	-	-	3 Year
D	civil eng.	2 passes A-level incl. maths	CCD	3 Year
E	geology	maths and two relevant	passes	3 Year

Table: Five Courses (A-E) in Environmental Education

The courses are designed to provide engineering education with a special emphasis on environmental matters. As far as achievements are measurable, the courses are successful, and graduate students find jobs in different branches of industry and the service economy.

The German course is different in both structure and purpose. It is designed for part-time study requiring at least four different modules taught for eight hours a week each semester. Attendance in each module is credited and successful completion of a project and passing of the final exam lead to the degree of "Umweltingenieur". The course is organised by a committee formed of representatives from all Faculties of the University; Engineering Faculties as well as Faculties of Medical Science, Human Science and Science are involved. The conception of the course is interdisciplinary in nature. The average age of the participants on the course is much higher than on the courses observed in the UK. The structure of the German course builds on lectures, seminars and a few group excursions. Substantial involvement of the participants is expected in the discussions after every lecture and during the seminars and excursions. The discussions benefit from the different backgrounds, experience and interests of the participants.

The German course is thus not readily comparable with the UK undergraduate courses in environmental engineering. Many of the differences between it and the British courses are characteristic of the differences between British and German higher education in general. The consideration of this German course is worthwhile though as it provides the interdisciplinary aspect which the authors consider as essential for a successful environmental education.

The range of UK undergraduate courses allows every potential student to find her or his most appropriate course. In designing a new environmental engineering course, intended to satisfy the demands of environmental education at a fundamental level, elements should be chosen from all the courses examined. Among these are interdisciplinary education, space for 'activity' based learning, for group discussions, evaluation, philosophical contemplation etc.. As the education process is aimed at young people of the age of 18+, the new course should provide learning exercises and opportunities to help the student to develop critical judgement and an ability to evaluate situations, in short, to develop their personality.

The Department of Manufacturing and Engineering Systems at Brunel University is currently designing an undergraduate course in environmental engineering which should satisfy the requisites of both personal development and the demand on interdisciplinarity.

4 Description of the SEE course under development at Brunel University

The new environmental engineering course, Special Environmental Engineering (SEE), at Brunel University is aimed at developing a kind of engineer whose focus is on creating clean and lean, effective and efficient, and humanly responsive environmental systems. It will run as a full-time and thick sandwich course and treat engineering not as a narrow discipline, but as a broad based activity appropriately reflecting human concerns of individuals, society and culture. The course will consist not only of the technology and management of environmental systems but also the ethical and legislative considerations underpinning their design.

The course is intended to meet the demand for broad based and well educated engineers. The development of the student's personality is part of the undergraduate course. Alternative teaching methods will be applied. 'Learning Through Discussion' and other kinds of group work carried and controlled by the students will be the major teaching models used for the integrated environmental subjects.

The course covers basic engineering subjects and provides basic knowledge in environmental matters. The engineering and science related environmental subjects are taught in an integrated way with subjects more related to human sciences, such as ethics, law, sociology and politics to enable students to make and take future decisions in full responsibility.

Built in the course there are different methods to help the student in developing her or his personality. Teaching and learning models as suggested in literature and tested in different classroom situations will be applied in the education process for the SEE course. Part of the learning process is student centred and led by the students. The syllabuses of integrated block of the course will describe the area which should be covered, while the route through the learning process in those subjects will be structured by the students. Personality teaching requires competent teachers who are able to guide the students through their learning.

A case in point is the Environmental Principles module. The primary aim of this course is to introduce the student to the major environmental issues and to an examination of the responses which are appropriate to the environmental problems identified. Firstly, the course is concerned with presenting factual information, secondly, the course is devoted to the consideration of the values germane to the issues involved. The two aspects of the course will be treated in an integrated way. The learning process will be based on case studies and group discussions. The students' work will be guided by staff.

The next few years will show whether the implementation of the theoretical ideas into the SEE course is successful. The evaluation of the course and its achievements on personality development will not be an easy task. Possible ways will be an evaluation by the students or a course evaluation through the assessment of the students' achievements. In the light of experiences the course will be modified to ensure that its aim can be reached and the students will develop their personality throughout the course. So that they may gain a general understanding to enable them to work effectively on engineering problems related to the environment.

5   Conclusions

The authors have identified the support of the students' personal development and the provision of interdisciplinary experience as the main demands on environmental engineering education. Several British environmental engineering courses and one from Germany have been examined for ability fulfilling the demand created by environmental education. The "Special Environmental Engineering" course at Brunel is designed to answer these demands. The course will be taught interdisciplinary. Furthermore, as described in the last chapter, alternative teaching methods based on learning and teaching theories will be employed. The next years will show whether the implementation of "student centred activities" and similar models of learning and teaching are successful.

6   References

Gagné, R.M. (1977) *The Conditions of Learning*, Holt, Rinehart and Winston, New York.
Hill, W.F. (1969) *Learning thru Discussion*, Sage Publication, Beverly Hill.
Jaques, D. (1984) *Learning in Groups*, Croom Helm, London.
Joyce, B., Weil, M. (1992) *Models of Teaching*, Allyn and Bacon, Boston.
Papanek, V. (1974) *Design for the Real World*, Thames and Hudson, London.
Schumacher,E.F. (1973) *small is beautiful*, Blond and Briggs Ltd., London.
Yorke, D.M. (1981) *Patterns of Teaching*, Council for Educational Technology, London.

Environmental Engineering - Bridging the Educational Divide

D.J. Blackwood, S. Sarkar
Department of Civil Engineering, Surveying and Building, Dundee Institute of Technology

Abstract
This paper describes an approach to environmental engineering education at undergraduate level which is concerned with the development, in students, of the ability to develop holistic strategies to the prevention and solution of environmental problems. The paper suggests that there are barriers within higher education in the United Kingdom, particularly the traditional educational divide between science and engineering, which hinder such an approach and describes a new programme of undergraduate study at Dundee Institute of Technology which is intended to bridge the divide.
Keywords: Environmental Technology, Environmental Engineering, Environmental Education, Curriculum Development.

1 Introduction

Over the past twenty years advanced Industrial societies have recognised more than ever before that the products of their development are fundamentally damaging both the natural and man-made environments and that the damage is reaching serious proportions. Public awareness of environmental issues is now international (Lowe et al 1985, ICIHI 1986, World Rainforest Movement 1990); and the international community openly supports the demand for solutions to existing problems and a proactive approach to future development of societies in an environmentally conscious way. More specifically, in the European Community several directives have been issued, or are being prepared, relating to environmental matters; and these together with associated United Kingdom legislation have brought to focus the importance of environmental pollution control as factors in commercial and industrial decision making, and in the establishment of public policies. It is essential that professional engineers are involved in this decision making process and it is therefore important that due consideration is given to environmental issues in the education and training of engineers. This has been recognised in the United Kingdom by the Engineering Council whose 1990 assembly stated that they would "like to bring environmental engineering issues into focus in schools, further education and higher education". One issue which must be addressed is the identification of the most appropriate form of environmental education, and this paper considers this in the context of undergraduate engineering education in the United Kingdom.

2 Requirements of Environmental Education

It is necessary to consider at a fundamental level the objectives of environmental education. In its broadest sense, Forbes (1987) describes the objective of environmental education as being to enable full public participation in the determination of policy on the major environmental issues. However society at large will always require proper guidance and accurate information from specialists and the engineering professions must play an important role in the provision of such information. This gives rise to a need for different levels of environmental education within society and consequently within the engineering professions. Stokes and Cranshaw (1986) proposed four groups which must exist within tertiary environmental education, the four major categories being:

    the lay group who would be concerned with developing attitudes and opinions on environmental issues (as described by Forbes);
    the technical group who would be concerned with measuring environmental parameters;
    the subject specialist group who would be concerned with a particular discipline and the environmental issues associated with that discipline; and,
    the management group who would be concerned with the identification of solutions to complex inter-disciplinary environmental issues.

Additionally, the need for a multi-disciplinary approach to the identification of, and the development of solutions to environmental problems has been widely discussed (Gloyna et al 1988, Tomlinson 1989).

3 Current Undergraduate Course Provision in Environmental Education

It is apparent then that there is a need for, firstly, a range of strategies for the introduction of environmental issues to education within the various disciplines and secondly, a multi-disciplinary approach to environmental education. Whilst there is evidence of the introduction of environmental issues to existing course curriculum, there is little evidence of a concerted effort towards a broader education of the various professions involved in environment related matters. They have generally been drawn from a broad spectrum of disciplines including engineers, chemists, biologists, physicists, architects, builders and planners. Traditionally, the education of these individual professions would produce graduates who are heavily biased to either:

    an "engineering" approach to the environment, being concerned with the design and operation of systems which can be employed to control environmental processes; or,
    a "scientific" approach to the environment, being concerned with an understanding of the processes and organisms in the environment

which cause environmental damage.
The separation of engineering and science faculties in higher education gives rise to this educational divide and a consequence of this is a lack of development of personnel equipped with the necessary skills to adopt the required holistic approach to the solution of environmental problems. Furthermore, although there has been a rapid expansion in the number of undergraduate courses in environment related subjects these generally reflect and continue to carry the schism between engineering and science. A recent guide to academic, professional and vocational courses related to the environment (Environmental Council 1992) identifies fifteen categories of courses, three of which contain the majority of engineering and science courses considered to be relevant to the environment. Of the engineering courses, eighty six were classified in a category entitled "Industry and the Environment" with most of these being Civil Engineering orientated. It is significant that few mechanical or electrical engineering courses appear in the guide. The majority of the science based courses was included in a category entitled "environmental studies" which contained one hundred and fifty three courses, mainly environmental and applied sciences and ecolgy. It is particularly significant that the entrance qualifications for the two categories of courses are in line with the normal entrance qualifications for science and engineering courses which suggest that the environmental issues are introduced to the courses through a change in emphasis within existing course provision rather than a radical review of course content and course philosophy. The third category of courses entitled "Environmental Engineering" list courses which are directly concerned with the application of engineering to the environment. Significantly there are only twelve courses listed in this category and, although there are known omissions from this list, the total number of such courses is very small. It would appear that the growth of environmental courses has further polarised engineering and scientific education. It can therefore be concluded that environmental engineering education is being provided at undergraduate level by introducing environmental issues to curriculum within the previously polarised specialist departments within Tertiary Education.
    In terms of Stoke's four categories of environmental education, the emphasis has been on the technical and subject specialist groups rather than the management group, and it is essential that due regard is given to the management, or multi-disciplinary approach to environmental engineering.

4   Environmental Management from an Engineering Perspective.

The overall objective of an environmental engineering management course must be the development, in students, of the ability to synthesise holistic strategies for the prevention and solution of environmental problems. The overall concept of a multi-disciplinary approach to environmental education is however constrained by the impracticable breadth of knowledge which would be required and the consequential lack of depth of understanding of the issues. It is therefore essential that a programme of study of environmental management would have to limit the overall breadth by an element of specialisation in either engineering or scientific study. Nevertheless whilst the course must be firmly based in one discipline, such as engineering, it would have to be radically different in content and philosophy from traditional engineering courses. The remainder of this paper describes the BSc Honours degree course in Environmental Technology at Dundee Institute of Technology as an example of such a radical approach.

The title Environmental Technology was considered to be appropriate as it best reflected the course content and objectives. There are many definitions of engineering but these are generally concerned with the production and operation of "machines and works of public utility" and this, in an environmental context, was not considered to be sufficiently broad. Technology however can be defined as the systematic application of engineering, science and management, which better describes the course philosophy and content. The Environmental Technologist can be considered to be the "general practitioner" of environmental issues being equipped with sufficient knowledge and expertise to lead multi-disciplinary teams in the analysis of environmental problems, the synthesis of solutions to these problems, and the installation and operation of these solutions as well as being able to communicate with, and demonstrate the importance of environmental issues to the public at large. The environmental technologist must also be able to communicate on a knowledgeable basis with engineers and other specialists, who would be involved in developing the details of the solutions to environmental problems. The Environmental Technology course is therefore engineering based and is underpinned by studies on the man-made environment being based in the department of civil engineering, surveying and building with significant inputs from staff of the other departments in the Faculty of Engineering and Construction. The total input from this faculty amounts to approximately 65% of the course content. This course differs from other engineering courses in the faculty because there is much more emphasis on the natural environment and its interface with the built environment and therefore there is a significant contribution from the department of molecular and life sciences amounting to some 25% of the course content. The analytical organisational and communications activities of the students are further enhanced by important contributions from the departments of mathematical and computer sciences, accountancy and economics and business studies.

The diverse nature of the course could present problems in establishing appropriate entrance qualifications. Fortunately the course is based on the Scottish pattern of four years full time study to honours level and therefore the first year has been designed as a foundation year which will introduce students to environmental issues through the two environmental units and ensure that the students have a common base of knowledge and skills for future years of the course through science foundation units, mathematics and statistics units, land surveying, materials, communications and information technology units. This course structure enabled the normally restrictive entrance qualifications associated with engineering courses to be broadened, the normal entrance qualifications for this course being five SCE passes including three at higher levels with mathematics and english being included at either grade.

In the second year of the course the basic knowledge and skills from year one are further developed with greater emphasis on environmental issues, through subjects such as environmental science, geomorphology and atmospheric systems. The technological aspects are developed in environmental engineering units and in a measurement and data collection units, which will also serve as an introduction to environmental management and energy management in years three and four. Students are also introduced this year to building science and services, and studies on the built environment will continue, thus ensuring that students will be able to make an informed choice of option subjects in years three and four.

The third and fourth years, which will operate as a continuum, are designed to consolidate the previous studies and to introduce an element of specialisation which would be related to the student's future career opportunities. An Environmental Management unit addresses important issues such as environmental impact analysis, environmental audits and, together with project management, will serve as a central focus of the studies and will integrate the skills and knowledge gained by students in the optional subjects. The need for graduates with a wide range of knowledge and abilities was discussed earlier in this paper but this must be associated with the ability to demonstrate an in-depth understanding of specific subjects. This has been accommodated in the course by allowing the students to select two major options and two minor options. The option subjects have themselves been designed to allow students to select complementary groups of subjects appropriate to the students' possible future career choices. For example a student who wishes to pursue a career related to the water industry might select major and minor combinations of Water Engineering, Pollution Technology, Environmental Biotechnology and Geotechnology. Alternatively, a student wishing to pursue a career related to the built environment might select major and minor combinations of Built Environment, Building Science and Services, Transportation and Geotechnology.

The course is strongly vocational and an optional sandwich year of supervised work experience is offered between the second and third years of academic study. Additionally, the course contains industry orientated group projects. In a third year group project students will assume the role consultants and will be required to develop a strategy for the solution of an environmental problem, selected from a range of topics proposed by members of the academic staff. Students will be required to present their findings during seminar sessions in the final stages of the project. A similar approach will also be adopted to energy management where the project will be case study based.

An individual project in year four is intended to allow students a period of sustained independent study at a high level on a particular aspect of Environmental Technology. This will take place substantially during term three of year four, and on completion of all other subjects. A secondary function of the individual project therefore will provide a period of active reflection on the taught subjects.

To summarise, the overall objective of the course is to enable students to develop an holistic view of environmental engineering, and through a combination of the breadth and depth of study in the course it is anticipated that graduates should be ideally suited to eventually lead and manage multi-disciplinary teams on environment related project. The graduates will be of value to a range of organisations including: environmental consultants, waste disposal contractors, local and central government departments, water companies and national rivers authorities.

5 Conclusions

This paper has identified a need to expand the course portfolio in environmental engineering education in an attempt to bridge the traditional educational divide between science and engineering. The Environmental Technology course at Dundee Institute of Technology is an example of an educational approach which should produce graduates with the necessary holistic view of the environment. It is difficult to predict the acceptability of the output from the course but the course team is very alive to the changing geo-political circumstances and are willing to adapt and fine tune the course as necessary. They are therefore confident that the course is a step in the right direction and that graduates will meet the expected needs and demands of industry and society. As a postscript, the environment is our most precious commodity - let us all preserve it; this course is a small contribution to that aim which we hope will be replicated or bettered elsewhere.

6 References

Environmental Council (1992), **Directory of Environmental Courses**, Environmental Council. London.

Forbes J (1987), Environmental education - implication for public policy, **The Environmentalist, 7, No 2, 131-142**.

Gloyna, E F & Sorber A S (1988), A look to the future: environmental education, **Journal of the Water Pollution Control Federation, 60, Pt 7, 1193-8**.

Independent Commission on International Humanitarian Issues (1986), **The Vanishing Forest**, Malayan Nature Society, Kuala Lumpur.

Lowe P & Godyer J (1986) **Environmental Groups in Politics**, Allen & Unwin, London.

Stokes D & Cranshaw B (1986), Teaching Strategies for Environmental Education, **The Environmentalist, 6, No 1, 35-43**.

Tomlinson P (1989), Environmental Statements: Guidance for Review and Audit, **The Planner, 3rd November 1989**.

World Rainforest Movement (1990), **Rainforest Destruction: Causes, Effects & False Solutions**, World Rainforest Movement, Penang.

Impact of Environmental Issues on Engineering Education

T.V. Duggan (*), K. McIvor (**), M.R.I. Purvis (***)
(*) Faculty of Engineering, The University of Portsmouth
(**) Environmental Education, Department of Chemical Engineering, The University of Queensland
(***) Environmental Engineering Programme, School of Systems Engineering, The University of Portsmouth

Abstract
Environmental issues are increasing in significance in the curricula of engineering courses in response to the needs of industry. Specialist environmental engineering courses are characterised by broad based curricula and a multi-disciplinary approach to courses development. These aspects are considered from examples of new Environmental Engineering courses offered at The University of Queensland and The University of Portsmouth. The attributes of network approaches are discussed as a means to integrate subject areas and to synergise complementary strengths across Departments and Institutions.
Keywords: Education, Engineering, Environment, Networks.

1 Introduction

Problems arising from material usage, high energy consumption, water quality and waste generation are the focus of a multiplicity of actions generated by Governments, International Agencies and Industry. Such actions reflect growing public concern on environmental issues and include the Environmental Protection Act (UK 1990), Directives from the European Community on resource sustainability and pollution control, statements on Third World expectations and resources from the United Nations and industrial strategies designed to assess and ameliorate the environmental impact of business practice. A consequence is pressure on the engineering education services of many countries to react to national environmental problems and to provide positive contributions toward the care and protection of the planet.

The Engineering Professional Associations of both the United Kingdom and Australia have taken a lead in formulating environmental codes of practice. Engineers and the Environment was the theme of the 1990 UK Engineering Council Assembly Conference. The conference made the recommendations that all engineers should:

- familiarise themselves with the current environmental issues.
- contribute in a positive way to debate on these issues.
- propose and implement sounding engineering solutions to safeguard the future.

Inputs to engineering courses to promote general awareness of environmental issues is one extreme of educational provision. The other extreme is the development of specialised curricula to allow the training of well rounded professional environmental engineers. The key issue is that environmental problems are wide ranging, complex and interactive. Thus specialised course curricula should accommodate a diversity of subjects concerning themes of engineering science, technology and management. In most cases this will involve collaboration and integration of expertise across Departments, Faculties and, perhaps, Institutions.

This paper presents for critical appraisal and comment examples of new courses development in environmental engineering at The University of Queensland, Australia and the University of Portsmouth, UK. Emphasis is placed on subject integration, international links and the value of network approaches as a means to share experiences and combine expertise for the solution of environmental problems.

2 Environmental Education Provision at The University of Queensland

The Department of Chemical Engineering at The University of Queensland has been funded by the Australian Government to develop courses in Environmental Engineering and Management as part of a national commitment to sustainable development and environmental research. The Department acts as a focus for integrated environmental education provision within the University of Queensland and deals with all matters concerning courses development, promotion and marketing. Table 1 gives a selection of the environmentally related subjects offered by Departments at The University of Queensland.

A courses review of all subjects offered by the University of Queensland demonstrated that appropriate groupings of subjects combined with new environmental management elective option subjects would allow the development of a high level programme of courses provision in the area of Environmental Engineering (Van Zeeland, Krol and Greenfield 1991). Collaborative arrangements were formalised between Departments and Campuses to make the interdisciplinary elective subjects available to any engineering or science degree course. The titles of the elective subjects are:

1. Elements of process Interaction.

2. Principles of Waste Management

3. Environment: Regulatory and legal issues.

4. Social Impact Assessment and Occupational Health.

5. Environmental Impact Assessment Project.

A Division of Environmental Engineering was established in the Faculty of Engineering in 1991 to support these initiatives and to promote courses developments in Environmental Engineering and Management. Details of these courses are contained in the paper by Van Zeeland, Krol and Greenfield (1991) and are summarised briefly as:

Bachelor of Environmental Engineering

This is a four year semester based study programme designed to develop professional engineers with an understanding of the principles of sustainable development and the knowledge of how theses principles apply to industry. A key focus is the consideration of environmental and ecological issues within the planning, design, operation and management of industrial processes. The curriculum comprises a common first year with other Engineering Divisions, core subjects and a wide range of elective options. The course is supported by inputs from specialists from Government, Industry and environmental groups.

Master of Environmental Management, (MEM)

This comprises full or part time Courses for industrially based entrants whose employers require specialists in the multi-discipline area of environmental engineering. The curriculum consists of a set of three general environmental overview subjects; a set of management subjects including communication and negotiation skills; project, evaluation and decision analysis; specialised environmental management subjects including economics, occupational health and environmental impact assessments; an environmental management project; electives covering such areas as mining, ecology and natural resources management.

Table 1

Departments and Selected Environmental Subjects at The University of Queensland

Department	Subject
Agriculture	Soil and water conservation nature resources management, land rehabilitation
Anthropology and Sociology	Aboriginal Australia
Management	Social Issues
Botany	Algae and the environment. Applied ecology
Chemical Engineering	Environmental engineering. Industrial waste water systems, modelling of environmental systems
Mechanical Engineering	Engineering Acoustics
Mining and metallurgical engineering	Mining environments
Economics	Environmental Economics
Geography	Landscape and environmental systems environmental control, remote sensing
Journalism	News communication theory
Microbiology	Industrial and public health microbiology
Physics	Environmental physics - atmosphere and oceans
Zoology	Aquatic ecology, behaviour ecology
Civil Engineering	Public health engineering, coastal engineering, hydrology and ground water
Interdisciplinary subjects	Coral reef biology and geology, science and the environment, tourism

Master of Engineering and Technology Management (METM)

This course is complementary to the MEM programme and is designed to develop management skills for people involved in engineering and technology related areas. It aims to produce graduates who will assume leadership roles in industrial and public service organisations.

The Masters Degree Courses incorporate a research project jointly supervised by University staff and personnel from industry. Industrial and commercially based staff also make contributions to courses management and delivery. The part time Masters Degree may be completed in 2 or 3 years while the full time courses may be pursued in 12 or 18 months. Assessment is on the basis of credit points and this includes a credit transfer facility from other Institutions. For the convenience of industrial entrants subjects are incorporated into one semester segments.

The total range of elective subjects provided by the University network is impressive. In the case of the MEM course there are 33 subjects covering areas of environmental science, technology and management. For the final year of the various streams of the undergraduate programme in the Environment Division some 45 elective subjects are available.

Although newly established, the Bachelors and Masters Degree courses are proving to be very popular with students and industry. It is interesting to note that 55% of the Bachelors Degree second year cohort are women. This early indication suggests that Environmental Engineering may be a mechanism for attracting more young women into the engineering profession.

3 Environmental Education Provision at the Faculty of Engineering, University of Portsmouth

As part of an overall strategy to provide Environmental Engineering courses at BEng/MEng and postgraduate levels the Faculty of Engineering has developed curricula for Postgraduate Certificate, Diploma and Masters Degree Courses to be offered from January 1993.

The courses build on the existing research strengths of the two Faculty Schools, School of Systems Engineering and School of Civil Engineering, in areas of Energy Engineering, Systems Engineering, Materials Engineering, Manufacturing Technology, Water Treatment and Quality, Public Health Engineering and Instrumentation and Control. Part time and full time study routes are available. Study Periods vary from 15 weeks for the full time Certificate to 5 years for the part time Masters Degree.

The courses overall rationale, aims and objectives and transferable skills have been formulated following substantial consultations with industry. These subsume the requirements for the supply of well rounded professionals having high level expertise in environmental engineering and management. The courses are timely in relation to the needs of UK industry where new and impending legislation places obligations upon companies for their discharges to land, air and water sources. Industrially based staff will make significant contributions to the management and delivery of the courses in order to ensure a rapid response to industrial need and state of the art applications in environmental technology.

The courses are structured to attract delegates from a wide range of backgrounds, including industrial entrants for the part time courses. This is achieved by provision of 30 hour study period modules, each of which relates to a well defined area of environmental engineering. Flexibility to accommodate personal delegate needs is supplied by choice of modules and assessment by credit accumulation and transfer. The latter includes a facility for exemption from certain modules by certificated or uncertificated prior learning. Each delegate will devise a personal study plan in consultation with the Courses Management Team.

Modules are divided into sets of core modules and expert modules. The core modules provide the essential background for the study of the specialisms associated with the expert modules. There are eight core modules which must be studies by all delegates. Expert modules are divided into sets corresponding to Energy (E modules), Water (W modules) and Materials/Manufacture (M modules) theme areas. Each of the E, W, M, sets comprise four modules. Choice of expert modules will be increased by the introduction of Elective Option Modules (O modules) depending upon demand from delegates.

Delegates for the postgraduate certificate will be required to complete successfully the eight core modules. Delegates for the Diploma or Masters Degree will be required to study 14 modules. These will comprise the eight core modules and six of the expert modules. In the case of the expert modules delegates must choose 4 modules from an individual theme area (E or M or W Sets). The remaining two modules may be selected from other theme areas or from the Elective Option modules. Seven Elective Option modules are currently available and these are expected to increase in number as the courses develop.

Delegates for the Diploma and Masters Degree are required to pursue a two week assignment on the topic of Integrated Pollution control. Masters degree delegates are also required to undertake a 23 week equivalent study period project jointly supervised by University and Industrially based staff.

All modules are allocated a credit rating of 5 points. The Integrated Pollution Control assignment and Masters Degree project contribute 20 points and 50 points respectively to the delegates assessment portfolio.

The part time Master of Science Degree benefits from UK Government sponsorship under the terms of the Integrated Graduate Development Scheme (IGDS). This programme, won on a competitive bid and peer assessment basis, is designed to increase the effectiveness of UK graduates already in industrial employment. Substantial funding has been obtained to pump prime the course for a period of seven years.

Inputs to the modular programme from other Departments and Faculties are being sought and developed. These inputs will complement the expertise available within the Faculty and will include Environmental Law and Business Opportunities in Environmental Engineering from the Business School; climate modelling from the Department of Geography and Energy Utilisation and Building Design from the Faculty of Environmental Studies.

External networking arrangements are an important and perhaps unique feature of the courses. The National IGDS Scheme currently comprises 18 programmes covering many areas of engineering. Holders of IGDS sponsorships are encouraged to use modules from other IGDS programmes in their Postgraduates Courses. In principle some 300 alternative modules are available as elective option subjects depending upon demand from delegates. Discussions are taking place with representatives of an IGDS programme held by Brighton Polytechnic and Sussex University concerning provision of a module entitled 'Management of Technology for the Environment'.

Procedures for international collaboration have been incorporated into the courses by means of Partnership agreements with overseas Institutions. The development of these agreements have been aided by British Council and ERASMUS scheme funded visits undertaken over several years. The aim of the Partnership agreement is to present delegates with an opportunity to study modules and carry out project work overseas in order to experience different and shared responses to environmental problems. Delegates who accrue 30 credit points of equivalent study in a European Partnership Institution as part of the overall assessment package may claim the title of European Masters Degree in Environmental Engineering. Structured Partnership Agreements are in an advanced stage of negotiation with Ecole Nationale d'Ingenieurs de St Etienne, France and Universitat Gasamthochschule, Siegen, Germany. Other links are being developed with Institutions in Hong Kong, China, Egypt and the Philippines. A TEMPUS bid in Environmental Engineering, involving 14 Institutions in Eastern and Western Europe, has been submitted as a Joint European Project proposal. If successful this will add considerable impetus to the establishment of an international Environmental Engineering Education network.

4 Discussion

The impact of environmental issues on Engineering Education has been considered from the viewpoint of the breadth of material required to satisfy complex environmental problems and the associated need to integrate subjects from a range of engineering, science and management disciplines. The examples of courses development at The University of Portsmouth and The University of Queensland exhibit surprising similarities in view of the fact that they were developed independently taking into account national circumstances on opposite sides of the globe. For example both courses programmes are Government sponsored, comprise a module or segmented structure, subsume assessment by means of credit accumulation and transfer, include industrial involvement in the design and delivery of the courses and use network approaches to synergise expertise and complementary strengths in response to environmental issues. It is apparent that The University of Queensland has developed a highly sophisticated internal network to assist courses development. The University of Portsmouth, on the other hand, has placed emphasis on the establishment of links with other Institutions in the UK and overseas.

The success of external network arrangements relies on operational details concerning common modular programmes and credit accumulation and transfer. To maintain academic credibility of courses, equivalence of standards need to be demonstrated across Institutions and this will require compliance with the aims and objectives of module syllabuses. The logistics of staff and delegate transfer between Institutions and the timing of modules can cause difficulties whose solution requires goodwill generated by long established contact. However the efforts involved in creating such networks is seen as having substantial benefits to academe and industry throughout the international community.

5 Conclusions

There are significant pressures on the developers of engineering course curricula to include environmental issues in programmes of study. Specialist environmental engineering courses at both undergraduate and postgraduate level are broad based and require multidisciplinary approaches to satisfy the training requirement of industry. Examples of new courses development at The University of Queensland and The University of Portsmouth show similar features which demonstrate a reassuring common purpose in environmental engineering education provision across an international dimension. The value of network arrangements as a means to synergise complementary strengths in institutions has been stressed. The sharing of values and experiences on the environment are seen as desirable attributes of International environmental engineering networks. The operation of such schemes for collaboration in courses development relies on modular study routes, assessment by credit accumulation and transfer and an understanding of the qualities of Partnership Institutions.

Reference

Van Zeeland K, Krol A, Greenfield P (1991) Environmental Education: An Engineering Essential. East-West Congress on Engineering Education Cracow, Poland, 264-268.

Educating Engineers for Positive Environmental Action

D.J. Hardy

Department of Mechanical Engineering, Design and Manufacture, Manchester Polytechnic

Abstract
Environmental issues are now on political agendas. Engineering courses should contain a compulsory subject whose **main** educational objective is to demonstrate that technical solutions to environmental decay will probably fail unless engineers consider how they might influence the formation of legislation, exploit mechanisms of technology uptake by motivation and understand their own accountability. Manchester Polytechnic have recently introduced an Energy and Environment subject in years 1 and 2 on the BEng Mech Eng course.
<u>Keywords</u>: Environmental Action, Engineering education, Accountability, Motivation, Environmental Legislation.

1 Introduction

Industry is one of the major causes behind environmental degradation. "I doubt that the dynamo of industrial activity can be stopped, and if it could the time taken to reverse its action is too long to prevent environmental catastrophe". So said a senior manager from a major car manufacturer. (annon 1988)

Moreover, scientific evidence for the causes of environmental destruction is enough for the World Climate Programme to require that CO_2 emissions are reduced by 20% before 2005 (WCP 1988).

Higher education for engineers, across Europe, is strongly biased towards technical training. The **weakness** is that engineers become strongly tempted to find purely technical solutions to problems.

A compulsory subject should be included in all engineering degrees and diplomas to help students assess the impact of technology on the environment. The course should aim to demonstrate that technical advance alone will not reverse environmental decay. It is, for example, technically possible to use ethanol as a substitute for petrol or diesel in motor vehicles. Engineers are well able to solve the problem of reduced range and possible engine damage. Ethanol is, environmentally, a cleaner fuel, but it is unlikely that this alternative technology will become widely available unless the public are given sufficient motivation to use it. Will legislation help to encourage this fuel switch ? Is it right that a government should force this on its population ? Should food supplies (grain) be diverted for fuel ?

The pyramid (fig 1) illustrates foundation topics that help to govern the uptake of technology that is not immediately acceptable to the public. Each are discussed in order of strategic priority and introduced with specific educational objectives.

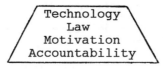

Figure 1

2 Accountability

Educational Objective : To encourage self examination. To reach personal conclusions as to whom individuals are accountable for preserving the environment.

Whose earth is it anyway ? The answer to this question will strategically determine our actions. Whilst every human being is given freewill to determine their ultimate destiny we are given the choice to believe that "The earth and everything in it belongs to the Lord." (psalm 24) Men and women are not just accountable to their grandchildren for the way they determine environmental recovery.

To illustrate that individuals can be personally accountable for environmental decay consider the poetry of Karl Marx, whose poem "Human Pride" (Rjazanov 1927) written as a student, dwells on an obsession to see the world ruined. Today, the economic collapse and environmental ruin of the Soviet Union ranks as a major disaster story.

3 Motivation

Educational Objective : To understand ways in which people can be motivated to positive action. Examine the relative effectiveness of market forces, government intervention, awareness programmes, incentive schemes.

Change best springs from the will. Enforcement may be necessary but choice has more lasting effects. The industrial 'dynamo' is driven by consumer demand. In an ideal world, motivation for choosing environmentally friendly products ought to arise from conscience, but in reality is more likely driven by financial incentives.

There is evidence, from market surveys, that whilst consumers feel inclined to go green they will not generally invest in environmentally friendly products because retail outlets do not encourage them to do so. (EEO 1990). Government, industry and business can provide legitimate motivation. The East Midlands Electricity Board recently chose to supply consumers with energy efficient lightbulbs as a business alternative to building a new substation. Financial returns are not always opposed to environmentally good decisions.

Reduction of tax on leaded petrol is an example where interventionist policies can be put to good effect for consumer and the environment.

Consumers have to be helped to make investments that are good for their pocket and the environment. Energy efficiency in the home, for example, always requires an initial expense. Government awareness programmes do little to motivate widespread investment unless financial incentives are offered.

4 Law

Educational Objective : To assess the effectiveness of recent environmental legislation throughout the UK and Europe. Appreciate how the costs of applying the law might be met. Understand how engineers can **influence** legislation that will encourage acceptance of new technology.

Where motivation fails to reverse environmental degradation, the law will have to provide a solution.

The Germans are often perceived as a nation committed to green issues. The policies of Wilhelm Brandt, a recent German Premier, no doubt accounts for this national reputation (Brandt 1983).

Despite perceptions that the German public are strongly motivated in green issues, the law still plays a very strong part in the German commitment to salvaging the environment. The government is showing strong determination to impose tight timetables and tough standards, that even industry and consumers fear they cannot meet, to deal with the unnecessary packaging of goods. From April 1 1992, for example, consumers will be allowed to dump packaging like the cardboard carton around toothpaste tubes at their local shop. This is proving a powerful irritant to shop keepers who say it is quite unrealistic, but they must do it because of the regulations.

Environmental law in the UK, in contrast, shows signs of caution, dither and compromise. Noises of concern have not necessarily turned into national intent.

The (UK) White Paper (HMSO 1990) was conceived as a radical blueprint for the future of the environment. In the event the Financial Times described the Act as a muted declaration of hesitant intent. (CEST 1990)

Critics of the Electricity Act (1988) have said that it has prejudiced the government against discussion of measures such as carbon tax. Supporters would argue that the Act, by action of market forces, encourages combined cycle gas turbine generation in favour of flue gas desulphurisation and that this solution not only cleans up emissions but is a more efficient generating solution.

The 1990 Environmental Protection Act and the 1974 Pollution Act set out a framework that requires industry to have authorisation from the Secretary of State before carrying out industrial processes that are recognised as polluting. Licences will be withheld from those whose processes, unnecessarily, pollute the environment. Nevertheless, in prescribing a legal framework to counter industrial pollution, the EPA devises an integrated pollution control (IPC) scheme which requires industry to find processes based on "best available techniques not entailing excessive costs" (BATNEEC) (EPA 1990). It is obvious that, if protecting the environment does entail a high cost, the EPA will not be a sufficient regulatory mechanism to protect the environment.

Fortunately, vigorous defence of the Montreal Protocol, to phase out CFC's, puts the UK in better light. UK law is tighter than the original protocol requires. (DOE 1991)

The Single European Act states that "action by the community, relating to to the environment, shall have the following objectives : (i) **to preserve, protect and improve the environment** (ii) to contribute towards protecting human health (iii) to ensure a prudent and rational utilisation of natural resources." (SEA 1986)

Furthermore, "action by the community, relating to the environment, shall be based on the principle that environmental damage should, as a priority, be rectified at source and that the **polluter should pay.**" (SEA 1986)

These observations give some indication that European Law will make every endeavour to protect the environment. Just how much it does so really depends on the commitment of member states to encourage the detailed formation of laws that prevent business, industry and consumers engaging in massive polluting activity.

Implementing environmental law in industry will be costly but in the long term may make the business more competitive (Cleasby 1992). Some companies will not be able to meet these costs. Should governments help ?

Engineers are in an informed position. They need to be shown the mechanisms available to them to lobby government through local MP's when they find the law weak in regard to important issues such as environmental protection.
Understanding the process of how law is made in the EC is an important prerequisite for engineers if they want to know how they can contribute to its shape and form.

5 Technology

Educational Objective : Review feasible technologies that could be applied to reverse environmental decay. Understand why technical solutions can fail if legislative and motivational mechanisms are not understood.

This is the strength of European engineers. Massive expertise across the EC is available to find technical solutions to almost any problem. In most EC countries the record of success in engineering enterprise is outstanding. And in environmental matters engineers are already showing their ability to design and manufacture products that show great potential. An example is the ability to generate electrical power from the wind at about the same cost per kwh as a modern coal fired power station, without major environmental impact (Musgrove and Lindley 1990). The recent Electricity Act has provided some incentive for commercialising wind power. (Non Fossil Fuel Obligation)

6 Teaching Experience

Energy and Environment is now taught, by lectures, as a subject in years 1 and 2 on the BEng Mech Eng degree course. It is a new course which is currently optional. It should be compulsory. The subject lends itself well to (sometimes heated) expression of opinion. This opportunity is sometimes suppressed by the lecture situation. A series of lectures is necessary to convey many facts **accurately** but more use could be made of discussion and debate. This is to be encouraged to allow personal convictions to develop.
    Assessment is currently by essay and examination. 15% of marks were given for student presentations in the form of a five minute talk on a specfic environmental topic.
    Case studies are used to illustrate the problem of introducing technology if no account is taken of accountability, motivation and law.

One case study looked at the issue of transport policy. The dilemma is that governments have to, at the same time, satisfy a balance between economic and social development alongside a form of control against pollution and excess.

Transport policy was looked at in the UK in relation to traffic congestion and pollution. The government White Paper "Roads for Prosperity" (1989) was cross examined against the Green Party transport policy statement "Roads to the Future" (1991). One group of students took this up for an assessed presentation.

7 Conclusion

Engineers who are being prepared for industrial, business and academic positions should understand the impact that their actions will have on the environment because these issues are now on political agendas.

The educational curriculum should include a compulsory subject on environmental issues that does not marginalise the importance of accountability, mechanisms of motivation and the law.

References

WCP (1988) The Changing Atmosphere : Implications for Global Security.Toronto Conference Statement

Psalm 24(1) New International Version of the Bible

Rjazanov (1927) Frankfurt-am-Main
    Marx Engels Archive 1927 I, i (2) p50

EEO (1990) Energy Efficiency Series No 13 : Energy Efficiency in Domestic Appliances p8 Energy Efficiency Office Department of Energy

Brandt (1983) Common Crisis : Brandt Commissioners Report

HMSO (1990) This Common Inheritance. A Summary of the White Paper on the Environment HMSO

CEST (1990) Industry and the Environment: A Strategic Overview section 3.3.2 The Centre for Exploitation of Science and Technology

EPA (1990) Environmental Protection Act section 7.2 (i) HMSO

DOE (1991) The Ozone Layer Department of the Environment

SEA (1986) Single European Act Art 130R par 1 & 2 HMSO

Cleasby (1992) Industrys' Green Burden : Professional Engineering Vol 5 No 4 p13 April 1992

Musgrove P and Lindley D (1990) Commercialisation of Wind Farms in Europe by 2000. Wind Energy Group European Wind Energy Conference Madrid

Engineering Education in the Philippines with Emphasis upon Energy and the Environment

J. Mabaylan (*), A.H. Pe (**), M.R.I. Purvis (**)
(*) College of Engineering, Xavier University
(**) Faculty of Engineering, University of Portsmouth

Abstract
In 1988 the President of the Philippines set up a Task Force to report on strategies and policies to enable the country to become a newly industrialised nation by the year 2000. This paper considers the relationship between the Task Force Plan and Engineering Education in the Philippines. Particular attention is paid to courses developments in energy and environmental studies at postgraduate level.
Keywords: Education, Engineering, Philippines, Energy, Environment.

1 Introduction

The Southern extremity of the 7000 islands forming the Philippine Archipelago lies 5° north of the equator and stretches some 15° of longitude. The thirteen largest islands, each having an area of more than 1000 square kilometres comprise about 94% of the total land mass. Total land area (30 million ha) and population (55 million) are very similar to the United Kingdom. Luzon in the north of the country and Mindanao in the south are the two largest islands. Manila is the capital of the Philippines and is located on Luzon. Average temperature is about 31°C and rainfall about 2600 mm, mostly obtained in the monsoon season. The two UK based authors have gained considerable experience of Filipino life and culture having visited the Philippines and collaborated with Filipino University staff involved in British Council Exchange programmes (Link Scheme).

Economic performance indicators for the Philippines are poor in comparison to other Association of South East Asia Nations (ASEAN) countries. Exports, labour productivity per capita GNP and investment in Research and Development are all low in relation to conditions in, for example, Malaysia, Hong Kong, South Korea, Singapore and Taiwan. To overcome these deficiencies the President of the Philippines created a Task Force on Science and Technology Development in August 1988. The Task Force was required to present a report within six months of the Presidential decree concerning strategies and policies to enable the Philippines to attain the status of a newly industrialised nation by the year 2000.

On the basis of the report (Presidential Task Force Report 1989) thirteen production and service sectors were identified for development. These covered such areas as Construction, Electronics, Metals and Engineering, Information Technology and Energy. Committees on Manpower Development and Policy were established to integrate the recommendations of individual sector working parties. Amongst other issues, these provided policies and objectives for the education and training of engineers at vocational, undergraduate and postgraduate levels. A subsequent report (Science and Technology (S and T) Master Plan 1990) gave background data, policy targets and actions related to the Task Force Plan aims. The purpose of the present paper is to provide a context for those features of the Task Force Plan dealing with engineering education and to consider issues associated with postgraduate training in energy and environmental studies.

2 Engineering Education Provision in the Philippines

All education provision in the Philippines must be approved by the Department of Education, Culture and Sports. Secondary education covers the age range 13-17 and is seen as a necessity by most Filipino families. A High School Diploma is conferred on successful students at the age 17 years. To enter College or University Students must pass a National College entrance examination. Most Universities also required success in their own entrance examinations. In July 1990 there were 184 engineering schools in the Philippines. Corresponding figures for 1989 and 1987 were 184 and 194 respectively. (Directory of Architecture and Engineering Schools 1990). The decline over these years is attributed to the high cost of engineering provision and significant unemployment amongst engineering graduates. Of the engineering schools total 75% are privately owned and 25% public (ie) State run. The division of schools between discipline areas is:

	% of total
Civil Engineering	39
Mechanical Engineering	24
Electrical Engineering	21
Chemical Engineering	12
Other (Including Postgraduate)	4

Engineering schools in the Philippines are closely monitored by the Technical Panel for Engineering Education. This prescribes general, unit based curricula outlined as follows:

Area 1: Technical Courses

	Units
Mathematics	24
Physical Sciences	16
Basic Engineering Sciences)	106
Professional and Allied Courses)	
Technical Elective subjects	6

Area 2: Non-Technical Courses

Languages, Humanities and Social Sciences	58
Total	210

Twenty one units represent the average number of units taken by a student per semester and the course length is five years. Success on the course allows a student to claim a Diploma or Certificate which is a prerequisite for entry to a National Board Examination. The National Board Examination is the route to Lisentiateship (ie) Chartered Status of an Institution and practice as a professional engineer.

Enrolment and success rate statistics show that some 40% of post secondary education students enter College (Directory of Architecture and Engineering Schools 1990). About 15% of this total choose engineering. Approximately one third of students fail to complete their engineering course. Half of the College graduates are successful in the National Board Examinations.

3 Engineering Education and the Task Force Plan

The Manpower Development component of the plan in Science and Engineering is designed to create a pool of scientists, engineers and technicians which the Philippines needs for industrialisation. The programme covers training for graduates, teachers administrators and technicians. Priority is given to personnel who will be engaged in Research and Development (R and D) activities. The target is to have 1000 PhD,

2000 Masters Degree and 5000 Bachelor Degree Graduates in the next ten years. (S and T Master Plan 1990). It is also intended to train two R and D technicians for each R and D professional worker.

Graduate Institutions will have the main responsibility for the development of the Masters Degree and PhD degree programmes. These developments will arise through

- Strengthening of the capabilities for training graduate manpower from faculty and facilities development and allocation of fellowships.
- Strengthening of R and D capabilities through research grants, facilities development and collaboration with R and D Institutions in the Philippines and abroad.
- Liaison with industry to identify R and D needs and to develop graduate programmes

Policy actions to aid the realisation of the above aims include the creation of new centres of excellence, provision for Science and Technology Parks and the promotion of private investment in R and D Institutions.

Twenty three engineering schools have been identified to form part of an initial engineering network to deal with priority discipline areas at Diploma and Higher Degree levels. On the basis of a five year programme for engineering (1990-95) it is anticipated that 55 PhD and 550 Masters Degree Graduates will be produced. Details of the outputs from particular Institutions are given in the Master Plan (1990).

4 Energy

The link between economic growth and increased energy consumption has been recognised by the inclusion of Energy as a high priority sector in the Task Force Plan. Policy objectives relate to the following:

- Supply – to provide adequate, timely and reasonably priced energy forms and distribute them to markets in support of production goals.
- Demand – to promote the judicious and efficient use of energy resources within techno-economic bounds.
- Environment – to accomplish supply and demand objectives in an environmentally acceptable manner.

To conform to the aims of the Task Force Plan, energy demand is expected to increase by 7.2% per year between 1992 and 2000. Current energy consumption is 106 million barrels of oil equivalent and comprises the mix of sources given in Table 1.

These figures show the dominance of imported oil on energy consumptions in the Philippines. In common with other ASEAN countries energy planning is aimed at the promotion of substitutes for imported oil by exploitation of indigenous resources (Ang 1990)

Research and development in geology and reservoir engineering are components of the Task Force Plan designed to exploit indigenous conventional energy sources. Impetus to activities has been provided by the recent discovery of a large offshore gas field (Hayes 1991). A steady rise in coal production from 1.4 MT in 1992 to 4.4 MT/annum in 2000 is anticipated as new mines become operable.

Developments in the use of non-conventional energy sources, (geothermal, biomass, solar, hydro and wind) are also included in the Energy component of the Task Force Plan. Geothermal energy has quite a short history of exploitation (from about 1979) but in 1990 some 20% of electricity was produced from geothermal sources. The Philippines is part of the so called 'Pacific Ring of Fire' and it has been estimated that there is sufficient underground energy to supply 2×10^5 MW for a Millenium (Energy Forum 1988). Fuel wood is an important source of energy in rural communities and there is interest in the

Table 1. Primary Energy consumption for the Philippines 1990 (Office of Energy Affairs 1990)

		% of total
Indigenous Energy		
1 Conventional		
	Oil	1.3
	Coal	3.8
	Hydro	8.7
	Geothermal	7.8
	Total	21.6
2 Non-conventional		
	Bagasse	4.5
	Coconut Husk/Shell	5.2
	Rice Husk	0.7
	Wood/Woodwaste	2.6
	Other	0.3
	Total	13.3
Imported Energy		
	Oil	63.2
	Coal	1.9
	Total	65.1
	Grand Total	100%

development of fuel wood plantations on problem soils serving no other useful agricultural purpose (Koffa 1991). Xavier University, located in the South of Mindinao, incorporates an appropriate technology centre with support from German Agro Action. Demonstration projects for rural communities have been initiated and include fuel wood supply, solar dryers, biogas and training schemes in the use of alternative energy technologies (Approtech 1990).

5 Environment

There is international concern that as Third World countries industrialise environmental issues will take a low priority in relation to economic growth (World Health Organisation 1990). In the case of the Philippines it is acknowledged that the Environment was not included as a high priority area in the initial Task Force Plan. However the omission is misleading since close inspection of the strategies associated with the priority areas denoted in the Master Plan show detailed attention to environmental hazards for the manufacturing and energy sectors.

Legislation for air pollution control in the Philippines is somewhat dated (Official Gazette 1978) and lags emissions directives currently being adopted in Europe (O'Riordan 1989). More attention is paid to water quality and new regulations came into force in March 1990 (Effluent Regulations of 1990). Experiences of the UK authors concerning visits to industrial plants in the Philippines suggested that overseas companies (mainly American and Japanese based) were sensitive to environmental protection and, in general, pursued pollution control in accordance with the best European practice.

Monitoring of the environment in the Philippines is the responsibility of the Department of the Environment and Natural Resources. A visit to their laboratories in Cagayan De Oro City indicated that able and well motivated staff were hampered in their work by an acute lack of resources. A particular area of need was the provision of trained technicians to operate sophisticated pollution monitoring equipment.

5 Postgraduate Energy and Environmental Studies

The College of Engineering, University of the Philippines, Quezon City, has been identified as a centre of excellence for training, research and development in energy and environmental engineering. A comprehensive set of courses is available at both Masters Degree and PhD level.

The Masters Degree courses cover the subject areas of Energy and Environmental Engineering. Both thesis and non thesis options are available. The programmes of study are based on sets of modules which allow a student to accumulate units of assessment. A choice of study route is available from elective option subjects. Each Masters Degree student must serve a one semester industrial training period. Unit requirements for the courses are summarised in the Table below.

	Thesis option units	Non Thesis option units
Major subjects	15	21
Applied Mathematics	6	6
Thesis	6	0
Electives	3	9
Total	30	36

The PhD programmes are dominated by unit based courses and include a minor thesis component. Fifty seven units are required for the award of PhD of which twelve units apply to the thesis. Up to nine units may be transferred into the assessment portfolio from study at other universities. The study programme options cover (i) Solar Energy, (ii) Wind, Hydro and Ocean energy, (iii) Applied Combustion, (iv) Bioenergy - Biological Processes, (v) Bioenergy - Thermal Processes, (vi) Energy Management. Students must serve a one semester period of industrial training. The thesis is defended at an oral examination and the work must be accepted for publication in a scientific journal. Normally the requirements for PhD should be satisfied within six years from inception.

It is interesting to note the similarities between the structure of the above courses and current curriculum developments in the United Kingdom, namely modular study programmes, unit based assessment and credit accumulation and transfer.

Conclusions

The Philippines Task Force Development Plan exhibits a remarkable degree of cooperation between government departments, academe and industry for the advancement of economic growth and engineering education. In many ways it is a model of coherence in contrast to UK experiences where ad hoc and disparate planning mitigate against constructive use of educational provision. Weaknesses in the Plan that can be detected relate to the lack of long term strategies to allow a poor, rural population to take advantage of new opportunities. The impression is gained that higher level education is aimed at the top few per cent of students and does not subsume concepts of wider access. High failure and completion rates in courses at both undergraduate and postgraduate level are cause for concern which beg attention to teaching methods. Although the Plan identifies general targeted investment in Science and Technology R and D there are no explicit figures for engineering education resources. It must be said, however, that these deficiencies are more apparent than real and will no doubt be addressed within the longer term of the Plan objectives.

The components of the Plan dealing with Energy place emphasis upon energy conservation programmes and indigenous resources at the expense of imported hydrocarbon fuels. The range of Postgraduate energy courses at the University of the Philippines is impressive and finely tuned to national need, particularly in the area of renewables. Less attention appears to be paid to studies in Environmental Engineering in Graduate Institutions despite the strategies contained in several high priority sector plans. Perhaps this is an area open to international collaboration involving shared values and technology transfer in pollution control.

The scope and vision of the Filippino Task Force Plan and its relationship to Engineering Education is arguably unique in the planning strategies of Nations. It is hoped that it will act as a vehicle for releasing the substantial energy and talents of the Filippino people.

References

Presidential Task Force Report on Science and Technology Development. March 1989. (*)
Science and Technology Master Plan, July 1990. Department of Science and Technology, Bicutan, Tagig, Metro Manila. (*)
Directory of Architecture and Engineering Schools, July 1990, Bureau of Higher Education, Department of Education, Culture and Sports. (*)
Office of Energy Affairs Data, Demand Management Division, Planning Services 1990.(*)
Ang BW (1990) Oil Substitution and the Changing Structure of Energy Demand in South East Asia: A case study. The Journal of Energy and Development 14, 1, 55-77
Hayes D (1991) Philippines - Another Leap for Gas in South East Asia Gas World International, March, 30
Energy Forum (1988) Publication of the Philippine National Oil Company, Manila (*)
Koffa SN (1991) Potential Use of Problem Soils for Energy Plantations in the Philippines, Bioresource Technology, 36, 101-111
Approtech (August 1990) Publication of the Appropriate Technology Centre, Xavier University, Cagayan Do Oro, Philippines, Vol 2 No 3 (*)
World Health Organisation (1990) Global Environment monitoring System, Assessment of Urban Air Quality
Official Gazette 1978, Rules and Regulations, Presidential Decree 1984, National Pollution Control Commission, Manila. (*)
O'Riordan 1989 Air Pollution Legislation and Regulation in the European Community, Atmospheric Environment, 23, 2, 293-306.
Revised Effluent Regulations of 1990, Department of the Environment and Natural Resources Administrative Order No 35.

Note:

Copies of references denoted * may be obtained either by application to:

    H E The Ambassador
    Embassy of the Philippines
    9A Palace Green
    London   W8 4QE
    United Kingdom

or Dr M R I Purvis
    School of Systems Engineering
    University of Portsmouth
    Portsmouth
    Hampshire
    PO1 3DJ
    United Kingdom

The Establishment and Operation of a MSc in Energy and Environmental Systems

C.U. Chisholm, S. Burek
School of Engineering, Glasgow Polytechnic

Abstract

The paper outlines the development of a MSc course which involves Energy Systems and Environmental Systems and their interaction in terms of operating complex and sophisticated technology against a base of diminishing natural resources. The course is developed to suit to a wide range of graduates and to give an understanding of energy and environmental systems in the context of economics and relevance to society. The course curriculum was evolved and developed on a much broader base than more conventional specialised courses in energy and environment. The operation and further development of the course over the past 3 years is reviewed.

The paper illustrates studies conducted in Hungary and Czechoslovakia through Academic Links Projects to establish a broad based course in energy and environmental systems against the economics of a evolving economy and a changing society demanding an improved quality of life. It is illustrated how the technology transfer will take place to the East European Universities in Hungary and Czechoslovakia and the paper discusses a number of the problems which have arisen.

The course is discussed in terms of completing a major project within industrial companies and the problems associated with achieving the standard of a MSc while at the same time satisfying the requirements of the company.

The paper illustrates how recent course design allows either full-time students, part-time students or distance learning students to study on this particular course.

It is concluded that the MSc course successfully provides a broad based education in energy and environment against an understanding of economics and legal aspects. It is further concluded that the broad based approach while virtually unknown in East Europe can be successfully transferred to these countries to educate graduates to achieve business growth through an understanding of energy and environmental systems.

Keywords: Postgraduate Study, Energy, Environment, Europe.

1 Introduction

The new course, hosted by the School of Engineering in Glasgow, is unique in that it aims to combine the energy aspects, the environmental aspects and the economic/legal aspects on a equal basis. The course was first offered in

January 1990 and has been taught as a one year full-time course with students entering from a wide variety of backgrounds ranging from engineering and science degrees through to degrees in economics. The graduates leaving the course develop an in-depth understanding of energy and environmental issues relating to their previous qualifications and professional backgrounds, in addition to establishing an in-depth understanding of problems in related areas.

The production and use of energy has a significant effect on the global and local environment, thus it is clear that energy technologies cannot be studied in isolation form the environment and conversely environmental studies need to be correlated to an understanding of energy technology. Increasingly the environment is being considered as an accountable economic resource with society becoming increasingly concerned over environment and environmental legislation beginning to dictate to a greater extent as to what can be allowed and what regulations are required.

Previously these topics have tended to be separated into specialist areas. However, the need for a broad based approach, particularly to engineering problems, has become increasingly clear, where such problems can no longer be considered within the isolation of traditional subject boundaries. It is considered that to study one subject without due regard to the others would produce an unbalanced and incomplete view of a far reaching and complicated subject area.

Therefore, in setting up the course, the strategic aim was to equip graduates from science, engineering and other disciplines with a unique approach to bringing together and integrating energy, environment and related economics and legal aspects.

2 Aims And Objectives

The course is designed to stretch students' educational horizons across traditional subject boundaries, encompassing areas normally associated with engineering, chemistry, biology, economics, social sciences and relevant law. It is aimed at those wishing to enter higher management posts in industry, commerce or public administration where decisions with respect to energy and environment can be made on sound analysis and advice. The emphasis is on analysis of data and information rather than its acquisition. Throughout the course, assessment is based on presentations where critical arguments need to be developed and creative solutions to problems sought, by making use of information from all aspects of the course. In particular the course provides an opportunity for graduates to develop or transfer to a career in energy and environment. In particular the course develops in students:
1 An approach within their chosen fields which gives high priority to considerations of energy use and environmental protection;
2 A critical awareness of advanced energy systems and the ability to use various techniques for the design and operation;
3 A mature understanding of current developments in environmental protection and pollution control;
4 A sound knowledge of social, legal, economic and political issues surrounding energy use and environmental protection;
5 An integrated perception of links between energy production and use, environmental control and protection, and social, legal, economic and political

issues;
6 Expanded perspectives so that they consider the consequences of action at local, national and global levels and at time spans beyond the immediate future.

The exchange of ideas between students of various disciplines is seen as an essential element, thus the course has been designed to appeal to students with widely differing backgrounds.

The course is designed for maximum flexibility where students can study either full-time or part-time or by distance learning. Each unit within the course is also made available to industry as part of a continuing professional development programme. The aim of the course is to provide maximum access to persons either unemployed who can attend the course full-time or to persons employed in industry and commerce who can attend on a part-time basis. The course at Glasgow Polytechnic is unique in providing the opportunity to study energy and environmental systems with a broad perspective and it has proved extremely popular with potential students and has opened up a number of employment opportunities to its recent graduates.

3 Review of Course

The course was initially validated under CNAA regulations in December 1989 and has had three intakes of students to date. These intakes have been supported under the High Technology National Training HTNT programme. This funding has now been discontinued and in the future the course will be operated as a full cost course available to graduates in the United Kingdom and in Europe and worldwide. It will also be available to suitable unemployed graduates who are able to find sponsorship to attend the course.

The demand for the course has been high, with the number of funded places vastly oversubscribed. Typically 150 applications have been received each year for 16 places. Most of the students attending the course have been in the mature age range from 22 to 59.

In reviewing the course to gain a wider participation it has been structured to operate on a 3 week cycle involving 1 week of intensive lectures followed by 2 weeks of tutorials, visits, seminars and other forms of consolidation. Where students are attending full-time the remaining 2 weeks of the cycle is used to consolidate the information presented in the first week. Where students are attending in the part-time mode, after the first week the students then receive a set of structured and directed studies to complete the unit. Organised in this way the course is thus suitable for full-time study, part-time study in block release mode, and is available for participants from industry who wish to attend individual units as 1 week short courses.

The course is divided into 2 parts with the first part involving 6 units, developing themes of Environmental Law, Environmental Economics, Thermal and Electric Power Systems, Environmental Pollution and Control, Renewable Energy Resources and Work Place Environment Control. A seventh unit is used to integrate the various aspects of the course by relevant case studies. The integrated case studies are based on extended work in groups and it is not formally taught. In addition to the normal 3 week period allowed for this unit a further number of hours is devoted to case studies spread among the previous 6 units. A major field trip is incorporated into the case study unit. This has been shown to be beneficial to the development of the students and the current cohort made a successful field trip to attend an International Conference on Fossil fuels and Environment in Prague where the students were able to benefit from listening to experts on energy and environment from many different countries and were able to study areas of Czechoslovakia and see environmental damage associated with fossil fuels and the protective measures being taken. The field trip is now considered to be a unique and essential feature of the course.

All the units are available on the credit accumulation and transfer scheme operated by the Polytechnic (SCOTCATS SCHEME).

The 6 units and the integrated case studies unit constitute an end point in terms of the award of a Postgraduate Diploma within the scheme for students wishing to terminate at this point.

The second part of the scheme comprises a further 6 units from which the students are normally required to choose 3. The units are Environmental Management, Risk Management, Environmental Measurement and Monitoring, Energy Law and Policy, Energy in Buildings and Technical Management of Power Plant.. The students can either opt for a specialised or more general selection of modules according to their requirements or entry qualifications. In addition all students are required to undertake a major project which is based either within the Polytechnic or preferably within an industrial company. Over the past year, the second cohort of students successfully completed their projects within industrial companies. This created a particular problem in terms of finding suitable places for the students where an MSc level project could be completed within a company. This situation was solved by producing a detailed set of guidelines of the types of projects which would be suitable and this was circulated to a number of companies with the help of the full-time Placements Officer within the School. Once a company had identified a possible project a meeting then took place with the Placement Officer and a Course Organiser and a project remit was then agreed. By carefully agreeing with the company prior to initiation of the project it was found that all the projects were successfully completed to the satisfaction of the academic supervisors and the host company.

Alternative routes through to the completion of the course have now been included where a part-time student can select to spend a greater amount of time on the major project with the student being exempted from 1, 2 or all 3 of the modules within the second part of the course. Where an extended project is agreed it will be completed through the establishment of a formal Learning Contract signed and agreed between the company, the academic institution and a student. The Learning Contract will ensure that a formal agreement is established with a set of predetermined goals laid out on the basis of the work programmes set up within the company. The academic supervisor, will determine the learning goals and thus will be able to establish the standard of the project at MSc level. It is believed that this unique feature of the course will be desirable for employers who are sponsoring part-time students and require a particular project area to be undertaken relating to the student's employment or alternatively where an overseas student requires to return to his/her home country to complete a project in industry or commerce. In such cases it would be the intention to negotiate a Learning Contract with the company involved and the academic supervisor would require to make visits to the home country of the student.

For part-time students the consolidation studies will require to take place away from the Polytechnic through a programme of directed studies and

tutorial sessions arranged to suit the individual student.
The course as originally conceived, was designed to act as a bridge between several disciplines, the major ones being that of energy systems and environmental systems. In reviewing the course it was decided that it was essential to maintain the broad based approach that also to ensure that the syllabi keeps abreast of local and global developments in energy and environmental systems. Since 1989 new relevant subject areas have been formally introduced involving Risk Management in Environmental Decision Making, Environmental Monitoring and Measurement Techniques and Methods of Environmental Impact Assessment and Environmental Auditing. The project has been found to be highly successful within the industrial and commercial environment and it is intended where possible to expand the project to provide a formal Learning Contract Framework particularly for part-time students in the United Kingdom and for overseas students where it will be possible for the student to complete a major project within their company.

4 Transferability of the Course to East European Institutions

The School of Engineering was awarded Academic Links Projects with Hungary and Czechoslovakia. The awarded projects were to develop an approach to business growth within the free economy through a better understanding of Energy and Environment. In the longer term it was decided that undergraduate students at the Technical University in Prague and the Technical University in Budapest could benefit from experience of the broad based course designed and operated at Glasgow Polytechnic. Over the past year discussions have taken place with a view to establishing the course at these Institutions. Neither the staff nor the students at these Universities are familiar with this type of broad based course and traditionally have been involved with undergraduate and postgraduate courses of a highly specialist nature. In particular staff are used to postgraduate courses of a highly specialist nature and directly related to a similar undergraduate course. Therefore the main problem in establishing the broad based course has been to illustrate the value of the broad based philosophy for training graduates to enable them to take forward energy and environmental issues within the developing free economy of these countries.

Another problem which emerged was the difficulty of sustaining the developments once a visit had taken place to the East European establishment. Essentially a significant time factor is required to allow the necessary culture and organisational changes required to support the establishment of such new ventures as a broad based MSc in Energy and Environmental Systems. It is believed that once the proper communication channels have been established through the Academic Links Project that a successful transfer will begin to take place with the establishment of the course at both the Technical University in Prague and the Technical University in Budapest. Essential to the successful transfer is the staff at Glasgow, Budapest and Prague being able to meet and understand the concepts of what is being transferred. In this respect while the level of English employed is of high quality it is still a barrier to quick communication in respect of all the various factors associated with the transfer and operation of the course at these Universities.

For both Hungary and Czechoslovakia there are large and significant problems to be solved in energy and environment and it is now accepted by the staff at both establishments that the development of this type of course would be desirable in terms of solving the future energy and environmental problems within the context of free economy economics and within the social context of these nations developing with the rest of Europe. In particular it is recognised that each unit within the course could be made available to industrial and commercial staff within Hungarian and Czechoslovakian Industry which would allow for professional updating of a number of key individuals required to take industry forward within the free economy.

In particular the part-time nature of the course in the form of intensive units was considered to be particularly suitable for returning industrialists.

Another model being examined is a joint course where students from Glasgow, Prague and Budapest would study units at each University leading to a joint award.

5 Summary and Conclusions

The large number of applications which were made each year for entry to the course demonstrate that it has already achieved a reputation in meeting the changing requirements for postgraduate education and training for the 1990's in energy and environmental systems. The emphasis on the broad based approach where energy and environment are considered against the social, legal, economic, and political issues combine to make the course unique in the United Kingdom. Recent modifications to provide a set of intensive study units provides for considerable flexibility in terms of individuals attending part-time from industry and provided grants could be made available, could accommodate the training of unemployed graduates for conversion to the energy and environment posts in industry and commerce.

The broad based course is ideal for providing graduate education in East Europe in energy and environment where East Europe faces many problems in relation to solving energy and environmental issues. The style of the course is also particularly suitable for East Europe where it will allow for the return of persons employed in industry to update in environmental and energy issues in the context of the free economy.

The course is unique in providing a Master's project through the use of a Learning Contract Framework where the project is completed within a company or other organisation by agreement of the company, the academic institution and the student.